Birkhäuser

Quantum Field Theory and Gravity

Conceptual and Mathematical Advances in the Search for a Unified Framework

Felix Finster
Olaf Müller
Marc Nardmann
Jürgen Tolksdorf
Eberhard Zeidler
Editors

Birkhäuser

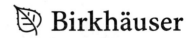

Editors
Felix Finster
Lehrstuhl für Mathematik
Universität Regensburg
Germany

Olaf Müller
Lehrstuhl für Mathematik
Universität Regensburg
Germany

Marc Nardmann
Fachbereich Mathematik
Universität Hamburg
Germany

Jürgen Tolksdorf
Max-Planck-Institut für
Mathematik in den Naturwissenschaften
Leipzig, Germany

Eberhard Zeidler
Max-Planck-Institut für
Mathematik in den Naturwissenschaften
Leipzig, Germany

ISBN 978-3-0348-0042-6 e-ISBN 978-3-0348-0043-3
DOI 10.1007/978-3-0348-0043-3
Springer Basel Dordrecht Heidelberg London New York

Library of Congress Control Number: 2012930975

Mathematical Subject Classification (2010): 81-06, 83-06

Printed on acid-free paper

Springer Basel AG is part of Springer Science+Business Media

www.birkhauser-science.com

Contents

Preface

The present volume arose from the conference on *"Quantum field theory and gravity – Conceptual and mathematical advances in the search for a unified framework"*, held at the University of Regensburg (Germany) from September 28 to October 1, 2010. This conference was the successor of similar conferences which took place at the *Heinrich Fabri Institut* in Blaubeuren in 2003 and 2005 and at the *Max Planck Institute for Mathematics in the Sciences* in Leipzig in 2007. The intention of this series of conferences is to bring together mathematicians and physicists to discuss profound questions within the non-empty intersection of mathematics and physics. More specifically, the series aims at discussing conceptual ideas behind different mathematical and physical approaches to quantum field theory and (quantum) gravity.

As its title states, the Regensburg conference was devoted to the search for a unified framework of quantum field theory and general relativity. On the one hand, the standard model of particle physics – which describes all physical interactions except gravitation – is formulated as a quantum field theory on a fixed Minkowski-space background. The affine structure of this background makes it possible for instance to interpret interacting quantum fields as asymptotically "free particles". On the other hand, the gravitational interaction has the peculiar property that all kinds of energy couple to it. Furthermore, since Einstein developed general relativity theory, gravity is considered as a dynamical property of space-time itself. Hence space-time does not provide a fixed background, and a back-reaction of quantum fields to gravity, i.e. to the curvature of space-time, must be taken into account. It is widely believed that such a back-reaction can be described consistently only by a (yet to be found) quantum version of general relativity, commonly called *quantum gravity*. Quantum gravity is expected to radically change our ideas about the structure of space-time. To find this theory, it might even be necessary to question the basic principles of quantum theory as well.

Similar to the third conference of this series, the intention of the conference held at the University of Regensburg was to provide a forum to discuss different mathematical and conceptual approaches to a quantum (field) theory including gravitational back-reactions. Besides the two well-known paths laid out by string theory and loop quantum gravity, also other ideas were presented. In particular, various functorial approaches were discussed, as well as the possibility that space-time emerges from discrete structures.

The present volume provides an appropriate cross-section of the conference. The refereed articles are intended to appeal to experts working in different fields of mathematics and physics who are interested in the subject of quantum field theory and (quantum) gravity. Together they give the reader some overview of new approaches to develop a quantum (field) theory taking a dynamical background into account.

As a complement to the invited talks which the articles in this volume are based on, discussion sessions were held on the second and the last day of the conference. We list some of the questions raised in these sessions:

1. **Can we expect to obtain a quantum theory of gravity by purely mathematical considerations?** What are the physical requirements to expect from a unified field theory? How can these be formulated mathematically? Are the present mathematical notions sufficient to formulate quantum gravity, or are new mathematical concepts needed? Are the criteria of mathematical consistency and simplicity promising guiding principles for finding a physical theory? Considering the wide variety of existing approaches, the use of gedanken experiments as guiding paradigms seems indispensable even for pure mathematicians in the field.
2. **Evolution or revolution?** Should we expect progress rather by small steps or by big steps? By "small steps" we mean a conservative approach towards a unified theory where one tries to keep the conventional terminology as far as possible. In contrast, proceeding in "big steps" often entails to replace the usual terminology and the conventional physical objects by completely new ones.

   In the discussion, the possibilities for giving up the following conventional structures were considered:
   - **Causality:** In what sense should it hold in quantum gravity?
   - **Superposition principle:** Should it hold in a unified field theory? More specifically, do we have to give up the Hilbert space formalism and its probabilistic interpretation?

   A related question is:
3. **Can we quantize gravity separately?** That is, does it make physical sense to formulate a quantum theory of pure gravity? Can such a formulation be mathematically consistent? Or is it necessary to include all other interactions to obtain a consistent theory?
4. **Background independence**: How essential is it, and which of the present approaches implement it? Which basic mathematical structure would be physically acceptable as implementing background independence?
5. **What are the relevant open problems in classical field theory?** One problem is the concept of charged point particles in classical electrodynamics (infinite self-energy). Other problems concern the notion of quasi-local mass in general relativity and the cosmic censorship conjectures.

6. **(How) can we test quantum gravity?** Can one hope to test quantum gravity in experiments whose initial conditions are controlled by humans, similar to tests of the standard model in particle accelerators? Or does one need to rely on astronomical observations (of events like supernovae or black hole mergers)?

Having listed some of the basic questions, we will now give brief summaries of the articles in this volume. They are presented in chronological order of the corresponding conference talks. Unfortunately, not all the topics discussed at the conference are covered in this volume, because a few speakers were unable to contribute; see also pp. xii–xiii below.

The volume begins with an overview by **Claus Kiefer** on the main roads towards quantum gravity. After a brief motivation why one should search for a quantum theory of gravitation, he discusses canonical approaches, covariant approaches like loop quantum gravity, and string theory. As two main problems that a theory of quantum gravity should solve, he singles out a statistical explanation of the Bekenstein–Hawking entropy and a description of the final stage of black-hole evaporation. He summarizes what the previously discussed approaches have found out about the first question so far.

Locally covariant quantum field theory is a framework proposed by Brunetti–Fredenhagen–Verch that replaces the Haag–Kastler axioms for a quantum field theory on a fixed Minkowski background, by axioms for a functor which describes the theory on a large class of curved backgrounds simultaneously. After reviewing this framework, **Klaus Fredenhagen*** [1] and **Katarzyna Rejzner** suggest that quantum gravity can be obtained from it via perturbative renormalization à la Epstein–Glaser of the Einstein–Hilbert action. One of the technical problems one encounters is the need for a global version of BRST cohomology related to diffeomorphism invariance. As a preliminary step, the authors discuss the classical analog of this quantum problem in terms of infinite-dimensional differential geometry.

Based on his work with Joel Smoller, **Blake Temple** suggests an alternative reason for the observed increase in the expansion rate of the universe, which in the standard model of cosmology is explained in terms of "dark energy" and usually assumed to be caused by a positive cosmological constant. He argues that since the moment when radiation decoupled from matter 379000 years after the big bang, the universe should be modelled by a wave-like perturbation of a Friedmann–Robertson–Walker space-time, according to the mathematical theory of Lax–Glimm on how solutions of conservation laws decay to self-similar wave patterns. The possible perturbations form a 1-parameter family. Temple proposes that a suitable member of this family describes the observed anomalous acceleration of the galaxies (without invoking a cosmological constant). He points out that his hypothesis makes testable predictions.

---

[1] In the cases where articles have several authors, the star marks the author who delivered the corresponding talk at the conference.

The term *"third quantization"* refers to the idea of quantum gravity as a quantum field theory on the space of geometries (rather than on space-time), which includes a dynamical description of topology change. **Steffen Gielen** and **Daniele Oriti*** explain how matrix models implement the third-quantization program for 2-dimensional Riemannian quantum gravity, via a rigorous continuum limit of discretized geometries. Group field theory (GFT) models, which originated in loop quantum gravity (LQG) but are also relevant in other contexts, implement third quantization for 3-dimensional Riemannian quantum gravity – but only in the discrete setting, without taking a continuum limit. The authors compare the GFT approach to the LQG-motivated idea of constructing, at least on a formal level, a continuum third quantization on the space of connections rather than geometries. They argue that the continuum situation should be regarded only as an effective description of a physically more fundamental GFT.

Andreas Döring* and **Rui Soares Barbosa** present the topos approach to quantum theory, an attempt to overcome some conceptual problems with the interpretation of quantum theory by using the language of category theory. One aspect is that physical quantities take their values not simply in the real numbers; rather, the values are families of real intervals. The authors describe a connection between the topos approach, noncommutative operator algebras and domain theory.

Many problems in general relativity, as well as the formulation of the AdS/CFT correspondence, involve assigning a suitable boundary to a given space-time. A popular choice is Penrose's *conformal boundary*, but it does not always exist, and it depends on non-canonical data and is therefore not always unique. **José Luis Flores**, **Jónatan Herrera** and **Miguel Sánchez*** explain the construction of a *causal boundary of space-time* which does not suffer from these problems. They describe its properties and the relation to the conformal boundary. Several examples are discussed, in particular pp-waves.

Dietrich Häfner gives a mathematically rigorous description of the Hawking effect for second-quantized spin-$\frac{1}{2}$ fields in the setting of the collapse of a rotating charged star. The result, which confirms physical expectations, is stated and proved using the language and methods of scattering theory.

One problem in constructing a background-free quantum theory is that the standard quantum formalism depends on a background metric: its operational meaning involves a background time, and its ability to describe physics *locally* in field theory arises dynamically, via metric concepts like causality and cluster decomposition. In his *general boundary formulation (GBF)* of quantum theory, **Robert Oeckl** tries to overcome this problem by using, instead of spacelike hypersurfaces, boundaries of arbitrary spacetime regions as carriers of quantum states. His article lists the basic GBF objects and the axioms they have to satisfy, and describes how the usual quantum states, observables and probabilities are recovered from a GBF setting. He proposes various quantization schemes to produce GBF theories from classical theories.

Felix Finster, **Andreas Grotz*** and **Daniela Schiefeneder** introduce causal fermion systems as a general mathematical framework for formulating relativistic quantum theory. A particular feature is that space-time is a secondary object which emerges by minimizing an action for the so-called universal measure. The setup provides a proposal for a "quantum geometry" in the Lorentzian setting. Moreover, numerical and analytical results on the support of minimizers of causal variational principles are reviewed which reveal a "quantization effect" resulting in a discreteness of space-time. A brief survey is given on the correspondence to quantum field theory and gauge theories.

Christian Bär* and **Nicolas Ginoux** present a systematic construction of bosonic and fermionic locally covariant quantum field theories on curved backgrounds in the case of free fields. In particular, they give precise mathematical conditions under which bosonic resp. fermionic quantization is possible. It turns out that fermionic quantization requires much more restrictive assumptions than bosonic quantization.

Christopher J. Fewster asks whether every locally covariant quantum field theory (cf. the article by Fredenhagen and Rejzner described above) represents "the same physics in all space-times". In order to give this phrase a rigorous meaning, he defines the "SPASs" property for families of locally covariant QFTs, which intuitively should hold whenever each member of the family represents the same physics in all space-times. But not every family of locally covariant QFTs has the SPASs property. However, for a "dynamical locality" condition saying that kinematical and dynamical descriptions of local physics coincide, every family of dynamically local locally covariant QFTs has SPASs.

Rainer Verch extends the concept of *local thermal equilibrium (LTE) states*, i.e. quantum states which are not in global thermal equilibrium but possess local thermodynamical parameters like temperature, to quantum field theory on curved space-times. He describes the ambiguities and anomalies that afflict the definition of the stress-energy tensor of QFT on curved space-times and reviews the work of Dappiaggi–Fredenhagen–Pinamonti which, in the setting of the semi-classical Einstein equation, relates a certain fixing of these ambiguities to cosmology. In this context, he applies LTE states and shows that the temperature behavior of a massless scalar quantum field in the very early history of the universe is more singular than the behavior of the usually considered model of classical radiation.

Inspired by a version of Mach's principle, **Julian Barbour** presents a framework for the construction of background-independent theories which aims at quantum gravity, but whose present culmination is a theory of classical gravitation called *shape dynamics.* Its dynamical variables are the elements of the set of compact 3-dimensional Riemannian manifolds divided by isometries and volume-preserving conformal transformations. It "eliminates time", involves a procedure called *conformal best matching*, and is equivalent to general relativity for space-times which admit a foliation by compact spacelike hypersurfaces of constant mean curvature.

Michael K.-H. Kiessling considers the old problem of finding the correct laws of motion for a joint evolution of electromagnetic fields and their point-charge sources. After reviewing the long history of proposals, he reports on recent steps towards a solution by coupling the Einstein–Maxwell–Born–Infeld theory for an electromagnetic space-time with point defects to a Hamilton–Jacobi theory of motion for these defects. He also discusses how to construct a "first quantization with spin" of the sources in this classical theory by replacing the Hamilton–Jacobi law with a de Broglie–Bohm–Dirac quantum law of motion.

Several theories related to quantum gravity postulate (large- or small-sized) extra dimensions of space-time. **Stefan Hollands**' contribution investigates a consequence of such scenarios, the possible existence of higher-dimensional black holes, in particular of stationary ones. Because of their large number, the possible types of such stationary black holes are much harder to classify than their 4-dimensional analogs. Hollands reviews some partial uniqueness results.

Since properties of general relativity, for instance the *Einstein equivalence principle (EEP)*, could conceivably fail to apply to quantum systems, experimental tests of these properties are important. **Domenico Giulini**'s article explains carefully which subprinciples constitute the EEP, how they apply to quantum systems, and to which accuracy they have been tested. In 2010, Müller–Peters–Chu claimed that the least well-tested of the EEP subprinciples, the *universality of gravitational redshift*, had already been verified with very high precision in some older atom-interferometry experiments. Giulini argues that this claim is unwarranted.

Besides the talks summarized above there were also presentations covering the "main roads" to quantum gravity and other topics related to quantum theory and gravity. PDF files of these presentations can be found at *www.uni-regensburg.de/qft2010.*

Dieter Lüst (LMU München) gave a talk with the title *The landscape of multiverses and strings: Is string theory testable?*. He argued that, despite the huge number of vacua that superstring/M-theory produces after compactification, it might still yield experimentally testable predictions. If the string mass scale, which can a priori assume arbitrary values in brane-world scenarios, is not much larger than 5 TeV, then effects like string Regge excitations will be seen at the Large Hadron Collider.

Christian Fleischhack from the University of Paderborn gave an overview of loop quantum gravity, emphasizing its achievements – e.g. the construction of geometric operators for area and volume, and the derivation of black hole entropy – but also its problems, in particular the still widely unknown dynamics of the quantum theory.

In her talk *New 'best hope' for quantum gravity*, **Renate Loll** from the University of Utrecht presented the motivation, the status and perspectives of "Quantum Gravity from Causal Dynamical Triangulation (CDT)" and how it is related to other approaches to a non-perturbative and mathematically rigorous formulation of quantum gravity.

Mu-Tao Wang from Columbia University gave a talk *On the notion of quasilocal mass in general relativity.* After explaining why it is difficult to define a satisfying notion of quasilocal mass, he presented a new proposal due to him and Shing-Tung Yau. This mass is defined via isometric embeddings into Minkowski space and has several desired properties, in particular a vanishing property that previous definitions were lacking.

Motivated by the question – asked by 't Hooft and others – whether quantum mechanics could be an emergent phenomenon that occurs on length scales sufficiently larger than the Planck scale but arises from different dynamics at shorter scales, **Thomas Elze** from the University of Pisa discussed in the talk *General linear dynamics: quantum, classical or hybrid* a path-integral representation of classical Hamiltonian dynamics which allows to consider direct couplings of classical and quantum objects. Quantum dynamics turns out to be rather special within the class of such general linear evolution laws.

In his talk on *Massive quantum gauge models without Higgs mechanism,* **Michael Dütsch** explained how to construct the S-matrix of a non-abelian gauge theory in Epstein–Glaser style, via the requirements of renormalizability and causal gauge invariance. These properties imply already the occurrence of Higgs fields in massive non-abelian models; the Higgs fields do not have to be put in by hand. He discussed the relation of this approach to model building via spontaneous symmetry breaking.

Jerzy Kijowski from the University of Warszawa spoke about *Field quantization via discrete approximations: problems and perspectives.* He explained how the set of discrete approximations of a physical theory is partially ordered, and that the observable algebras form an inductive system for this partially ordered set, whereas the states form a projective system. Then he argued that loop quantum gravity is the best existing proposal for a quantum gravity theory, but suffers from the unphysical property that its states form instead an inductive system.

Acknowledgments

It is a great pleasure for us to thank all participants for their contributions, which have made the conference so successful. We are very grateful to the staff of the Department of Mathematics of the University of Regensburg, especially to Eva Rütz, who managed the administrative work before, during and after the conference excellently.

We would like to express our deep gratitude to the German Science Foundation (DFG), the Leopoldina – National Academy of Sciences, the Alfried Krupp von Bohlen und Halbach Foundation, the International Association of Mathematical Physics (IAMP), the private Foundation Hans Vielberth of the University of Regensburg, and the Institute of Mathematics at the University of Regensburg for their generous financial support.

Also, we would like to thank the Max Planck Institute of Mathematics in the Sciences, Leipzig, for the financial support of the conference volume. Finally, we thank Barbara Hellriegel and her team from Birkhäuser for the excellent cooperation.

Bonn, Leipzig, Regensburg
September 2011

Felix Finster
Olaf Müller
Marc Nardmann
Jürgen Tolksdorf
Eberhard Zeidler

Quantum Gravity: Whence, Whither?

Claus Kiefer

Abstract. I give a brief summary of the main approaches to quantum gravity and highlight some of the recent developments.

Mathematics Subject Classification (2010). Primary 83-02; Secondary 83C45, 83C47, 83E05, 83E30.

Keywords. Quantum gravity, string theory, quantum geometrodynamics, loop quantum gravity, black holes, quantum cosmology.

1. Why quantum gravity?

Quantum theory provides a universal framework that encompasses so far all particular interactions – with one exception: gravitation. The question whether gravity must also be described by a quantum theory at the most fundamental level and, if yes, how such a theory can be constructed, is perhaps the deepest unsolved problem of theoretical physics. In my contribution I shall try to give a general motivation and a brief overview of the main approaches as well as of some recent developments and applications. A comprehensive presentation can be found in [1], where also many references are given; an earlier short overview is [2].

The main obstacle so far in constructing a theory of quantum gravity is the lack of experimental support. Physics is an empirical science, and it is illusory to expect that a new fundamental physical theory can be found without the help of data. This difficulty is connected with the fact that the fundamental quantum-gravity scale – the Planck scale – is far from being directly accessible. The Planck scale (Planck length, Planck time, and Planck mass or energy) follows upon combining the gravitational constant, G, the speed of light, c, and the quantum of action, $\hbar$,

$$l_{\rm P} = \sqrt{\frac{\hbar G}{c^3}} \approx 1.62 \times 10^{-33} \text{ cm} , \tag{1}$$

$$t_{\rm P} = \frac{l_{\rm P}}{c} = \sqrt{\frac{\hbar G}{c^5}} \approx 5.39 \times 10^{-44} \text{ s} , \tag{2}$$

$$m_{\rm P} = \frac{\hbar}{l_{\rm P} c} = \sqrt{\frac{\hbar c}{G}} \approx 2.18 \times 10^{-5} \text{ g} \approx 1.22 \times 10^{19} \text{ GeV}/c^2 . \tag{3}$$

To probe the Planck scale with present technology, for example, one would need a storage ring of galactic size, something beyond any imagination. So why should one be interested in looking for a quantum theory of gravity?

The reasons are of conceptual nature. The current edifice of theoretical physics cannot be complete. First, Einstein's theory of general relativity (GR) breaks down in certain situations, as can be inferred from the singularity theorems. Such situations include the important cases of big bang (or a singularity in the future) and the interior of black holes. The hope is that a quantum theory can successfully deal with such situations and cure the singularities. Second, present quantum (field) theory and GR use concepts of time (and spacetime) that are incompatible with each other. Whereas current quantum theory can only be formulated with a rigid external spacetime structure, spacetime in GR is dynamical; in fact, even the simplest features of GR (such as the gravitational redshift implemented e.g. in the GPS system) cannot be understood without a dynamical spacetime. This is often called the *problem of time*, since non-relativistic quantum mechanics is characterized by the absolute Newtonian time t as opposed to the dynamical configuration space. A fundamental quantum theory of gravity is therefore assumed to be fully background-independent. And third, the hope that all interactions of Nature can be unified into one conceptual framework will only be fulfilled if the present hybrid character of the theoretical structure is overcome.

In the following, I shall first review the situations where quantum effects are important in a gravitational context. I shall then give an overview of the main approaches and end with some applications.

2. Steps towards quantum gravity

The first level of connection between gravity and quantum theory is quantum mechanics in an external Newtonian gravitational field. This is the only level where experiments exist so far. The quantum-mechanical systems are mostly neutrons or atoms. Neutrons, like any spin-1/2 system, are described by the Dirac equation, which for the experimental purposes is investigated in a non-relativistic approximation ('Foldy–Wouthuysen approximation'). One thereby arrives at

$$\mathrm{i}\hbar \frac{\partial \psi}{\partial t} \approx H_{\rm FW} \psi$$

with (in a standard notation)

$$
\begin{aligned}
H_{\mathrm{FW}} = & \underbrace{\beta mc^2}_{\text{rest mass}} + \underbrace{\frac{\beta}{2m}\mathbf{p}^2}_{\text{kinetic energy}} - \underbrace{\frac{\beta}{8m^3c^2}\mathbf{p}^4}_{\text{SR correction}} + \underbrace{\beta m(\mathbf{a}\,\mathbf{x})}_{\text{COW}} \\
& - \underbrace{\omega\mathbf{L}}_{\text{Sagnac effect}} - \underbrace{\omega\mathbf{S}}_{\text{Mashhoon effect}} \\
& + \frac{\beta}{2m}\mathbf{p}\frac{\mathbf{a}\,\mathbf{x}}{c^2}\mathbf{p} + \frac{\beta\hbar}{4mc^2}\vec{\Sigma}(\mathbf{a}\times\mathbf{p}) + \mathcal{O}\left(\frac{1}{c^3}\right) .
\end{aligned} \tag{4}
$$

The underbraced terms have been experimentally tested directly or indirectly. ('COW' stands for the classic neutron interferometry experiment performed by Colella, Overhauser, and Werner in 1975.)

The next level on the way to quantum gravity is quantum field theory in an external curved spacetime (or, alternatively, in a non-inertial system in Minkowski spacetime). Although no experimental tests exist so far, there are definite predictions.

One is the Hawking effect for black holes. Black holes radiate with a temperature proportional to $\hbar$,

$$
T_{\mathrm{BH}} = \frac{\hbar\kappa}{2\pi k_{\mathrm{B}}c} , \tag{5}
$$

where κ is the surface gravity. In the important special case of a Schwarzschild black hole with mass M, one has for the Hawking temperature,

$$
\begin{aligned}
T_{\mathrm{BH}} &= \frac{\hbar c^3}{8\pi k_{\mathrm{B}}GM} \\
&\approx 6.17\times 10^{-8}\left(\frac{M_\odot}{M}\right)\ \mathrm{K} .
\end{aligned}
$$

Due to the smallness of this temperature, the Hawking effect cannot be observed for astrophysical black holes. One would need for this purpose primordial black holes or small black holes generated in accelerators.

Since black holes are thermodynamical systems, one can associate with them an entropy, the Bekenstein–Hawking entropy

$$
S_{\mathrm{BH}} = k_{\mathrm{B}}\frac{A}{4l_{\mathrm{P}}^2} \overset{\text{Schwarzschild}}{\approx} 1.07\times 10^{77}k_{\mathrm{B}}\left(\frac{M}{M_\odot}\right)^2 . \tag{6}
$$

Among the many questions for a quantum theory of gravity is the microscopic foundation of S_{BH} in the sense of Boltzmann.

There exists an effect analogous to (5) in flat spacetime. An observer linearly accelerated with acceleration a experiences a temperature

$$
T_{\mathrm{DU}} = \frac{\hbar a}{2\pi k_{\mathrm{B}}c} \approx 4.05\times 10^{-23}\ a\left[\frac{\mathrm{cm}}{\mathrm{s}^2}\right]\ \mathrm{K} , \tag{7}
$$

the 'Unruh' or 'Davies–Unruh' temperature. The analogy to (5) is more than obvious. An experimental confirmation of (7) is envisaged with higher-power, short-pulse lasers [3].

The fact that black holes behave like thermodynamical systems has led to speculations that the gravitational field might not be fundamental, but is instead an effective macroscopic variable like in hydrodynamics, see e.g. [4] for a discussion. If this were true, the search for a quantum theory of the gravitational field would be misleading, since one never attempts to quantize effective (e.g. hydrodynamic) variables. So far, however, no concrete 'hydrodynamic' theory of gravity leading to a new prediction has been formulated.

The third, and highest, level is full quantum gravity. At present, there exist various approaches about which no consensus is in sight. The most conservative class of approaches is quantum general relativity, that is, the direct application of quantization rules to GR. Methodologically, one distinguishes between covariant and canonical approaches. A more radical approach is string theory (or M-theory), which starts with the assumption that a quantum description of gravity can only be obtained within a unified quantum theory of all interactions. Out of these approaches have grown many other ones, most of them building on discrete structures. Among them are quantum topology, causal sets, group field theory, spin-foam models, and models implementing non-commutative geometry. In the following, I shall restrict myself to quantum general relativity and to string theory. More details on discrete approaches can be found in [5] and in other contributions to this volume.

3. Covariant quantum gravity

The first, and historically oldest, approach is covariant perturbation theory. For this purpose one expands the four-dimensional metric $g_{\mu\nu}$ around a classical background given by $\bar{g}_{\mu\nu}$,

$$g_{\mu\nu} = \bar{g}_{\mu\nu} + \sqrt{\frac{32\pi G}{c^4}} f_{\mu\nu} , \tag{8}$$

where $f_{\mu\nu}$ denotes the perturbation. This is similar to the treatment of weak gravitational waves in GR. Associated with $f_{\mu\nu}$ is a massless 'particle' of spin 2, the graviton. The strongest observational constraint on the mass of the graviton comes from investigating gravity over the size of galaxy clusters and leads to $m_{\rm g} \lesssim 10^{-29}$ eV, cf. [6] for a discussion of this and other constraints. This mass limit would correspond to a Compton wavelength of 2×10^{22} m.

One can now insert the expansion (8) into the Einstein–Hilbert action and develop Feynman rules as usual. This can be done [1], but compared to Yang–Mills theories an important difference occurs: perturbative quantum gravity is non-renormalizable, that is, one would need infinitely many parameters to absorb the divergences. As has been shown by explicit calculations, the expected divergences indeed occur from two loops on. Recent progress in this direction was made in the context of $N = 8$ supergravity [7], see also [8]. $N = 8$ supergravity, which has maximal supersymmetry, is *finite* up to four loops, as was shown by an explicit calculation using powerful new methods. There are arguments that it is finite even at five and six loops and perhaps up to eight loops. If this is true, the question will arise whether there exists a hitherto unknown symmetry that prevents the occurrence of divergences at all.

Independent of this situation, one must emphasize that there exist theories at the non-perturbative level that are perturbatively non-renormalizable. One example is the non-linear σ model for dimension $D > 2$, which exhibits a non-trivial UV fixed point at some coupling g_c ('phase transition'). An expansion in $D - 2$ and use of renormalization-group (RG) techniques gives information about the behaviour in the vicinity of the non-trivial fixed point. The specific heat exponent of superfluid helium as described by this model was measured in a space shuttle experiment, and the results are in accordance with the calculations; the details are described, for example, in [9].

Another covariant approach that makes heavy use of RG techniques is asymptotic safety. A theory is called asymptotically safe if all essential coupling parameters g_i of the theory approach for $k \to \infty$ a non-trivial (i.e. non-vanishing) fixed point. This approach has recently attracted a lot of attention, see, for example, [9, 10] and the references therein. The paper [10] puts particular emphasis on the role of background independence in this approach.

Most modern covariant approaches make use of path integrals. Formally, one has to integrate over all four-dimensional metrics,

$$Z[g] = \int \mathcal{D}g_{\mu\nu}(x)\ \mathrm{e}^{\mathrm{i}S[g_{\mu\nu}(x)]/\hbar}\ ,$$

and, if needed, non-gravitational fields. The expression is formal, since for a rigorous definition one would have to specify the details of the measure and the regularization procedure. Except for general manipulations, the path integral has therefore been used mainly in a semiclassical expansion or for discretized approaches. An example for the first is Hawking's use of the Euclidean path integral in quantum cosmology, while examples for the second application are Regge calculus and dynamical triangulation. In dynamical triangulation, for example, one decomposes spacetime into simplices whose edge lengths remain fixed. The sum in the path integral is then performed over all possible combinations with equilateral simplices, and heavy use of Monte-Carlo simulations is made, see, for example [11] for a review. Among the many interesting results of this approach, I want to mention here the fact that the (expected) four-dimensionality of spacetime emerges at macroscopic scales, but that spacetime appears two-dimensional at small scales. Surprisingly, this microscopic two-dimensionality is also a result of the asymptotic-safety approach.

In spite of being perturbatively non-renormalizable, quantum general relativity can be used in the limit of small energies as an effective field theory. One can obtain, for example, one-loop corrections to non-relativistic potentials from the scattering amplitude by calculating the non-analytic terms in the momentum transfer. In this way one can find one-loop corrections to the

Newton potential [12],

$$V(r) = -\frac{Gm_1m_2}{r}\left(1 + 3\frac{G(m_1+m_2)}{rc^2} + \frac{41}{10\pi}\frac{G\hbar}{r^2c^3}\right) ,$$

as well as to the Coulomb potential [13],

$$V(r) = \frac{Q_1Q_2}{r}\left(1 + 3\frac{G(m_1+m_2)}{rc^2} + \frac{6}{\pi}\frac{G\hbar}{r^2c^3}\right) + \dots ,$$

The first correction terms, which do not contain $\hbar$, describe, in fact, effects of classical GR. The quantum gravitational corrections themselves are too small to be measurable in the laboratory, but they are at least definite predictions from quantum gravity.

4. Canonical quantum gravity

Canonical quantum gravity starts from a Hamiltonian formulation for GR and uses quantization rules to arrive at a wave functional Ψ that depends on the configuration space of the theory [1]. A central feature of all canonical theories is the presence of constraints,

$$\hat{H}\Psi = 0 , \tag{9}$$

where (9) stands for both the Hamiltonian and the diffeomorphism (momentum) constraints, which arise as a consequence of the presence of redundancies ('coordinate freedom') in GR. The various canonical versions of GR can be distinguished by the choice of canonical variables. The main approaches are

Geometrodynamics. The canonical variables are the 3-dimensional metric h_{ab} and a linear combination p^{cd} of the components of the extrinsic curvature.

Connection dynamics. The canonical variables are a connection A_a^i and a coloured electric field E_i^a.

Loop dynamics. The canonical variables are a holonomy constructed from A_a^i and the flux of E_i^a through a two-dimensional surface.

I shall give a brief review of the first and the third approach.

4.1. Quantum geometrodynamics

Quantum geometrodynamics is a very conservative approach [14]. One arrives inevitably at the relevant equations if one proceeds analogously to Schrödinger in 1926. In that year Schrödinger found his famous equation by looking for a wave equation that leads to the Hamilton–Jacobi equation in the (as we now say) semiclassical limit. As already discussed by Peres in 1962, the Hamilton–Jacobi equation(s)[1] for GR reads (here presented for

[1]The second equation states that S be invariant under infinitesimal three-dimensional coordinate transformations.

simplicity in the vacuum case)

$$16\pi G\, G_{abcd}\frac{\delta S}{\delta h_{ab}}\frac{\delta S}{\delta h_{cd}} - \frac{\sqrt{h}}{16\pi G}({}^{(3)}R - 2\Lambda) = 0\ , \tag{10}$$

$$D_a\frac{\delta S}{\delta h_{ab}} = 0\ , \tag{11}$$

where G_{abcd} is a local function of the three-metric and is called 'DeWitt metric', since it plays the role of a metric on the space of all three-metrics.

The task is now to find a functional wave equation that yields the Hamilton–Jacobi equation(s) in the semiclassical limit given by

$$\Psi[h_{ab}] = C[h_{ab}]\exp\left(\frac{\mathrm{i}}{\hbar}S[h_{ab}]\right)\ ,$$

where the variation of the prefactor C with respect to the three-metric is much smaller than the corresponding variation of S. From (10) one then finds the Wheeler–DeWitt equation (Hamiltonian constraint)

$$\hat{H}\Psi \equiv \left(-16\pi G\hbar^2 G_{abcd}\frac{\delta^2}{\delta h_{ab}\delta h_{cd}} - (16\pi G)^{-1}\sqrt{h}({}^{(3)}R - 2\Lambda)\right)\Psi = 0, \tag{12}$$

and from (11) the quantum diffeomorphism (momentum) constraints

$$\hat{D}^a\Psi \equiv -2\nabla_b\frac{\hbar}{\mathrm{i}}\frac{\delta\Psi}{\delta h_{ab}} = 0\ . \tag{13}$$

The latter equations guarantee that the wave functional Ψ is independent of infinitesimal three-dimensional coordinate transformations.

A detailed discussion of this equation and its applications can be found in [1]. We emphasize here only a central conceptual issue: the wave functional does not depend on any external time parameter. This is a direct consequence of the quantization procedure, which treats the three-metric and the extrinsic curvature (which can be imagined as the 'velocity' of the three-metric) as canonically conjugated, similar to position and momentum in quantum mechanics. By its local hyperbolic form, however, one can introduce an intrinsic timelike variable that is constructed out of the three-metric itself; in simple quantum cosmological models, the role of intrinsic time is played by the scale factor a of the Universe.

By its very construction, it is obvious that one can recover quantum field theory in an external spacetime from (12) and (13) in an appropriate limit [1]. The corresponding approximation scheme is similar to the Born–Oppenheimer approximation in molecular physics. In this way one finds the equations (10) and (11) together with a functional Schrödinger equation for non-gravitational fields on the background defined by the Hamilton–Jacobi equation. The time parameter in this Schrödinger equation is a many-fingered time and emerges from the chosen solution S.

The next order in this Born–Oppenheimer approximation gives corrections to the Hamiltonian $\hat{H}^{\mathrm{m}}$ that occurs in the Schrödinger equation for the non-gravitational fields. They are of the form

$$\hat{H}^{\mathrm{m}} \to \hat{H}^{\mathrm{m}} + \frac{1}{m_{\mathrm{P}}^2}\,(\text{various terms}) \ ,$$

see [1] for details. From this one can calculate, for example, the quantum gravitational correction to the trace anomaly in de Sitter space. The result is [15]

$$\delta\epsilon \approx -\frac{2G\hbar^2 H_{\mathrm{dS}}^6}{3(1440)^2\pi^3 c^8} \ .$$

A more recent example is the calculation of a possible contribution to the CMB anisotropy spectrum [16]. The terms lead to an enhancement of power at small scales; from the non-observation of such an enhancement one can then get a weak upper limit on the Hubble parameter of inflation, $H_{\mathrm{dS}} \lesssim 10^{17}$ GeV.

One may ask whether there is a connection between the canonical and the covariant approach. Such a connection exists at least at a formal level: the path integral satisfies the Wheeler–DeWitt equation and the diffeomorphism constraints. At the one-loop level, this connection was shown in a more explicit manner. This means that the full path integral with the Einstein–Hilbert action (if defined rigorously) should be equivalent to the constraint equations of canonical quantum gravity.

4.2. Loop quantum gravity

An alternative and inequivalent version of canonical quantum gravity is loop quantum gravity [17]. The development started with the introduction of Ashtekar's New Variables in 1986, which are defined as follows. The new momentum variable is the densitized version of the triad,

$$E_i^a(x) := \sqrt{h(x)}e_i^a(x) \ ,$$

and the new configuration variable is the connection defined by

$$GA_a^i(x) := \Gamma_a^i(x) + \beta K_a^i(x) \ ,$$

where $\Gamma_a^i(x)$ is the spin connection, and $K_a^i(x)$ is related to the extrinsic curvature. The variable β is called the Barbero–Immirzi parameter and constitutes an ambiguity of the theory; its meaning is still mysterious. The variables are canonically conjugated,

$$\{A_a^i(x), E_j^b(y)\} = 8\pi\beta\delta_j^i\delta_a^b\delta(x,y) \ ,$$

and define the connection representation mentioned above.

In loop gravity, one uses instead the following variables derived from them. The new configuration variable is the holonomy around a loop (giving the theory its name),

$$U[A,\alpha] := \mathcal{P}\exp\left(G\int_\alpha A\right) \ ,$$

and the new momentum variable is the densitized triad flux through the surface $\mathcal{S}$ enclosed by the loop,

$$E_i[\mathcal{S}] := \int_{\mathcal{S}} \mathrm{d}\sigma_a \ E_i^a \ .$$

In the quantum theory, these variables obey canonical commutation rules. It was possible to prove a theorem analogous to the Stone–von Neumann theorem in quantum mechanics [18]: under some mild assumption, the holonomy–flux representation is unique. The kinematical structure of loop quantum gravity is thus essentially unique. As in quantum geometrodynamics, one finds a Hamiltonian constraint and a diffeomorphism constraint, although their explicit forms are different from there. In addition, a new constraint appears in connection with the use of triads instead of metrics ('Gauss constraint').

A thorough presentation of the many formal developments of loop quantum gravity can be found in [17], see also [19] for a critical review. A main feature is certainly the discrete spectrum of geometric operators. One can associate, for example, an operator $\hat{A}$ with the surface area of a classical two-dimensional surface $\mathcal{S}$. Within the well-defined and essentially unique Hilbert space structure at the kinematical level one can find the spectrum

$$\hat{A}(\mathcal{S})\Psi_S[A] = 8\pi\beta l_{\mathrm{P}}^2 \sum_{P\in S\cap\mathcal{S}} \sqrt{j_P(j_P+1)}\Psi_S[A] \ ,$$

where the j_P denote integer multiples of $1/2$, and P denotes an intersection point between the fundamental discrete structures of the theory (the 'spin networks') and $\mathcal{S}$. Area is thus quantized and occurs as a multiple of a fundamental quantum of area proportional to l_{P}^2. It must be emphasized that this (and related) results are found at the kinematical level, that is, before all quantum constraints are solved. It is thus an open problem whether they survive the solution of the constraints, which would be needed in order to guarantee physical meaning. Moreover, in contrast to quantum geometrodynamics, it is not yet clear whether loop quantum gravity has the correct semiclassical limit.

5. String theory

String theory is fundamentally different from the approaches described above. The aim is not to perform a direct quantization of GR, but to construct a quantum theory of all interactions (a 'theory of everything') from where quantum gravity can be recovered in an appropriate limit. The inclusion of gravity in string theory is, in fact, unavoidable, since no consistent theory can be constructed without the presence of the graviton.

String theory has many important features such as the presence of gauge invariance, supersymmetry, and higher dimensions. Its structure is thus much more rigid than that of quantum GR which allows but does not demand these features. The hope with string theory is that perturbation theory is finite at all orders, although the sum diverges. The theory contains only three fundamental dimensionful constants, $\hbar, c, l_s$, where l_s is the string length. The expectation is (or was) that all other parameters (couplings, masses, ...) can be derived from these constants, once the path from the higher-dimensional (10- or 11-dimensional) spacetime to four dimensions is found. Whether this goal can ever be reached is far from clear. It is even claimed that there are so many possibilities available that a sensible selection can only be made on the basis of the anthropic principle. This is the idea of the 'string landscape' in which at least 10^{500} 'vacua' corresponding to a possible world are supposed to exist, cf. [20]. If this were true, much of the original motivation for string theory would have gone.

Since string theory contains GR in some limit, the above arguments that lead to the Wheeler–DeWitt equation remain true, that is, this equation should also follow as an approximate equation in string theory if one is away from the Planck (or string) scale. The disappearance of external time should thus also hold in string theory, but has not yet been made explicit.

It is not the place here to give a discussion of string theory. An accessible introduction is, for example, [21]; some recent developments can be found in [5] as well as in many other sources. In fact, current research focuses on issues such as AdS/CFT correspondence and holographic principle, which are motivated by string theory but go far beyond it [22]. Roughly, this correspondence states that non-perturbative string theory in a background spacetime that is asymptotically anti-de Sitter (AdS) is dual to a conformal field theory (CFT) defined in a flat spacetime of one less dimension, a conjecture made by Maldacena in 1998. This is often considered as a mostly background-independent definition of string theory, since information about the background metric enters only through boundary conditions at infinity.[2]

AdS/CFT correspondence is considered to be a realization of the holographic principle which states that all the information needed for the description of a spacetime region is already contained in its boundary. If the Maldacena conjecture is true, laws including gravity in three space dimensions will be equivalent to laws excluding gravity in two dimensions. In a sense, space has then vanished, too. It is, however, not clear whether this equivalence is a statement about the reality of Nature or only (as I suspect) about the formal properties of two descriptions describing a world with gravity.

6. Black holes and cosmology

As we have seen, effects of quantum gravity in the laboratory are expected to be too small to be observable. The main applications of quantum gravity should thus be found in the astrophysical realm – in cosmology and the physics of black holes.

---

[2]But it is also claimed that string *field* theory is the only truly background-independent approach to string theory, see the article by Taylor in [5].

As for black holes, at least two questions should be answered by quantum gravity. First, it should provide a statistical description of the Bekenstein–Hawking entropy (6). And second, it should be able to describe the final stage of black-hole evaporation when the semiclassical approximation used by Hawking breaks down.

The first problem should be easier to tackle because its solution should be possible for black holes of arbitrary size, that is, also for situations where the quantum effects of the final evaporation are negligible. In fact, preliminary results have been found in all of the above approaches and can be summarized as follows:

Loop quantum gravity. The microscopic degrees of freedom are the spin networks; S_{BH} only follows for a specific choice of the Barbero–Immirzi parameter β: $\beta = 0.237532\ldots$ [23].

String theory. The microscopic degrees of freedom are the 'D-branes'; S_{BH} follows for special (extremal or near-extremal) black holes. More generally, the result follows for black holes characterized by a near-horizon region with an AdS_3-factor [24].

Quantum geometrodynamics. One can find $S \propto A$ in various models, for example the LTB model describing a self-gravitating spherically-symmetric dust cloud [25].

A crucial feature is the choice of the correct state counting [26]. One must treat the fundamental degrees of freedom either as distinguishable (e.g. loop quantum gravity) or indistinguishable (e.g. string theory) in order to reproduce (6).

The second problem (final evaporation phase) is much more difficult, since the full quantum theory of gravity is in principle needed. At the level of the Wheeler–DeWitt equation, one can consider oversimplified models such as the one presented in [27], but whether the results have anything in common with the results of the full theory is open.

The second field of application is cosmology. Again, it is not the place here to give an introduction to quantum cosmology and its many applications, see, for example, [1, 28] and the references therein. Most work in this field is done in the context of canonical quantum gravity. For example, the Wheeler–DeWitt equation assumes the following form for a Friedmann universe with scale factor a and homogeneous scalar field ϕ,

$$\frac{1}{2}\left(\frac{G\hbar^2}{a^2}\frac{\partial}{\partial a}\left(a\frac{\partial}{\partial a}\right) - \frac{\hbar^2}{a^3}\frac{\partial^2}{\partial\phi^2} - G^{-1}a + G^{-1}\frac{\Lambda a^3}{3} + m^2a^3\phi^2\right)\psi(a,\phi) = 0 \,.$$

In loop quantum cosmology, the Wheeler–DeWitt equation is replaced by a difference equation [29]. Important issues include the possibility of singularity avoidance, the semiclassical limit including decoherence, the justification of an inflationary phase in the early Universe, the possibility of observational confirmation, and the origin of the arrow of time.

Acknowledgment
I thank the organizers of this conference for inviting me to a most inspiring meeting.

References

[1] C. Kiefer, *Quantum gravity.* 2nd edition, Oxford University Press, Oxford, 2007.

[2] C. Kiefer, Quantum gravity — a short overview, in: *Quantum gravity*, edited by B. Fauser, J. Tolksdorf, and E. Zeidler. Birkhäuser Verlag, Basel, 2006, pp. 1–13.

[3] P. G. Thirolf *et al.*, Signatures of the Unruh effect via high-power, short-pulse lasers, Eur. Phys. J. D **55**, 379–389 (2009).

[4] T. Padmanabhan, Thermodynamical aspects of gravity: new insights, Rep. Prog. Phys. **73**, 046901 (2010).

[5] *Approaches to quantum gravity*, edited by D. Oriti. Cambridge University Press, Cambridge, 2009.

[6] A. S. Goldhaber and M. N. Nieto, Photon and graviton mass limits, Rev. Mod. Phys. **82**, 939–979 (2010).

[7] Z. Bern *et al.*, Ultraviolet behavior of $N = 8$ supergravity at four loops, Phys. Rev. Lett. **103**, 081301 (2009).

[8] H. Nicolai, Vanquishing infinity, Physics **2**, 70 (2009).

[9] H. W. Hamber, *Quantum gravitation – The Feynman path integral approach.* Springer, Berlin, 2009.

[10] M. Reuter and H. Weyer, The role of background independence for asymptotic safety in quantum Einstein gravity, Gen. Relativ. Gravit. **41**, 983–1011 (2009).

[11] J. Ambjørn, J. Jurkiewicz, and R. Loll, Quantum gravity as sum over spacetimes, in: Lect. Notes Phys. **807**, 59–124 (2010).

[12] N. E. J. Bjerrum-Bohr, J. F. Donoghue, and B. R. Holstein, Quantum gravitational corrections to the nonrelativistic scattering potential of two masses, Phys. Rev. D **67**, 084033 (2003).

[13] S. Faller, Effective field theory of gravity: leading quantum gravitational corrections to Newton's and Coulomb's law, Phys. Rev. D **77**, 12409 (2008).

[14] C. Kiefer, Quantum geometrodynamics: whence, whither?, Gen. Relativ. Gravit. **41**, 877–901 (2009).

[15] C. Kiefer, Quantum gravitational effects in de Sitter space, in: *New frontiers in gravitation*, edited by G. A. Sardanashvily. Hadronic Press, Palm Harbor, 1996, see also `arXiv:gr-qc/9501001v1 (1995)`.

[16] C. Kiefer and M. Krämer, Quantum gravitational contributions to the CMB anisotropy spectrum. `arXiv:1103.4967v1 [gr-qc]`.

[17] T. Thiemann, *Modern canonical quantum general relativity.* Cambridge University Press, Cambridge, 2007.

[18] C. Fleischhack, Representations of the Weyl algebra in quantum geometry, Commun. Math. Phys. **285**, 67–140 (2009); J. Lewandowski *et al.*, Uniqueness of diffeomorphism invariant states on holonomy-flux algebras, Commun. Math. Phys. **267**, 703–733 (2006).

[19] H. Nicolai, K. Peeters, and M. Zamaklar, Loop quantum gravity: an outside view, Class. Quantum Grav. **22**, R193–R247 (2005).

[20] *Universe or multiverse?*, edited by B. Carr (Cambridge University Press, Cambridge, 2007).

[21] B. Zwiebach, *A first course in string theory*, 2nd Edition. Cambridge University Press, Cambridge, 2009.

[22] O. Aharony *et al.*, Large N field theories, string theory and gravity, Phys. Rep. **323**, 183–386 (2000).

[23] M. Domagala and J. Lewandowski, Black-hole entropy from quantum geometry, Class. Quantum Grav. **21**, 5233–5243 (2004); K. A. Meissner, Black-hole entropy in loop quantum gravity, Class. Quantum Grav. **21**, 5245–5251 (2004).

[24] A. Strominger, Black hole entropy from near-horizon microstates, JHEP 02 (1998) 009.

[25] C. Vaz, S. Gutti, C. Kiefer, and T. P. Singh, Quantum gravitational collapse and Hawking radiation in 2+1 dimensions, Phys. Rev. D **76**, 124021 (2007).

[26] C. Kiefer and G. Kolland, Gibbs' paradox and black-hole entropy, Gen. Relativ. Gravit. **40**, 1327–1339 (2008).

[27] C. Kiefer, J. Marto, and P. V. Moniz, Indefinite oscillators and black-hole evaporation, Annalen der Physik **18**, 722–735 (2009).

[28] M. Bojowald, C. Kiefer, and P. V. Moniz, Quantum cosmology for the 21st century: a debate. `arXiv:1005.2471 [gr-qc]` (2010).

[29] M. Bojowald, *Canonical gravity and applications.* Cambridge University Press, Cambridge, 2011.

Claus Kiefer
Institut für Theoretische Physik
Universität zu Köln
Zülpicher Straße 77
D 50937 Köln
Germany
e-mail: `kiefer@thp.uni-koeln.de`

Local Covariance and Background Independence

Klaus Fredenhagen and Katarzyna Rejzner

Abstract. One of the many conceptual difficulties in the development of quantum gravity is the role of a background geometry for the structure of quantum field theory. To some extent the problem can be solved by the principle of local covariance. The principle of local covariance was originally imposed in order to restrict the renormalization freedom for quantum field theories on generic spacetimes. It turned out that it can also be used to implement the request of background independence. Locally covariant fields then arise as background-independent entities.

Mathematics Subject Classification (2010). 83C45, 81T05, 83C47.

Keywords. Local covariance principle, quantum gravity, background independence, algebraic quantum field theory.

1. Introduction

The formulation of a theory of quantum gravity is one of the most important unsolved problems in physics. It faces not only technical but, above all, conceptual problems. The main one arises from the fact that, in quantum physics, space and time are a priori structures which enter the definition of the theory as well as its interpretation in a crucial way. On the other hand, in general relativity, spacetime is a dynamical object, determined by classical observables. To solve this apparent discrepancy, radical new approaches were developed. Among these the best-known are string theory and loop quantum gravity. Up to now all these approaches meet the same problem: It is extremely difficult to establish the connection to actual physics.

Instead of following the standard approaches to quantum gravity we propose a more conservative one. We concentrate on the situation when the influence of the gravitational field is weak. This idealization is justified in a large scope of physical situations. Under this assumption one can approach the problem of quantum gravity from the field-theoretic side. In the first step we consider spacetime to be a given Lorentzian manifold, on which quantum fields live. In the second step gravitation is quantized around a given background. This is where the technical problems start. The resulting theory is nonrenormalizable, in the sense that infinitely many counterterms arise in the process of renormalization. Furthermore, the causal structure of the theory is determined by the background metric. Before discussing these difficulties we want to point out that also the first step is by no means trivial. Namely, the standard formalism of quantum field theory is based on the symmetries of Minkowski space. Its generalization even to the most symmetric spacetimes (de Sitter, anti-de Sitter) poses problems. There is no vacuum, no particles, no S-matrix, etc. A solution to these difficulties is provided by concepts of algebraic quantum field theory and methods from microlocal analysis.

One starts with generalizing the Haag-Kastler axioms to generic spacetimes. We consider algebras $\mathfrak{A}(\mathcal{O})$ of observables which can be measured within the spacetime region $\mathcal{O}$, satisfying the axioms of isotony, locality (commutativity at spacelike distances) and covariance. Stability is formulated as the existence of a vacuum state (spectrum condition). The existence of a dynamical law (field equation) is understood as fulfilling the timeslice axiom (primitive causality) which says that the algebra of a timeslice is already the algebra of the full spacetime. This algebraic framework, when applied to generic Lorentzian manifolds, still meets a difficulty. The causal structure is well defined, but the absence of nontrivial symmetries raises the question: What is the meaning of repeating an experiment? This is a crucial point if one wants to keep the probability interpretation of quantum theory. A related issue is the need of a generally covariant version of the spectrum condition. These problems can be solved within *locally covariant quantum field theory*, a new framework for QFT on generic spacetimes proposed in [8].

2. Locally covariant quantum field theory

The framework of locally covariant quantum field theory was developed in [8, 17, 18]. The idea is to construct the theory simultaneously on all spacetimes (of a given class) in a coherent way. Let $\mathcal{M}$ be a globally hyperbolic, oriented, time-oriented Lorentzian 4d spacetime. Global hyperbolicity means that $\mathcal{M}$ is diffeomorphic to $\mathbb{R}\times\Sigma$, where Σ is a Cauchy surface of $\mathcal{M}$. Between spacetimes one considers a class of admissible embeddings. An embedding $\chi:\mathcal{N}\to\mathcal{M}$ is called admissible if it is isometric, time-orientation and orientation preserving, and causally convex in the following sense: If γ is a causal curve in $\mathcal{M}$ with endpoints $p,q\in\chi(\mathcal{N})$, then $\gamma=\chi\circ\gamma'$ with a causal curve γ' in $\mathcal{N}$. A locally covariant QFT is defined by assigning to spacetimes $\mathcal{M}$ corresponding unital C^*-algebras $\mathfrak{A}(\mathcal{M})$. This assignment has to fulfill a set of axioms, which generalize the Haag-Kastler axioms:

1. $\mathcal{M}\mapsto\mathfrak{A}(\mathcal{M})$ unital C^*-algebra (*local observables*).
2. If $\chi:\mathcal{N}\to\mathcal{M}$ is an admissible embedding, then $\alpha_\chi:\ \mathfrak{A}(\mathcal{N})\to\mathfrak{A}(\mathcal{M})$ is a unit-preserving C^*-homomorphism (*subsystems*).

3. Let $\chi : \mathcal{N} \to \mathcal{M}$, $\chi' : \mathcal{M} \to \mathcal{L}$ be admissible embeddings; then $\alpha_{\chi' \circ \chi} = \alpha_{\chi'} \circ \alpha_\chi$ (*covariance*).
4. If $\chi_1 : \mathcal{N}_1 \to \mathcal{M}$, $\chi_2 : \mathcal{N}_2 \to \mathcal{M}$ are admissible embeddings such that $\chi_1(\mathcal{N}_1)$ and $\chi_2(\mathcal{N}_2)$ are spacelike separated in $\mathcal{M}$, then
$$[\alpha_{\chi_1}(\mathfrak{A}(\mathcal{N}_1)), \alpha_{\chi_2}(\mathfrak{A}(\mathcal{N}_2))] = 0 \quad (\textit{locality}).$$
5. If $\chi(\mathcal{N})$ contains a Cauchy surface of $\mathcal{M}$, then $\alpha_\chi(\mathfrak{A}(\mathcal{N})) = \mathfrak{A}(\mathcal{M})$ (*time-slice axiom*).

Axioms 1-3 have a natural interpretation in the language of category theory. Let **Loc** be the category of globally hyperbolic Lorentzian spacetimes with admissible embeddings as morphisms and **Obs** the category of unital C^*-algebras with homomorphisms as morphisms. Then a locally covariant quantum field theory is defined as a covariant functor $\mathfrak{A}$ between **Loc** and **Obs**, with $\mathfrak{A}\chi := \alpha_\chi$.

The fourth axiom is related to the tensorial structure of the underlying categories. The one for the category **Loc** is given in terms of disjoint unions. It means that objects in $\mathbf{Loc}^\otimes$ are all elements $\mathcal{M}$ that can be written as $\mathcal{M}_1 \otimes \ldots \otimes \mathcal{M}_N := \mathcal{M}_1 \coprod \ldots \coprod \mathcal{M}_n$ with the unit provided by the empty set $\varnothing$. The admissible embeddings are maps $\chi : \mathcal{M}_1 \coprod \ldots \coprod \mathcal{M}_n \to \mathcal{M}$ such that each component satisfies the requirements mentioned above and additionally all images are spacelike to each other, i.e., $\chi(\mathcal{M}_1) \perp \ldots \perp (\mathcal{M}_n)$. The tensorial structure of the category **Obs** is a more subtle issue. Since there is no unique tensor structure on general locally convex vector spaces, one has to either restrict to some subcategory of **Obs** (for example nuclear spaces) or make a choice of the tensor structure based on some physical requirements. The functor $\mathfrak{A}$ can be then extended to a functor $\mathfrak{A}^\otimes$ between the categories $\mathbf{Loc}^\otimes$ and $\mathbf{Obs}^\otimes$. It is a covariant tensor functor if the following conditions hold:

$$\mathfrak{A}^\otimes\left(\mathcal{M}_1 \coprod \mathcal{M}_2\right) = \mathfrak{A}(\mathcal{M}_1) \otimes \mathfrak{A}(\mathcal{M}_2) \tag{2.1}$$
$$\mathfrak{A}^\otimes(\chi \otimes \chi') = \mathfrak{A}^\otimes(\chi) \otimes \mathfrak{A}^\otimes(\chi') \tag{2.2}$$
$$\mathfrak{A}^\otimes(\varnothing) = \mathbb{C} \tag{2.3}$$

It can be shown that if $\mathfrak{A}$ is a tensor functor, then the causality follows. To see this, consider the natural embeddings $\iota_i : \mathcal{M}_i \to \mathcal{M}_1 \coprod \mathcal{M}_2$, $i = 1, 2$, for which $\mathfrak{A}\iota_1(A_1) = A_1 \otimes \mathbb{1}$, $\mathfrak{A}\iota_2(A_2) = \mathbb{1} \otimes A_2$, $A_i \in \mathfrak{A}(\mathcal{M}_i)$. Now let $\chi_i : \mathcal{M}_i \to \mathcal{M}$ be admissible embeddings such that the images of χ_1 and χ_2 are causally disjoint in $\mathcal{M}$. We define now an admissible embedding $\chi : \mathcal{M}_1 \coprod \mathcal{M}_2 \to \mathcal{M}$ by

$$\chi(x) = \begin{cases} \chi_1(x) & , \quad x \in \mathcal{M}_1 \\ \chi_2(x) & , \quad x \in \mathcal{M}_2 \end{cases} . \tag{2.4}$$

Since $\mathfrak{A}^\otimes$ is a covariant tensor functor, it follows that

$$[\mathfrak{A}\chi_1(A_1), \mathfrak{A}\chi_2(A_2)] = \mathfrak{A}\chi[\mathfrak{A}\iota_1(A_1), \mathfrak{A}\iota_2(A_2)] = \mathfrak{A}\chi[A_1 \otimes \mathbb{1}, \mathbb{1} \otimes A_2] = 0. \tag{2.5}$$

This proves the causality. With a little bit more work it can be shown that also the opposite implication holds, i.e., the causality axiom implies that the functor $\mathfrak{A}$ is tensorial.

The last axiom is related to cobordisms of Lorentzian manifolds. One can associate to a Cauchy surface $\Sigma \subset \mathcal{M}$ a family of algebras $\{\mathfrak{A}(\mathcal{N})\}_{\mathcal{N}\in I}$, where the index set consists of all admissibly embedded subspacetimes $\mathcal{N}$ of $\mathcal{M}$ that contain the Cauchy surface Σ. On this family we can introduce an order relation $\geq$, provided by the inclusion. Let $\mathcal{N}_i, \mathcal{N}_j \in I$, such that $\mathcal{N}_i \subset \mathcal{N}_j \in I$; then we say that $\mathcal{N}_i \geq \mathcal{N}_j$. Clearly Σ is the upper limit with respect to the order relation $\geq$, hence we obtain a directed system of algebras $(\{\mathfrak{A}(\mathcal{N})\}_{\mathcal{N}\in I}, \geq)$. Now let $\chi_{ij} : \mathcal{N}_i \hookrightarrow \mathcal{N}_j$ be the canonical isometric embedding of $\mathcal{N}_i \geq \mathcal{N}_j$. From the covariance it follows that there exists a morphism of algebras $\alpha_{\chi_{ji}} : \mathfrak{A}(\mathcal{N}_i) \hookrightarrow \mathfrak{A}(\mathcal{N}_j)$. We can now consider a family of all such mappings between the elements of the directed system $(\{\mathfrak{A}(\mathcal{N})\}_{\mathcal{N}\in I}, \geq)$. Clearly $\alpha_{\chi_{ii}}$ is the identity on $\mathfrak{A}(\mathcal{N}_i)$, and $\alpha_{\chi_{ik}} = \alpha_{\chi_{ij}} \circ \alpha_{\chi_{jk}}$ for all $\mathcal{N}_i \leq \mathcal{N}_j \leq \mathcal{N}_k$. This means that the family of mappings $\alpha_{\chi_{ij}}$ provides the transition morphisms for the directed system $(\{\mathfrak{A}(\mathcal{N})\}_{\mathcal{N}\in I}, \geq)$ and we can define the projective (inverse) limit of the inverse system of algebras $(\{\mathfrak{A}(\mathcal{N})\}_{\mathcal{N}\in I}, \geq, \{\alpha_{\chi_{ij}}\})$, i.e.:

$$\mathfrak{A}(\Sigma) := \varprojlim_{\mathcal{N}\supset\Sigma} \mathfrak{A}(\mathcal{N}) = \Big\{ \text{germ of } (a)_I \in \prod_{\mathcal{N}\in I} \mathfrak{A}(\mathcal{N}) \;\Big|\; a_{\mathcal{N}_i} = \alpha_{\chi_{ij}}(a_{\mathcal{N}_j})\ \forall\, \mathcal{N}_i \leq \mathcal{N}_j \Big\}. \tag{2.6}$$

The algebra $\mathfrak{A}(\Sigma)$ obtained in this way depends in general on the germ of Σ in $\mathcal{M}$. If we consider natural embeddings of Cauchy surfaces Σ in $\mathcal{M}$, then acting with the functor $\mathfrak{A}$ we obtain homomorphisms of algebras, which we denote by $\alpha_{\mathcal{M}\Sigma}$. The timeslice axiom implies that these homomorphisms are in fact isomorphisms. It follows that the propagation from Σ_1 to another Cauchy surface Σ_2 is described by the isomorphism

$$\alpha^{\mathcal{M}}_{\Sigma_1\Sigma_2} := \alpha^{-1}_{\mathcal{M}\Sigma_1} \alpha_{\mathcal{M}\Sigma_2} \,. \tag{2.7}$$

Given a cobordism, i.e. a Lorentzian manifold $\mathcal{M}$ with future/past boundary $\Sigma_\pm$, we obtain an assignment: $\Sigma_\pm \mapsto \mathfrak{A}(\Sigma_\pm)$, $\mathcal{M} \mapsto \alpha^{\mathcal{M}}_{\Sigma_-\Sigma_+}$. The concept of relative Cauchy evolution obtained in this way realizes the notion of dynamics in the locally covariant quantum field theory framework. This provides a solution to the old problem of Schwinger to formulate the functional evolution of the quantum state. The original idea to understand it as a unitary map between Hilbert spaces turned out not to be a viable concept even in Minkowski spacetime [25]. Nevertheless one can understand the dynamical evolution on the algebraic level as an isomorphism of algebras corresponding to the Cauchy surfaces. This idea was already applied by Buchholz and Verch [26] to some concrete examples, and the locally covariant quantum theory provides a more general framework in which this approach is justified. Note also the structural similarity to topological field theory [24]. There, however, the objects are finite-dimensional vector spaces, so the functional-analytic obstructions which are typical for quantum field theory do not arise.

3. Perturbative quantum gravity

After the brief introduction to the locally covariant QFT framework we can now turn back to the problem of quantum gravity seen from the point of view of perturbation theory. First we split off the metric:

$$g_{ab} = g^{(0)}_{ab} + h_{ab}\,, \tag{3.1}$$

where $g^{(0)}$ is the background metric, and h is a quantum field. Now we can renormalize the Einstein-Hilbert action by the Epstein-Glaser method (interaction restricted to a compact region between two Cauchy surfaces) and construct the functor $\mathfrak{A}$. Next we compute (2.7) for two background metrics which differ by κ_{ab} compactly supported between two Cauchy surfaces. Let $\mathcal{M}_1 = (M, g^{(0)})$ and $\mathcal{M}_2 = (M, g^{(0)} + \kappa)$. Following [8, 7], we assume that there are two causally convex neighbourhoods $\mathcal{N}_\pm$ of the Cauchy surfaces $\Sigma_\pm$ which can be admissibly embedded both in $\mathcal{M}_1$ and $\mathcal{M}_2$, and that κ is supported in a compact region between $\mathcal{N}_-$ and $\mathcal{N}_+$. We denote the corresponding embeddings by $\chi_i^\pm : \mathcal{N}_\pm \to \mathcal{M}_i$, $i = 1, 2$. We can now define an automorphism of $\mathcal{M}_1$ by:

$$\beta_\kappa := \alpha_{\chi_1^-} \circ \alpha^{-1}_{\chi_2^-} \circ \alpha_{\chi_2^+} \circ \alpha^{-1}_{\chi_1^+}\,. \tag{3.2}$$

This automorphism corresponds to a change of the background between the two Cauchy surfaces. Under the geometrical assumptions given in [8] one can calculate a functional derivative of β_κ with respect to κ. If the metric is not quantized, it was shown in [8] that this derivative corresponds to the commutator with the stress-energy tensor. In the case of quantum gravity, $\frac{\delta\beta_\kappa}{\delta\kappa_{ab}(x)}$ involves in addition also the Einstein tensor. Therefore the background independence may be formulated as the condition that $\frac{\delta\beta_\kappa}{\delta\kappa_{ab}(x)} = 0$, i.e., one requires the validity of Einstein's equation for the quantized fields. This can be translated into a corresponding renormalization condition.

The scheme proposed above meets some technical obstructions. The first of them is the nonrenormalizability. This means that in every order new counterterms appear. Nevertheless, if these terms are sufficiently small, we can still have a predictive power of the resulting theory, as an effective theory. The next technical difficulty is imposing the constraints related to the gauge (in this case: diffeomorphism) invariance. In perturbation theory this can be done using the BRST method [4, 5] (or more generally Batalin-Vilkovisky formalism [3, 1]). Since this framework is based on the concept of local objects, one encounters another problem. Local BRST cohomology turns out to be trivial [14], hence one has to generalize the existing methods to global objects. Candidates for global quantities are *fields*, considered as natural transformations between the functor of test function spaces $\mathfrak{D}$ and the quantum field theory functor $\mathfrak{A}$. A quantum field $\Phi : \mathfrak{D} \to \mathfrak{A}$ corresponds therefore to a family of mappings $(\Phi_\mathcal{M})_{\mathcal{M}\in\mathrm{Obj}(\mathbf{Loc})}$, such that $\Phi_\mathcal{M}(f) \in \mathfrak{A}(\mathcal{M})$ for $f \in \mathfrak{D}(M)$ and given a morphism $\chi : \mathcal{N} \to \mathcal{M}$ we have $\alpha_\chi(\Phi_\mathcal{N}(f)) = \Phi_\mathcal{M}(\chi_* f)$.

4. BRST cohomology for classical gravity

While quantum gravity is still elusive, classical gravity is (to some extent) well understood. Therefore one can try to test concepts for quantum gravity in a suitable framework for classical gravity. Such a formalism is provided by the algebraic formulation, where classical field theory occurs as the $\hbar = 0$ limit of quantum field theory [9, 10, 12, 11]. In this approach (the *functional approach*) one replaces the associative involutive algebras by Poisson algebras. In the case of gravity, to obtain a suitable Poisson structure one has to fix the gauge. In the BRST method this is done by adding a gauge-fixing term and a ghost term to the Einstein-Hilbert action. The so-called *ghost fields* have a geometrical interpretation as Maurer-Cartan forms on the diffeomorphism group. This can be made precise in the framework of infinite-dimensional differential geometry. The notion of infinite-dimensional manifolds and, in particular, infinite-dimensional Lie groups is known in mathematics since the reviews of Hamilton [16] and Milnor [21]. Because one needs to consider manifolds modeled on general locally convex vector spaces, an appropriate calculus has to be chosen. Unfortunately the choice is not unique when we go beyond Banach spaces. Historically the earliest works concerning such generalizations of calculus are those of Michal [19] (1938) and Bastiani [2] (1964). At present there are two main frameworks in which the problems of infinite-dimensional differential geometry can be approached: the convenient setting of global analysis [15, 20] and the locally convex calculus [16, 22]. Up to now both calculi coincide in the examples which were considered.

First we sketch the BRST construction performed on the fixed background $\mathcal{M}$. The basic objects of the classical theory are:

- S, a diffeomorphism-invariant action;
- field content: configuration space $\mathfrak{E}(\mathcal{M})$, considered as an infinite-dimensional manifold: scalar, vector, tensor and spinor fields (including the metric), gauge fields;
- ghost fields (fermions): forms on the gauge algebra ΓTM, i.e. elements of $(\Gamma TM)^*$;
- antifields (fermions): vector fields $\Gamma T\mathfrak{E}(\mathcal{M})$ on the configuration space;
- antifields of ghosts (bosons): compactly supported vector fields $\Gamma_c TM$.

The fields listed above constitute the minimal sector of the theory. To impose a concrete gauge, one can also introduce further fields, the so-called nonminimal sector. For the harmonic gauge it consists of Nakanishi-Lautrup fields (bosonic) and antighosts (fermionic). The minimal sector of the BRST-extended functional algebra takes the form

$$\mathfrak{BV}(\mathcal{M}) = Sym(\Gamma_c T\mathcal{M}) \mathbin{\widehat{\otimes}} \Lambda(\Gamma T\mathfrak{E}(\mathcal{M})) \mathbin{\widehat{\otimes}} \Lambda(\Gamma T\mathcal{M})^* , \tag{4.1}$$

where $\widehat{\otimes}$ denotes the sequentially completed tensor product, and *Sym* is the symmetric algebra. The algebra (4.1) is equipped with a grading called *the ghost number* and a graded differential s consisting of two terms $s = \delta + \gamma$. Both δ and γ are graded differentials. The natural action of $\Gamma T\mathcal{M}$ by the

Lie derivative on $\mathfrak{E}(\mathcal{M})$ induces in a natural way an action on $T\mathfrak{E}(\mathcal{M})$. Together with the adjoint action on $\Gamma_c T\mathcal{M}$ we obtain an action of $\Gamma T\mathcal{M}$ on $Sym(\Gamma_c T\mathcal{M}) \otimes \Lambda(\Gamma T\mathfrak{E}(\mathcal{M}))$ which we denote by ρ. We can now write down how δ and γ act on the basic fields $a \in \Gamma_c T\mathcal{M}$, $Q \in \Gamma T\mathfrak{E}(\mathcal{M})$, $\omega \in (\Gamma T\mathcal{M})^*$:

- $\langle \gamma(a \otimes Q \otimes \mathbb{1}), X \rangle := \rho_X(a \otimes Q \otimes \mathbb{1})$,
- $\langle \gamma(a \otimes Q \otimes \omega), X \wedge Y \rangle :=$
  $$\rho_X(a \otimes Q \otimes \langle \omega, Y \rangle) - \rho_Y(a \otimes Q \otimes \langle \omega, X \rangle) - a \otimes Q \otimes \langle \omega, [X,Y] \rangle,$$
- $\delta(\mathbb{1} \otimes Q \otimes \omega) := \mathbb{1} \otimes \partial_Q S \otimes \omega$,
- $\delta(a \otimes \mathbb{1} \otimes \omega) := \mathbb{1} \otimes \rho(a) \otimes \omega$.

Up to now the construction was done on the fixed spacetime, but it is not difficult to see that the assignment of the graded algebra $\mathfrak{BV}(\mathcal{M})$ to spacetime $\mathcal{M}$ can be made into a covariant functor [14].

As indicated already, the BRST method when applied to gravity has to be generalized to global objects. Otherwise the cohomology of the BRST operator s turns out to be trivial. This corresponds to the well-known fact that there are no local on-shell observables in general relativity. It was recently shown in [14] that one can introduce the BRST operator on the level of natural transformations and obtain in this way a nontrivial cohomology. Fields are now understood as natural transformations. Let $\mathfrak{D}^k$ be a functor from the category **Loc** to the product category $\mathbf{Vec}^k$, that assigns to a manifold M a k-fold product of the test section spaces $\mathfrak{D}(M) \times \ldots \times \mathfrak{D}(M)$. Let $\mathrm{Nat}(\mathfrak{D}^k, \mathfrak{BV})$ denote the set of natural transformations from $\mathfrak{D}^k$ to $\mathfrak{BV}$. We define the extended algebra of fields as

$$Fld = \bigoplus_{k=0}^{\infty} \mathrm{Nat}(\mathfrak{D}^k, \mathfrak{BV}) . \tag{4.2}$$

It is equipped with a graded product defined by

$$(\Phi\Psi)_M(f_1, ..., f_{p+q}) = \frac{1}{p!q!} \sum_{\pi \in P_{p+q}} \Phi_M(f_{\pi(1)}, ..., f_{\pi(p)}) \Psi_M(f_{\pi(p+1)}, ..., f_{\pi(p+q)}), \tag{4.3}$$

where the product on the right-hand side is the product of the algebra $\mathfrak{BV}(M)$. Let Φ be a field; then the action of the BRST differential on it is defined by

$$(s\Phi)_M(f) := s(\Phi_M(f)) + (-1)^{|\Phi|} \Phi_M(\pounds_{(.)} f) , \tag{4.4}$$

where $|.|$ denotes the ghost number and the action of s on $\mathfrak{BV}(\mathcal{M})$ is given above. The physical fields are identified with the 0-th cohomology of s on *Fld*. Among them we have for example scalars constructed covariantly from the metric.

5. Conclusions

It was shown that a construction of quantum field theory on generic Lorentzian spacetimes is possible, in accordance with the principle of general covariance. This framework can describe a wide range of physical situations. Also a consistent incorporation of the quantized gravitational field seems to be possible. Since the theory is invariant under the action of an infinite-dimensional Lie group, the framework of infinite-dimensional differential geometry plays an important role. It provides the mathematical setting in which the BV method has a clear geometrical interpretation. The construction of a locally covariant theory of gravity in the proposed setting was already performed for the classical theory. Based on the gained insight it seems to be possible to apply this treatment also in the quantum case. One can then investigate the relations to other field-theoretical approaches to quantum gravity (Reuter [23], Bjerrum-Bohr [6], ...). As a conclusion we want to stress that quantum field theory should be taken serious as a third way to quantum gravity.

References

[1] G. Barnich, F. Brandt, M. Henneaux, Phys. Rept. **338** (2000) 439 [arXiv:hep-th/0002245].

[2] A. Bastiani, J. Anal. Math. **13**, (1964) 1-114.

[3] I. A. Batalin, G. A. Vilkovisky, Phys. Lett. **102B** (1981) 27.

[4] C. Becchi, A. Rouet, R. Stora, Commun. Math. Phys. **42** (1975) 127.

[5] C. Becchi, A. Rouet, R. Stora, Annals Phys. **98** (1976) 287.

[6] N. E. J. Bjerrum-Bohr, Phys. Rev. **D 67** (2003) 084033.

[7] R. Brunetti, K. Fredenhagen, proceedings of Workshop on Mathematical and Physical Aspects of Quantum Gravity, Blaubeuren, Germany, 28 Jul - 1 Aug 2005. In: B. Fauser et al. (eds.), Quantum gravity, 151-159 [arXiv:gr-qc/0603079v3].

[8] R. Brunetti, K. Fredenhagen, R. Verch, Commun. Math. Phys. **237** (2003) 31-68.

[9] R. Brunetti, M. Dütsch, K. Fredenhagen, Adv. Theor. Math. Phys. **13**(5) (2009) 1541-1599 [arXiv:math-ph/0901.2038v2].

[10] M. Dütsch and K. Fredenhagen, Proceedings of the Conference on Mathematical Physics in Mathematics and Physics, Siena, June 20-25 2000 [arXiv:hep-th/0101079].

[11] M. Dütsch and K. Fredenhagen, Rev. Math. Phys. **16**(10) (2004) 1291-1348 [arXiv:hep-th/0403213].

[12] M. Dütsch and K. Fredenhagen, Commun. Math. Phys. **243** (2003) 275 [arXiv:hep-th/0211242].

[13] K. Fredenhagen, *Locally Covariant Quantum Field Theory*, Proceedings of the XIVth International Congress on Mathematical Physics, Lisbon 2003 [arXiv:hep-th/0403007].

[14] K. Fredenhagen, K. Rejzner, [arXiv:math-ph/1101.5112].

[15] A. Frölicher, A. Kriegl, *Linear spaces and differentiation theory*, Pure and Applied Mathematics, J. Wiley, Chichester, 1988.

[16] R. S. Hamilton, Bull. Amer. Math. Soc. (N.S.) **7**(1) (1982) 65-222.

[17] S. Hollands, R. Wald, Commun. Math. Phys. **223** (2001) 289.

[18] S. Hollands, R. M. Wald, Commun. Math. Phys. **231** (2002) 309.

[19] A. D. Michal, Proc. Nat. Acad. Sci. USA **24** (1938) 340-342.

[20] A. Kriegl, P. Michor, *The convenient setting of global analysis*, Mathematical Surveys and Monographs **53**, American Mathematical Society, Providence 1997. `www.ams.org/online_bks/surv53/1`.

[21] J. Milnor, *Remarks on infinite-dimensional Lie groups*, In: B. DeWitt, R. Stora (eds.), Les Houches Session XL, Relativity, Groups and Topology II, North-Holland (1984), 1007-1057.

[22] K.-H. Neeb, *Monastir lecture notes on infinite-dimensional Lie groups*, `www.math.uni-hamburg.de/home/wockel/data/monastir.pdf1`.

[23] M. Reuter, Phys. Rev. **D 57** (1998) 971 [arXiv:hep-th/9605030].

[24] G. Segal, In: Proceedings of the IXth International Congress on Mathematical Physics (Bristol, Philadelphia), eds. B. Simon, A. Truman and I. M. Davies, IOP Publ. Ltd. (1989), 22-37.

[25] C. Torre, M. Varadarajan, Class. Quant. Grav. **16**(8) (1999) 2651.

[26] R. Verch, Ph.D. Thesis, University of Hamburg, 1996.

Klaus Fredenhagen and Katarzyna Rejzner
II Inst. f. Theoretische Physik
Universität Hamburg
Luruper Chaussee 149
D-22761 Hamburg
Germany
e-mail: `klaus.fredenhagen@desy.de`
`katarzyna.rejzner@desy.de`

The "Big Wave" Theory for Dark Energy

Blake Temple

(Joint work with Joel Smoller)

Abstract. We explore the author's recent proposal that the anomalous acceleration of the galaxies might be due to the displacement of nearby galaxies by a wave that propagated during the radiation phase of the Big Bang.

Mathematics Subject Classification (2010). Primary 83B05; Secondary 35L65.

Keywords. Anomalous acceleration, self-similar waves, conservation laws.

1. Introduction

By obtaining a linear relation between the recessional velocities of distant galaxies (redshift) and luminosity (distance), the American astronomer Edwin Hubble showed in 1927 that the universe is expanding. This confirmed the so-called *standard model of cosmology*, that the universe, on the largest scale, is evolving according to a Friedmann-Robertson-Walker (FRW) spacetime. The starting assumption in this model is the *Cosmological Principle*—that on the largest scale, we are not in a special place in the universe—that, in the words of Robertson and Walker, the universe is *homogeneous* and *isotropic* about every point like the FRW spacetime. In 1998, more accurate measurements of the recessional velocity of distant galaxies based on new Type 1a supernova data made the surprising discovery that the universe was actually accelerating relative to the standard model. This is referred to as the *anomalous acceleration of the galaxies.* The only way to preserve the FRW framework and the Cosmological Principle is to modify the Einstein equations by adding an extra term called the *cosmological constant. Dark Energy*, the physical interpretation of the cosmological constant, is then an unknown source of anti-gravitation that, for the model to be correct, must account for some 70 percent of the energy density of the universe.

Supported by NSF Applied Mathematics Grant Number DMS-070-7532.

In [14] the authors introduced a family of self-similar expanding wave solutions of the Einstein equations of General Relativity (GR) that contain the standard model during the radiation phase of the Big Bang. Here I discuss our cosmological interpretation of this family, and explore the possibility that waves in the family might account for the anomalous acceleration of the galaxies without the cosmological constant or Dark Energy (see [14, 16] for details). In a nutshell, our premise is that the Einstein equations of GR during the radiation phase form a highly nonlinear system of wave equations that support the propagation of waves, and [14] is the culmination of our program to discover waves that perturb the uniform background Friedmann universe (the setting for the standard model of cosmology), something like water waves perturb the surface of a still pond. I also use this as a vehicle to record our unpublished *Answers to reporter's questions* which appeared on the author's website the week our PNAS paper [14] appeared, August 17, 2009.

In Einstein's theory of General Relativity, gravitational *forces* turn out to be just anomalies of spacetime *curvature*, and the propagation of curvature through spacetime is governed by the *Einstein equations*. The Einstein equations during the radiation phase (when the equation of state simplifies to $p = \rho c^2/3$) form a highly nonlinear system of conservation laws that support the propagation of waves, including compressive shock waves and self-similar expansion waves. Yet since the 1930s, the modern theory of cosmology has been based on the starting assumption of the Copernican Principle, which restricts the whole theory to the Friedmann spacetimes, a special class of solutions of the Einstein equations which describe a uniform three-space of constant curvature and constant density evolving in time. Our approach has been to look for general-relativistic waves that could perturb a uniform Friedmann background. The GR self-similar expanding waves in the family derived in [14] satisfy two important conditions: they perturb the standard model of cosmology, and they are the kind of waves that more complicated solutions should settle down to according to the quantitative theories of Lax and Glimm on how solutions of conservation laws decay in time to self-similar wave patterns. The great accomplishment of Lax and Glimm was to explain and quantify how *entropy*, *shock-wave dissipation* and *time-irreversibility* (concepts that originally were understood only in the context of ideal gases) could be given meaning in general systems of *conservation laws*, a setting much more general than gas dynamics. (This viewpoint is well expressed in the celebrated works [10, 5, 6].) The conclusion: Shock-waves introduce dissipation and increase of entropy into the dynamics of solutions, and this provides a mechanism by which complicated solutions can settle down to orderly self-similar wave patterns, *even when dissipative terms are neglected in the formulation of the equations.* A rock thrown into a pond demonstrates how the mechanism can transform a chaotic "plunk" into a series of orderly outgoing self-similar waves moments later. As a result, our new construction of a family of GR self-similar waves that apply when this decay mechanism should be in place received a good deal of media attention when it came out in PNAS, August 2009. (A sampling of press releases and articles can be found on my homepage.[1]

At the value of the *acceleration parameter* $a = 1$ (the free parameter in our family of self-similar solutions), the solution reduces exactly to the critical FRW spacetime of the standard model with pure radiation sources, and solutions look remarkably similar to FRW when $a \neq 1$. When $a \neq 1$, we prove that the spacetimes in the family are distinct from all the other non-critical FRW spacetimes, and hence it follows that the critical FRW during the radiation phase is characterized as the unique spacetime lying at the intersection of these two one-parameter families. Since adjustment of the free parameter a speeds up or slows down the expansion rate relative to the standard model, we argue they can account for the leading-order quadratic correction to redshift vs luminosity observed in the supernova data, without the need for Dark Energy. I first proposed the idea that the anomalous acceleration might be accounted for by a wave in the talk *Numerical Shockwave Cosmology*, New Orleans, January 2007,[2] and set out to simulate such a wave numerically. While attempting to set up the numerical simulation, we discovered that the standard model during the radiation phase admits a coordinate system (Standard Schwarzschild Coordinates (SSC)) in which the Friedmann spacetime is *self-similar*. That is, it took the form of a non-interacting time-asymptotic wave pattern according to the theory of Lax and Glimm. This was the key. Once we found this, we guessed that the Einstein equations in these coordinates must close to form a new system of ODEs in the same self-similar variable. After a struggle, we derived this system of equations, and showed that the standard model was one point in a family of solutions parameterized by four initial conditions. Symmetry and regularity at the center then reduced the four-parameter family to an implicitly defined one-parameter family, one value of which gives the critical Friedmann spacetime of the standard model during the radiation phase of the Big Bang. Our idea then: an expansion wave that formed during the radiation epoch, when the Einstein equations obey a highly nonlinear system of conservation laws for which we must expect self-similar non-interacting waves to be the end state of local fluctuations, could account for the anomalous acceleration of the galaxies without Dark Energy. Since we have explicit formulas for such waves, it is a verifiable proposition.

2. Statement of results

In this section we state three theorems which summarize our results in [14, 16]. (Unbarred coordinates (t, r) refer to FRW co-moving coordinates, and barred coordinates $(\bar{t}, \bar{r})$ refer to (SSC).)

---

[1]see *Media Articles* on my homepage `http://www.math.ucdavis.edu/~temple/`

[2]the fourth entry under *Conference/Seminar Talks* on my homepage

Theorem 2.1. *Assume $p = \frac{1}{3}\rho c^2$, $k = 0$ and $R(t) = \sqrt{t}$. Then the FRW metric*

$$ds^2 = -dt^2 + R(t)^2 dr^2 + \bar{r}^2 d\Omega^2,$$

under the change of coordinates

$$\bar{t} = \psi_0 \left\{ 1 + \left[\frac{R(t) r}{2t} \right]^2 \right\} t, \tag{2.1}$$

$$\bar{r} = R(t) r, \tag{2.2}$$

transforms to the SSC-metric

$$ds^2 = -\frac{d\bar{t}^2}{\psi_0^2 \left(1 - v^2(\xi)\right)} + \frac{d\bar{r}^2}{1 - v^2(\xi)} + \bar{r}^2 d\Omega^2, \tag{2.3}$$

where

$$v = \frac{1}{\sqrt{AB}} \frac{\bar{u}^1}{\bar{u}^0} \tag{2.4}$$

is the SSC velocity, which also satisfies

$$v = \frac{\zeta}{2}, \tag{2.5}$$

$$\psi_0 \xi = \frac{2v}{1 + v^2}. \tag{2.6}$$

Theorem 2.2. *Let ξ denote the self-similarity variable*

$$\xi = \frac{\bar{r}}{\bar{t}}, \tag{2.7}$$

and let

$$G = \frac{\xi}{\sqrt{AB}}. \tag{2.8}$$

Assume that $A(\xi)$, $G(\xi)$ and $v(\xi)$ solve the ODEs

$$\xi A_\xi = -\left[\frac{4(1-A)v}{(3+v^2)G - 4v} \right] \tag{2.9}$$

$$\xi G_\xi = -G \left\{ \left(\frac{1-A}{A} \right) \frac{2(1+v^2)G - 4v}{(3+v^2)G - 4v} - 1 \right\} \tag{2.10}$$

$$\xi v_\xi = -\left(\frac{1-v^2}{2\{\cdot\}_D} \right) \left\{ (3+v^2)G - 4v + \frac{4\left(\frac{1-A}{A}\right)\{\cdot\}_N}{(3+v^2)G - 4v} \right\}, \tag{2.11}$$

where

$$\{\cdot\}_N = \left\{ -2v^2 + 2(3-v^2)vG - (3-v^4)G^2 \right\} \tag{2.12}$$

$$\{\cdot\}_D = \left\{ (3v^2 - 1) - 4vG + (3-v^2)G^2 \right\}, \tag{2.13}$$

and define the density by

$$\kappa\rho = \frac{3(1-v^2)(1-A)G}{(3+v^2)G - 4v} \frac{1}{\bar{r}^2}. \tag{2.14}$$

Then the metric

$$ds^2 = -B(\xi)d\bar{t}^2 + \frac{1}{A(\xi)}d\bar{r}^2 + \bar{r}^2 d\Omega^2 \tag{2.15}$$

solves the Einstein-Euler equations $G = \kappa T$ with velocity $v = v(\xi)$ and equation of state $p = \frac{1}{3}\rho c^2$. In particular, the FRW metric (2.3) *solves equations* (2.9)–(2.11).

Note that it is not evident from the FRW metric in standard co-moving coordinates that self-similar variables even exist, and if they do exist, by what ansatz one should extend the metric in those variables to obtain nearby self-similar solutions that solve the Einstein equations exactly. The main point is that our coordinate mapping to SSC explicitly identifies the self-similar variables as well as the metric ansatz that together accomplish such an extension of the metric.

In [14, 16] we prove that the three-parameter family (2.9)–(2.11) (parameterized by three initial conditions) reduces to an (implicitly defined) one-parameter family by removing time-scaling invariance and imposing regularity at the center. The remaining parameter a changes the expansion rate of the spacetimes in the family, and thus we call it the *acceleration parameter*. Transforming back to (approximate) co-moving coordinates, the resulting one-parameter family of metrics is amenable to the calculation of a redshift vs luminosity relation, to third order in the redshift factor z, leading to the following theorem which applies during the radiation phase of the expansion, cf. [14, 16]:

Theorem 2.3. *The redshift vs luminosity relation, as measured by an observer positioned at the center of the expanding wave spacetimes (metrics of form* (2.15)*), is given up to fourth order in redshift factor z by*

$$d_\ell = 2ct\left\{z + \frac{a^2-1}{2}z^2 + \frac{(a^2-1)(3a^2+5)}{6}z^3 + O(1)|a-1|z^4\right\}, \tag{2.16}$$

where d_ℓ is luminosity distance, ct is invariant time since the Big Bang, and a is the acceleration parameter that distinguishes expanding waves in the family.

When $a = 1$, (2.16) reduces to the correct linear relation of the standard model, [8]. Assuming redshift vs luminosity evolves continuously in time, it follows that the leading-order part of any (small) anomalous correction to the redshift vs luminosity relation of the standard model observed *after* the radiation phase could be accounted for by suitable adjustment of parameter a.

3. Discussion

These results suggest an interpretation that we might call a *conservation law* explanation of the anomalous acceleration of the galaxies. That is, the theory of Lax and Glimm explains how highly interactive oscillatory solutions of conservation laws will decay in time to non-interacting waves (shock waves and expansion waves), by the mechanisms of wave interaction and shock-wave dissipation. The subtle point is that even though dissipation terms are neglected in the formulation of the equations, there is a canonical dissipation and consequent loss of information due to the *nonlinearities*, and this can be modeled by shock-wave interactions that drive solutions to non-interacting wave patterns. Since the one fact most certain about the standard model is that our universe arose from an earlier hot dense epoch in which all sources of energy were in the form of radiation, and since it is approximately uniform on the largest scale but highly oscillatory on smaller scales[3], one might reasonably conjecture that decay to a non-interacting expanding wave occurred during the radiation phase of the standard model, via the highly nonlinear evolution driven by the large sound speed, and correspondingly large modulus of *genuine nonlinearity*[4], present when $p = \rho c^2/3$, cf. [11]. Our analysis has shown that FRW is just one point in a family of non-interacting, self-similar expansion waves, and as a result we conclude that some further explanation is required as to why, on some length scale, decay during the radiation phase of the standard model would not proceed to a member of the family satisfying $a \neq 1$. If decay to $a \neq 1$ did occur, then the galaxies that formed from matter at the end of the radiation phase (some $379,000$ years after the Big Bang) would be displaced from their anticipated positions in the standard model at present time, and this displacement would lead to a modification of the observed redshift vs luminosity relation. In short, the displacement of the fluid particles (i.e., the displacement of the co-moving frames in the radiation field) by the wave during the radiation epoch leads to a displacement of the galaxies at a later time. In principle such a mechanism could account for the anomalous acceleration of the galaxies as observed in the supernova data. Of course, if $a \neq 1$, then the spacetime within the expansion wave has a center, and this would violate the so-called *Copernican Principle*, a simplifying assumption generally accepted in cosmology, at least on the scale of the wave (cf. the discussions in [17] and [1]). Moreover, if our Milky Way galaxy did not lie within some threshold of the center of expansion, the expanding wave theory would imply unobserved angular variations in the expansion rate. In fact, all of these observational issues have already been discussed recently in [2, 1, 3] (and references therein), which explore the possibility that the anomalous acceleration of the galaxies might be due to a local *void* or under-density of galaxies in the vicinity of the Milky Way.[5] Our proposal then is

---

[3]In the standard model, the universe is approximated by uniform density on a scale of a billion light years or so, about a tenth of the radius of the visible universe, [18]. The stars, galaxies and clusters of galaxies are then evidence of large oscillations on smaller scales.

[4]Again, *genuine nonlinearity* is, in the sense of Lax, a measure of the magnitude of nonlinear compression that drives decay, cf. [10].

[5]The size of the center, consistent with the angular dependence that has been observed in the actual supernova and microwave data, has been estimated to be about 15 megaparsecs, approximately the distance between clusters of galaxies, roughly $1/200$ the distance across the visible universe, cf. [1, 2, 3].

that the one-parameter family of general-relativistic self-similar expansion waves derived here is a family of possible end-states that could result after dissipation by wave interactions during the radiation phase of the standard model is completed, and such waves could thereby account for the appearance of a local under-density of galaxies at a later time.

In any case, the expanding wave theory is testable. For a first test, we propose next to evolve the quadratic and cubic corrections to redshift vs luminosity recorded here in relation (2.16), valid at the end of the radiation phase, up through the $p \approx 0$ stage to present time in the standard model, to obtain the present-time values of the quadratic and cubic corrections to redshift vs luminisity implied by the expanding waves, as a function of the acceleration parameter a. Once accomplished, we can look for a best fit value of a via comparison of the quadratic correction at present time to the quadratic correction observed in the supernova data, leaving the third-order correction at present time as a prediction of the theory. That is, in principle, the predicted third-order correction term could be used to distinguish the expanding wave theory from other theories (such as Dark Energy) by the degree to which they match an accurate plot of redshift vs luminosity from the supernove data (a topic of the authors' current research). The idea that the anomalous acceleration might be accounted for by a local under-density in a neighborhood of our galaxy was expounded in the recent papers [2, 3]. Our results here might then give an accounting for the source of such an under-density.

The expanding wave theory could in principle give an explanation for the observed anomalous acceleration of the galaxies within classical General Relativity, with classical sources. In the expanding wave theory, the so-called anomalous acceleration is not an acceleration at all, but is a correction to the standard model due to the fact that we are looking outward into an expansion wave. The one-parameter family of non-interacting, self-similar, general-relativistic expansion waves derived here contains all possible end-states that could result by wave interaction and dissipation due to nonlinearities back when the universe was filled with pure radiation sources. And when $a \neq 1$, they introduce an anomalous acceleration into the standard model of cosmology. Unlike the theory of Dark Energy, this provides a possible explanation for the anomalous acceleration of the galaxies that is not *ad hoc* in the sense that it is derivable exactly from physical principles and a mathematically rigorous theory of general-relativistic expansion waves. In particular, this explanation does not require the *ad hoc* assumption of a universe filled with an as yet unobserved form of energy with anti-gravitational properties (the standard physical interpretation of the cosmological constant) in order to fit the data.

In summary, these new general-relativistic expanding waves provide a new paradigm to test against the standard model. Even if they do not in the end explain the anomalous acceleration of the galaxies, one has to believe they are present and propagating on some scale, and their presence represents an instability in the standard model in the sense that an explanation is required as to why small-scale oscillations have to settle down to large-scale $a = 1$ expansions instead of $a \neq 1$ expansions (either locally or globally) during the radiation phase of the Big Bang.

We now use this proceedings to record the *Answers to reporter's questions* which appeared on our websites shortly after our PNAS paper came out in August 2009.

4. Answers to reporter's questions: Blake Temple and Joel Smoller, August 17, 2009.

To Begin: Let us say at the start that what is definitive about our work is the construction of a new one-parameter family of exact, self-similar expanding wave solutions to Einstein's equations of General Relativity. They apply during the radiation phase of the Big Bang, and approximate the standard model of cosmology arbitrarily well. For this we have complete mathematical arguments that are not controvertible. Our intuitions that led us to these, and their physical significance to the anomalous acceleration problem, are based on lessons learned from the mathematical theory of nonlinear conservation laws, and only this interpretation is subject to debate.

1) *Could you explain—in simple terms—what an expanding wave solution is and what other phenomena in nature can be explained through this mathematics?*

To best understand what an expanding wave is, imagine a stone thrown into a pond, making a splash as it hits the water. The initial "plunk" at the start creates chaotic waves that break every which way, but after a short time the whole disturbance settles down into orderly concentric circles of waves that radiate outward from the center—think of the resulting final sequence of waves as the "expanding wave". In fact, it is the initial breaking of waves that dissipates away all of the disorganized motion, until all that is left is the orderly expansion of waves. For us, the initial "plunk" of the stone is the chaotic Big Bang at the start of the radiation phase, and the expansion wave is the orderly expansion that emerges at the end of the radiation phase. What we have found is that the standard model of cosmology is not the only expanding wave that could emerge from the initial "plunk". In fact, we constructed a whole family of possible expanding waves that could emerge; and we argue that which one would emerge depends delicately on the nature of the chaos in the initial "plunk". That is, one expanding wave in the family is equally likely to emerge as another. Our family depends on a freely assignable number a which we call the *acceleration parameter*, such that if we pick $a = 1$, then we get the standard model of cosmology, but if $a > 1$ we get an expanding wave that looks a lot like the standard model, but expands faster, and if $a < 1$, then it expands slower. So an "anomalous acceleration" would result if $a > 1$.

Summary: By "expanding wave" we mean a wave that expands outward in a "self-similar" orderly way in the sense that at each time the wave looks like it did at an earlier time, but more "spread out". The importance of an expanding wave is that it is the end state of a chaotic disturbance because it is what remains after all the complicated breaking of waves is over... one part of the expanding solution no longer affects the other parts. Our thesis, then, is that we can account for the anomalous acceleration of the galaxies without Dark Energy by taking $a > 1$.

2) *Could you explain how and why you decided to apply expanding wave solutions to this particular issue?*

We (Temple) got the idea that the anomalous acceleration of the galaxies might be explained by a secondary expansion wave reflected backward from the shock wave in our earlier construction of a shock wave in the standard model of cosmology, and proposed to numerically simulate such a wave. Temple got this idea while giving a public lecture to the National Academy of Sciences in Bangalore India, in 2006. We set out together to simulate this wave while Temple was Gehring Professor in Ann Arbor in 2007, and in setting up the simulation, we subsequently discovered exact formulas for a family of such waves, without the need for the shock wave model.

3) *Do you think this provides the strongest evidence yet that Dark Energy is a redundant idea?*

At this stage we personally feel that this gives the most plausible explanation for the anomalous acceleration of the galaxies that does not invoke Dark Energy. Since we don't believe in "Dark Energy"... [more detail in (12) below].

We emphasize that our model implies a verifiable prediction, so it remains to be seen whether the model fits the red-shift vs luminosity data better than the Dark Energy theory. (We are working on this now.)

4) *Is this the first time that expanding wave solutions of the Einstein equations have been realized?*

As far as we know, this is the first time a family of self-similar expanding wave solutions of the Einstein equations has been constructed for the radiation phase of the Big Bang, such that the members of the family can approximate the standard model of cosmology arbitrarily well. Our main point is *not* that we have self-similar expanding waves, but that we have self-similar expanding waves during the radiation phase when (1) decay to such waves is possible because $p \neq 0$, and (2) they are close to the standard model. We are not so interested in self-similar waves when $p = 0$ because we see no reason to believe that self-similar waves during the time when $p = \rho c^2/3 \neq 0$ will evolve into exact self-similar waves in the present era when $p = 0$. That is, they should evolve into some sort of expanding spacetime when $p = 0$, but not a pure (self-similar) expansion wave.

5) *How did you reach the assumption that* $p = [\rho][c]^2/3$, *a wise one?*

We are mathematicians, and in the last several decades, a theory for how highly nonlinear equations can decay to self-similar waves was worked out by mathematicians, starting with fundamental work of Peter Lax and Jim Glimm. The theory was worked out for model equations much simpler than the Einstein equations. We realized that only during the radiation phase of the expansion were the equations "sufficiently nonlinear" to expect sufficient breaking of waves at the start to create enough dissipation to drive a chaotic disorganized disturbance into an orderly self-similar expansion wave at the end. The subtle point is that even though no mechanisms for dissipation are put into the model, the nonlinearities alone can cause massive dissipation via the breaking of waves that would drive a chaotic disturbance into an orderly expansion wave.

6) *How does your suggestion—that the observed anomalous acceleration of the galaxies could be due to our view into an expansion wave—compare with an idea that I heard Subir Sarkar describe recently: that the Earth could be in a void that is expanding faster than the outer parts of the universe?*

We became aware of this work in the fall of '08, and forwarded our preprints. Our view here is that after the radiation phase is over, and the pressure drops to zero, there is no longer any nonlinear mechanism that can cause the breaking of waves that can cause dissipation into an expansion wave. Thus during the recent $p = 0$ epoch (after some 300,000 years after the Big Bang), you might model the evolution of the remnants of such an expanding wave or under-density (in their terms a local "void"), but there is no mechanism in the $p = 0$ phase to explain the constraints under which such a void could form. (When $p = 0$, everything is in "freefall", and there can be no breaking of waves.) The expanding wave theory we present provides a possible quantitative explanation for the formation of such a void.

7) *How do you intend to develop your research from here?*

Our present paper demonstrates that there is some choice of the number a (we proved it exists, but still do not know its precise value) such that the member of our family of expanding waves corresponding to that value of the acceleration parameter a will account for the leading-order correction of the anomalous acceleration. That is, it can account for how the plot of redshift vs luminosity of the galaxies curves away from a straight line at the center. But once the correct value of a is determined exactly, that value will give a prediction of how the plot should change beyond the first breaking of the curve. (There are no more free parameters to adjust!) We are currently working on finding that exact value of a consistent with the observed anomalous acceleration, so that from this we can calculate the next-order correction it predicts, all with the goal of comparing the expanding wave prediction to the observed redshift vs luminosity plot, to see if it does better than the prediction of Dark Energy.

8) *What is your view on the relevance of the Copernican Principle to these new expanding waves?*

These self-similar expanding waves represent possible end states of the expansion of the Big Bang that we propose could emerge at the end of the radiation phase when there exists a mechanism for their formation. We imagine that decay to such an expanding wave could have occurred locally in the vicinity of the Earth, over some length scale, but we can only conjecture as to what length scale that might be—the wave could extend out to some fraction of the distance across the visible universe or it could extend even beyond, we cannot say, but to explain the anomalous acceleration the Earth must lie within some proximity of the center. That is, for the $a > 1$ wave to account for the anomalous acceleration observed in the galaxies, we would have to lie in some proximity of the center of such a wave to be consistent with no observed angular dependence in the redshift vs luminosity plots. (The void theory has the same implication.)

Now one might argue that our expanding waves violate the so-called *Copernican* or *Cosmological Principle* which states that *on the largest length scale* the universe looks the same everywhere. This has been a simplifying assumption taken in cosmology since the mid thirties when Howard Robertson and Geoffrey Walker proved that the Friedmann spacetimes of the standard model (constructed by Alexander Friedmann a decade earlier) are the unique spacetimes that are spatially *homogeneous* and *isotropic* about every point—a technical way of saying there is no special place in the universe. The introduction of Dark Energy via the cosmological constant is the only way to preserve the Copernican Principle *and* account for the anomalous acceleration on the largest scale, *everywhere.* The stars, galaxies and clusters of galaxies are evidence of small-scale variations that violate the Copernican Principle on smaller length scales. We are arguing that there could be an even larger length scale than the clusters of galaxies on which local decay to one of our expanding waves has occurred, and we happen to be near the center of one. This would violate the Copernican Principle if these expanding waves describe the entire universe—but our results allow for the possibility that on a scale even larger than the scale of the expanding waves, the universe may look everywhere the same like the standard model. Thus our view is that the Copernican Principle is really a moot issue here. But it does beg the question as to how big the effective center can be for the value of a that accounts for the anomalous acceleration. This is a problem we hope to address in the future.

Another way to look at this is, if you believe there is no cosmological constant or Dark Energy [see (12) below], then the anomalous acceleration may really be the first definitive evidence that in fact, by accident, we just happen to lie near the center of a great expansion wave of cosmic dimensions. We believe our work at this stage gives strong support for this possibility.

9) *How large would the displacement of matter caused by the expanding wave be, and how far out would it extend?*

For our model, the magnitude of the displacement depends on the value of the acceleration parameter a. It can be very large or very small, and we argue that somewhere in between it can be right on for the first breaking of the observed redshift vs luminosity curve near the center. To meet the observations, it has to displace the position of a distant galaxy the right amount to displace the straight line redshift vs luminosity plot of the standard model into the curved graph observed. In their article referenced in our paper (exposition of this appeared as the cover article in Scientific American a few months ago), Clifton and Ferreira quote that the bubble of under-density observed today should extend out to about one billion lightyears, about a tenth of the distance across the visible universe, and the size of the center consistent with no angular variation is about 15 megaparsecs, about 50 million lightyears, and this is approximately the distance between clusters of galaxies, a distance about 1/200 across the visible universe.[6]

10) *How do the spacetimes associated with the expanding waves compare to the spacetime of the standard model of cosmology?*

Interestingly, we prove that the spacetimes associated with the expanding waves when $a \neq 1$ actually have properties surprisingly similar to the standard model $a = 1$. Firstly, the expanding spacetimes ($a \neq 1$) look more and more like the standard model $a = 1$ as you approach the center of expansion. (That is why you have to go far out to see an anomalous acceleration.)

[6] The following back-of-the-envelope calculation provides a ballpark estimate for what we might expect the extent of the remnants of one of our expanding waves might be today. Our thesis is that the self-similar expanding waves that can exist during the pure radiation phase of the standard model can emerge at the end of the radiation phase by the dissipation created by the strong nonlinearities. Now matter becomes transparent with radiation at about 300,000 years after the Big Bang, so we might estimate that our wave should have emerged by about $t_{\text{endrad}} \approx 10^5$ years after the Big Bang. At this time, the distance of light-travel since the Big Bang is about 10^5 lightyears. Since the sound speed $c/\sqrt{3} \approx .58c$ during the radiation phase is comparable to the speed of light, we could estimate that dissipation that drives decay to the expanding wave might reasonably be operating over a scale of 10^5 lightyears by the end of the radiation phase. Now in the $p = 0$ expansion that follows the radiation phase, the scale factor (that gives the expansion rate) evolves like

$$R(t) = t^{2/3},$$

so a distance of 10^5 lightyears at $t = t_{\text{endrad}}$ years will expand to a length L at present time $t_{\text{present}} \approx 10^{10}$ years by a factor of

$$\frac{R(t_{\text{present}})}{R(t_{\text{endrad}})} \approx \frac{(10^{10})^{2/3}}{(10^5)^{2/3}} = 10^{4.7} \geq 5 \times 10^4.$$

It follows then that we might expect the scale of the wave at present time to extend over a distance of about

$$L = 5 \times 10^5 \times 10^4 = 5 \times 10^9 \text{ lightyears}.$$

This is a third to a fifth of the distance across the visible universe, and agrees with the extent of the under-density void region quoted in the Clifton-Ferreira paper, with room to spare.

Moreover, out to a great distance from the center, say out to about 1/3 to 1/2 the distance across the visible universe, (where the anomalous acceleration is apparent), we prove that (to within negligible errors) there is a time coordinate t such that the 3-space at each fixed t has zero curvature, just like the standard model of cosmology, and observers fixed in time or at a fixed distance from the center will measure distances and times exactly the same as in a Friedmann universe, the spacetime of the standard model of cosmology. In technical terms, only line elements changing in space *and* time will measure dilation of distances and times relative to the standard model. This suggests that it would be easy to mistake one of these expanding waves for the Friedmann spacetime itself until you did a measurement of redshift vs luminosity far out where the differences are highly apparent (that is, you measured the anomalous acceleration).

11) *Your expanding wave theory is more complicated than a universe filled with Dark Energy, and we have to take into account the Occam's razor principle. What do you think about this assertion?*

To quote Wikipedia, Occam's razor states: "The explanation of any phenomenon should make as few assumptions as possible, eliminating those that make no difference in the observable predictions of the explanatory hypothesis or theory."

We could say that our theory does not require the extra hypothesis of Dark Energy or a cosmological constant to explain the anomalous acceleration. Since there is no obvious reason why an expansion wave with one value of a over another would come out locally at any given location at the end of the radiation phase, and since we don't need Dark Energy in the expanding wave explanation, we could argue that the expanding wave explanation of the anomalous acceleration is simpler than Dark Energy. But a better answer is that our theory has an observable prediction, and only experiments, not the 14th-century principle of Occam, can resolve the physics. Occam's razor will have nothing whatsoever to say about whether we are, or are not, near the center of a cosmic expansion wave.

12) *If, as you suggest, Dark Energy doesn't exist, what is the ingredient of 75% of the mass-energy in our universe?*

In short, nothing is required to replace it. The term "anomalous acceleration" of the galaxies begs the question "acceleration relative to what?". The answer is that the anomalous acceleration of the galaxies is an acceleration relative to the prediction of the standard model of cosmology. In the expanding wave theory, we prove that there is no "acceleration" because the anomalous acceleration can be accounted for in redshift vs luminosity by the fact that the galaxies in the expanding wave are displaced from their anticipated position in the standard model. So the expanding wave theory requires only classical sources of mass-energy for the Einstein equations.

13) *If Dark Energy doesn't exist, it would be just an invention. What do you think about Dark Energy theory?*

Keep in mind that Einstein's equations have been confirmed without the need for the cosmological constant or Dark Energy, in every physical setting except in cosmology.

Dark Energy is the physical interpretation of the cosmological constant. The cosmological constant is a source term with a free parameter (similar to but different from our a) that can be added to the original Einstein equations and still preserve the frame independence, the "general relativity" if you will, of Einstein's equations. Einstein's equations express that mass-energy is the source of spacetime curvature. So if you interpret the cosmological constant as the effect of some exotic mass-energy, then you get Dark Energy. For the value of the cosmological constant required to fit the anomalous acceleration observed in the redshift vs luminosity data, this Dark Energy must account for some 73 percent of the mass-energy of the universe, and it has to have the physical property that it *anti-gravitates*—that is, it gravitationally repels instead of attracts. Since no one has ever observed anything that has this property (it would not fall to Earth like an apple, it would fly up like a balloon), it seems rather suspect that such mass-energy could possibly exist. If it does exist, then it also is not like any other mass-energy in that the density of it stays constant, stuck there at the same value forever, even as the universe expands and spreads all the other mass-energy out over larger and larger scales—and there is no principle that explains why it has the value it has.[7] On the other hand, if you put the cosmological constant on the other side of the equation with the curvature then there is always some (albeit very small) baseline curvature permeating spacetime, and the zero-curvature spacetime is no longer possible; that is, the empty-space Minkowski spacetime of Special Relativity no longer solves the equations. So when the cosmological constant is over on the curvature side of Einstein's equation, the equations no longer express the physical principle that led Einstein to discover them in the first place—that mass-energy should be the sole source of spacetime curvature.

Einstein put the cosmological constant into his equations shortly after he discovered them in 1915, because this was the only way he could get the possibility of a static universe. (*Anti-gravity* holds the static universe up!) After Hubble proved that the universe was expanding in 1929, Einstein took back the cosmological constant, declaring it was the greatest blunder of his career, as he could have predicted the expansion ahead of time without it. At the time, taking out the cosmological constant was interpreted as a great victory for General Relativity. Since then, cosmologists have become more comfortable putting the cosmological constant back in. There are many respected scientists who see no problem with Dark Energy.

[7]In the expanding wave theory, the principle for determining a is that all values of a near $a = 1$ should be (roughly) equally likely to appear, and one of them did...

14) *How does the coincidence in the value of the cosmological constant in the Dark Energy theory compare to the coincidence that the Milky Way must lie near a local center of expansion in the expanding wave theory?*

The Dark Energy explanation of the anomalous acceleration of the galaxies requires a value of the cosmological constant that accounts for some 73 percent of the mass-energy of the universe. That is, to correct for the anomalous acceleration in the supernova data, you need a value of the cosmological constant that is just three times the energy-density of the rest of the mass-energy of the universe. Now there is no principle that determines the value of the cosmological constant ahead of time, so its value could apriori be anything. Thus it must be viewed as a great coincidence that it just happens to be so close to the value of the energy density of the rest of the mass-energy of the universe. (Keep in mind that the energy-density of all the classical sources decreases as the universe spreads out, while the cosmological constant stays *constant.*) So why does the value of the cosmological constant come out so close to, *just 3 times*, the value of the rest of the mass-energy of the universe, instead of 10^{10} larger or 10^{-10} smaller? This raises a very suspicious possibility. Since the magnitude of the sources sets the scale for the overall *oomph* of the solution, when you need to adjust the equations by an amount on the order of the sources present in order to fit the data, that smacks of the likelihood that you are really just adding corrections to the wrong underlying solution. So to us it looks like the coincidence in the value of the cosmological constant in the Dark Energy theory may well be greater than the coincidence that we lie near a local center of expansion in the expanding wave theory.

In summary: Our view is that the Einstein equations make more physical sense without Dark Energy or the cosmological constant, and Dark Energy is most likely an unphysical *fudge factor*, if you will, introduced into the theory to meet the data. But ultimately, whether Dark Energy or an expanding wave correctly explains the anomalous acceleration of the galaxies can only be decided by experiments, not the Copernican Principle or Occam's razor.

References

[1] C. Copi, D. Huterer, D.J. Schwarz, G.D. Starkman, *On the large-angle anomalies of the microwave sky*, Mon. Not. R. Astron. Soc. **367**, 79–102 (2006).

[2] T. Clifton, P.G. Ferreira, K. Land, *Living in a void: testing the Copernican principle with distant supernovae*, Phys. Rev. Lett. **101**, 131302 (2008). arXiv:0807.1443v2 [astro-ph].

[3] T. Clifton and P.G. Ferreira, *Does dark energy really exist?*, Sci. Am., April 2009, 48–55.

[4] D. Eardley, *Self-similar spacetimes: geometry and dynamics*, Commun. Math. Phys. **37**, 287–309 (1974).

[5] J. Glimm, *Solutions in the large for nonlinear hyperbolic systems of equations*, Comm. Pure Appl. Math. **18**, 697–715 (1965).

[6] J. Glimm, P.D. Lax, *Decay of solutions of systems of nonlinear hyperbolic conservation laws*, Memoirs Amer. Math Soc. **101** (1970).

[7] J. Groah and B. Temple, *Shock-wave solutions of the Einstein equations: existence and consistency by a locally inertial Glimm scheme*, Memoirs Amer. Math Soc. **172** (2004), no. 813.

[8] O. Gron and S. Hervik, *Einstein's general theory of relativity with modern applications in cosmology*, Springer, 2007.

[9] P. Hartman, *A lemma in the structural stability of ordinary differential equations*, Proc. Amer. Math. Soc. **11**, 610–620 (1960).

[10] P.D. Lax, *Hyperbolic systems of conservation laws, II*, Comm. Pure Appl. Math. **10**, 537–566 (1957).

[11] J. Smoller and B. Temple, *Global solutions of the relativistic Euler equations*, Comm. Math. Phys. **157**, 67–99 (1993).

[12] J. Smoller and B. Temple, *Cosmology, black holes, and shock waves beyond the Hubble length*, Meth. Appl. Anal. **11**, 77–132 (2004).

[13] J. Smoller and B. Temple, *Shock-wave cosmology inside a black hole*, Proc. Nat. Acad. Sci. **100**, 11216–11218 (2003), no. 20.

[14] B. Temple and J. Smoller, *Expanding wave solutions of the Einstein equations that induce an anomalous acceleration into the Standard Model of Cosmology*, Proc. Nat. Acad. Sci. **106**, 14213–14218 (2009), no. 34.

[15] B. Temple and J. Smoller, *Answers to questions posed by reporters: Temple–Smoller GR expanding waves*, August 19, 2009. `http://www.math.ucdavis.edu/~temple/`.

[16] J. Smoller and B. Temple, *General relativistic self-similar waves that induce an anomalous acceleration into the Standard Model of Cosmology*, Memoirs Amer. Math. Soc., to appear.

[17] B. Temple, *Numerical refinement of a finite mass shock-wave cosmology*, AMS National Meeting, Special Session *Numerical Relativity*, New Orleans (2007). `http://www.math.ucdavis.edu/~temple/talks/NumericalShockWaveCosTalk.pdf`.

[18] S. Weinberg, *Gravitation and cosmology: principles and applications of the general theory of relativity*, John Wiley & Sons, 1972.

Blake Temple
Dept of Mathematics
1 Shields Avenue
Davis, CA
USA 95618
e-mail: `temple@math.ucdavis.edu`

Discrete and Continuum Third Quantization of Gravity

Steffen Gielen and Daniele Oriti

Abstract. We give a brief introduction to matrix models and group field theory (GFT) as realizations of the idea of a third quantization of gravity, and present the basic features of a continuum third quantization formalism in terms of a field theory on the space of connections. Building up on the results of loop quantum gravity, we explore to what extent one can rigorously define such a field theory. We discuss the relation between GFT and this formal continuum third-quantized gravity, and what it can teach us about the continuum limit of GFTs.

Mathematics Subject Classification (2010). 83C45, 83C27, 81T27.

Keywords. Group field theory, third quantization, loop quantum gravity, lattice gravity, spin foam models.

1. Introduction

Several approaches to quantum gravity have been developed over the last years [40], with a remarkable convergence, in terms of mathematical structures used and basic ideas shared. Group field theories have been proposed [15, 38, 37, 34] as a kind of second quantization of canonical loop quantum gravity, in the sense that one turns into a dynamical (quantized) field its canonical wave function. They can be also understood as second quantizations of simplicial geometry [37]. Because both loop quantum gravity and simplicial quantum gravity are supposed to represent the quantization of a (classical) field theory, General Relativity, this brings GFT into the conceptual framework of "third quantization", a rather inappropriate label for a rather appealing idea: a field theory on the space of geometries, rather than spacetime, which also allows for a dynamical description of topology change. This idea has been brought forward more than 20 years ago in the context of canonical geometrodynamics, but has never been seriously developed due to huge mathematical difficulties, nor recast in the language of connection dynamics, which allowed so much progress in loop quantum gravity.

In this contribution, we give a brief introduction to the group field theory formalism, and we present in some more detail the idea and basic features of a continuum third-quantization formalism in the space of connections. Building up on the results of loop quantum gravity, we explore to what extent one can rigorously define such a field theory. As a concrete example, we shall restrict ourselves to the simple case of 3d Riemannian GR. Finally, we discuss the relation between GFT and this formal continuum third-quantized gravity, and what it can teach us about the continuum limit of GFTs.

Our purpose is partly pedagogical, partly technical, partly motivational. The third-quantized framework in the continuum is not a well-defined approach to quantum gravity. Still, its initial motivations remain valid, in our opinion, and it is worth keeping them in mind and thus presenting this framework in some detail. Most importantly, group field theories can realize this third-quantization program, even if at the cost of using a language that is farther away from that of General Relativity. The GFT formalism itself is still in its infancy, so that many conceptual aspects are not often stressed. By comparison with continuum third quantization we clarify here some of these conceptual issues: the "level of quantization" adopted, the consequent interpretation of the GFT classical action, the role of topology change, how the usual canonical quantum theory should be recovered, the difference between the "global" nature of the traditional quantization scheme (dynamics of a quantum universe) versus the more "local" nature of the GFT approach (quantizing building blocks of the same universe), and the implementation of spacetime symmetries into the GFT context.

At the more technical level, our analysis of the third-quantized framework, even if incomplete and confined to the classical field theory on the space of connections, will be new and potentially useful for further developments. Among them, we have in mind both the study of the continuum limit of GFT models, which we will discuss in some detail, and the construction of simplified "third-quantized" models, in contexts where they can be made amenable to actual calculations, for example in the context of quantum cosmology.

Finally, one more goal is to stimulate work on the continuum approximation of GFTs. We will sketch possible lines of research and speculate on how the continuum limit of GFTs can be related to third-quantized continuum gravity, what such a continuum limit may look like, and on the emergence of classical and quantum General Relativity from it. In the end, the hope is to learn something useful about GFT from the comparison with a more formal framework, but one that is also closer to the language of General Relativity.

1.1. Canonical quantization, loop quantum gravity and third quantization

The geometrodynamics program [49] was an early attempt to apply the canonical quantization procedure to General Relativity, in metric variables, on a manifold of given topology $\mathcal{M} = \Sigma \times \mathbb{R}$. The space of "coordinates" is *superspace*, the space of Riemannian geometries on Σ, *i.e.* metrics modulo spatial diffeomorphisms. One is led to the infamous Wheeler-DeWitt equation

[12],

$$\mathcal{H}(x)\Psi[h_{ij}] \equiv \left[\mathcal{G}^{ijkl}[h](x)\frac{\delta^2}{\delta h_{ij}(x)\delta h_{kl}(x)} - (R[h](x) - 2\Lambda)\right]\Psi[h_{ij}] = 0, \tag{1.1}$$

which is the analogue of the classical Hamiltonian constraint generating reparametrization of the time coordinate. Here $\mathcal{G}^{ijkl} = \sqrt{h}h^{ij}h^{kl} + \ldots$ is the DeWitt supermetric, and the wave functional Ψ depends on the 3-dimensional metric h_{ij} which encodes the geometry of Σ. (1.1) is mathematically ill-defined as an operator equation, and suffers from severe interpretational problems. Geometrodynamics therefore never made much progress as a physical theory. Nevertheless, quite a lot is known about geometrical and topological properties of superspace itself; for a nice review see [23].

While from the conceptual point of view *superspace* is a kind of "meta-space", or "a space of spaces", in the sense that each of its "points", a 3-metric, represents a possible physical space, from the mathematical point of view it is a manifold in its own right, with a given fixed metric and a given topology. This makes it possible to combine background independence with respect to physical spacetime required by GR with the use of the background-dependent tools of (almost) ordinary quantum field theory in a third-quantized field theory formalism. This key aspect is shared also by GFTs.

With the reformulation of General Relativity in connection variables [1], the canonical quantization program experienced a revival in the form of loop quantum gravity (LQG) [44, 47]. Superspace is replaced by the space of $\mathfrak{g}$-connections on Σ, where G is the gauge group of the theory (usually, $G = \mathrm{SU}(2)$). One then takes the following steps towards quantization:

- Reparametrize the classical phase space, going from connections to (G-valued) holonomies, exploiting the fact that a connection can be reconstructed if all of its holonomies along paths are known. The conjugate variable to the connection, a triad field, is replaced by its ($\mathfrak{g}$-valued) fluxes through surfaces; one finds Poisson-commutativity among functions of holonomies and non-commutativity among fluxes. In the simplest case, one considers a fixed graph Γ embedded into Σ and takes as elementary variables cylindrical functions Φ_f of holonomies along the edges of the graph and fluxes E^i_e through surfaces intersecting the graph only at a single edge e. One then finds [2]

$$\{\Phi_f, \Phi_{f'}\} = 0, \qquad \{E^i_e, E^j_{e'}\} = \delta_{e,e'}C^{ij}{}_k E^k_e,$$

where $C^{ij}{}_k$ are the structure constants of the Lie algebra.
One hence has two possible representations, one where wave functions are functionals of the connection, and a non-commutative flux representation [3]. A third representation arises by decomposing a functional of the connection into representations of G (these three representations also exist for GFT).

- In the connection formulation, one defines a Hilbert space of functionals of (generalized) $\mathfrak{g}$-connections by decomposing such functionals into sums of cylindrical functions which only depend on a finite number of holonomies, associated to a given graph, each: $\mathfrak{H} \sim \bigoplus_\Gamma \mathfrak{H}_\Gamma$. Holonomy operators act by multiplication and flux operators as left-invariant vector fields on G, hence implementing their non-commutativity.

• Define the action of diffeomorphism and Gauss constraints on cylindrical functions, and pass to a reduced Hilbert space of gauge-invariant, (spatially) diffeomorphism-invariant states.

While this procedure implements the kinematics of the theory rigorously, the issue of dynamics, *i.e.* the right definition of the analogue of the Hamiltonian constraint (1.1) on the kinematical Hilbert space and hence of the corresponding space of physical states, is to a large extent still an open issue, also in its covariant sum-over-histories formulation [45].

What we have discussed so far is a "first quantization", where the wave function gives probabilities for states of a single hypersurface Σ. One can draw an analogy to the case of a relativistic particle, where the mass-shell constraint $p^2 + m^2 = 0$ leads to the wave equation

$$\left[g^{\mu\nu}(x)\frac{\partial}{\partial x^\mu}\frac{\partial}{\partial x^\nu} - m^2\right]\Psi(x) = 0, \tag{1.2}$$

the Klein-Gordon equation. The straightforward interpretation of $\Psi(x)$ as a single-particle wave function fails: In order to define on the kinematical Hilbert space a projection to the solutions of (1.2) which has the correct composition properties and which on the solutions reduces to a positive definite inner product (for an overview of possible definitions for inner products and their composition laws, see [26]; for analogous expressions in loop quantum cosmology see [10]), one needs to define a splitting of the solutions to (1.2) into positive- and negative-frequency solutions. This splitting relies on the existence of a timelike Killing vector k and hence a conserved quantity $k \cdot p$ ("energy") on Minkowski space; as is well known, for generic metrics without isometries there is no unambiguous particle concept. This leads to "second quantization" where the particle concept is secondary.

The close analogy between (1.1) and (1.2) suggests that for a meaningful "one-universe" concept in quantum geometrodynamics, one would need a conserved quantity on all solutions of the constraints. It was argued in [29] that no such quantity exists, so that one has to go to a many-geometries formalism in which "universes," 3-manifolds of topology Σ, can be created and annihilated, and hence to a QFT on superspace. The analogy was pushed further, on a purely formal level, by Teitelboim [46], who gave analogues of the Feynman propagator, QFT perturbation theory etc. for such a theory.

The general idea is to define a (scalar) field theory on superspace $\mathcal{S}$ for a given choice of spatial manifold topology Σ, *e.g.* the 3-sphere, essentially turning the wave function of the canonical (first-quantized) theory into an operator $\Phi[{}^3h]$, whose dynamics is defined by an action with a kinetic term of the type

$$S_{\text{free}}(\Phi) = \int_{\mathcal{S}} \mathcal{D}^3h\, \Phi[{}^3h]\mathcal{H}\Phi[{}^3h]\,, \tag{1.3}$$

with $\mathcal{H}$ being the Wheeler-DeWitt differential operator of canonical gravity (1.1) here defining the free propagation of the theory. One thinks of a quantum field which is a functional $\Phi[^3h]$ of the 3-metric defined on Σ; the operator $\Phi[^3h]$ creates a 3-manifold of topology Σ with metric h.

The quantum theory would be "defined" by the perturbative expansion of the partition function $Z = \int \mathfrak{D}\Phi \, e^{-S(\Phi)}$ in "Feynman diagrams". Adding a term cubic in Φ would give diagrams corresponding to processes such as the following:

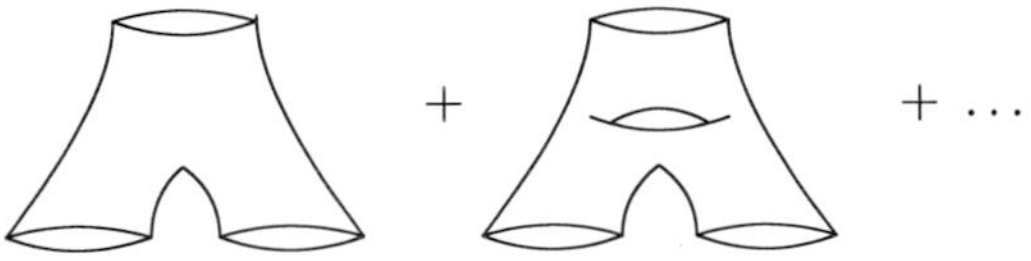

Thus such a formalism also has the attractive aspect of incorporating topology change. In the simplified setting of homogeneous 3-spheres (*i.e.* a third-quantized minisuperspace model), it was explored by Giddings and Strominger in [20] in the hope to find a dynamical mechanism determining the value of the cosmological constant, under the name "third quantization."

The Feynman amplitudes are given by the quantum-gravity path integral for each spacetime topology, identified with a particular interaction of universes, with the one for trivial topology representing a one-particle propagator, a Green function for the Wheeler-DeWitt equation. Other features of this very formal setting are: 1) the classical equations of motion for the scalar field on superspace are a non-linear extension of the Wheeler-DeWitt equation, due to the interaction term in the action, *i.e.* the inclusion of topology change; 2) the perturbative third-quantized vacuum of the theory is the "no-space" state, and not any state with a semiclassical geometric interpretation.

In the third-quantization approach one has to deal with the quantization of a field theory defined on an infinite-dimensional manifold, clearly a hopeless task. In a connection formulation, there is more hope of at least defining the classical theory, due to the work done in LQG. To make more progress, one will have to reduce the complexity of the system. One possibility is to pass to a symmetry-reduced sector of GR before quantization, obtaining a third-quantized minisuperspace model, as done in metric variables in [20] and for connection variables in [9]. Another possibility is to consider, instead of a continuous manifold Σ, a discrete structure such as a simplicial complex where one is only interested in group elements characterizing the holonomies along links. This is the idea behind GFTs. Before getting to the GFT setting, we will show how the third-quantization idea is implemented rigorously in a simpler context, that of 2d quantum gravity, by means of matrix models.

2. Matrix models: a success story

A simple context in which the idea of "third quantization" of gravity can be realized rigorously is that of 2d Riemannian quantum gravity. The quantization is achieved by using matrix models, in the same "go local, go discrete" way that characterizes group field theories, which indeed can be seen as a generalization of the same formalism.

Define a simple action for an $N \times N$ Hermitian matrix M, given by

$$\begin{aligned} S(M) &= \frac{1}{2}\,\mathrm{Tr}M^2 - \frac{g}{\sqrt{N}}\,\mathrm{Tr}M^3 = \frac{1}{2}M^i{}_j M^j{}_i - \frac{g}{\sqrt{N}}M^i{}_j M^j{}_k M^k{}_i \\ &= \frac{1}{2}M^i{}_j K_{jkli} M^k{}_l - \frac{g}{\sqrt{N}}M^i{}_j M^m{}_n M^k{}_l\, V_{jmknli} \end{aligned} \tag{2.1}$$

with propagator $(K^{-1})_{jkli} = K_{jkli} = \delta^j{}_k\,\delta^l{}_i$ and vertex $V_{jmknli} = \delta^j{}_m\,\delta^n{}_k\,\delta^l{}_i$.

Feynman diagrams are constructed, representing matrices by two points corresponding to their indices, out of: lines of propagation (made of two strands), non-local "vertices" of interaction (providing a re-routing of strands) and faces (closed loops of strands) obtained after index contractions. This combinatorics of indices can be given a simplicial interpretation by viewing a matrix as representing an edge in a 2-dimensional (dual) simplicial complex:

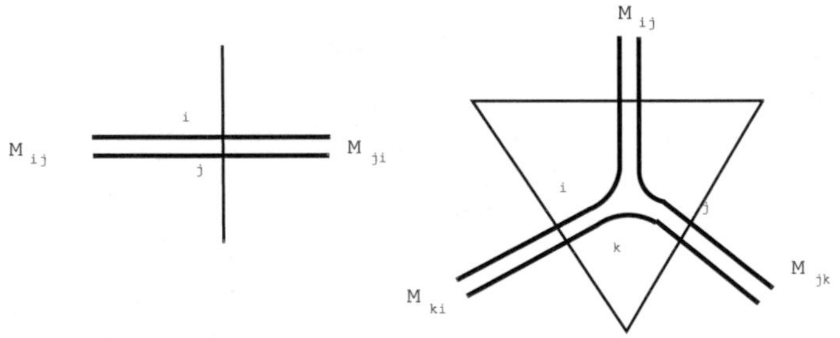

The diagrams used in evaluating the partition function $Z = \int dM_{ij}\, e^{-S(M)}$ correspond to complexes of arbitrary topology, obtained by arbitrary gluing of edges to form triangles (in the interaction) and of triangles along common edges (dictated by the propagator). A discrete spacetime emerges as a virtual construction, encoding the interaction processes of fundamental space quanta.

The partition function, expressed in terms of Feynman amplitudes, is

$$Z = \sum_\Gamma g^{V_\Gamma} N^{F_\Gamma - \frac{1}{2}V_\Gamma} = \sum_\Gamma g^{V_\Gamma}\, N^\chi,$$

where V_Γ is the number of vertices and F_Γ the number of faces of a graph Γ, and χ is the Euler characteristic of the simplicial complex.

Each Feynman amplitude is associated to a simplicial path integral for gravity discretized on the associated simplicial complex Δ. The action for 2d GR with cosmological constant on a 2d manifold is $\frac{1}{G}\int d^2x\sqrt{g}\,(-R(g) + \Lambda) = -\frac{4\pi}{G}\chi + \frac{\Lambda a}{G}t$, where the surface is discretized into t equilateral triangles of area a. We can now identify ($t_\Delta = V_\Gamma$, each vertex is dual to a triangle)

$$Z = \sum_\Gamma g^{V_\Gamma}\, N^\chi \equiv \sum_\Delta e^{+\frac{4\pi}{G}\chi(\Delta) - \frac{a\Lambda}{G}\,t_\Delta}, \qquad g \equiv e^{-\frac{\Lambda a}{G}}, \qquad N \equiv e^{+\frac{4\pi}{G}},$$

which is a sum over histories of discrete 2d GR, trivial as the only geometric variable associated to each surface is its area. In addition to this sum over geometries, we obtain a sum over 2d complexes of all topologies; the matrix model defines a discrete third quantization of 2d GR! We can control both sums [11, 21] by expanding in a topological parameter, the genus h,

$$Z = \sum_{\Delta} g^{t_\Delta} N^{2-2h} = \sum_{h} N^{2-2h} Z_h(g) = N^2 Z_0(g) + Z_1(g) + \dots .$$

In the limit $N \to \infty$, only spherical simplicial complexes (of genus 0) contribute. As $N \to \infty$ one can also define a continuum limit and match the results of the continuum GR path integral: Expanding $Z_0(g)$ in powers of g,

$$Z_0(g) = \sum_t t^{\gamma-3} \left(\frac{g}{g_c}\right)^t \simeq_{t\to\infty} (g-g_c)^{2-\gamma},$$

we see that as $t \to \infty$, $g \to g_c$ ($\gamma > 2$), $Z_0(g)$ diverges, signaling a phase transition. In order to identify it as a continuum limit we compute the expectation value for the area: $\langle A \rangle = a\frac{\partial}{\partial g} \ln Z_0(g) \simeq_{t\to\infty} \frac{a}{g-g_c}$. Thus we can send $a \to 0$, $t \to \infty$, tuning at the same time the coupling constant to g_c, to get finite macroscopic areas. This continuum limit reproduces [11, 21] the results from a continuum 2d gravity path integral when this can be computed; otherwise it can be taken to *define* the continuum path integral.

One can compute the contribution from all topologies in the continuum limit, defining a *continuum* third quantization of 2d gravity, using the so-called double-scaling limit [11, 21]. One finds that in the continuum limit

$$Z \simeq_{t\to\infty} \sum_h \kappa^{2h-2} f_h, \qquad \kappa^{-1} := N (g-g_c)^{\frac{(2-\beta)}{2}}$$

for some constant β. We can then take the limits $N \to \infty$, $g \to g_c$, holding κ fixed. The result is a continuum theory to which all topologies contribute.

3. Group field theory: a sketchy introduction

One can construct combinatorial generalizations of matrix models to higher tensor models; however such models do not possess any of the nice scaling limits (large-N limit, continuum and double-scaling limit) that allow to control the sum over topologies in matrix models and to relate the quantum amplitudes to those of continuum gravity. One needs to generalize further by defining corresponding field theories, replacing the indices of the tensor models by continuous variables. For the definition of good models for quantum gravity, the input from other approaches to quantum gravity is crucial.

We now describe in some detail the GFT formalism for 3d Riemannian GR, clarifying the general features of the formalism in this specific example.

Consider a triangle in $\mathbb{R}^3$ and encode its kinematics in a (real) field φ, a function on the space of geometries for the triangle, parametrized by three $\mathfrak{su}(2)$ elements x_i attached to its edges which are interpreted as triad discretized along the edges. Using the non-commutative group Fourier transform [17, 18, 5], based on plane waves $e_g(x) = e^{i\vec{p}_g \cdot \vec{x}}$ on $\mathfrak{g} \sim \mathbb{R}^n$ (with coordinates

$\vec{p}_g$ on SU(2)), the field can alternatively be seen as a function on $\mathrm{SU}(2)^3$,

$$\varphi(x_1, x_2, x_3) = \int [\mathrm{d}g]^3 \, \varphi(g_1, g_2, g_3) \, e_{g_1}(x_1) e_{g_2}(x_2) e_{g_3}(x_3)$$

where the $g_i \in \mathrm{SU}(2)$ are thought of as parallel transports of the gravity connection along links dual to the edges of the triangle represented by φ.

In order to define a geometric triangle, the vectors x_i have to sum to zero. We thus impose the constraint ($\star$ is a non-commutative product reflecting the non-commutativity of the group multiplication in algebra variables)

$$\varphi = C \star \varphi, \quad C(x_1, x_2, x_3) = \delta_0(x_1 + x_2 + x_3) := \int \mathrm{d}g \, e_g(x_1 + x_2 + x_3). \tag{3.1}$$

In terms of the dual field $\varphi(g_1, g_2, g_3)$, the closure constraint (3.1) implies invariance under the diagonal (left) action of the group SU(2),

$$\varphi(g_1, g_2, g_3) = P\varphi(g_1, g_2, g_3) = \int_{\mathrm{SU}(2)} \mathrm{d}h \, \phi(hg_1, hg_2, hg_3). \tag{3.2}$$

In group variables, the field can be best depicted graphically as a 3-valent vertex with three links, dual to the three edges of the closed triangle (Fig. 1). This object will be the GFT building block of our quantum space.

A third representation is obtained by decomposition into irreducible representations (compare with the LQG construction, sect. 5),

$$\varphi(g_1, g_2, g_3) = \sum_{j_1, j_2, j_3} \varphi^{j_1 j_2 j_3}_{m_1 m_2 m_3} \, D^{j_1}_{m_1 n_1}(g_1) D^{j_2}_{m_2 n_2}(g_2) D^{j_3}_{m_3 n_3}(g_3) \, C^{j_1 j_2 j_3}_{n_1 n_2 n_3} \,, \tag{3.3}$$

where $C^{j_1 j_2 j_3}_{n_1 n_2 n_3}$ is the Wigner invariant 3-tensor, the 3j-symbol.

Graphically, one can think of the GFT field in any of the three representations (Lie algebra, group, representation), as appropriate:

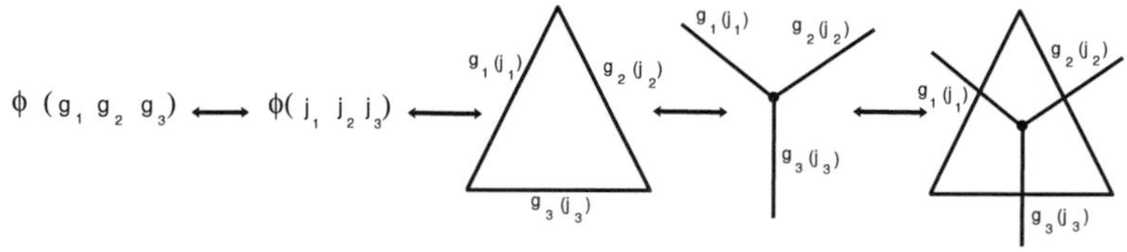

FIGURE 1. Different representations of the GFT field φ.

The convolution (in the group or Lie algebra picture) or tracing (in the representation picture) of multiple fields with respect to a common argument represents the gluing of triangles along common edges, and thus the formation of more complex simplicial structures, or more complex dual graphs:

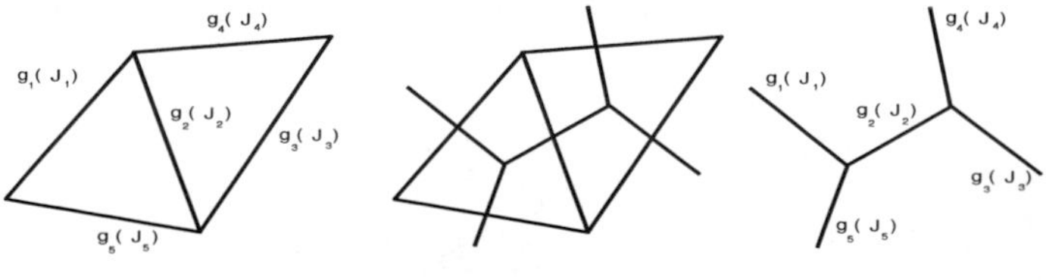

The corresponding field configurations represent extended chunks of space; a generic polynomial observable is associated with a particular quantum space.

We now define a classical dynamics for the GFT field. In the interaction term four geometric triangles should be glued along common edges to form a tetrahedron. The kinetic term should encode the gluing of two tetrahedra along common triangles. With $\varphi_{123} := \varphi(x_1, x_2, x_3)$, we define the action

$$S = \frac{1}{2} \int [\mathrm{d}x]^3 \, \varphi_{123} \star \varphi_{123} - \frac{\lambda}{4!} \int [\mathrm{d}x]^6 \, \varphi_{123} \star \varphi_{345} \star \varphi_{526} \star \varphi_{641} \, , \tag{3.4}$$

where $\star$ relates repeated indices as $\phi_i \star \phi_i := (\phi \star \phi_-)(x_i)$, with $\phi_-(x) = \phi(-x)$.

One can generalize the GFT field to a function of n arguments which when satisfying (3.1) or (3.2) can be taken to represent a general n-gon (dual to an n-valent vertex) and glued to other fields, giving a polygonized quantum space just as for triangles. There is also no difficulty in considering a more general action in which, for given $\varphi(x_i)$, one adds other interaction terms corresponding to the gluing of triangles (polygons) to form general polyhedra or more pathological configurations (*e.g.* with multiple identifications among triangles). The only restriction may come from the symmetries of the action.

The projection (3.2) takes into account parallel transport between different frames; from (3.2) and (3.4) we can identify a propagator and a vertex:

$$\begin{aligned}
\mathcal{K}(x_i, y_i) &= \int \mathrm{d}h_t \prod_{i=1}^{3} (\delta_{-x_i} \star e_{h_t})(y_i), \\
\mathcal{V}(x_i, y_i) &= \int \prod_t \mathrm{d}h_t \prod_{i=1}^{6} (\delta_{-x_i} \star e_{h_{tt'}})(y_i);
\end{aligned} \tag{3.5}$$

the variables h_t and $h_{t\tau}$ arising from (3.2) are interpreted as parallel transports through the triangle t and from the center of the tetrahedron τ to triangle t, respectively. We may represent the propagator and vertex as follows:

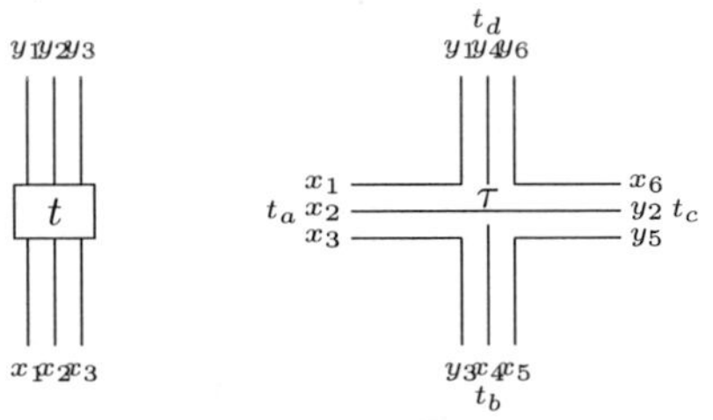

(3.6)

The integrands in (3.5) factorize into products of functions associated to strands (one for each field argument), with a geometrical meaning: The variables (x_i, y_i) associated to the edge i correspond to the edge vectors in the frames associated to the triangles t, t' sharing it; opposite edge orientations in different triangles and a mismatch between reference frames associated to the same triangle in two different tetrahedra are taken into account.

Using the group Fourier transform we obtain the form of Boulatov [8],

$$S[\phi] = \frac{1}{2}\int [\mathrm{d}g]^3 \varphi(g_1,g_2,g_3)\varphi(g_3,g_2,g_1)$$
$$-\frac{\lambda}{4!}\int [\mathrm{d}g]^6 \varphi(g_{12},g_{13},g_{14})\varphi(g_{14},g_{24},g_{34})\varphi(g_{34},g_{13},g_{23})\varphi(g_{23},g_{24},g_{12}). \tag{3.7}$$

In this group representation, the kinetic and vertex functions are

$$\mathcal{K}(g_i,\tilde{g}_i) = \int \mathrm{d}h \prod_{k=1}^{3} \delta(g_k h \tilde{g}_k^{-1}),$$
$$\mathcal{V}(g_{ij},g_{ji}) = \int \prod_{i=1}^{4} \mathrm{d}h_i \prod_{i<j} \delta(g_{ij} h_i h_j^{-1} g_{ji}^{-1}). \tag{3.8}$$

Also in group variables, the geometric content of the model is apparent: the six delta functions in $\mathcal{V}$ encode the flatness of each "wedge," *i.e.* of the portion of each dual face inside a single tetrahedron. This flatness is characteristic of the piecewise-flat context in which the GFT models are best understood.

There is also a form of the action in representation variables, using (3.3),

$$S(\varphi) = \frac{1}{2}\sum_{\{j\},\{m\}} \varphi^{j_1j_2j_3}_{m_1m_2m_3}\varphi^{j_3j_2j_1}_{m_3m_2m_1}$$
$$-\frac{\lambda}{4!}\sum_{\{j\},\{m\}} \varphi^{j_1j_2j_3}_{m_1m_2m_3}\varphi^{j_3j_4j_5}_{m_3m_4m_5}\varphi^{j_5j_2j_6}_{m_5m_2m_6}\varphi^{j_6j_4j_1}_{m_6m_4m_1}\begin{Bmatrix} j_1 & j_2 & j_3 \\ j_4 & j_5 & j_6 \end{Bmatrix}. \tag{3.9}$$

The classical equations of motion for this model are, in group space:

$$0 = \int \mathrm{d}h\, \phi(g_1h,g_2h,g_3h) - \frac{\lambda}{3!}\int \prod_{i=1}^{3} \mathrm{d}h_i \int \prod_{j=4}^{6} \mathrm{d}g_j\, \phi(g_3h_1,g_4h_1,g_5h_1)$$
$$\times\, \phi(g_5h_2,g_6h_2,g_2h_2)\phi(g_6h_3,g_4h_3,g_1h_3). \tag{3.10}$$

These equations define the classical dynamics of the theory, allowing the identification of classical background solutions. Considering the interpretation of the GFT as a "third quantization" of gravity, the classical GFT equations encode fully the quantum dynamics of the underlying canonical quantum gravity theory, where quantum gravity wave functions are constructed, or the quantum dynamics of first-quantized spin networks implementing the constraints of canonical GR. This should be clearer once we have presented the formal third quantization of gravity in connection variables.

The quantum dynamics is defined by the expansion of the partition function in Feynman diagrams, viewed as dual to 3d simplicial complexes (where propagators correspond to triangles and the vertices to tetrahedra):

$$Z = \int \mathcal{D}\varphi \, e^{-S[\varphi]} = \sum_{\Gamma} \frac{\lambda^N}{\mathrm{sym}[\Gamma]} Z(\Gamma),$$

where N is the number of interaction vertices in the Feynman graph Γ, $\mathrm{sym}[\Gamma]$ is a symmetry factor for Γ, and $Z(\Gamma)$ the corresponding Feynman amplitude. In Lie algebra variables, the amplitude for a generic Feynman diagram is [5]

$$Z(\Gamma) = \int \prod_p \mathrm{d}h_p \prod_f \mathrm{d}x_f \, e^{i \sum_f \mathrm{Tr}\, x_f H_f}, \tag{3.11}$$

where f denotes loops in the graph which bound faces of the 2-complex formed by the graph; such faces are dual to edges in the simplicial complex, and are associated a Lie algebra variable x_f. For each propagator p we integrate over a group element h_p interpreted as a discrete connection. (3.11) corresponds to the simplicial path integral of 3d gravity in first-order form with action

$$S(e,\omega) = \int_{\mathcal{M}} \mathrm{Tr}\,(e \wedge F(\omega)), \tag{3.12}$$

with $\mathfrak{su}(2)$-valued triad $e^i(x)$ and $\mathfrak{su}(2)$-connection $\omega^j(x)$ with curvature $F(\omega)$. Introducing the simplicial complex Δ and its dual cellular complex Γ, we can discretize e^i along edges of Δ as $x_e = \int_e e(x) = x^i \tau_i \in \mathfrak{su}(2)$, and ω^j along links of Γ, dual to triangles of Δ, as $h_L = e^{\int_L \omega} \in \mathrm{SU}(2)$. The discrete curvature is given by the holonomy around the face f in Γ dual to an edge e of Δ, obtained as an ordered product $H_f = \prod_{L \in \partial f} h_L \in \mathrm{SU}(2)$. The discrete counterpart of (3.12) is then the action $\sum_e \mathrm{Tr}\, x_e H_e$ in (3.11).

The GFT model we have introduced succeeds in at least one point where tensor models failed: in defining amplitudes for its Feynman diagrams (identified with discrete spacetimes), arising in a perturbative expansion around the "no-space state," that correctly encode classical and quantum simplicial geometry and that can be nicely related to a simplicial gravity action.

The Feynman amplitudes can be computed in the other representations we have at our disposal. In the group picture one obtains a delta function $\delta(\prod_{L \in \partial f} h_L)$ in the overall amplitude, making the flatness constraint explicit; the expression of $Z(\Gamma)$ in terms of representations is an assignment of an irreducible $\mathrm{SU}(2)$ representation j_f to each face of Γ, and of a group intertwiner to each link of the complex, a *spin foam* [35, 42, 36]:

$$Z(\Gamma) = \left(\prod_f \sum_{j_f} \right) \prod_f (2j_f + 1) \prod_v \begin{Bmatrix} j_1 & j_2 & j_3 \\ j_4 & j_5 & j_6 \end{Bmatrix}.$$

This is the Ponzano-Regge model [43] for 3d quantum gravity. The correspondence between spin-foam models and GFT amplitudes is generic.

The above constructions can be generalized to the computation of GFT observables, in particular n-point functions, which translate into the calculation of Feynman amplitudes for diagrams/simplicial complexes of arbitrary topology and with boundaries. These in turn take again the form of simplicial gravity path integrals on the corresponding topology.

Once more, we have a discrete realization of the third-quantization idea. We refer to the literature (*e.g.* the recent more extended introduction in [34]) for more details and for an account of recent results, in particular for the definition of GFT models of 4d quantum gravity, for the identification of (discrete) diffeomorphism symmetry in GFT, and for the proof that a generalization of the large-N limit of matrix models holds true also in (some) GFT models, leading to the suppression of non-trivial topologies, as well as for more work on the topological properties of the GFT Feynman expansion.

4. Continuum third quantization: heuristics

Going back to the idea of a third quantization of continuum gravity, we would like to define a field theory on the configuration space of GR that reproduces the constraints of canonical GR through its equations of motion. Since such constraints have to be satisfied at each point in Σ, the action should reproduce equations of the form (γ_α are the variables of GR, metric or connection)

$$\mathcal{C}_i\left[\gamma_\alpha, \frac{\delta}{\delta\gamma_\alpha}; x\right) \Phi[\gamma_\alpha] = 0 , \tag{4.1}$$

where we have introduced a field which is classically a functional of γ_α, *i.e.* a canonical quantum gravity wave function. The naive action

$$S = \int \mathcal{D}\gamma_\alpha \sum_i \Phi[\gamma_\alpha]\, \mathcal{C}_i\left[\gamma_\alpha, \frac{\delta}{\delta\gamma_\alpha}\right] \Phi[\gamma_\alpha] \tag{4.2}$$

reproduces constraints $\mathcal{C}_i = 0$ independent of the point on Σ. This was noted by Giddings and Strominger in [20], who took $\mathcal{C}$ to be the Hamiltonian constraint integrated over $\Sigma = S^3$.

To get the right number of constraints, one has to increase the number of degrees of freedom of Φ. One possibility, for only a single constraint per point $\mathcal{C}$, would be to add an explicit dependence of Φ on points in Σ:

$$S = \int \mathcal{D}\gamma_\alpha \int_\Sigma \mathrm{d}^D x\, \Phi[\gamma_\alpha; x)\, \mathcal{C}\left[\gamma_\alpha, \frac{\delta}{\delta\gamma_\alpha}; x\right) \Phi[\gamma_\alpha; x). \tag{4.3}$$

Defining a formal differential calculus by

$$\mathfrak{d}\Phi[\gamma_\alpha; x)/\mathfrak{d}\Phi[\delta_\beta; y) = \delta[\gamma_\alpha - \delta_\beta]\delta(x-y) ,$$

the variation $\mathfrak{d}S/\mathfrak{d}\Phi$ would reproduce (4.1). However, such a dependence on points seems unnatural from the point of view of canonical gravity; points in Σ are not variables on the phase space. An alternative and more attractive possibility arises by realizing that in canonical GR one really has an integrated Hamiltonian involving non-dynamical quantities which are originally part of the phase space. We thus integrate the constraints (4.1) using

Lagrange multipliers, and extend Φ to be a functional of all original phase space variables. An action is then defined as

$$S = \int \mathcal{D}\gamma_\alpha \, \mathcal{D}\Lambda^i \, \Phi[\gamma_\alpha, \Lambda^i] \left(\int_\Sigma \mathrm{d}^D x \, \Lambda^i(x) \, \mathcal{C}_i \left[\gamma_\alpha, \frac{\delta}{\delta\gamma_\alpha}; x \right] \right) \Phi[\gamma_\alpha, \Lambda^i]. \quad (4.4)$$

Variation with respect to the scalar field Φ then yields the equations of motion

$$0 = \mathcal{C}\left[\gamma_\alpha, \frac{\delta}{\delta\gamma_\alpha}, \Lambda^i\right] \Phi[\gamma_\alpha, \Lambda^i] \equiv \left(\int_\Sigma \mathrm{d}^D x \, \Lambda^i(x) \, \mathcal{C}_i \left[\gamma_\alpha, \frac{\delta}{\delta\gamma_\alpha}; x \right] \right) \Phi[\gamma_\alpha, \Lambda^i]. \quad (4.5)$$

$\mathcal{C}$ is now the Hamiltonian of canonical gravity; since (4.5) has to hold for arbitrary values of $\Lambda^i(x)$, it is equivalent to (4.1). The fact that Λ^i are non-dynamical is apparent from the action which does not contain functional derivatives with respect to them. Introducing smeared constraints is also what one does in LQG since equations of the form (4.1) are highly singular operator equations, just as (1.1). We are now treating *all* the constraints of canonical gravity on equal footing, and expect all of them to result from the equations of motion of the theory; if we are working in metric variables, our field is defined on the space of metrics on Σ before any constraint is imposed.

Implicit in the very idea of a continuum third quantization of gravity is the use of a non-trivial kinetic term, corresponding to *quantum* constraints. On the one hand this has to be immediately contrasted with the standard GFT action which has instead a trivial kinetic term; we will discuss this point more extensively in the concluding section. On the other hand, since the *classical* action of the field theory on superspace encodes and makes use of the *quantum* constraint operators, choices of operator ordering and related issues enter prominently also at this level; this should clarify why the *classical* GFT action makes use, for example, of $\star$-products in its definition.

One more important issue is that one should expect the symmetries of classical (and presumably quantum) GR – spatial diffeomorphisms, time reparametrizations etc. – to be manifest as symmetries of (4.4) in some way. The kind of symmetries one normally considers, transformations acting on spacetime that do not explicitly depend on the dynamical fields, will from the viewpoint of the third-quantized framework defined on some form of superspace appear as *global* rather than local symmetries[1]. This basic observation is in agreement with recent results in GFT [4], where one can identify a global symmetry of the GFT action that has the interpretation of diffeomorphisms.

If we now focus on GR in connection variables ($\gamma_\alpha = \omega_i^{ab}$), the variable conjugate to ω_i^{ab}, some function of the frame field, would be represented as a functional derivative. We then follow the same strategy as in LQG, redefining variables on the phase space to obtain a field which depends on holonomies only, and rewriting the Hamiltonian in terms of holonomies. Issues of operator ordering and regularization will be dealt with in the way familiar from LQG. As an example which will allow us to implement this procedure to some extent

[1]In extensions of standard GR where one allows for topology change at the classical level by allowing degenerate frame fields, the situation may be different [27].

and to make connections to the GFT formalism, we specialize to the case of 3d Riemannian GR in first-order formulation (3.12), heuristically defining

$$S = \int \mathcal{D}\omega_i^{ab}\, \mathcal{D}\chi^a\, \mathcal{D}\Omega^{ab}\, \Phi[\omega_i^{ab}, \chi^a, \Omega^{ab}] H \Phi[\omega_i^{ab}, \chi^a, \Omega^{ab}], \qquad (4.6)$$

where H is the Hamiltonian of 3d GR (without cosmological constant),

$$H = -\frac{1}{2}\int_\Sigma \mathrm{d}^2x \left\{\epsilon_{abc}\chi^a\epsilon^{ij}R_{ij}^{bc} + \Omega^{ab}\nabla_j^{(\omega)}\pi^j_{ab}\right\}, \qquad (4.7)$$

involving R_{ij}^{bc}, the curvature of the $\mathfrak{su}(2)$ connection ω_i^{ab}, the conjugate momentum $\pi^j_{bc} \sim \epsilon^{ij}\epsilon_{abc}e^a_i$, the covariant derivative $\nabla^{(\omega)}$, and an $\mathfrak{su}(2)$-valued scalar $\epsilon_{abc}\chi^a$. Alternatively one could go to a reduced configuration space by dividing out gauge transformations generated by $\nabla_j^{(\omega)}\pi^j_{ab} = 0$, and define

$$S = \int \mathcal{D}\omega_i^{ab}\, \mathcal{D}\chi^a\, \Phi[\omega_i^{ab}, \chi^a] \left(-\frac{1}{2}\int_\Sigma \mathrm{d}^2x\, \epsilon_{abc}\chi^a(x)\epsilon^{ij}R_{ij}^{bc}(x)\right) \Phi[\omega_i^{ab}, \chi^a], \qquad (4.8)$$

where Φ is now understood as a gauge-invariant functional. At this formal level, an analogous "construction" would be possible in metric variables, but the construction of a Hilbert space for GR in connection variables done in LQG will allow us to make more progress towards a rigorous definition of the action if we work with an SU(2) connection. Let us investigate this in detail.

5. Towards a rigorous construction: passing to holonomies, decomposing in graphs

Connection variables allow the use of technology developed for LQG (summarized in [47]) to give a more rigorous meaning to (4.4). One changes phase space variables to holonomies, and decomposes functionals on the space of (generalized) connections into functions depending on a finite number of holonomies. A measure in (4.4) can be defined from a normalized measure on the gauge group, extended to arbitrary numbers of holonomies by so-called projective limits; the Hilbert space is defined as the space of square-integrable functionals on the space of (generalized) connections with respect to this measure. This strategy works well for connections with compact gauge group; here we require functionals that also depend on Lagrange multipliers. One way of dealing with scalars is by defining "point holonomies" $U_{\lambda,x}(\chi) := \exp(\lambda\chi(x)) \in G$ where the field χ is in the Lie algebra $\mathfrak{g}$ (for (4.8), $\mathfrak{g} = \mathfrak{su}(2)$). One can reconstruct χ from its point holonomies, taking derivatives or the limit $\lambda \to 0$ [47, sect. 12.2.2.2]. The Lagrange multiplier and connection can then be treated on the same footing. We will in the following focus on 3d Riemannian GR but a similar procedure would be applicable in any setting where one has a connection with compact gauge group.

We detail the construction step by step. Consider a graph Γ embedded into the surface Σ, consisting of E edges and V vertices. Define a functional of a $\mathfrak{g}$-connection ω and a $\mathfrak{g}$-valued scalar χ, called a *cylindrical function*, by[2]

$$\Phi_{\Gamma,f}[\omega_i^{ab},\chi^a] = f(h_{e_1}(\omega),\ldots,h_{e_E}(\omega),U_{\lambda,x_1}(\chi),\ldots,U_{\lambda,x_V}(\chi)), \tag{5.1}$$

where f is a square-integrable function on G^{E+V} whose arguments are the holonomies of the connection along the edges e_i of Γ and the point holonomies of χ at the vertices x_k. Functional integration of such functions is defined as

$$\int \mathcal{D}\omega_i^{ab}\,\mathcal{D}\chi^a\;\Phi_{\Gamma,f}[\omega_i^{ab},\chi^a] := \int [\mathrm{d}h]^E[\mathrm{d}U]^V\, f(h_1,\ldots,h_E,U_1,\ldots,U_V). \tag{5.2}$$

Now the linear span of the functionals (5.1) is dense in the required space of functionals of generalized connections [44]. An orthonormal basis for the functions associated to each graph, *i.e.* for square-integrable functions on G^{E+V}, is provided by the entries of the representation matrices for group elements in (irreducible) representations (due to the Peter-Weyl theorem),

$$\begin{aligned} T_{\Gamma,j_e,C_v,m_e,n_e,p_v,q_v}[\omega_i^{ab},\chi^a] \;\sim\; & D^{j_1}_{m_1n_1}(h_{e_1}(\omega))\ldots D^{j_E}_{m_En_E}(h_{e_E}(\omega)) \\ & \times D^{C_1}_{p_1q_1}(U_{\lambda,x_1}(\chi))\ldots D^{C_V}_{p_Vq_V}(U_{\lambda,x_V}(\chi)), \end{aligned} \tag{5.3}$$

where we omit normalization factors. Note that one associates one such representation j_e to each edge and another one C_v to each vertex in the graph.

If we go to *gauge-invariant* functionals of the connection ω_i^{ab} and the field χ^a, passing to a reduced phase space where one has divided out gauge orbits, a basis is provided by functions associated to generalized spin networks, graphs whose edges start and end in vertices, labeled by irreducible representations at edges and vertices and intertwiners at vertices projecting onto singlets. For (4.8), there is only a single SU(2), and schematically we have (an analogous decomposition would be possible at the level of (5.1) already)

$$\Phi[\omega_i^{ab},\chi^a] = \sum_{\{\Gamma,j_e,C_v,i_v\}} \Phi(\Gamma,j_e,C_v,i_v)\,T_{\Gamma,j_e,C_v,i_v}[\omega_i^{ab},\chi^a]\,, \tag{5.4}$$

where the spin-network functions T_{Γ,j_e,C_v,i_v} are (again up to normalization)

$$\begin{aligned} T_{\Gamma,j_e,C_v,i_v}[\omega_i^{ab},\chi^a] \;\sim\; & D^{j_1}_{m_1n_1}(h_{e_1}(\omega))\ldots D^{j_E}_{m_En_E}(h_{e_E}(\omega))D^{C_1}_{p_1q_1}(U_{\lambda,x_1}(\chi)) \\ & \ldots D^{C_V}_{p_Vq_V}(U_{\lambda,x_V}(\chi))(i_1)^{n_1n_2\ldots q_1}_{m_1m_2\ldots p_1}\cdots(i_V)^{q_V\ldots n_E}_{p_V\ldots m_E}; \end{aligned} \tag{5.5}$$

one takes the functions (5.3) in the representations j_e and C_v and contracts with intertwiners i_v at the vertices. We may represent a spin network as follows:

[2](5.1) bears a close resemblance to "projected cylindrical functions" used in [31] to obtain a Lorentz-covariant 4d formalism, both functions of connection and Lagrange multiplier.

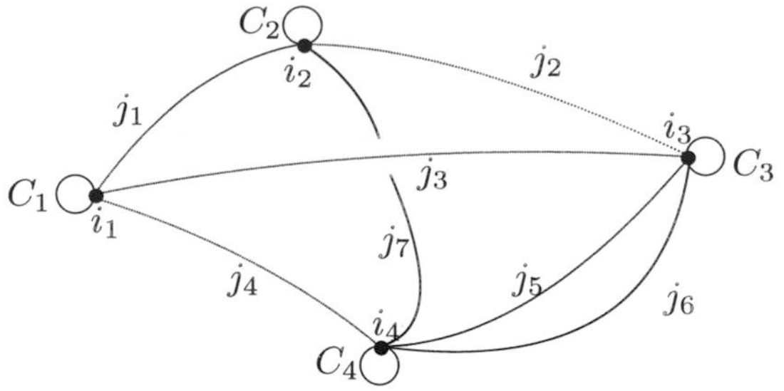

Using the expansion (5.4) of the field Φ, (4.8) can be formally written as

$$S = \sum_{s,s' \in S} \Phi(s)\Phi(s') \int \mathcal{D}\omega_i^{ab}\, \mathcal{D}\chi^a\, T_s[\omega, \chi] \left(-\frac{1}{2} \int_\Sigma \mathrm{d}^2x\, \epsilon_{abc}\chi^a \epsilon^{ij} R_{ij}^{bc} \right) T_{s'}[\omega, \chi], \tag{5.6}$$

where we use $s = \{\Gamma, j_e, C_v, i_v\}$ to describe a spin network, and S is the set of all spin networks embedded into Σ. The task is then to determine the action of the Hamiltonian on a spin-network function and to use the orthogonality property $\langle T_s, T_{s'} \rangle = \delta_{s,s'}$ for normalized spin-network functions to collapse the sums to a single sum. Let us again be more explicit on the details.

The Hamiltonian can be expressed in terms of holonomies, following the strategy for 3d LQG [33]: For given s and s', consider a cellular decomposition of Σ, chosen so that the graphs of s and s' are contained in the 0-cells and 1-cells, into plaquettes p with sides of coordinate length ϵ; approximate

$$\int_\Sigma \mathrm{d}^2x\, \epsilon_{abc}\chi^a(x)\epsilon^{ij} R_{ij}^{bc}(x) \approx \sum_p \epsilon^2 \mathrm{Tr}(\chi_p R_p[\omega]) \approx \frac{2}{\lambda} \sum_p \mathrm{Tr}(U_{\lambda,p}(\chi) h_p[\omega]) , \tag{5.7}$$

where $(\chi_p)_{bc} = \epsilon_{abc}\chi^a(x_p)$ for some $x_p \in p$ and $(R_p)^{bc} = \epsilon^{ij} R_{ij}^{bc}$ which we approximate by the point holonomy and the holonomy h_p around p:

$$U_{\lambda,p} = 1 + \lambda\chi_p + \ldots, \qquad h_p = 1 + \frac{1}{2}\epsilon^2 R_p + \ldots . \tag{5.8}$$

Instead of treating λ as a parameter taken to zero at the end, one could take $\lambda = \lambda(\epsilon)$ and then a single limit $\epsilon \to 0$. The action becomes, formally,

$$S = \lim_{\substack{\epsilon \to 0 \\ \lambda \to 0}} \sum_{s,s'} \Phi(s)\Phi(s') \int \mathcal{D}\omega \mathcal{D}\chi\, T_s[\omega, \chi] \left(-\frac{1}{\lambda} \sum_p \mathrm{Tr}\big(U_{\lambda,p}(\chi) h_p[\omega]\big) \right) T_{s'}[\omega, \chi]. \tag{5.9}$$

Next one computes the action of the Hamiltonian on spin-network functions. Since the latter form a basis for gauge-invariant functionals one can expand

$$-\frac{1}{\lambda} \sum_p \mathrm{Tr}(U_{\lambda,p}(\chi) h_p[\omega]) T_{s'}[\omega_i^{ab}, \chi^a] = \sum_{s'' \in S} c^H_{s',s''} T_{s''}[\omega_i^{ab}, \chi^a] . \tag{5.10}$$

Then one uses the orthogonality property for spin-network functions,

$$\begin{aligned}&\int \mathcal{D}\omega_i^{ab}\,\mathcal{D}\chi^a\; T_s[\omega_i^{ab},\chi^a]T_{s''}[\omega_i^{ab},\chi^a]\\ \equiv &\int [\mathrm{d}h]^E[\mathrm{d}U]^V\; T_s[h_i,U_k]T_{s''}[h_i,U_k] = \delta_{s,s''}\;,\end{aligned} \tag{5.11}$$

where integration of spin-network functions is defined as in (5.2); one takes the union of the graphs Γ and Γ' used to define s and s' and extends the spin-network functions trivially to the union, integrating over $E+V$ copies of SU(2) (with normalized Haar measure), where E is the number of edges and V the number of vertices in the union of Γ and Γ'. The action becomes

$$S = \lim_{\substack{\epsilon\to 0\\ \lambda\to 0}} \sum_{s,s'\in S} \Phi(s)\Phi(s')c^H_{s',s}, \tag{5.12}$$

with the action of the Hamiltonian now hidden in the matrix elements $c^H_{s',s}$, and one has to worry about $\epsilon\to 0$ and $\lambda\to 0$. (In LQG the only limit is $\epsilon\to 0$; the Hamiltonian is a finite and well-defined operator in this limit [33].)

In the traditional LQG formulation [44, 47] the effect of $\mathrm{Tr}(U_{\lambda,p}(\chi)h_p[\omega])$ on a spin network is to add p as a loop with $j=1/2$, a vertex at x_p with $C_v=1/2$ and an intertwiner contracting the two SU(2) matrices, so that

$$-\frac{1}{\lambda}\sum_p \mathrm{Tr}(U_{\lambda,p}(\chi)h_p[\omega])T_{s'}[\omega_i^{ab},\chi^a] = -\frac{1}{\lambda}\sum_p T_{s'\cup p_{j=C_{x_p}=1/2}}[\omega_i^{ab},\chi^a]\;, \tag{5.13}$$

so that $c^H_{s',s''} = -\frac{1}{\lambda}\sum_p \delta_{s'',s'\cup p_{j=C_{x_p}=1/2}}$, and hence

$$S = \lim_{\substack{\epsilon\to 0\\ \lambda\to 0}} \sum_{s\in S} -\frac{1}{\lambda}\sum_p \Phi(s\cup p_{j=C_{x_p}=1/2})\Phi(s). \tag{5.14}$$

There have been suggestions for non-graph-changing holonomy operators, which would mean that $c^H_{s',s''}=0$ unless s' and s'' define the same graph, so that one would have a (free) action "local in graphs". Here a more concrete definition relies on input from canonical quantum gravity (here LQG).

The expansion of Φ into spin-network functions, imposed by the structure of the LQG Hilbert space, suggests a change of perspective: While a quantum field $\Phi[\omega_i^{ab},\chi^a]$ would be thought of as creating a hypersurface of topology Σ with a continuous geometry, the operators $\Phi(s)$ would create labeled *graphs* of arbitrary valence and complexity. One would consider a Fock space of "many-graph states", a huge extension of the LQG Hilbert space:

$$\mathfrak{H}_{\mathrm{LQG}} \sim \bigoplus_\Gamma \mathfrak{H}_\Gamma \quad\longrightarrow\quad \mathfrak{H}_{\text{3rd quant.}} \sim \bigoplus_n \mathfrak{H}^n_{\mathrm{LQG}} \sim \bigoplus_\Gamma\bigoplus_n \mathfrak{H}^n_\Gamma. \tag{5.15}$$

This (we stress again, purely formal) picture in terms of graphs is closer to the discrete structures in the interpretation of GFT Feynman graphs, and suggests to take the generalization of GFT to fields with arbitrary number of arguments seriously, as a vertex of arbitrary valence n would be described by a GFT field with n arguments. In (5.12), the graphs forming the spin networks are embedded into Σ, but the next step towards a GFT interpretation in which one views them as abstract is conceivably straightforward. A crucial difference between the continuum third quantization and GFT, however, is that between entire graphs represented by $\Phi(s)$, and building blocks of graphs represented by the GFT field, where entire graphs emerge as Feynman graphs in the quantum theory. We will discuss this issue further in the next section.

If one wants to add an interaction term to the action, there is not much guidance from the canonical theory. Nevertheless, the addition of such a term would in principle pose no additional difficulty. One would, in adding such a term, presumably impose conservation laws associated to the geometric interpretation of the quantities imposed, *e.g.* conservation of the total area/volume represented by two graphs merged into one, which when expressed in terms of the canonically conjugate connection would mean requiring "locality" in the connection. This strategy is proposed for a minisuperspace model in [9].

6. Lessons for group field theories and some speculations

Let us now conclude with a discussion of the relation between the two formulations of third-quantized gravity we have presented.

One could try to derive the GFT setting from the formal continuum third-quantization framework, by somehow discretizing (5.12). The relevant questions are then: What is the limit in which a suitable GFT arises from (5.12)? What is the expression for the action if one adapts the Hamiltonian constraint to a generic triangulation, dual to the spin network graph? One possibility seems to be to assume that $\Phi(s) = 0$ except on a fixed graph γ_0. One would then hope to be able to identify the nontrivial contributions as $\epsilon \to 0$, and then do a summation over the variables C_v to get a GFT model. Here we immediately encounter a basic issue: whatever graph γ_0 chosen, it has to be one of those appearing in the decomposition of a gauge-invariant functional coming from LQG, and so it will have to be a closed graph, while the fundamental GFT field is typically associated to an open graph.

However, not much faith can be placed on the third quantization in the continuum as a consistent theory of quantum gravity. It is more sensible to view the problem in the opposite direction, and consider such a continuum formulation as an effective description of the dynamics of a more fundamental GFT in some approximation. It makes more sense, therefore, to tackle the big issue of the continuum approximation of the GFT dynamics. It may be helpful to use the idea and general properties of the formal continuum third quantization to gain some insight on how the continuum limit of GFTs should be approached and on what it may look like. We see two main possibilities:

- The first possibility is to try to define a continuum limit of GFTs at the level of the classical action, trying to obtain the classical action of a third-quantized field on continuum superspace.

This would mean recovering, from the GFT field, the LQG functional as an infinite combination of cylindrical functions each depending on a finite number of group elements. This entails a generalization of the GFT framework to allow for all possible numbers of arguments of the GFT field(s), and combinatorial patters among them in the interaction term. This generalization is possible, but will certainly be very difficult to handle. Even this huge action would still codify the dynamics of *building blocks* of graphs, rather than graphs themselves, falling still a step shorter of the continuum third-quantized action. Let us discuss a few more technical points.

First, in (5.14), we have assumed a graph-changing Hamiltonian. From the point of view of a dynamics of *vertices*, this implies a non-conservation of the same, and thus suggests an underlying interacting field theory of such vertices, with the canonical quantum dynamics encoded in both the kinetic and interaction terms of the underlying action, and the same interaction term that produces graph changing also produces topology change. This graph dynamics is in contrast with the formal continuum one: The continuum third-quantized action features a clear distinction between the topology-preserving contribution to the dynamics, corresponding to the canonical (quantum) gravitational dynamics, encoded in a non-trivial kinetic term (Hamiltonian constraint), and the topology-changing contribution, encoded in the interaction term. This distinction would remain true also in the case of a graph-preserving continuum Hamiltonian constraint. Only, the kinetic term would decompose into a sum of kinetic terms each associated to a different graph. Both the distinction between topology-preserving and topology-changing contribution to the dynamics and a non-trivial kinetic term are not available in (standard) GFTs. Concerning the issue of a restriction of the GFT dynamics to a given (trivial) topology, there have been encouraging results extending the large-N limit of matrix models [24, 25], but these clearly relate to the quantum GFT dynamics, and a similar approximation would not lead to any significant modification at the level of the GFT action. Still, the study of GFT perturbative renormalization [16, 19, 32, 6] is relevant because it may reveal (signs of this are already in [6]) that the GFT action has to be extended to include non-trivial kinetic terms, in order to achieve renormalizability. Another strategy is the expansion of the GFT action around a non-trivial background configuration. What one gets generically [14, 22, 41, 30] is an effective dynamics for "perturbations" characterized by a non-trivial kinetic term and thus a non-trivial topology- (and graph-)preserving dynamics, encoded in the linear part of the GFT equations of motion, and a topology- (and graph-)changing dynamics encoded in the non-linearities, *i.e.* in the GFT interaction. This observation suggests that the GFT dynamics which, in a continuum limit, would correspond to the (quantum) GR dynamics, is to be looked for in a different phase, and not around the trivial $\Phi = 0$ (Fock, no-space) vacuum.

Notwithstanding these considerations, it seems unlikely that a matching between discrete and continuum dynamics can be obtained working purely at the level of the action. The discretization of the continuum third-quantized framework represented by the GFT formalism seems to operate at a more radical, fundamental level, in terms of the very dynamical objects chosen to constitute a discrete quantum space: its microscopic building blocks.

- The second main possibility is that the continuum limit is to be defined at the level of the quantum theory, at the level of its transition amplitudes.

This seems much more likely, for several reasons. One is simply that, in a quantum GFT, the classical action will be relevant only in some limited regime. A second one is the analogy with matrix models, where the contact between discrete and continuum quantization is made at the level of the transition amplitudes. Moreover, the continuum limit of the theory corresponds to a phase transition in the thermodynamic limit that can not be seen at the level of the action. A deeper implication of this is that the emergence of a classical spacetime is the result of a purely quantum phenomenon.

When considering such a limit at the level of the quantum GFT theory, two main mathematical and conceptual issues have to be solved, already mentioned above: the restriction to the dynamics of geometry for given (trivial) topology, and the step from a description of the dynamics of (gravitational) degrees of freedom associated to elementary building blocks of graphs, thus of a quantum space, to that of entire graphs, thus quantum states for the whole of space (as in LQG). On the first issue, the recent results on the large-N approximation of the GFT dynamics [24, 25] (for topological models) are certainly going to be crucial, and represent a proof that the needed restriction can be achieved. The other example of a procedure giving, in passing, the same restriction is the mean-field approximation of the GFT dynamics around a non-trivial background configuration.

The second issue is, in a sense, more thorny, and does not seem to require only the ability to solve a mathematical problem, but some new conceptual ingredient, some new idea. Let us speculate on what this idea could be. Looking back at the LQG set-up, one sees that the continuum nature of the theory, with infinitely many degrees of freedom, is encoded in the fact that a generic wave function of the connection decomposes into a sum over cylindrical functions associated to arbitrarily complicated graphs. In terms of graph vertices, the continuum nature of space is captured in the regime of the theory in which an arbitrary high number of such graph vertices is interacting. The best way to interpret and use the GFT framework, in this respect, seems then to be that of many-particle and statistical physics. From this point of view, quantum space is a sort of weird condensed-matter system, where the continuum approximation would play the role of the hydrodynamic approximation, and the GFT represents the theory of its "atomic" building blocks. This perspective has been advocated already in [39] and it resonates with other ideas about spacetime as a condensate [28, 48] (see also [41] for recent results in this direction). In practice, this possibility leads immediately to the need for ideas and techniques from statistical field theory, in particular to the issue of phase transitions in GFTs, the idea being that the discrete-continuum transition is one such phase transition and that continuum spacetime physics applies only in some of the phases of the GFT system.

It becomes even more apparent, then, that the study of GFT renormalization [16, 19, 32], both perturbative and non-perturbative, will be crucial in the longer run for solving the problem of the continuum. The analogy with matrix models, where the idea of discrete-continuum phase transition is realized explicitly, and the generalization of ideas and tools developed for them to the GFT context, will be important as well. Probably a crucial role will be played by diffeomorphism invariance, in both GFT and simplicial gravity, as a guiding principle for recovering a good continuum limit and even for devising the appropriate procedure to do so [4, 13]. Similarly, classical and quantum simplicial gravity will provide insights about what sort of approximation of variables and which regime of the quantum dynamics give the continuum.

In both cases above, the theory one would recover/define from the GFT dynamics would be a form of quantum gravitational dynamics, either encoded in a classical action which would give equations of motion corresponding to canonical quantum gravity, or in a quantum gravity path integral. An extra step would then be needed to recover *classical* gravitational dynamics, presumably some modified form of GR, from it. This is where semi-classical approximations would be needed, in a GFT context. Their role will be that of "de-quantizing" the formalism, *i.e.* to go from what would still be a third-quantized formalism, albeit now in the continuum, to a second-quantized one. This semi-classical approximation, which is a very different approximation with respect to the continuum one, would be approached most naturally using WKB techniques or coherent states. Here it is worth distinguishing further between third-quantized coherent states, coherent states for the quantized GFT field, within a Fock-space construction for GFTs still to be defined properly, and second-quantized ones, coherent states for the canonical wave function and defining points in the canonical classical phase space, as defined and used in LQG. The last type of coherent states have been used to extract an effective dynamics using mean-field-theory techniques and de-quantize the system in one stroke in [41]. This issue could also be fruitfully studied in a simplified minisuperspace third quantization; work on such a model, which can be seen both as a truncation of the GFT dynamics and as a generalization of the well-developed loop quantum cosmology [7] to include topology change, is in progress [9].

Acknowledgments

This work is funded by the A. von Humboldt Stiftung, through a Sofja Kovalevskaja Prize, which is gratefully acknowledged.

References

[1] Abhay Ashtekar. New variables for classical and quantum gravity. *Physical Review Letters*, 57:2244–2247, November 1986.

[2] Abhay Ashtekar, Alejandro Corichi, and José A. Zapata. Quantum theory of geometry: III. Non-commutativity of Riemannian structures. *Classical and Quantum Gravity*, 15:2955–2972, October 1998.

[3] Aristide Baratin, Bianca Dittrich, Daniele Oriti, and Johannes Tambornino. Non-commutative flux representation for loop quantum gravity. 2010.

[4] Aristide Baratin, Florian Girelli, and Daniele Oriti. Diffeomorphisms in group field theories. 2011.

[5] Aristide Baratin and Daniele Oriti. Group field theory with noncommutative metric variables. *Physical Review Letters*, 105(22):221302, November 2010.

[6] Joseph Ben Geloun and Valentin Bonzom. Radiative corrections in the Boulatov-Ooguri tensor model: the 2-point function. January 2011.

[7] Martin Bojowald. Loop quantum cosmology. *Living Reviews in Relativity*, 11:4, July 2008.

[8] D. V. Boulatov. A model of three-dimensional lattice gravity. *Modern Physics Letters A*, 7:1629–1646, 1992.

[9] Gianluca Calcagni, Steffen Gielen, and Daniele Oriti. Work in progress.

[10] Gianluca Calcagni, Steffen Gielen, and Daniele Oriti. Two-point functions in (loop) quantum cosmology. 2010.

[11] F. David. Planar diagrams, two-dimensional lattice gravity and surface models. *Nuclear Physics B*, 257:45–58, 1985.

[12] Bryce S. DeWitt. Quantum theory of gravity. I. The canonical theory. *Physical Review*, 160:1113–1148, August 1967.

[13] Bianca Dittrich. Diffeomorphism symmetry in quantum gravity models. 2008.

[14] Winston J. Fairbairn and Etera R. Livine. 3D spinfoam quantum gravity: matter as a phase of the group field theory. *Classical and Quantum Gravity*, 24:5277–5297, October 2007.

[15] L. Freidel. Group field theory: an overview. *International Journal of Theoretical Physics*, 44:1769–1783, October 2005.

[16] Laurent Freidel, Razvan Gurau, and Daniele Oriti. Group field theory renormalization - the 3d case: Power counting of divergences. *Phys. Rev. D*, 80:044007, 2009.

[17] Laurent Freidel and Etera R. Livine. Ponzano-Regge model revisited III: Feynman diagrams and effective field theory. *Class. Quant. Grav.*, 23:2021–2062, 2006.

[18] Laurent Freidel and Shahn Majid. Noncommutative harmonic analysis, sampling theory and the Duflo map in 2+1 quantum gravity. *Classical and Quantum Gravity*, 25(4):045006, February 2008.

[19] Joseph Ben Geloun, Thomas Krajewski, Jacques Magnen, and Vincent Rivasseau. Linearized group field theory and power counting theorems. *Class. Quant. Grav.*, 27:155012, 2010.

[20] Steven B. Giddings and Andrew Strominger. Baby universes, third quantization and the cosmological constant. *Nuclear Physics B*, 321:481–508, July 1989.

[21] P. Ginsparg. Matrix models of 2d gravity. 1991.

[22] Florian Girelli, Etera R. Livine, and Daniele Oriti. 4d deformed special relativity from group field theories. *Phys. Rev. D*, 81:024015, 2010.

[23] Domenico Giulini. The superspace of geometrodynamics. *General Relativity and Gravitation*, 41:785–815, April 2009.

[24] Razvan Gurau. The 1/N expansion of colored tensor models. 2010.

[25] Razvan Gurau and Vincent Rivasseau. The 1/N expansion of colored tensor models in arbitrary dimension. 2011.

[26] Jonathan J. Halliwell and Miguel E. Ortiz. Sum-over-histories origin of the composition laws of relativistic quantum mechanics and quantum cosmology. *Physical Review D*, 48:748–768, July 1993.

[27] G. T. Horowitz. Topology change in classical and quantum gravity. *Classical and Quantum Gravity*, 8:587–601, April 1991.

[28] B. L. Hu. Can spacetime be a condensate? *Int. J. Theor. Phys.*, 44:1785–1806, 2005.

[29] Karel Kuchař. General relativity: dynamics without symmetry. *Journal of Mathematical Physics*, 22:2640–2654, November 1981.

[30] E. Livine, D. Oriti, and J. Ryan. Effective Hamiltonian constraint from group field theory, in preparation.

[31] Etera R. Livine. Projected spin networks for Lorentz connection: linking spin foams and loop gravity. *Classical and Quantum Gravity*, 19:5525–5541, November 2002.

[32] Jacques Magnen, Karim Noui, Vincent Rivasseau, and Matteo Smerlak. Scaling behaviour of three-dimensional group field theory. *Class. Quant. Grav.*, 26:185012, 2009.

[33] Karim Noui and Alejandro Perez. Three-dimensional loop quantum gravity: physical scalar product and spin-foam models. *Classical and Quantum Gravity*, 22:1739–1761, May 2005.

[34] D. Oriti. The microscopic dynamics of quantum space as a group field theory, in: Foundations of space and time, G. Ellis, J. Murugan (eds.). Cambridge: Cambridge University Press, 2011.

[35] Daniele Oriti. Spacetime geometry from algebra: spin foam models for non-perturbative quantum gravity. *Reports on Progress in Physics*, 64:1489–1543, December 2001.

[36] Daniele Oriti. Spin foam models of quantum spacetime. PhD thesis, University of Cambridge, 2003.

[37] Daniele Oriti. Quantum gravity as a quantum field theory of simplicial geometry, in: Mathematical and Physical Aspects of Quantum Gravity, B. Fauser, J. Tolksdorf and E. Zeidler (eds.). Birkhäuser, Basel, 2006.

[38] Daniele Oriti. The group field theory approach to quantum gravity. 2006.

[39] Daniele Oriti. Group field theory as the microscopic description of the quantum spacetime fluid: a new perspective on the continuum in quantum gravity. 2007.

[40] Daniele Oriti. Approaches to quantum gravity: toward a new understanding of space, time and matter. Cambridge: Cambridge Univ. Press, 2009.

[41] Daniele Oriti and Lorenzo Sindoni. Towards classical geometrodynamics from group field theory hydrodynamics. *New J. Phys.*, 13:025006, 2011.

[42] Alejandro Perez. Spin foam models for quantum gravity. *Classical and Quantum Gravity*, 20(6):R43, 2003.

[43] G. Ponzano and T. Regge. Semiclassical limit of Racah coefficients, in: Spectroscopic and Group Theoretical Methods in Physics, F. Block (ed.), 1–58. New York: John Wiley and Sons, Inc., 1968.

[44] Carlo Rovelli. Quantum gravity. Cambridge University Press, 2004.

[45] Carlo Rovelli. A new look at loop quantum gravity. April 2010.

[46] Claudio Teitelboim. Quantum mechanics on the gravitational field. *Phys. Rev. D*, 25:3159–3179, June 1982.

[47] Thomas Thiemann. Modern canonical quantum general relativity. Cambridge University Press, 2007.

[48] Grigory Volovik. From quantum hydrodynamics to quantum gravity. 2006.

[49] J. A. Wheeler. Superspace and the nature of quantum geometrodynamics, in: Quantum cosmology, L. Z. Fang, R. Ruffini (eds.), 27–92. Singapore: World Scientific, 1987.

Steffen Gielen and Daniele Oriti
Max Planck Institute for Gravitational Physics (Albert Einstein Institute)
Am Mühlenberg 1
D-14476 Golm
Germany
e-mail: gielen@aei.mpg.de
doriti@aei.mpg.de

Unsharp Values, Domains and Topoi

Andreas Döring and Rui Soares Barbosa

Abstract. The so-called topos approach provides a radical reformulation of quantum theory. Structurally, quantum theory in the topos formulation is very similar to classical physics. There is a state object $\underline{\Sigma}$, analogous to the state space of a classical system, and a quantity-value object $\underline{\mathbb{R}^{\leftrightarrow}}$, generalising the real numbers. Physical quantities are maps from the state object to the quantity-value object – hence the 'values' of physical quantities are not just real numbers in this formalism. Rather, they are families of real intervals, interpreted as 'unsharp values'. We will motivate and explain these aspects of the topos approach and show that the structure of the quantity-value object $\underline{\mathbb{R}^{\leftrightarrow}}$ can be analysed using tools from domain theory, a branch of order theory that originated in theoretical computer science. Moreover, the base category of the topos associated with a quantum system turns out to be a domain if the underlying von Neumann algebra is a matrix algebra. For general algebras, the base category still is a highly structured poset. This gives a connection between the topos approach, noncommutative operator algebras and domain theory. In an outlook, we present some early ideas on how domains may become useful in the search for new models of (quantum) space and space-time.

Mathematics Subject Classification (2010). Primary 81P99; Secondary 06A11, 18B25, 46L10.

Keywords. Topos approach, domain theory, intervals, unsharp values, von Neumann algebras.

You cannot depend on your eyes
when your imagination is out of focus.
Mark Twain (1835–1910)

1. Introduction

The search for a theory of quantum gravity is ongoing. There is a range of approaches, all of them differing significantly in scope and technical content. Of course, this is suitable for such a difficult field of enquiry. Most approaches accept the Hilbert space formalism of quantum theory and try to find extensions, additional structures that would capture gravitational aspects, or reproduce them from the behaviour of underlying, more fundamental entities.

Yet, standard quantum theory is plagued with many conceptual difficulties itself. Arguably, these problems get more severe when we try to take the step to quantum gravity and quantum cosmology. For example, in the standard interpretations of quantum theory, measurements on a quantum system by an external classical observer play a central rôle. This concept clearly becomes meaningless if the whole universe is to be treated as a quantum system. Moreover, standard quantum theory and quantum field theory are based on a continuum picture of space-time. Mathematically, the continuum in the form of the real numbers underlies all structures like Hilbert spaces, operators, manifolds, and differential forms (and also strings and loops). If space-time fundamentally is not a smooth continuum, then it may be wrong to base our mathematical formalism on the mathematical continuum of the real numbers.

These considerations motivated the development of the *topos approach* to the formulation of physical theories, and in particular the topos approach to quantum theory. In a radical reformulation based upon structures in suitable, physically motivated topoi, all aspects of quantum theory – states, physical quantities, propositions, quantum logic, etc. – are described in a novel way. As one aspect of the picture emerging, physical quantities take their values not simply in the real numbers. Rather, the formalism allows to describe 'unsharp', generalised values in a systematic way. In this article, we will show that the mathematical structures used to formalise unsharp values can be analysed using techniques from domain theory. We will take some first steps connecting the topos approach with domain theory.

Domain theory is a branch of order theory and originated in theoretical computer science, where it has become an important tool. Since domain theory is not well-known among physicists, we will present all necessary background here. Recently, domain theory has found some applications in quantum theory in the work of Coecke and Martin [3] and in general relativity in the work of Martin and Panangaden [27]. Domain theory also has been connected with topos theory before, in the form of synthetic domain theory, but this is technically and conceptually very different from our specific application.

The plan of the paper is as follows: in section 2, we will present a sketch of the topos approach to quantum theory, with some emphasis on how generalised, 'unsharp' values for physical quantities arise. In section 3, we present some background on domain theory. In section 4, it will be shown that the structure of the quantity-value object $\underline{\mathbb{R}^{\leftrightarrow}}$, a presheaf whose global elements are the unsharp values, can be analysed with the help of domain-theoretical techniques. Section 5 shows that the base category $\mathcal{V}(\mathcal{N})$ of the topos $\mathbf{Set}^{\mathcal{V}(\mathcal{N})^{\mathrm{op}}}$ associated with a quantum system is a directed complete poset, and moreover an algebraic domain if $\mathcal{N}$ is a matrix algebra. Physically, $\mathcal{V}(\mathcal{N})$ is the collection of all classical perspectives on the quantum system. In section 6, we show that the poset $\mathcal{V}(\mathcal{N})$ is not continuous, and hence not a domain, for non-matrix algebras $\mathcal{N}$, and in section 7, we present some speculative ideas on how domains may become useful in the construction of new models of space and space-time adequate for quantum theory and theories 'beyond quantum theory' in the context of the topos approach. Section 8 concludes.

2. The topos approach, contexts and unsharp values

Basic ideas. In recent years, the *topos approach* to the formulation of physical theories has been developed by one of us (AD), largely in collaboration with Chris Isham [9, 10, 11, 12, 5, 13, 6, 7, 8]. This approach originates from works by Isham [21] and Isham/Butterfield [22, 23, 24, 25, 26]. Landsman et al. have presented a closely related scheme for quantum theory [17, 2, 18, 19], developing some aspects topos-internally, and Flori has developed a history version [15].

The main goal of the topos approach is to provide a framework for the formulation of 'neo-realist' physical theories. Such a theory describes a physical system by (i) a state space, or more generally, a state object Σ, (ii) a quantity-value object $\mathcal{R}$, where physical quantities take their values, and (iii) functions, or more generally, arrows $f_A : \Sigma \to \mathcal{R}$ from the state object to the quantity-value object corresponding to physical quantities like position, momentum, energy, spin, etc. Both the state object and the quantity-value object are objects in a *topos*, and the arrows between them representing physical quantities are arrows in the topos. Roughly speaking, a topos is a mathematical structure, more specifically a category, that can be seen as a universe of generalised sets and generalised functions between them.

Each topos has an *internal logic* that is of intuitionistic type. In fact, one typically has a multivalued, intuitionistic logic and not just two-valued Boolean logic as in the familiar topos **Set** of sets and functions. One main aspect of the topos approach is that it makes use of the internal logic of a given topos to provide a logic for a physical system. More specifically, the subobjects of the state object, or a suitable subfamily of these, are the representatives of *propositions* about the values of physical quantities. In the simplest case, one considers propositions of the form "$A \varepsilon \Delta$", which stands for "the physical quantity A has a value in the (Borel) set Δ of real numbers".

As an example, consider a classical system: the topos is **Set**, the state object is the usual state space, a symplectic manifold $\mathcal{S}$, and a physical quantity A is represented by a function f_A from $\mathcal{S}$ to the set of real numbers $\mathbb{R}$, which in this case is the quantity-value object. The subset $T \subseteq \mathcal{S}$ representing a proposition "$A \varepsilon \Delta$" consists of those states $s \in \mathcal{S}$ for which $f_A(s) \in \Delta$ holds (i.e., $T = f_A^{-1}(\Delta)$). If we assume that the function f_A representing the physical quantity A is (at least) measurable, then the set T is a Borel subset of the state space $\mathcal{S}$. Hence, in classical physics the representatives of propositions are the Borel subsets of $\mathcal{S}$. They form a σ-complete *Boolean algebra*, which ultimately stems from the fact that the topos **Set** in which classical physics is formulated has the familiar two-valued Boolean logic as its internal logic. Of course, we rarely explicitly mention **Set** and its internal logic – it is just the usual mathematical universe in which we formulate our theories. As an underlying structure, it usually goes unnoticed.

A topos for quantum theory. For non-relativistic quantum theory, another topos is being used though. The details are explained elsewhere, see [5, 8] for an introduction to the topos approach and [13] for a more detailed description. The main idea is to use *presheaves* over the set $\mathcal{V}(\mathcal{N})$ of abelian subalgebras of the nonabelian von Neumann algebra $\mathcal{N}$ of physical quantities.[1] $\mathcal{V}(\mathcal{N})$ is partially ordered under inclusion; the topos of presheaves over $\mathcal{V}(\mathcal{N})$ is denoted as $\mathbf{Set}^{\mathcal{V}(\mathcal{N})^{\mathrm{op}}}$. The poset $\mathcal{V}(\mathcal{N})$, also called the *context category*, is interpreted as the collection of all classical perspectives on the quantum system: each *context* (abelian subalgebra) $V \in \mathcal{V}(\mathcal{N})$ provides a set of commuting self-adjoint operators, representing compatible physical quantities. The poset $\mathcal{V}(\mathcal{N})$ keeps track of how these classical perspectives overlap, i.e., to which degree they are mutually compatible. Presheaves over $\mathcal{V}(\mathcal{N})$, which are contravariant functors from $\mathcal{V}(\mathcal{N})$ to **Set**, automatically encode this information, too. A presheaf is not a single set, but a 'varying set': a family $\underline{P} = (\underline{P}_V)_{V \in \mathcal{V}(\mathcal{N})}$ of sets indexed by elements from $\mathcal{V}(\mathcal{N})$, together with functions $\underline{P}(i_{V'V}) : \underline{P}_V \to \underline{P}_{V'}$ between the sets whenever there is an inclusion $i_{V'V} : V' \to V$ in $\mathcal{V}(\mathcal{N})$.

The state object for quantum theory is the so-called *spectral presheaf* $\underline{\Sigma}$ that is given as follows:

- To each abelian subalgebra $V \in \mathcal{V}(\mathcal{N})$, one assigns the Gel'fand spectrum $\underline{\Sigma}_V$ of the algebra V;
- to each inclusion $i_{V'V} : V' \to V$, one assigns the function $\underline{\Sigma}(i_{V'V}) : \underline{\Sigma}_V \to \underline{\Sigma}_{V'}$ that sends each $\lambda \in \underline{\Sigma}_V$ to its restriction $\lambda|_{V'} \in \underline{\Sigma}_{V'}$.

One can show that propositions of the form "$A \varepsilon \Delta$" correspond to so-called *clopen subobjects* of $\underline{\Sigma}$. The set $\mathrm{Sub_{cl}}(\underline{\Sigma})$ of clopen subobjects is the analogue of the set of Borel subsets of the classical state space. Importantly, $\mathrm{Sub_{cl}}(\underline{\Sigma})$ is a complete *Heyting algebra*, which relates to the fact that the internal logic of the presheaf topos $\mathbf{Set}^{\mathcal{V}(\mathcal{N})^{\mathrm{op}}}$ is intuitionistic (and not just Boolean). Note that unlike in Birkhoff-von Neumann quantum logic, which is based on the non-distributive lattice $\mathcal{P}(\mathcal{N})$ of projections in the algebra $\mathcal{N}$, in the topos scheme propositions are represented by elements in a distributive lattice $\mathrm{Sub_{cl}}(\underline{\Sigma})$. This allows to give a better interpretation of this form of quantum

[1] To be precise, we only consider abelian von Neumann subalgebras V of $\mathcal{N}$ that have the same unit element as $\mathcal{N}$. In the usual presentation of this approach, the trivial algebra $\mathbb{C}\hat{1}$ is excluded from $\mathcal{V}(\mathcal{N})$. However, here we will keep the trivial algebra as a bottom element of $\mathcal{V}(\mathcal{N})$. We will occasionally point out which results depend on $\mathcal{V}(\mathcal{N})$ having a bottom element.

logic [13, 7]. The map from the $\mathcal{P}(\mathcal{N})$ to $\mathrm{Sub_{cl}}(\underline{\Sigma})$ is called *daseinisation of projections.* Its properties are discussed in detail in [8].

Unsharp values. In this article, we will mostly focus on the quantity-value object for quantum theory and its properties. Like the state object $\underline{\Sigma}$, the quantity-value object, which will be denoted $\underline{\mathbb{R}^{\leftrightarrow}}$, is an object in the presheaf topos $\mathbf{Set}^{\mathcal{V}(\mathcal{N})^{\mathrm{op}}}$. The topos approach aims to provide models of quantum systems that can be interpreted as realist (or as we like to call them, neo-realist, because of the richer intuitionistic logic coming from the topos). One aspect is that physical quantities should have values at all times, independent of measurements. Of course, this immediately meets with difficulties: in quantum theory, we cannot expect physical quantities to have sharp, definite values. In fact, the Kochen-Specker theorem shows that under weak and natural assumptions, there is no map from the self-adjoint operators to the real numbers that could be seen as an assignment of values to the physical quantities represented by the operators.[2] The Kochen-Specker theorem holds for von Neumann algebras [4].

The simple idea is to use *intervals*, interpreted as 'unsharp values', instead of sharp real numbers. The possible (generalised) values of a physical quantity A are real intervals, or more precisely, real intervals intersected with the spectrum of the self-adjoint operator $\hat{A}$ representing A. In our topos approach, each self-adjoint operator $\hat{A}$ in the algebra $\mathcal{N}$ of physical quantities is mapped to an arrow $\breve{\delta}(\hat{A})$ from the state object $\underline{\Sigma}$ to the quantity-value object $\underline{\mathbb{R}^{\leftrightarrow}}$. (The latter object will be defined below.)

We will not give the details of the construction of the arrow $\breve{\delta}(\hat{A})$ (see [13, 8]), but we present some physical motivation here. For this, consider two contexts $V, V' \in \mathcal{V}(\mathcal{N})$ such that $V' \subset V$, that is, V' is a smaller context than V. 'Smaller' means that there are fewer physical quantities available from the classical perspective described by V' than from V, hence V' gives a more limited access to the quantum system. The step from V to $V' \subset V$ is interpreted as a process of coarse-graining.

For example, we may be interested in the value of a physical quantity A. Let us assume that the corresponding self-adjoint operator $\hat{A}$ is contained in a context V, but not in a context $V' \subset V$. For simplicity, let us assume furthermore that the state of the quantum system is an eigenstate of $\hat{A}$. Then, from the perspective of V, we will get a sharp value, namely the eigenvalue of $\hat{A}$ corresponding to the eigenstate. But from the perspective of V', the operator $\hat{A}$ is not available, so we have to approximate $\hat{A}$ by self-adjoint operators in V'. One uses one approximation from below and one from above, both taken with respect to the so-called spectral order. These approximations always exist. In this way, we obtain two operators $\delta^i(\hat{A})_{V'}, \delta^o(\hat{A})_{V'}$ in V' which, intuitively speaking, contain as much information about $\hat{A}$ as is

[2] The conditions are: (a) each self-adjoint operator $\hat{A}$ is assigned an element of its spectrum; and (b) if $\hat{B} = f(\hat{A})$ for two self-adjoint operators $\hat{A}, \hat{B}$ and a Borel function f, then the value $v(\hat{B}) = v(f(\hat{A}))$ assigned to $\hat{B}$ is $f(v(\hat{A}))$, where $v(\hat{A})$ is the value assigned to $\hat{A}$.

available from the more limited classical perspective V'. If we now ask for the value of A in the given state from the perspective of V', then we will get two real numbers, one from $\delta^i(\hat{A})_{V'}$ and one from $\delta^o(\hat{A})_{V'}$. By the properties of the spectral order, these two numbers lie in the spectrum of $\hat{A}$. We interpret them as the endpoints of a real interval, which is an 'unsharp value' for the physical quantity A from the perspective of V'. Note that we get 'unsharp values' for each context $V \in \mathcal{V}(\mathcal{N})$ (and for some V, we may get sharp values, namely in an eigenstate-eigenvalue situation).

In a nutshell, this describes the idea behind *daseinisation of self-adjoint operators*, which is a map from self-adjoint operators in the nonabelian von Neumann algebra $\mathcal{N}$ of physical quantities to arrows in the presheaf topos $\mathbf{Set}^{\mathcal{V}(\mathcal{N})^{\mathrm{op}}}$, sending $\hat{A} \in \mathcal{N}_{\mathrm{sa}}$ to $\breve{\delta}(\hat{A}) \in \mathrm{Hom}(\underline{\Sigma}, \underline{\mathbb{R}^{\leftrightarrow}})$. The arrow $\breve{\delta}(\hat{A})$ is the topos representative of the physical quantity A. We will now consider the construction of the quantity-value object $\underline{\mathbb{R}^{\leftrightarrow}}$, the codomain of $\breve{\delta}(\hat{A})$.

As a first step, we formalise the idea of unsharp values as real intervals. Define

$$\mathbf{I}\mathbb{R} := \{[a,b] \mid a,b \in \mathbb{R},\ a \le b\}. \tag{2.1}$$

Note that we consider closed intervals and that the case $a = b$ is included, which means that the intervals of the form $[a,a]$ are contained in $\mathbf{I}\mathbb{R}$. Clearly, these intervals can be identified with the real numbers. In this sense, $\mathbb{R} \subset \mathbf{I}\mathbb{R}$.

It is useful to think of the presheaf $\underline{\mathbb{R}^{\leftrightarrow}}$ as being given by one copy of $\mathbf{I}\mathbb{R}$ for each classical perspective $V \in \mathcal{V}(\mathcal{N})$. Each observer hence has the whole collection of 'unsharp' values available. The task is to fit all these copies of $\mathbf{I}\mathbb{R}$ together into a presheaf. In particular, whenever we have $V, V' \in \mathcal{V}(\mathcal{N})$ such that $V' \subset V$, then we need a function from $\mathbf{I}\mathbb{R}_V$ to $\mathbf{I}\mathbb{R}_{V'}$ (here, we put an index on each copy of $\mathbf{I}\mathbb{R}$ to show to which context the copy belongs). The simplest idea is to send each interval $[a,b] \in \mathbf{I}\mathbb{R}_V$ to the same interval in $\mathbf{I}\mathbb{R}_{V'}$. But, as we saw, in the topos approach the step from the larger context V to the smaller context V' is seen as a process of coarse-graining. Related to that, we expect to get an even more unsharp value, corresponding to a bigger interval, in $\mathbf{I}\mathbb{R}_{V'}$ than in $\mathbf{I}\mathbb{R}_V$ in general. In fact, we want to be flexible and define a presheaf such that we can map $[a,b] \in \mathbf{I}\mathbb{R}_V$ either to the same interval in $\mathbf{I}\mathbb{R}_{V'}$, or to any bigger interval $[c,d] \supset [a,b]$, depending on what is required.

We note that as we go from a larger context V to smaller contexts $V' \subset V$, $V'' \subset V'$, ..., the left endpoints of the intervals will get smaller and smaller, and the right endpoints will get larger and larger in general. The idea is to formalise this by two functions $\mu_V, \nu_V : \downarrow\! V \to \mathbb{R}$ that give the left resp. right endpoints of the intervals for all $V' \subseteq V$. Here, $\downarrow\! V := \{V' \in \mathcal{V}(\mathcal{N}) \mid V' \subseteq V\}$ denotes the downset of V in $\mathcal{V}(\mathcal{N})$. Physically, $\downarrow\! V$ is the collection of all subcontexts of V, that is, all smaller classical perspectives than V. This leads to the following definition.

Definition 2.1. The *quantity-value object* $\underline{\mathbb{R}^{\leftrightarrow}}$ for quantum theory is given as follows:

- To each $V \in \mathcal{V}(\mathcal{N})$, we assign the set
$$\underline{\mathbb{R}^{\leftrightarrow}}_V := \{(\mu,\nu) \mid \mu,\nu : {\downarrow}V \to \mathbb{R}, \\ \mu \text{ order-preserving}, \nu \text{ order-reversing}, \mu \leq \nu\}; \tag{2.2}$$
- to each inclusion $i_{V'V} : V' \to V$, we assign the map
$$\underline{\mathbb{R}^{\leftrightarrow}}(i_{V'V}) : \underline{\mathbb{R}^{\leftrightarrow}}_V \longrightarrow \underline{\mathbb{R}^{\leftrightarrow}}_{V'} \\ (\mu,\nu) \longmapsto (\mu|_{{\downarrow}V'}, \nu|_{{\downarrow}V'}). \tag{2.3}$$

A *global element* γ of the presheaf $\underline{\mathbb{R}^{\leftrightarrow}}$ is a choice of one pair of functions $\gamma_V = (\mu_V, \nu_V)$ for every context $V \in \mathcal{V}(\mathcal{N})$ such that, whenever $V' \subset V$, one has $\gamma_{V'} = (\mu_{V'}, \nu_{V'}) = (\mu_V|_{V'}, \nu_V|_{V'}) = \gamma_V|_{V'}$. Clearly, a global element γ gives a pair of functions $\mu, \nu : \mathcal{V}(\mathcal{N}) \to \mathbb{R}$ such that μ is order-preserving (smaller contexts are assigned smaller real numbers) and ν is order-reversing (smaller contexts are assigned larger numbers). Note that μ and ν are defined on the whole poset $\mathcal{V}(\mathcal{N})$. Conversely, each such pair of functions determines a global element of $\underline{\mathbb{R}^{\leftrightarrow}}$. Hence we can identify a global element γ with the corresponding pair of functions (μ,ν). We see that $\gamma = (\mu,\nu)$ provides one closed interval $[\mu(V), \nu(V)]$ for each context V. Moreover, whenever $V' \subset V$, we have $[\mu(V'), \nu(V')] \supseteq [\mu(V), \nu(V)]$, that is, the interval at V' is larger than or equal to the interval at V. We regard a global element γ of $\underline{\mathbb{R}^{\leftrightarrow}}$ as one unsharp value. Each interval $[\mu(V), \nu(V)]$, $V \in \mathcal{V}(\mathcal{N})$, is one component of such an unsharp value, associated with a classical perspective/context V. The set of global elements of $\underline{\mathbb{R}^{\leftrightarrow}}$ is denoted as $\Gamma\underline{\mathbb{R}^{\leftrightarrow}}$.

In the following, we will show that the set $\Gamma\underline{\mathbb{R}^{\leftrightarrow}}$ of unsharp values for physical quantities that we obtain from the topos approach is a highly structured poset, and that a subset of $\Gamma\underline{\mathbb{R}^{\leftrightarrow}}$ naturally can be seen as a so-called *domain* if the context category $\mathcal{V}(\mathcal{N})$ is a domain (see section 4). Domains play an important rôle in theoretical computer science. The context category $\mathcal{V}(\mathcal{N})$ turns out to be a domain, even an algebraic domain, if $\mathcal{N}$ is a matrix algebra. This leads to a first, simple connection between noncommutative operator algebras, the topos approach and domain theory (see section 5). For more general von Neumann algebras $\mathcal{N}$, $\mathcal{V}(\mathcal{N})$ is a not a domain, see section 6.

3. Domain theory

Basics. This section presents some basic concepts of domain theory. Standard references are [16, 1]. Since domain theory is not well-known among physicists, we give some definitions and motivation. Of course, we barely scratch the surface of this theory here.

The study of domains was initiated by Dana Scott [28, 29], with the aim of finding a denotational semantics for the untyped λ-calculus. Since then, it has undergone significant development and has become a mathematical theory in its own right, as well as an important tool in theoretical computer science.

Domain theory is a branch of order theory, yet with a strong topological flavour, as it captures notions of convergence and approximation. The basic concepts are easy to grasp: the idea is to regard a partially ordered set as a (qualitative) hierarchy of information content or knowledge. Under this interpretation, we think of $x \sqsubseteq y$ as meaning 'y is more specific, carries more information than x'. Therefore, non-maximal elements in the poset represent incomplete/partial knowledge, while maximal elements represent complete knowledge. In a more computational perspective, we can see the non-maximal elements as intermediate results of a computation that proceeds towards calculating some maximal element (or at least a larger element with respect to the information order).

For the rest of this section, we will mainly be considering a single poset, which we will denote by $\langle P, \sqsubseteq \rangle$.

Convergence – directed completeness of posets. The first important concept in domain theory is that of *convergence*. We start by considering some special subsets of P.

Definition 3.1. A nonempty subset $S \subseteq P$ is *directed* if

$$\forall x, y \in S \ \exists z \in S : x, y \sqsubseteq z. \tag{3.1}$$

A directed set can be seen as a consistent specification of information: the existence of a $z \sqsupseteq x, y$ expresses that x and y are compatible, in the sense that it is possible to find an element which is larger, i.e., contains more information, than both x and y. Alternatively, from the computational viewpoint, directed sets describe computations that converge in the sense that for any pair of intermediate results (that can be reached in a finite number of steps), there exists a better joint approximation (that can also be reached in a finite number of steps). This is conceptually akin to converging sequences in a metric space. Hence the natural thing to ask of directed sets is that they possess a suitable kind of limit – that is, an element containing all the information about the elements of the set, but not more information. This limit, if it exists, can be seen as the ideal result the computation approximates. This leads to the concept of directed-completeness:

Definition 3.2. A *directed-complete poset* (or *dcpo*) is a poset in which any directed set has a supremum (least upper bound).

Following [1], we shall write $\bigsqcup^{\uparrow} S$ to denote the supremum of a directed set S, instead of simply $\bigsqcup S$. Note that $\bigsqcup^{\uparrow} S$ means 'S is a directed set, and we are considering its supremum'.

The definition of morphisms between dcpos is the evident one:

Definition 3.3. A function $f : P \longrightarrow P'$ between dcpos $\langle P, \sqsubseteq \rangle$ and $\langle P', \sqsubseteq' \rangle$ is *Scott-continuous* if

- f is order-preserving (monotone);
- for any directed set $S \subseteq P$, $f(\bigsqcup^{\uparrow} S) = \bigsqcup^{\uparrow} f^{\rightarrow}(S)$, where $f^{\rightarrow}(S) = \{f(s) \mid s \in S\}$.

Clearly, dcpos with Scott-continuous functions form a category. The definition of a Scott-continuous function can be extended to posets which are not dcpos, by carefully modifying the second condition to say 'for any directed set S that has a supremum'. The reference to 'continuity' is not fortuitous, as there is the so-called *Scott topology*, with respect to which these (and only these) arrows are continuous. The Scott topology will be defined below.

Approximation - continuous posets. The other central notion is sometimes called approximation. This is captured by the following relation on elements of P.

Definition 3.4. We say that x *approximates* y or x *is way below* y, and write $x \ll y$, whenever for any directed set S with a supremum, we have

$$y \sqsubseteq \bigsqcup^{\uparrow} S \Rightarrow \exists s \in S : x \sqsubseteq s. \tag{3.2}$$

The 'way-below' relation captures the fact that x is much simpler than y, yet carries essential information about y. In the computation analogy, we could say that x is an unavoidable step in any computation of y, in the sense that any computation that tends to (i.e., successively approximates) y must reach or pass x in a finite amount of steps.

In particular, one can identify certain elements which are 'finite' or 'simple', in the sense that they cannot be described by (i.e., given as the supremum of) any set of smaller elements that does not contain the element itself already.

Definition 3.5. An element $x \in P$ such that $x \ll x$ is called a *compact* or a *finite* element. $K(P)$ stands for the set of compact elements of P.

Another interpretation of a compact element x is to say that any computation that tends to x eventually reaches x in a finite number of steps.

In a poset $\langle \mathcal{P}A, \subseteq \rangle$ of subsets of a set A, the compact elements are exactly the finite subsets of A: if one covers a finite set F by a directed collection $(S_i)_{i \in I}$, F will be contained in one of the S_i already. Also, the definition of the way-below relation (particularly of $x \ll x$) has a striking similarity with that of a compact set in topology. Indeed, in the poset $\langle \mathcal{O}(X), \subseteq \rangle$ of open subsets of a topological space X, the compact elements are simply the compact open sets.

Given an element x in a poset P, we write $\downarrow x$ for the *downset* $\{y \in P \mid y \leq x\}$ of x in P. If $X \subseteq P$, then $\downarrow X := \{y \in P \mid \exists x \in X : y \leq x\}$. The sets $\uparrow x$ and $\uparrow X$ are defined analogously. Similarly, we write $\twoheaddownarrow x, \twoheaddownarrow X, \twoheaduparrow x, \twoheaduparrow X$ for the corresponding sets with respect to the way-below relation $\ll$, e.g.

$$\twoheaddownarrow x := \{y \in P \mid y \ll x\}. \tag{3.3}$$

We now come to another requirement that is usually imposed on the posets of interest, besides directed-completeness.

Definition 3.6. A poset P is a *continuous poset* if, for any $y \in P$, one has $\bigsqcup^{\uparrow} \twoheaddownarrow y = y$, and P is an *algebraic poset* if, for any $y \in P$, $\bigsqcup^{\uparrow}(\twoheaddownarrow y \cap K(P)) = y$ holds.

Recall that $\bigsqcup^{\uparrow}\Downarrow y = y$ means that $\Downarrow y$ is directed and has supremum y. Continuity basically says that the elements 'much simpler' than y carry all the information about y, when taken together. Algebraicity further says that the 'primitive' (i.e., compact) elements are enough.

Bases and domains. The continuity and algebraicity requirements are often expressed in terms of the notion of a basis. The definition is slightly more involved, but the concept of a basis is useful in its own right.

Definition 3.7. A subset $B \subseteq P$ is a *basis* for P if, for all $x \in X$, $\bigsqcup^{\uparrow}(B \cap \Downarrow x) = x$.

It is immediate that continuity implies that P itself is a basis. Conversely, the existence of a basis implies continuity.

Definition 3.8. A *domain* (or *continuous domain*) $\langle D, \sqsubseteq\rangle$ is a dcpo which is continuous. Equivalently, a domain is a dcpo that has a basis. $\langle D, \sqsubseteq\rangle$ is an *algebraic domain* if it is a domain and algebraic, that is, if the set $K(D)$ of compact elements is a basis for $\langle D, \sqsubseteq\rangle$. An *$\omega$-continuous* (resp. *$\omega$-algebraic*) domain is a continuous (resp. algebraic) domain with a countable basis.

Note that a domain always captures the notions of convergence and of approximation as explained above.

Bounded complete posets. Later on, we will need another completeness property that a poset P may or may not have.

Definition 3.9. A poset is *bounded complete* (or a *bc-poset*) if all subsets S with an upper bound have a supremum. It is *finitely bounded complete* if all finite subsets with an upper bound have a supremum. It is *almost (finitely) bounded complete* if all nonempty (finite) subsets with an upper bound have a supremum.[3]

We state the following result without proof.

Proposition 3.10. *A(n almost) finitely bounded complete dcpo is (almost) bounded complete.*[4]

Another property, stronger than bounded completeness, will be needed later on:

Definition 3.11. An *L-domain* is a domain D in which, for each $x \in D$, the principal ideal $\downarrow x$ is a complete lattice.

The Scott topology. We now define the appropriate topology on dcpos and domains, called the Scott topology, and present some useful results. In fact, the Scott topology can be defined on any poset, but we are mostly interested in dcpos and domains.

Definition 3.12. Let $\langle P, \leq\rangle$ be a poset. A subset G of P is said to be *Scott-open* if

---

[3]Note that the 'almost' versions don't require a least element $\bot$.

[4]Bounded-complete dcpos are the same as complete semilattices (see [16]).

- G is an upper set, that is

$$x \in G \wedge x \leq y \Rightarrow y \in G; \tag{3.4}$$

- G is inaccessible by directed suprema, i.e. for any directed set S with a supremum,

$$\bigsqcup^{\uparrow} S \in G \Rightarrow \exists s \in S : s \in U. \tag{3.5}$$

The complement of a Scott-open set is called a *Scott-closed* set. Concretely, this is a lower set closed for all existing directed suprema.

The name Scott-open is justified by the following result.

Proposition 3.13. *The Scott-open subsets of P are the opens of a topology on P, called the* Scott topology.

Proposition 3.14. *If P is a continuous poset, the collection*

$$\{\twoheaduparrow x \mid x \in P\} \tag{3.6}$$

is a basis for the Scott topology.

The Scott topology encodes a lot of information about the domain-theoretical properties of P relating to convergence and (in the case of a continuous poset) approximation. The following is one of its most important properties, relating the algebraic and topological aspects of domain theory.

Proposition 3.15. *Let P and Q be two posets. A function $f : P \longrightarrow Q$ is Scott-continuous if and only if it is (topologically) continuous with respect to the Scott topologies on P and Q.*

The results above (and proofs for the more involved ones) can be found in [1, §1.2.3]. More advanced results can be found in [1, §4.2.3].

As for separation properties, the Scott topology satisfies only a very weak axiom in all interesting cases.

Proposition 3.16. *The Scott topology on P gives a $T0$ topological space. It is $T2$ if and only if the order in P is trivial.*

The real interval domain. As we saw in section 2, the collection of real intervals can serve as a model for 'unsharp values' of physical quantities (at least if we consider only one classical perspective V on a quantum system). We will now see that the set $\mathbf{I}\mathbb{R}$ of closed real intervals defined in equation (2.1) actually is a domain, the so-called *interval domain*. This domain was introduced by Scott [30] as a computational model for the real numbers.

Definition 3.17. The *interval domain* is the poset of closed intervals in $\mathbb{R}$ (partially) ordered by reverse inclusion,

$$\mathbf{I}\mathbb{R} := \langle \{[a,b] \mid a,b \in \mathbb{R}\}, \sqsubseteq := \supseteq \rangle . \tag{3.7}$$

The intervals are interpreted as approximations to real numbers, hence the ordering by reverse inclusion: we think of $x \sqsubseteq y$ as 'y is sharper than x'. Clearly, the maximal elements are the real numbers themselves (or rather, more precisely, the intervals of the form $[a,a]$).

We shall denote by x_- and x_+ the left and right endpoints of an interval $x \in \mathbf{I}\mathbb{R}$. That is, we write $x = [x_-, x_+]$. Also, if $f : X \longrightarrow \mathbf{I}\mathbb{R}$ is a function to the interval domain, we define the functions $f_-, f_+ : X \longrightarrow \mathbb{R}$ given by $f_\pm(x) := (f(x))_\pm$, so that, for any $x \in X$, $f(x) = [f_-(x), f_+(x)]$. Clearly, one always has $f_- \leq f_+$ (the order on functions being defined pointwise). Conversely, any two functions $g, h : X \longrightarrow \mathbb{R}$ with $g \leq h$ determine a function f such that $f_- = g$ and $f_+ = h$.

Writing $x = [x_-, x_+]$ amounts to regarding $\mathbf{I}\mathbb{R}$ as being embedded in $\mathbb{R} \times \mathbb{R}$ (as a set). The decomposition for functions can then be depicted as follows:

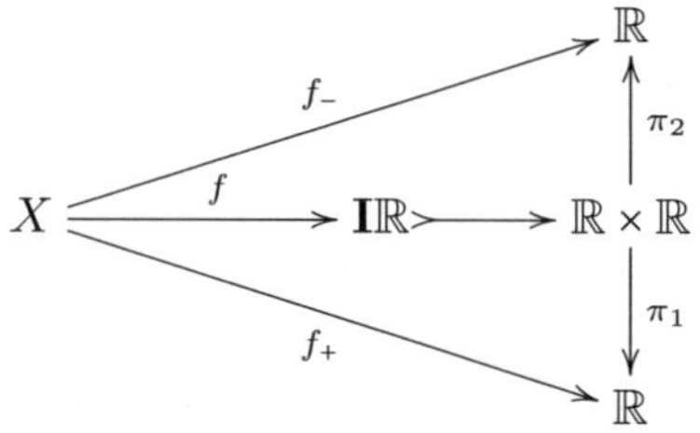

$$(f_- \leq f_+)\,, \tag{3.8}$$

and it is nothing more than the universal property of the (categorical) product $\mathbb{R} \times \mathbb{R}$ restricted to $\mathbf{I}\mathbb{R}$, the restriction being reflected in the condition $f_- \leq f_+$.

This diagram is more useful in understanding $\mathbf{I}\mathbb{R}$ than it may seem at first sight. Note that we can place this diagram in the category Pos (of posets and monotone maps) if we make a judicious choice of order in $\mathbb{R}^2$. This is achieved by equipping the first copy of $\mathbb{R}$ with its usual order and the second copy with the opposite order $\geq$. Adopting this view, equations 3.9 and 3.10 below should become apparent.

Domain-theoretic structure on $\mathbf{I}\mathbb{R}$. The way-below relation in $\mathbf{I}\mathbb{R}$ is given by:

$$x \ll y \ \text{ iff } \ (x_- < y_-) \wedge (y_+ < x_-). \tag{3.9}$$

Suprema of directed sets exist and are given by intersection. This can be written directly as an interval: let S be a directed set, then

$$\bigsqcup{}^{\uparrow} S = \bigcap S = [\sup\{x_- \mid x \in S\}, \inf\{x_+ \mid x \in S\}]. \tag{3.10}$$

One observes easily that $\mathbf{I}\mathbb{R}$ is an ω-continuous dcpo and hence a domain. (To show ω-continuity, one can consider the basis $\mathbf{I}\mathbb{Q}$ for $\mathbf{I}\mathbb{R}$.)

Moreover, $\mathbf{I}\mathbb{R}$ is a meet-semilattice. Also, we observe that $\mathbf{I}\mathbb{R}$ is an almost-bounded-complete poset: if $S \subset \mathbf{I}\mathbb{R}$ is a non-empty subset with an upper bound (which just means that all intervals in S overlap), then S has a supremum, clearly given by the intersection of the intervals. The related poset $\mathbf{I}\mathbb{R}_\perp$ (where we add a least element, which can be interpreted as $\perp = \mathbb{R} = [-\infty, +\infty]$) is then bounded complete. Also, it is easy to see that $\mathbf{I}\mathbb{R}_\perp$ is an L-domain.

We now consider the Scott topology on $\mathbf{I}\mathbb{R}$. The basic open sets of the Scott topology on $\mathbf{I}\mathbb{R}$ are of the form

$$\Uparrow[a,b] = \{[c,d] \mid a < c \leq d < b\} \tag{3.11}$$

for each $[a,b] \in \mathbf{I}\mathbb{R}$. The identity

$$\Uparrow[a,b] = \{t \in \mathbf{I}\mathbb{R} \mid t \subseteq (a,b)\} \tag{3.12}$$

allows us to see $\Uparrow[a,b]$ as a kind of open interval (a,b). More precisely, it consists of closed intervals contained in the real interval (a,b).

Recall that we can see the poset $\mathbf{I}\mathbb{R}$ as sitting inside $\mathbb{R}^{\leq} \times \mathbb{R}^{\geq}$. In topological terms, this means that the Scott topology is inherited from the Scott topologies in $\mathbb{R}^{\leq}$ and $\mathbb{R}^{\geq}$. These are simply the usual lower and upper semicontinuity topologies on $\mathbb{R}$, with basic open sets respectively (a,∞) and $(-\infty,b)$. Interpreting diagram 3.8 in Top gives the following result.

Proposition 3.18. *Let X be a topological space and $f : X \longrightarrow \mathbf{I}\mathbb{R}$ be a function. Then f is continuous iff f_- is lower semicontinuous and f_+ is upper semicontinuous.*

Hence, the subspace topology on $\mathbf{I}\mathbb{R}$ inherited from $\mathbb{R}^{LSC} \times \mathbb{R}^{USC}$, where LSC and USC stand for the lower and upper semicontinuity topologies, is the Scott topology.

Generalising $\mathbb{R}$. We want to regard $\mathbf{I}\mathbb{R}$ as a generalisation of $\mathbb{R}$. Note that the set $\max \mathbf{I}\mathbb{R}$ consists of degenerate intervals $[x,x] = \{x\}$. This gives an obvious way of embedding the usual continuum $\mathbb{R}$ in $\mathbf{I}\mathbb{R}$. What is more interesting, this is actually a homeomorphism.

Proposition 3.19. *$\mathbb{R} \cong \max \mathbf{I}\mathbb{R}$ as topological spaces, where $\mathbb{R}$ is equipped with its usual topology and $\max \mathbf{I}\mathbb{R}$ with the subspace topology inherited from $\mathbf{I}\mathbb{R}$.*

This result is clear from the following observation that identifies basic open sets:

$$\Uparrow[a,b] \cap \max \mathbf{I}\mathbb{R} = \{t \in \mathbf{I}\mathbb{R} \mid t \subseteq (a,b)\} \cap \max \mathbf{I}\mathbb{R} = \{\{t\} \mid t \in (a,b)\} = (a,b).$$

Another (maybe more informative) way of seeing this is by thinking of $\mathbf{I}\mathbb{R}$ as sitting inside $\mathbb{R}^{LSC} \times \mathbb{R}^{USC}$. The well-known fact that the topology of $\mathbb{R}$ is given as the join of the two semicontinuity topologies can be stated as follows: the diagonal map $diag : \mathbb{R} \longrightarrow \mathbb{R}^{LSC} \times \mathbb{R}^{USC}$ gives an isomorphism between $\mathbb{R}$ and its image Δ. This is because basic opens in Δ are

$$\Delta \cap ((a,\infty) \times (-\infty,b)) = (a,\infty) \cap (-\infty,b) = (a,b). \tag{3.13}$$

The result above just says that this diagonal map factors through $\mathbf{I}\mathbb{R}$, with image $\max \mathbf{I}\mathbb{R}$.

The only separation axiom satisfied by the topology of $\mathbf{I}\mathbb{R}$ is the T0 axiom, whereas $\mathbb{R}$ is a T6 space. However, this topology still keeps some properties of the topology on $\mathbb{R}$. It is second-countable and locally compact (the general definition of local compactness for non-Hausdorff spaces is given in [31]). Note that, obviously, adding a least element $\bot = [-\infty,+\infty]$ to the domain would make it compact.

4. The quantity-value object and domain theory

We now study the presheaf $\underline{\mathbb{R}^{\leftrightarrow}}$ of (generalised) values that shows up in the topos approach in the light of domain theory. First of all, one can consider each component $\underline{\mathbb{R}^{\leftrightarrow}}_V$ individually, which is a set. More importantly, we are interested in the set $\Gamma\underline{\mathbb{R}^{\leftrightarrow}}$ of global elements of $\underline{\mathbb{R}^{\leftrightarrow}}$. We will relate these two sets with the interval domain $\mathbf{I}\mathbb{R}$ introduced in the previous section.

The authors of [10], where the presheaf $\underline{\mathbb{R}^{\leftrightarrow}}$ was first introduced as the quantity-value object for quantum theory, were unaware of domain theory at the time. The idea that $\underline{\mathbb{R}^{\leftrightarrow}}$ (or rather, a closely related co-presheaf) is related to the interval domain was first presented by Landsman et al. in [17]. These authors considered $\underline{\mathbb{R}^{\leftrightarrow}}$ as a topos-internal version of the interval domain. Here, we will focus on topos-external arguments.

Rewriting the definition of $\underline{\mathbb{R}^{\leftrightarrow}}$. We start off by slightly rewriting the definition of the presheaf $\underline{\mathbb{R}^{\leftrightarrow}}$.

Recall from the previous section that a function $f : X \longrightarrow \mathbf{I}\mathbb{R}$ to the interval domain can be decomposed into two functions $f_-, f_+ : X \longrightarrow \mathbb{R}$ giving the left and right endpoints of intervals. In case X is a poset, it is immediate from the definitions that such an f is order-preserving if and only if f_- is order-preserving and f_+ is order-reversing with respect to the usual order on the real numbers. This allows us to rewrite the definition of $\underline{\mathbb{R}^{\leftrightarrow}}$:

Definition 4.1. The quantity-value presheaf $\underline{\mathbb{R}^{\leftrightarrow}}$ is given as follows:

- To each $V \in \mathcal{V}(\mathcal{N})$, we assign the set

$$\underline{\mathbb{R}^{\leftrightarrow}}_V = \{f : \downarrow\!V \longrightarrow \mathbf{I}\mathbb{R} \mid f \text{ order-preserving}\}; \tag{4.1}$$

- to each inclusion $i_{V'V}$, we assign the function

$$\begin{aligned} \underline{\mathbb{R}^{\leftrightarrow}}(i_{V'V}) : \underline{\mathbb{R}^{\leftrightarrow}}_V &\longrightarrow \underline{\mathbb{R}^{\leftrightarrow}}_{V'} \\ f &\longmapsto f|_{\downarrow V'} . \end{aligned}$$

This formulation of $\underline{\mathbb{R}^{\leftrightarrow}}$ brings it closer to the interval domain.

Global elements of $\underline{\mathbb{R}^{\leftrightarrow}}$. In section 2, we stated that in the topos approach, the generalised values of physical quantities are given by global elements of the presheaf $\underline{\mathbb{R}^{\leftrightarrow}}$. We remark that the global elements of a presheaf are (analogous to) points if the presheaf is regarded as a generalised set. Yet, the set of global elements may not contain the full information about the presheaf. There are non-trivial presheaves that have no global elements at all – the spectral presheaf $\underline{\Sigma}$ is an example. In contrast, $\underline{\mathbb{R}^{\leftrightarrow}}$ has many global elements.

We give a slightly modified characterisation of the global elements of $\underline{\mathbb{R}^{\leftrightarrow}}$ (compare end of section 2):

Proposition 4.2. *Global elements of $\underline{\mathbb{R}^{\leftrightarrow}}$ are in bijective correspondence with order-preserving functions from $\mathcal{V}(\mathcal{N})$ to $\mathbf{I}\mathbb{R}$.*

Proof. Let $\underline{\mathbf{1}}$ be the terminal object in the topos $\mathbf{Set}^{\mathcal{V}(\mathcal{N})^{\mathrm{op}}}$, that is, the constant presheaf with a one-element set $\{*\}$ as component for each $V \in \mathcal{V}(\mathcal{N})$. A global element of $\underline{\mathbb{R}^{\leftrightarrow}}$ is an arrow in $\mathbf{Set}^{\mathcal{V}(\mathcal{N})^{\mathrm{op}}}$, i.e., a natural transformation

$$\eta : \underline{\mathbf{1}} \longrightarrow \underline{\mathbb{R}^{\leftrightarrow}}. \tag{4.2}$$

For each object V in the base category $\mathcal{V}(\mathcal{N})$, this gives a function

$$\eta_V : \{\star\} \longrightarrow \underline{\mathbb{R}^{\leftrightarrow}}_V \tag{4.3}$$

which selects an element $\gamma_V := \eta_V(\star)$ of $\underline{\mathbb{R}^{\leftrightarrow}}_V$. Note that each γ_V is an order-preserving function $\gamma_V : \downarrow\! V \longrightarrow \mathbf{I}\mathbb{R}$. The naturality condition, expressed by the diagram

$$\begin{array}{ccc}
\mathbf{1}_V = \{*\} & \xrightarrow{\ \eta_V\ } & \underline{\mathbb{R}^{\leftrightarrow}}_V \\
\Vert & & \downarrow{\scriptstyle \underline{\mathbb{R}^{\leftrightarrow}}(i_{V'V})} \\
\mathbf{1}_{V'} = \{*\} & \xrightarrow{\ \eta_V\ } & \underline{\mathbb{R}^{\leftrightarrow}}_{V'}
\end{array}$$

then reads $\gamma_{V'} = \underline{\mathbb{R}^{\leftrightarrow}}(i_{V'V})(\gamma_V) = f_V|_{\downarrow V'}$. Thus, a global element of $\underline{\mathbb{R}^{\leftrightarrow}}$ determines a unique function

$$\begin{aligned}
\tilde{\gamma} : \mathcal{V}(\mathcal{N}) &\longrightarrow \mathbf{I}\mathbb{R} \\
V' &\longmapsto \gamma_V(V'),
\end{aligned}$$

where V is some context such that $V' \subseteq V$. The function $\tilde{\gamma}$ is well-defined: if we pick another $W \in \mathcal{V}(\mathcal{N})$ such that $V' \subseteq W$, then the naturality condition guarantees that $\gamma_W(V') = \gamma_V(V')$. The monotonicity condition for each γ_V forces $\tilde{\gamma}$ to be a monotone (order-preserving) function from $\mathcal{V}(\mathcal{N})$ to $\mathbf{I}\mathbb{R}$.

Conversely, given an order-preserving function $\tilde{\gamma} : \mathcal{V}(\mathcal{N}) \to \mathbf{I}\mathbb{R}$, we obtain a global element of $\underline{\mathbb{R}^{\leftrightarrow}}$ by setting

$$\forall V \in \mathcal{V}(\mathcal{N}) : \gamma_V := \tilde{\gamma}|_{\downarrow V}. \tag{4.4}$$

□

Global elements as a dcpo. So far, we have seen that each $\underline{\mathbb{R}^{\leftrightarrow}}_V$, $V \in \mathcal{V}(\mathcal{N})$, and the set of global elements $\Gamma\underline{\mathbb{R}^{\leftrightarrow}}$ are sets of order-preserving functions from certain posets to the interval domain $\mathbf{I}\mathbb{R}$. Concretely,

$$\underline{\mathbb{R}^{\leftrightarrow}}_V = \mathcal{OP}(\downarrow\! V, \mathbf{I}\mathbb{R}), \tag{4.5}$$

$$\Gamma\underline{\mathbb{R}^{\leftrightarrow}} = \mathcal{OP}(\mathcal{V}(\mathcal{N}), \mathbf{I}\mathbb{R}), \tag{4.6}$$

where $\mathcal{OP}(P, \mathbf{I}\mathbb{R})$ denotes the order-preserving functions from the poset P to $\mathbf{I}\mathbb{R}$.

We now want to apply the following result (for a proof of a more general result, see Prop. II-4.20 in [16]):

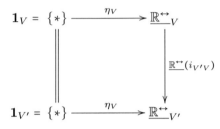

Proposition 4.3. *Let X be a topological space. If P is a dcpo (resp. bounded complete dcpo, resp. almost bounded complete dcpo) equipped with the Scott topology, then $C(X,P)$ with the pointwise order is a dcpo (resp. bounded complete dcpo, resp. almost bounded complete dcpo).*

The problem is that if we want to apply this result to our situation, then we need *continuous* functions between the posets, but neither $\downarrow V$ nor $\mathcal{V}(\mathcal{N})$ have been equipped with a topology so far. (On $\mathbf{I}\mathbb{R}$, we consider the Scott topology.) The following result shows what topology to choose:

Proposition 4.4. *Let P, Q be posets, and let $f : P \longrightarrow Q$ be a function. If the poset Q is continuous, the following are equivalent:*

1. *f is order-preserving;*
2. *f is continuous with respect to the upper Alexandroff topologies on P and Q;*
3. *f is continuous with respect to the upper Alexandroff topology on P and the Scott topology on Q.*

Hence, we put the upper Alexandroff topology on $\downarrow V$ and $\mathcal{V}(\mathcal{N})$ to obtain the following equalities:

- For each $V \in \mathcal{V}(\mathcal{N})$, we have $\underline{\mathbb{R}^{\leftrightarrow}}_V = C((\downarrow V)^{UA}, \mathbf{I}\mathbb{R})$;
- for the global elements of $\underline{\mathbb{R}^{\leftrightarrow}}$, we have $\Gamma\underline{\mathbb{R}^{\leftrightarrow}} = C(\mathcal{V}(\mathcal{N})^{UA}, \mathbf{I}\mathbb{R})$.

By Prop. 4.3, since $\mathbf{I}\mathbb{R}$ is an almost bounded complete dcpo, both $\underline{\mathbb{R}^{\leftrightarrow}}_V$ (for each V), and $\Gamma\underline{\mathbb{R}^{\leftrightarrow}}$ also are almost bounded complete dcpos.

A variation of the quantity-value object. The following is a slight variation of the quantity-value presheaf where we allow for completely undetermined values (and not just closed intervals $[a,b]$). This is achieved by including a bottom element in the interval domain $\mathbf{I}\mathbb{R}$, this bottom element of course being interpreted as the whole real line. Using the L-domain $\mathbf{I}\mathbb{R}_\perp$, let $\underline{\mathbb{R}^{\leftrightarrow}_\perp}$ be the presheaf defined as follows:

- To each $V \in \mathcal{V}(\mathcal{N})$, we assign the set
$$\underline{\mathbb{R}^{\leftrightarrow}_\perp}_V = \{f : \downarrow V \longrightarrow \mathbf{I}\mathbb{R}_\perp \mid f \text{ order-preserving}\}; \tag{4.7}$$
- to each inclusion $i_{V'V}$, we assign the function
$$\begin{aligned}\underline{\mathbb{R}^{\leftrightarrow}_\perp}(i_{V'V}) : \underline{\mathbb{R}^{\leftrightarrow}_\perp}_V &\longrightarrow \underline{\mathbb{R}^{\leftrightarrow}_\perp}_{V'}\\ f &\longmapsto f|_{\downarrow V'}.\end{aligned}$$

Clearly, we have:

- For each $V \in \mathcal{V}(\mathcal{N})$, $\underline{\mathbb{R}^{\leftrightarrow}_\perp}_V = C((\downarrow V)^{UA}, \mathbf{I}\mathbb{R}_\perp)$;
- $\Gamma\underline{\mathbb{R}^{\leftrightarrow}_\perp} = C(\mathcal{V}(\mathcal{N})^{UA}, \mathbf{I}\mathbb{R}_\perp)$.

Hence, by Prop. 4.3, $\underline{\mathbb{R}^{\leftrightarrow}_\perp}_V$ (for each V) and $\Gamma\underline{\mathbb{R}^{\leftrightarrow}}$ are bounded complete dcpos. Note that $\underline{\mathbb{R}^{\leftrightarrow}}$ is a subpresheaf of $\underline{\mathbb{R}^{\leftrightarrow}_\perp}$.

Also, one can consider the presheaves $\underline{S\mathbb{R}^{\leftrightarrow}}$ and $\underline{S\mathbb{R}^{\leftrightarrow}_\perp}$ defined analogously to $\underline{\mathbb{R}^{\leftrightarrow}}$ and $\underline{\mathbb{R}^{\leftrightarrow}_\perp}$, but requiring the functions to be Scott-continuous rather than simply order-preserving. Again, $\underline{S\mathbb{R}^{\leftrightarrow}}$ is a subpresheaf of $\underline{S\mathbb{R}^{\leftrightarrow}_{\perp}}$. Moreover, note that $\underline{S\mathbb{R}^{\leftrightarrow}}$ (resp. $\underline{S\mathbb{R}^{\leftrightarrow}_{\perp}}$) is a subpresheaf of $\underline{\mathbb{R}^{\leftrightarrow}}$ (resp. $\underline{\mathbb{R}^{\leftrightarrow}_{\perp}}$). For a finite-dimensional $\mathcal{N}$, since the poset $\mathcal{V}(\mathcal{N})$ has finite height, any order-preserving function is Scott-continuous, hence there is no difference between $\underline{\mathbb{R}^{\leftrightarrow}}$ and $\underline{S\mathbb{R}^{\leftrightarrow}}$ (resp. $\underline{\mathbb{R}^{\leftrightarrow}_{\perp}}$ and $\underline{S\mathbb{R}^{\leftrightarrow}_{\perp}}$). The presheaves $\underline{S\mathbb{R}^{\leftrightarrow}}$ and $\underline{S\mathbb{R}^{\leftrightarrow}_{\perp}}$ using Scott-continuous functions are interesting since the arrows of the form $\breve{\delta}(\hat{A}) : \underline{\Sigma} \to \underline{\mathbb{R}^{\leftrightarrow}}$ that one obtains from daseinisation of self-adjoint operators [10, 13] actually have image in $\underline{S\mathbb{R}^{\leftrightarrow}}$. We will not prove this result here, since this would lead us too far from our current interest. We have:

- For each $V \in \mathcal{V}(\mathcal{N})$, $\underline{S\mathbb{R}^{\leftrightarrow}}_V = C((\downarrow V)^S, \mathbf{I}\mathbb{R})$ and $\underline{S\mathbb{R}^{\leftrightarrow}_{\perp}}_V = C((\downarrow V)^S, \mathbf{I}\mathbb{R}_{\perp})$;
- $\Gamma\underline{S\mathbb{R}^{\leftrightarrow}} = C(\mathcal{V}(\mathcal{N})^S, \mathbf{I}\mathbb{R})$ and $\Gamma\underline{S\mathbb{R}^{\leftrightarrow}_{\perp}} = C(\mathcal{V}(\mathcal{N})^S, \mathbf{I}\mathbb{R}_{\perp})$.

Domain-theoretic structure on global sections. We now consider continuity. The following result (from theorem A in [32]) helps to clarify things further:

Proposition 4.5. *If D is a continuous L-domain and X a core-compact space (i.e. its poset of open sets is continuous), then $C(X, D)$ (where D has the Scott topology) is a continuous L-domain.*

In particular, any locally compact space is core-compact ([14]). Note that $\mathcal{V}(\mathcal{N})$ with the Alexandroff topology is always locally compact; it is even compact if we consider $\mathcal{V}(\mathcal{N})$ to have a least element. Moreover, we saw before that $\mathbf{I}\mathbb{R}_{\perp}$ is an L-domain. Hence, we can conclude that the global sections of $\underline{\mathbb{R}^{\leftrightarrow}_{\perp}}$ form an L-domain. Similarly, the sections of $\underline{\mathbb{R}^{\leftrightarrow}_{\perp}}_V$ (over $\downarrow V$) form an L-domain.

For the presheaves $\underline{\mathbb{R}^{\leftrightarrow}}$, $\underline{S\mathbb{R}^{\leftrightarrow}}$, and $\underline{S\mathbb{R}^{\leftrightarrow}_{\perp}}$, it is still an open question whether their global sections form a domain or not.

5. The category of contexts as a dcpo

We now turn our attention to the poset $\mathcal{V}(\mathcal{N})$. In this section, we investigate it from the perspective of domain theory. We will show that $\mathcal{V}(\mathcal{N})$ is a dcpo, and that the assignment $\mathcal{N} \mapsto \mathcal{V}(\mathcal{N})$ gives a functor from von Neumann algebras to the category of dcpos.

From a physical point of view, the fact that $\mathcal{V}(\mathcal{N})$ is a dcpo shows that the information contained in a coherent set of physical contexts is captured by a larger (limit) context. If $\mathcal{N}$ is a finite-dimensional algebra, that is, a finite direct sum of matrix algebras, then $\mathcal{V}(\mathcal{N})$ is an algebraic domain. This easy fact will be shown below. We will show in section 6 that, for other types of von Neumann algebras, $\mathcal{V}(\mathcal{N})$ is not continuous.

In fact, most of this section is not concerned with $\mathcal{V}(\mathcal{N})$ itself, but with a more general kind of posets of which $\mathcal{V}(\mathcal{N})$ is but one example.

The fact that $\mathcal{V}(\mathcal{N})$ is an algebraic domain in the case of matrix algebras $\mathcal{N}$ was suggested to us in private communication by Chris Heunen.

Domains of subalgebras. Common examples of algebraic domains are the posets of subalgebras of an algebraic structure, e.g. the poset of subgroups of a group. (Actually, this is the origin of the term 'algebraic'.) We start by formalising this statement and prove that posets of this kind are indeed domains. This standard result will be the point of departure for the following generalisations to posets of subalgebras modulo equations and topological algebras.

Mathematically, we will be using some simple universal algebra. A cautionary remark for the reader familiar with universal algebra: the definitions of some concepts were simplified. For example, for a fixed algebra $\mathcal{A}$, its subalgebras are simply defined to be subsets (and not algebras in their own right).

Definition 5.1. A *signature* is a set Σ of so-called *operation symbols* together with a function $ar : \Sigma \longrightarrow \mathbb{N}$ designating the *arity* of each symbol.

Definition 5.2. A Σ-*algebra* (or an algebra with signature Σ) is a pair $\mathcal{A} = \langle A; \mathcal{F} \rangle$ consisting of a set A (the *support*) and a set of operations $\mathcal{F} = \{f^{\mathcal{A}} \mid f \in \Sigma\}$, where $f^{\mathcal{A}} : A^{ar(f)} \longrightarrow A$ is said to realise the operation symbol f.

Note that the signature describes the operations an algebra is required to have. Unless the distinction is necessary, we will omit the superscript and therefore not make a distinction between operator symbols and operations themselves.

As an example, a monoid $\langle M; \cdot, 1 \rangle$ is an algebra with two operations, $\cdot$ and 1, of arities $ar(\cdot) = 2$ and $ar(1) = 0$.

Definition 5.3. Given an algebra $\mathcal{A} = \langle A; \mathcal{F} \rangle$, a *subalgebra* of $\mathcal{A}$ is a subset of A which is closed under all operations $f \in \mathcal{F}$. We denote the set of subalgebras of $\mathcal{A}$ by $Sub_{\mathcal{A}}$.

Definition 5.4. Let $\mathcal{A} = \langle A; \mathcal{F} \rangle$ be an algebra and $G \subseteq A$. The *subalgebra of* $\mathcal{A}$ *generated by* G, denoted $\langle G \rangle$, is the smallest algebra containing G. This is given explicitly by the closure of G under the operations, i.e. given

$$\begin{aligned} G_0 &= G, \\ G_{k+1} &= G_k \cup \bigcup_{f \in \mathcal{F}} \{ f(x_1, \ldots, x_{ar(f)}) \mid x_1, \ldots, x_{ar(f)} \in G_k \}, \end{aligned}$$

we obtain

$$\langle G \rangle = \bigcup_{k \in \mathbb{N}} G_k. \tag{5.1}$$

A subalgebra $B \subseteq A$ is said to be *finitely generated* whenever it is generated by a finite subset of A.

We will consider the poset $(Sub_{\mathcal{A}}, \subseteq)$ in some detail. The following results are well-known and rather easy to prove:

Proposition 5.5. $(Sub_{\mathcal{A}}, \subseteq)$ *is a complete lattice with the operations*

$$\begin{aligned} \bigwedge \mathcal{S} &:= \bigcap \mathcal{S}, \\ \bigvee \mathcal{S} &:= \langle \bigcup \mathcal{S} \rangle. \end{aligned}$$

Proposition 5.6. *If $\mathcal{S} \subseteq Sub_{\mathcal{A}}$ is directed, then $\bigvee \mathcal{S} = \langle \bigcup \mathcal{S} \rangle = \bigcup \mathcal{S}$.*

Let $B \in Sub_{\mathcal{A}}$. We write

$$\begin{aligned} Sub_{fin}(B) &:= \{C \in Sub_{\mathcal{A}} \mid C \subseteq B \,\text{and}\, C \,\text{is finitely generated}\} \\ &= \{\langle x_1, \ldots, x_n \rangle \mid n \in \mathbb{N}, \{x_1, \ldots, x_n\} \subseteq B\} \end{aligned} \tag{5.2}$$

for the finitely generated subalgebras of B.

Lemma 5.7. *For all $B \in Sub_{\mathcal{A}}$, we have $B = \bigsqcup^{\uparrow} Sub_{fin}(B) = \bigcup Sub_{fin}(B)$.*

We now characterise the way-below relation for the poset $Sub_{\mathcal{A}}$.

Lemma 5.8. *For $B, C \in Sub_{\mathcal{A}}$, one has $C \ll B$ if and only if $C \subseteq B$ and C is finitely generated.*

Proposition 5.9. *$Sub_{\mathcal{A}}$ is an algebraic complete lattice (i.e. a complete lattice which is an algebraic domain).*

Domains of subalgebras with additional algebraic properties. We are interested in subalgebras that satisfy certain algebraic properties that are not present in the algebra $\mathcal{A}$. For example, if $\mathcal{A}$ is a monoid, we may be interested in the set of abelian submonoids of $\mathcal{A}$. To formalise this, we must be able to incorporate 'equational properties'.

Definition 5.10. The set $\Sigma[x_1, \ldots, x_n]$ of terms (or polynomials) over the signature Σ in the variables $x_1, \ldots, x_n$ is defined inductively as follows:

- For each $i \in \{1, \ldots, n\}$, we have $x_i \in \Sigma[x_1, \ldots, x_n]$;
- for each $f \in \Sigma$, if $p_1, \ldots, p_{ar(f)} \in \Sigma[x_1, \ldots, x_n]$, then $f(p_1, \ldots, p_{ar(f)}) \in \Sigma[x_1, \ldots, x_n]$.

Note that $f(p_1, \ldots, p_{ar(f)})$ does not denote the application of the function f, but simply a formal string of symbols. The variables are also just symbols.

Definition 5.11. Let $\mathcal{A}$ be a Σ-algebra. Let $p \in \Sigma[x_1, \ldots, x_n]$, and let $\nu : \{x_1, \ldots, x_n\} \longrightarrow A$ (called a valuation of the variables). Then one extends ν to a function $|.|_\nu : \Sigma[x_1, \ldots, x_n] \longrightarrow A$ by the following inductive rules:

- $|x_i|_\nu := \nu(x_i)$;
- for each $f \in \Sigma$, $|f(p_1, \ldots, p_n)|_\nu := f^{\mathcal{A}}(|p_1|_\nu, \ldots, |p_n|_\nu)$.

Definition 5.12. A *polynomial equation* over $\mathcal{A}$ is a pair $\langle p, q \rangle$ of polynomials in the same variables. A *system of polynomial equations* over $\mathcal{A}$ is a set of such pairs.

Definition 5.13. A subalgebra $B \in Sub_{\mathcal{A}}$ is said to *satisfy an equation* $\langle p, q \rangle$ in the variables $\{x_1, \ldots, x_n\}$, whenever, for all valuations $\nu : \{x_1, \ldots, x_n\} \longrightarrow B \subseteq A$, we have $|p|_\nu = |q|_\nu$. B is said to *satisfy a system E of equations* if it satisfies all $\langle p, q \rangle \in E$. Further, we define

$$Sub_{\mathcal{A}}/E = \{B \in Sub_{\mathcal{A}} \mid B \,\text{satisfies}\, E\}. \tag{5.3}$$

We will now consider the poset $\langle Sub_{\mathcal{A}}/E, \subseteq\rangle$, which is a subposet of $\langle Sub_{\mathcal{A}}, \subseteq\rangle$ considered before. Clearly, if $\mathcal{A}$ satisfies E, we have $Sub_{\mathcal{A}}/E = Sub_{\mathcal{A}}$, which is not very interesting. In all other cases $Sub_{\mathcal{A}}/E$ has no top element. In fact, it is not even a lattice in most cases. However, we still have some weakened form of completeness:

Proposition 5.14. *$Sub_{\mathcal{A}}/E$ is a bounded-complete algebraic domain, i.e., it is a bc-dcpo and it is algebraic.*

Proof. To prove that it is an algebraic domain, it is enough to show that $Sub_{\mathcal{A}}/E$ is a Scott-closed subset of $Sub_{\mathcal{A}}$, i.e., that it is closed for directed suprema (hence a dcpo) and downwards closed (hence, given the first condition, $\mathop{\downarrow\!\!\!\downarrow} x$ is the same in both posets, and algebraicity follows from the same property on $Sub_{\mathcal{A}}$). Downwards-closedness follows immediately, since a subset of a set satisfying an equation also satisfies it.

For directed completeness, let $\mathcal{S}$ be a directed subset of $Sub_{\mathcal{A}}/E$. We want to show that its supremum, the union $\bigcup \mathcal{S}$, is still in $Sub_{\mathcal{A}}/E$. Let $\langle p, q\rangle \in E$ be an equation over the variables $\{x_1, \ldots, x_n\}$ and $\nu : \{x_1, \ldots, x_n\} \longrightarrow \bigcup \mathcal{S}$ any valuation on $\bigcup \mathcal{S}$. Let us write $a_1, \ldots, a_n$ for $\nu(x_1), \ldots, \nu(x_n)$. There exist $S_1, \ldots, S_n \in \mathcal{S}$ such that $a_1 \in S_1, \ldots, a_n \in S_n$. By directedness of $\mathcal{S}$, there is an $S \in \mathcal{S}$ such that $\{a_1, \ldots, a_n\} \subseteq S$. Since $S \in Sub_{\mathcal{A}}/E$ and the valuation ν is defined on S, one must have $|p|_\nu = |q|_\nu$. Therefore, $\bigcup \mathcal{S}$ satisfies the equations in E, and so $Sub_{\mathcal{A}}/E$ is closed under directed suprema.

For bounded completeness, note that if $\mathcal{S} \subseteq Sub_{\mathcal{A}}/E$ is bounded above by a subalgebra S' that also satisfies the equations, then the subalgebra generated by $\mathcal{S}$, $\langle \bigcup \mathcal{S}\rangle$, being a subset of S', must also satisfy the equations. So $\langle \bigcup \mathcal{S}\rangle$ is in $Sub_{\mathcal{A}}/E$ and is a supremum for $\mathcal{S}$ in this poset. □

Topologically closed subalgebras. The results of the previous section apply to any kind of algebraic structure. In particular, this is enough to conclude that, given a $*$-algebra $\mathcal{N}$, the set of its abelian $*$-subalgebras forms an algebraic domain. If we restrict attention to the finite-dimensional situation, i.e., matrix algebras $\mathcal{N}$, or finite direct sums of matrix algebras, then the result shows that the poset $\mathcal{V}(\mathcal{N})$ of abelian von Neumann subalgebras of a matrix algebra $\mathcal{N}$ is an algebraic domain, since every algebraically closed abelian $*$-subalgebra is also weakly closed (and hence a von Neumann algebra) in this case.

But for a general von Neumann algebra $\mathcal{N}$, not all abelian $*$-subalgebras need to be abelian von Neumann subalgebras. We need to consider the extra condition that each given subalgebra is closed with respect to a certain topology, namely the weak operator topology (or the strong operator topology, or the σ-weak topology for that matter).

Again, we follow a general path, proving what assertions can be made about posets of subalgebras of any kind of algebraic structures, where its subalgebras are also topologically closed.

For the rest of this section, we only consider Hausdorff topological spaces. This condition is necessary for our proofs to work. The reason is that, in a Hausdorff space, a net $(a_i)_{i\in I}$ converges to at most one point a. We shall write this as $(a_i)_{i\in I} \longrightarrow a$.

Definition 5.15. A *topological algebra* is an algebra $\mathcal{A} = \langle A; \mathcal{F}\rangle$ where A is equipped with a Hausdorff topology.

We single out the substructures of interest:

Definition 5.16. Given a *topological algebra* $\mathcal{A} = \langle A; \mathcal{F}\rangle$, a *closed subalgebra* is a subset B of A which is simultaneously a subalgebra and a topologically closed set. The set of closed subalgebras of $\mathcal{A}$ is denoted by $CSub_{\mathcal{A}}$. Moreover, for a system of polynomial equations E over $\mathcal{A}$, $CSub_{\mathcal{A}}/E$ denotes $CSub_{\mathcal{A}} \cap Sub_{\mathcal{A}}/E$.

Our goal is to extend the results about $Sub_{\mathcal{A}}/E$ to $CSub_{\mathcal{A}}/E$. For this, we need to impose a topological condition on the behaviour of the operations of $\mathcal{A}$. Usually, one requires the algebraic operations to be continuous, which would allow us to prove results regarding completeness which are similar to those in previous subsections. An example would then be the poset of (abelian) closed subgroups of a topological group. However, multiplication in a von Neumann algebra is not continuous with respect to the weak operator topology, only separately continuous in each argument. Thus, in order to capture the case of $\mathcal{V}(\mathcal{N})$, we need to weaken the continuity assumption. The following condition will suffice:

Definition 5.17. Let $\mathcal{A} = \langle A; \mathcal{F}\rangle$ be a topological algebra and $f \in \mathcal{F}$ an operation with arity n. We say that f is *separately continuous* if, for any $k = 1, \ldots, n$ and elements $b_1, \ldots, b_{k-1}, b_{k+1}, \ldots, b_n \in A$, the function

$$\begin{aligned} \mathcal{A} &\longrightarrow \mathcal{A} \\ a &\longmapsto f(b_1, \ldots, b_{k-1}, a, b_{k+1}, \ldots, b_n) \end{aligned}$$

is continuous. Equivalently, one can say that for any net $(a_i)_{i\in I}$ such that $(a_i)_{i\in I} \longrightarrow a$, we have

$$(f(b_1, \ldots, b_{k-1}, a_i, b_{k+1}, \ldots, b_n))_{i\in I} \longrightarrow f(b_1, \ldots, b_{k-1}, a, b_{k+1}, \ldots, b_n).$$

The fact that we allow a weaker form of continuity than it is costumary on the algebraic operations forces us to impose a condition on the allowed equations.

Definition 5.18. A polynomial over $\mathcal{A}$ in the variables $x_1, \ldots, x_n$ is *linear* if each of the variables occurs at most once.

Lemma 5.19. *Let $\mathcal{A} = \langle A; \mathcal{F}\rangle$ be a topological algebra with separately continuous operations, and p a linear polynomial over $\mathcal{A}$ in the variables $x_1, \ldots, x_n$. Then the function*

$$\begin{aligned} \widetilde{p} : \mathcal{A}^n &\longrightarrow \mathcal{A} \\ (a_1, \ldots, a_n) &\longmapsto |p|_{\nu: x_i \longmapsto a_i} \end{aligned}$$

is separately continuous.

Proof. The proof goes by induction on (linear) polynomials (refer back to definitions 5.10 and 5.11):

- If $p = x_j$, then
$$\widetilde{p}(a_1, \ldots, a_n) = |x_j|_{\nu : x_i \longmapsto a_i} = \nu(x_j) = a_j. \tag{5.4}$$
So $\widetilde{p} = \pi_j$, which is clearly separately continuous (either constant or the identity when arguments taken separately).
- If $p = f(p_1, \ldots, p_l)$, assume as the induction hypothesis that $\widetilde{p_1}, \ldots, \widetilde{p_l}$ are separately continuous. Then
$$\widetilde{p} = f \cdot \langle \widetilde{p_1}, \ldots, \widetilde{p_l} \rangle. \tag{5.5}$$
Let us see that this is separately continuous in the first argument x_1 (the other arguments can be treated analogously). Let $b_2, \ldots, b_n \in A$. Since p is linear, the variable x_1 occurs on at most one subpolynomial p_k (with $k = 1, \ldots, l$). Without loss of generality, say it occurs on p_1. Then the functions
$$t_k = a \longmapsto \widetilde{p_k}(a, b_2, \ldots, b_n)$$
for $k = 2, \ldots, l$ are constant (since x_1 does on occur in p_k), whereas the function
$$t_1 = a \longmapsto \widetilde{p_1}(a, b_2, \ldots, b_n)$$
is continuous (by the induction hypothesis).

  Now, we consider the function $\widetilde{p}$ with only the first argument varying. This is the function $f \cdot \langle t_1, \ldots, t_l \rangle$. But $t_2, \ldots, t_l$ are constant functions, which means that only the first argument of f varies. Since this is given by a continuous function t_1 and f is continuous in the first argument (because it is separately continuous), $f \cdot \langle t_1, \ldots, t_l \rangle : A \longrightarrow A$ is continuous, meaning that $\widetilde{p}$ is (separately) continuous in the first argument. For the other arguments, the proof is similar. □

We shall denote by $cl(-)$ the (Kuratowski) closure operator associated with the topology of $\mathcal{A}$.

Lemma 5.20. *For a topological algebra $\mathcal{A} = \langle A; \mathcal{F} \rangle$ with separately continuous operations:*

1. *If $B \in Sub_{\mathcal{A}}$, then $cl(B) \in Sub_{\mathcal{A}}$.*
2. *Given a system of linear polynomial equations E over $\mathcal{A}$, if $B \in Sub_{\mathcal{A}}/E$, then $cl(B) \in Sub_{\mathcal{A}}/E$.*

Proof. (1) We need to show that $cl(B)$ is closed under the operations in $\mathcal{F}$. We consider only the case of a binary operation $f \in \mathcal{F}$. It will be apparent that the general case follows from a simple inductive argument.

Let $a, b \in cl(B)$. Then there exist nets $(a_i)_{i \in I}$ and $(b_j)_{j \in J}$ consisting of elements of B such that $(a_i)_{i \in I} \longrightarrow a$ and $(b_j)_{j \in J} \longrightarrow b$. By fixing an index j at a time, we conclude, by separate continuity, that
$$\forall j \in J : (f(a_i, b_j))_{i \in I} \longrightarrow f(a, b_j). \tag{5.6}$$

Since all the elements of the net $(f(a_i, b_j))_{i \in I}$ are in B (for $a_i, b_j \in B$ and B is closed under f), all elements of the form $f(a, b_j)$ are in $cl(B)$, as they are limits of nets of elements of B. Now, letting j vary again, by separate continuity we obtain

$$(f(a, b_j))_{j \in J} \longrightarrow f(a, b). \tag{5.7}$$

But $(f(a, b_j))_{j \in J}$ is a net of elements of $cl(B)$, thus $f(a, b) \in cl(cl(B)) = cl(B)$. This completes the proof that $cl(B)$ is closed under f.

(2) The procedure is very similar and again we consider only a polynomial equation $\langle p, q \rangle$ in two variables x and y. We will prove that $cl(B)$ satisfies the equation whenever B does.

Let $a, b \in cl(B)$, with sets $(a_i)_{i \in I}$ and $(b_j)_{j \in J}$ as before. We use the function $\widetilde{p} : A^2 \longrightarrow A$ from lemma 5.19, which is given by

$$\widetilde{p}(k_1, k_2) = |p|_{[x_1 \mapsto k_1, x_2 \mapsto k_2]}. \tag{5.8}$$

Because of Lemma 5.19, we can follow the same procedure as in the first part (for the separately continuous function f) to conclude that

$$\forall j \in J : (\widetilde{p}(a_i, b_j))_{i \in I} \longrightarrow \widetilde{p}(a, b_j) \quad \wedge \quad (\widetilde{q}(a_i, b_j))_{i \in I} \longrightarrow \widetilde{q}(a, b_j). \tag{5.9}$$

But $a_i, b_j \in B$ and B satisfies the equations, so the nets $(\widetilde{p}(a_i, b_j))_{i \in I}$ and $(\widetilde{q}(a_i, b_j))_{i \in I}$ are the same. Since we have Hausdorff spaces by assumption, a net has at most one limit. This implies that

$$\forall j \in J : \widetilde{p}(a, b_j) = \widetilde{q}(a, b_j). \tag{5.10}$$

As before, we take (again by separate continuity of $\widetilde{p}$ and $\widetilde{q}$)

$$(\widetilde{p}(a, b_j))_{j \in J} \longrightarrow \widetilde{p}(a, b) \quad \wedge \quad (\widetilde{q}(a, b_j))_{j \in J} \longrightarrow \widetilde{q}(a, b). \tag{5.11}$$

Because of 5.10 and 5.11, the nets are the same again, yielding

$$\widetilde{p}(a, b) \ = \ \widetilde{q}(a, b). \tag{5.12}$$

This proves that $cl(B)$ satifies $\langle p, q \rangle$. □

We now can state our result. As far as completeness (or convergence) properties are concerned, we have a similar situation as in the non-topological case.

Proposition 5.21. *Let $\mathcal{A} = \langle A; \mathcal{F} \rangle$ be a topological algebra with separately continuous operations. Let E be a system of linear polynomial equations. Then:*

1. *$CSub_{\mathcal{A}}$ is a complete lattice.*
2. *$CSub_{\mathcal{A}}/E$ is a bounded-complete dcpo.*

Proof. If S is a set of closed subalgebras, then, by 5.20-1, there is a least closed subalgebra containing S, given by $cl(\langle \bigcup S \rangle)$. This proves the first claim.

Now, suppose that all $B \in S$ satisfy E and that S is directed. By directedness, the supremum is given by $cl(\langle \bigcup S \rangle) = cl(\bigcup S)$. By Lemma 5.14, $\bigcup S$ satisfies the equations E. Then, by 5.20-2, $cl(\bigcup S)$ also does.

Similarly, suppose all $B \in S$ satisfy E and that S is bounded above by $C \in CSub_{\mathcal{A}}/E$. Then the supremum of S in $CSub_{\mathcal{A}}$, namely $cl(\langle \bigcup S \rangle)$, is a closed subalgebra smaller than C. Therefore, $cl(\langle\bigcup S\rangle)$ must also satisfy E, hence it is a supremum for S in $CSub_{\mathcal{A}}/E$. □

$\mathcal{V}(\mathcal{N})$ as a dcpo. The results above apply easily to the situation we are interested in, namely the poset of abelian von Neumann subalgebras of a non-abelian von Neumann algebra $\mathcal{N}$. We can describe $\mathcal{N}$ as the algebra[5]

$$\langle \mathcal{N}; \mathbf{1}, +, \cdot_c, \times, * \rangle \tag{5.13}$$

whose operations have arities 0, 2, 1, 2 and 1, respectively. The operations are just the usual $*$-algebra operations. Note that $\cdot_c$ (scalar multiplication by $c \in \mathbb{C}$) is in fact an uncountable set of operations indexed by $\mathbb{C}$. Note that the inclusion of $\mathbf{1}$ as an operation allows to restrict our attention to unital subalgebras with the same unit and to unital homomorphisms. Recall that a C^*-subalgebra of $\mathcal{N}$ is a $*$-subalgebra that is closed with respect to the norm topology. A von Neumann subalgebra is a $*$-subalgebra that is closed with respect to the weak operator topology. Since both these topologies are Hausdorff and the operations are separately continuous with respect to both, the results from the previous sections apply.

Proposition 5.22. *Let $\mathcal{N}$ be a von Neumann algebra. Then*

- *the set of unital $*$-subalgebras (respectively, C^*-subalgebras, and von Neumann subalgebras) of $\mathcal{N}$ with the same unit element as $\mathcal{N}$ is a complete lattice.*
- *The set of abelian unital $*$-subalgebras (respectively, C^*-subalgebras, and von Neumann subalgebras) of $\mathcal{N}$ with the same unit element as $\mathcal{N}$ is a bounded complete dcpo (or complete semilattice).*

The set of abelian von Neumann subalgebras of $\mathcal{N}$ with the same unit element as $\mathcal{N}$ is simply $\mathcal{V}(\mathcal{N})$. Hence, we have shown that $\mathcal{V}(\mathcal{N})$ is a dcpo. Moreover, it is clear that $\mathcal{V}(\mathcal{N})$ is bounded complete if we include the trivial algebra as the bottom element. Note that this holds for *all* types of von Neumann algebras. Furthermore, our proof shows that analogous results hold for the poset of abelian C^*-subalgebras of unital C^*-algebras (and other kinds of topological algebras and equations, such as associative closed Jordan subalgebras, or closed subgroups of a topological group).

As already remarked at the beginning of this subsection on topologically closed subalgebras, $\mathcal{V}(\mathcal{N})$ is an algebraic domain in the case that $\mathcal{N}$ is a finite-dimensional algebra (finite direct sum of matrix algebras over $\mathbb{C}$). This can be seen directly from the fact that, in the finite-dimensional situation, any subalgebra is weakly (resp. norm) closed. Hence, the topological aspects do not play a rôle and proposition 5.14 applies. The result that $\mathcal{V}(\mathcal{N})$ is an algebraic domain if $\mathcal{N}$ is a finite-dimensional matrix algebra can also be deduced from the fact that $\mathcal{V}(\mathcal{N})$ has finite height in this case. This clearly means that any directed set has a maximal element and, consequently, the

[5] "Algebra" here has two different meanings. The first is the (traditional) meaning, of which the $*$-algebras, C^*-algebras and von Neumann algebras are special cases. When we say just algebra, though, we mean it in the sense of universal algebra as in the previous two sections.

way-below relation coincides with the order relation, thus $\mathcal{V}(\mathcal{N})$ is trivially an algebraic domain.

One can also see directly that the context category $\mathcal{V}(\mathcal{N})$ is a dcpo for an arbitrary von Neumann algebra $\mathcal{N}$: let $S \subset \mathcal{V}(\mathcal{N})$ be a directed subset, and let

$$A := \left\langle \bigcup_{V \in S} V \right\rangle \tag{5.14}$$

be the algebra generated by the elements of S. Clearly, A is a self-adjoint abelian algebra that contains $\hat{1}$, but not a von Neumann algebra in general, since it need not be weakly closed. By von Neumann's double commutant theorem, the weak closure $\tilde{V} := \overline{A}^w$ of A is given by the double commutant A'' of A. The double commutant construction preserves commutativity, so $A'' = \tilde{V}$ is an *abelian* von Neumann subalgebra of $\mathcal{N}$, namely the smallest von Neumann algebra containing all the $V \in S$, so

$$\tilde{V} = \bigsqcup{}^{\uparrow} S. \tag{5.15}$$

Hence, every directed subset S has a supremum in $\mathcal{V}(\mathcal{N})$.

A functor from von Neumann algebras to dcpos. We have seen that to each von Neumann algebra $\mathcal{N}$, we can associate a dcpo $\mathcal{V}(\mathcal{N})$. We will now show that this assignment is functorial. Again, this result arises as a special case of a more general proposition, which is quite easy to show.

Definition 5.23. Let $\mathcal{A}$ and $\mathcal{B}$ be two algebras with the same signature Σ (that is, they have the same set of operations). A function $\phi : A \longrightarrow B$ is called a *homomorphism* if it preserves every operation. That is, for each operation f of arity n,

$$\phi(f^{\mathcal{A}}(a_1, \dots, a_n)) = f^{\mathcal{B}}(\phi(a_1), \dots, \phi(a_n)). \tag{5.16}$$

Lemma 5.24. *Let $\mathcal{A}$ and $\mathcal{B}$ be topological algebras and $\phi : A \longrightarrow B$ a continuous homomorphism that preserves closed subalgebras. Then, for any E, the function*

$$\begin{aligned} \phi^{\rightarrow}|_{CSub_{\mathcal{A}}/E} : CSub_{\mathcal{A}}/E &\longrightarrow CSub_{\mathcal{B}}/E \\ S &\longmapsto \phi^{\rightarrow} S \end{aligned} \tag{5.17}$$

is a Scott-continuous function.

Proof. First, we check that the function is well-defined, that is, each element of the domain determines one element in the codomain. Let $S \in CSub_{\mathcal{A}}/E$. Then $S' := \phi^{\rightarrow}(S)$ is in $CSub_{\mathcal{B}}$ by the assumption that ϕ preserves closed subalgebras. The homomorphism condition implies that it also preserves the satisfaction of equations E.

As for the result itself, note that monotonicity is trivial. So we only need to show that this map preserves directed joins. We know that the directed join is given as $\bigsqcup^{\uparrow} D = cl(\bigcup D)$. Thus, we want to prove that for any directed set D:

$$\phi^{\rightarrow}(cl(\bigcup_{S \in D} S)) \subseteq cl(\bigcup_{S \in D} \phi^{\rightarrow}(S)) .$$

This follows easily since $f^{\rightarrow}$ commutes with union of sets and the inequality $\phi^{\rightarrow}cl(S) \subseteq cl(\phi^{\rightarrow}S)$ is simply the continuity of ϕ. □

We can combine the previous lemma with Prop. 5.21 as follows:

Proposition 5.25. *Let $\mathcal{C}$ be a category such that*

- *objects of $\mathcal{C}$ are topological Σ-algebras (for a fixed signature Σ) that fulfil the conditions stated in Prop.* 5.21*,*
- *morphisms of $\mathcal{C}$ are continuous homomorphisms preserving closed subalgebras.*

Then, for any system of linear polynomial equations E, the assignments

$$A \longmapsto CSub_A/E \tag{5.18}$$

$$(\phi : A \longrightarrow B) \longmapsto (\phi^{\rightarrow}|_{CSub_A/E} : CSub_A/E \longrightarrow CSub_B/E) \tag{5.19}$$

define a functor $F_E : \mathcal{C} \longrightarrow \mathsf{DCPO}$ from $\mathcal{C}$ to the category of dcpos and Scott-continuous functions.

Proof. 5.21 and 5.24 imply that this is a well-defined map between the categories. Functoriality is immediate from the properties of images $f^{\rightarrow}$ of a function f: $id_A^{\rightarrow} = id_{\mathcal{P}A}$ and $f^{\rightarrow} \circ g^{\rightarrow} = (f \circ g)^{\rightarrow}$. □

This result can be applied to our situation. Let $\mathcal{N}$ be a von Neumann algebra. Recall that a von Neumann subalgebra is a $*$-subalgebra closed with respect to the σ-weak topology.[6] This topology is Hausdorff, and σ-weakly continuous (or normal) $*$-homomorphisms map von Neumann subalgebras to von Neumann subalgebras. Hence, we get

Theorem 5.26. *The assignments*

$$\mathcal{V} : \mathsf{vNAlg} \longrightarrow \mathsf{DCPO}$$

$$\mathcal{N} \longmapsto \mathcal{V}(\mathcal{N}) \tag{5.20}$$

$$\phi \longmapsto \phi^{\rightarrow}|_{\mathcal{V}(\mathcal{N})} \tag{5.21}$$

define a functor from the category vNAlg *of von Neumann algebras and σ-weakly continuous, unital algebra homomorphisms to the category* DCPO *of dcpos and Scott-continuous functions.*

It is easy to see that the functor $\mathcal{V}$ is not full. It is currently an open question whether $\mathcal{V}$ is faithful.

6. The context category $\mathcal{V}(\mathcal{N})$ in infinite dimensions – lack of continuity

The case of $\mathcal{N} = \mathcal{B}(\mathcal{H})$ with infinite-dimensional $\mathcal{H}$. We now show that for an infinite-dimensional Hilbert space $\mathcal{H}$ and the von Neumann algebra $\mathcal{N} =$

[6] Equivalently, a von Neumann algebra is closed in the weak operator topology, as mentioned before.

$\mathcal{B}(\mathcal{H})$, the poset $\mathcal{V}(\mathcal{N})$ is *not* continuous, and hence not a domain. This counterexample is due to Nadish de Silva, whom we thank.

Let $\mathcal{H}$ be a separable, infinite-dimensional Hilbert space, and let $(e_i)_{i\in\mathbb{N}}$ be an orthonormal basis. For each $i \in \mathbb{N}$, let $\hat{P}_i$ be the projection onto the ray $\mathbb{C}e_i$. This is a countable set of pairwise-orthogonal projections.

Let $\hat{P}_e := \sum_{i=1}^{\infty} \hat{P}_{2i}$, that is, $\hat{P}_e$ is the sum of all the $\hat{P}_i$ with even indices. Let $V_e := \{\hat{P}_e, \hat{1}\}'' = \mathbb{C}\hat{P}_e + \mathbb{C}\hat{1}$ be the abelian von Neumann algebra generated by $\hat{P}_e$ and $\hat{1}$. Moreover, let V_m be the (maximal) abelian von Neumann subalgebra generated by all the $\hat{P}_i$, $i \in \mathbb{N}$. Clearly, we have $V_e \subset V_m$.

Now let $V_j := \{\hat{P}_1, ..., \hat{P}_j, \hat{1}\}'' = \mathbb{C}\hat{P}_1 + ... + \mathbb{C}\hat{P}_j + \mathbb{C}\hat{1}$ be the abelian von Neumann algebra generated by $\hat{1}$ and the first j projections in the orthonormal basis, where $j \in \mathbb{N}$. The algebras V_j form a directed set (actually a chain) $(V_j)_{j\in\mathbb{N}}$ whose directed join is V_m, obviously. But note that none of the algebras V_j contains V_e, since the projection $\hat{P}_e$ is not contained in any of the V_j. Hence, we have the situation that V_e is contained in $\bigsqcup^{\uparrow}(V_j)_{j\in\mathbb{N}} = V_m$, but $V_e \nsubseteq V_j$ for any element V_j of the directed set, so V_e is not way below itself.

Since V_e is an atom in the poset $\mathcal{V}(\mathcal{N})$ (i.e. only the bottom element, the trivial algebra, is below V_e in the poset), we have that $\mathord{\downarrow\!\!\!\downarrow} V_e = \{\bot\} = \{\mathbb{C}\hat{1}\}$, so $\bigsqcup^{\uparrow}\mathord{\downarrow\!\!\!\downarrow} V_e = \bigsqcup^{\uparrow}\{\bot\} = \bot \neq V_e$, which implies that $\mathcal{V}(\mathcal{N})$ is not continuous.

Other types of von Neumann algebras. The proof that $\mathcal{V}(\mathcal{N}) = \mathcal{V}(\mathcal{B}(\mathcal{H}))$ is not a continuous poset if $\mathcal{H}$ is infinite-dimensional can be generalised to other types of von Neumann algebras. ($\mathcal{B}(\mathcal{H})$ is a type I_∞ factor.) We will use the well-known fact that an abelian von Neumann algebra V is of the form

$$V \simeq \ell^\infty(X, \mu) \tag{6.1}$$

for some measure space (X, μ). It is known that the following cases can occur: $X = \{1, ..., n\}$, for each $n \in \mathbb{N}$, or $X = \mathbb{N}$, equipped with the counting measure μ_c; or $X = [0,1]$, equipped with the Lebesgue measure μ; or the combinations $X = \{1, ..., n\} \sqcup [0,1]$ and $X = \mathbb{N} \sqcup [0,1]$.

A maximal abelian subalgebra V_m of $\mathcal{B}(\mathcal{H})$ generated by the projections onto the basis vectors of an orthonormal basis $(e_i)_{i\in\mathbb{N}}$ of an infinite-dimensional, separable Hilbert space $\mathcal{H}$ corresponds to the case of $X = \mathbb{N}$, so $V_m \simeq \ell^\infty(\mathbb{N})$. The proof that in this case $\mathcal{V}(\mathcal{B}(\mathcal{H}))$ is not continuous is based on the fact that $X = \mathbb{N}$ can be decomposed into a countable family $(S_i)_{i\in\mathbb{N}}$ of measurable, pairwise disjoint subsets with $\bigcup_i S_i = X$ such that $\mu(S_i) > 0$ for each i. Of course, the sets S_i can just be taken to be the singletons $\{i\}$, for $i \in \mathbb{N}$, in this case. These are measurable and have measure greater than zero since the counting measure on $X = \mathbb{N}$ is used.

Yet, also for $X = ([0,1], \mu)$ we can find a countable family $(S_i)_{i\in\mathbb{N}}$ of measurable, pairwise disjoint subsets with $\bigcup_i S_i = X$ such that $\mu(S_i) > 0$ for each i: for example, take $S_1 := [0, \frac{1}{2})$, $S_2 := [\frac{1}{2}, \frac{3}{4})$, $S_3 := [\frac{3}{4}, \frac{7}{8})$, etc. Each measurable subset S_i corresponds to a projection $\hat{P}_i$ in the abelian von Neumann algebra $V \simeq \ell^\infty([0,1], \mu)$, so we obtain a countable family $(\hat{P}_i)_{i\in\mathbb{N}}$

of pairwise orthogonal projections with sum $\hat{1}$ in V. Let $\hat{P}_e := \sum_{i=1}^{\infty} \hat{P}_{2i}$, and let $V_e := \{\hat{P}_e, \hat{1}\}'' = \mathbb{C}\hat{P}_e + \mathbb{C}\hat{1}$.

Moreover, let $V_j := \{\hat{P}_1, ..., \hat{P}_j, \hat{1}\}''$ be the abelian von Neumann algebra generated by $\hat{1}$ and the first j projections, where $j \in \mathbb{N}$. Clearly, $(V_j)_{j \in \mathbb{N}}$ is a directed set, and none of the algebras V_j, $j \in \mathbb{N}$, contains the algebra V_e, since $\hat{P}_e \notin V_j$ for all j. Let $V := \bigsqcup^{\uparrow} V_j$ be the abelian von Neumann algebra generated by all the V_j. Then we have $V_e \subset V$, so V_e is not way below itself and $\bigsqcup^{\uparrow} \twoheaddownarrow V_e \neq V_e$.

This implies that $\mathcal{V}(\mathcal{N})$ is not continuous if $\mathcal{N}$ contains an abelian subalgebra of the form $V \simeq \ell^{\infty}([0,1], \mu)$. Clearly, analogous results hold if $\mathcal{N}$ contains any abelian subalgebra of the form $V = \ell^{\infty}(\{1, ..., n\} \sqcup [0,1])$ or $V \simeq \ell^{\infty}(\mathbb{N} \sqcup [0,1])$ – in none of these cases is $\mathcal{V}(\mathcal{N})$ a domain.

A von Neumann algebra $\mathcal{N}$ contains only abelian subalgebras of the form $V \simeq \ell^{\infty}(\{1, ..., n\}, \mu_c)$ if and only if $\mathcal{N}$ is either (a) a finite-dimensional matrix algebra $M_n(\mathbb{C}) \otimes \mathcal{C}$ with $\mathcal{C}$ an abelian von Neumann algebra (i.e., $\mathcal{N}$ is a type I_n-algebra) such that the centre $\mathcal{C}$ does not contain countably many pairwise orthogonal projections,[7] or (b) a finite direct sum of such matrix algebras. Equivalently, $\mathcal{N}$ is a *finite-dimensional von Neumann algebra*, that is, $\mathcal{N}$ is represented (faithfully) on a finite-dimensional Hilbert space. We have shown:

Theorem 6.1. *The context category $\mathcal{V}(\mathcal{N})$, that is, the partially ordered set of abelian von Neumann subalgebras of a von Neumann algebra $\mathcal{N}$ which share the unit element with $\mathcal{N}$, is a continuous poset and hence a domain if and only if $\mathcal{N}$ is finite-dimensional.*

7. Outlook: Space and space-time from interval domains?

There are many further aspects of the link between the topos approach and domain theory to be explored. In this brief section, we indicate some very early ideas about an application to concepts of space and space-time.[8] The ideas partly go back to discussions with Chris Isham, but we take full responsibility.

Let us take seriously the idea suggested by the topos approach that in quantum theory it is not the real numbers $\mathbb{R}$ where physical quantities take their values, but rather the presheaf $\underline{\mathbb{R}^{\leftrightarrow}}$. This applies in particular to the physical quantity position. We know from ordinary quantum theory that depending on its state, a quantum particle will be localised more or less well. (Strictly speaking, there are no eigenstates of position, so a particle can never be localised perfectly.) The idea is that we can now describe 'unsharp positions' as well as sharp ones by using the presheaf $\underline{\mathbb{R}^{\leftrightarrow}}$ and its global elements. A quantum particle will not 'see' a point in the continuum $\mathbb{R}$ as its

---

[7]This of course means that $\mathcal{C} \simeq \ell^{\infty}(\{1, ..., k\}, \mu_c)$, that is, $\mathcal{C}$ is isomorphic to a finite-dimensional algebra of diagonal matrices with complex entries.

[8]The reader not interested in speculation may well skip this section!

position, but, depending on its state, it will 'see' some global element of $\underline{\mathbb{R}^{\leftrightarrow}}$ as its (generalised) position.

These heuristic arguments suggest the possibility of building a model of space based not on the usual continuum $\mathbb{R}$, but on the quantity-value presheaf $\underline{\mathbb{R}^{\leftrightarrow}}$. The real numbers are embedded (topologically) in $\underline{\mathbb{R}^{\leftrightarrow}}$ as a tiny part, they correspond to global elements $\gamma = (\mu, \nu)$ such that $\mu = \nu = \mathrm{const}_a$, where $a \in \mathbb{R}$ and const_a is the constant function with value a.

It is no problem to form $\underline{\mathbb{R}^{\leftrightarrow}}^3$ or $\underline{\mathbb{R}^{\leftrightarrow}}^n$ within the topos, and also externally it is clear what these presheaves look like. It is also possible to generalise notions of distance and metric from $\mathbb{R}^n$ to $\mathbf{I}\mathbb{R}^n$, i.e., powers of the interval domain. This is an intermediate step to generalising metrics from $\mathbb{R}^n$ to $\underline{\mathbb{R}^{\leftrightarrow}}^n$. Yet, there are many open questions.

The main difficulty is to find a suitable, physically sensible group that would act on $\underline{\mathbb{R}^{\leftrightarrow}}^n$. Let us focus on space, modelled on $\underline{\mathbb{R}^{\leftrightarrow}}^3$, for now. We picture an element of $\underline{\mathbb{R}^{\leftrightarrow}}^3_V$ as a 'box': there are three intervals in three independent coordinate directions. We see that in this interpretation, there implicitly is a spatial coordinate system given. Translations are unproblematic, they will map a box to another box. But rotations will map a box to a rotated box that in general cannot be interpreted as a box with respect to the same (spatial) coordinate system. This suggests that one may want to consider a *space of (box) shapes* rather than the boxes themselves. These shapes would then represent generalised, or 'unsharp', positions. Things get even more difficult if we want to describe space-time. There is no time operator, so we have to take some liberty when interpreting time intervals as 'unsharp moments in time'. More crucially, the Poincaré group acting on classical special relativistic space-time is much more complicated than the Galilei group that acts separately on space and time, so it will be harder to find a suitable 'space of shapes' on which the Poincaré group, or some generalisation of it, will act.

There is ongoing work by the authors to build a model of space and space-time based on the quantity-value presheaf $\underline{\mathbb{R}^{\leftrightarrow}}$. It should be mentioned that classical, globally hyperbolic space-times can be treated as domains, as was shown by Martin and Panangaden [27]. These authors consider generalised interval domains over so-called bicontinuous posets. This very interesting work serves as a motivation for our ambitious goal of defining a notion of generalised space and space-time suitable for quantum theory and theories 'beyond quantum theory', in the direction of quantum gravity and quantum cosmology. Such a model of space and space-time would not come from a discretisation, but from an embedding of the usual continuum into a much richer structure. It is very plausible that domain-theoretic techniques will still apply, which could give a systematic way to define a classical limit of these quantum space-times. For now, we merely observe that the topos approach, which so far applies to non-relativistic quantum theory, provides suggestive structures that may lead to new models of space and space-time.

8. Conclusion

We developed some connections between the topos approach to quantum theory and domain theory. The topos approach provides new 'spaces' of physical relevance, given as objects in a presheaf topos. Since these new spaces are not even sets, many of the usual mathematical techniques do not apply, and we have to find suitable tools to analyse them. In this article, we focused on the quantity-value object $\underline{\mathbb{R}^{\leftrightarrow}}$ in which physical quantities take their values, and showed that domain theory provides tools for understanding the structure of this space of generalised, unsharp values. This also led to some preliminary ideas about models of space and space-time that would incorporate unsharp positions, extending the usual continuum picture of space-time.

Acknowledgements

We thank Chris Isham, Prakash Panangaden and Samson Abramsky for valuable discussions and encouragement. Many thanks to Nadish de Silva for providing the counterexample in Section 6, and to the anonymous referee for a number of valuable suggestions and hints. RSB gratefully acknowledges support by the Marie Curie Initial Training Network in Mathematical Logic - MALOA - From MAthematical LOgic to Applications, PITN-GA-2009-238381, as well as previous support by Santander Abbey bank. We also thank Felix Finster, Olaf Müller, Marc Nardmann, Jürgen Tolksdorf and Eberhard Zeidler for organising the very enjoyable and interesting "Quantum Field Theory and Gravity" conference in Regensburg in September 2010.

References

[1] S. Abramsky, A. Jung. Domain Theory. In *Handbook of Logic in Computer Science*, eds. S. Abramsky, D.M. Gabbay, T.S.E. Maibaum, Clarendon Press, 1–168 (1994).

[2] M. Caspers, C. Heunen, N.P. Landsman, B. Spitters. Intuitionistic quantum logic of an n-level system. *Found. Phys.* **39**, 731–759 (2009).

[3] B. Coecke, K. Martin. A partial order on classical and quantum states. In *New Structures for Physics*, ed. B. Coecke, Springer, Heidelberg (2011).

[4] A. Döring. Kochen-Specker theorem for von Neumann algebras. *Int. Jour. Theor. Phys.* **44**, 139–160 (2005).

[5] A. Döring. Topos theory and 'neo-realist' quantum theory. In *Quantum Field Theory, Competitive Models*, eds. B. Fauser, J. Tolksdorf, E. Zeidler, Birkhäuser (2009).

[6] A. Döring. Quantum states and measures on the spectral presheaf. *Adv. Sci. Lett.* **2**, special issue on "Quantum Gravity, Cosmology and Black Holes", ed. M. Bojowald, 291–301 (2009).

[7] A. Döring. Topos quantum logic and mixed states. In *Proceedings of the 6th International Workshop on Quantum Physics and Logic (QPL 2009), Oxford*, eds. B. Coecke, P. Panangaden, and P. Selinger. *Electronic Notes in Theoretical Computer Science* **270**(2) (2011).

[8] A. Döring. The physical interpretation of daseinisation. In *Deep Beauty*, ed. Hans Halvorson, 207–238, Cambridge University Press, New York (2011).

[9] A. Döring, and C.J. Isham. A topos foundation for theories of physics: I. Formal languages for physics. *J. Math. Phys* **49**, 053515 (2008).

[10] A. Döring, and C.J. Isham. A topos foundation for theories of physics: II. Daseinisation and the liberation of quantum theory. *J. Math. Phys* **49**, 053516 (2008).

[11] A. Döring, and C.J. Isham. A topos foundation for theories of physics: III. Quantum theory and the representation of physical quantities with arrows $\breve{A} : \underline{\Sigma} \to \mathcal{P}\mathbb{R}$. *J. Math. Phys* **49**, 053517 (2008).

[12] A. Döring, and C.J. Isham. A topos foundation for theories of physics: IV. Categories of systems. *J. Math. Phys* **49**, 053518 (2008).

[13] A. Döring, and C.J. Isham. 'What is a thing?': topos theory in the foundations of physics. In *New Structures for Physics*, ed. B. Coecke, Springer (2011).

[14] T. Erker, M. H. Escardó and K. Keimel. The way-below relation of function spaces over semantic domains. *Topology and its Applications* **89**, 61–74 (1998).

[15] C. Flori. A topos formulation of consistent histories. *Jour. Math. Phys* **51** 053527 (2009).

[16] G. Gierz, K.H. Hofmann, K. Keimel, J.D. Lawson, M.W. Mislove, D.S. Scott. *Continuous Lattices and Domains.* Encyclopedia of Mathematics and its Applications **93**, Cambridge University Press (2003).

[17] C. Heunen, N.P. Landsman, B. Spitters. A topos for algebraic quantum theory. *Comm. Math. Phys.* **291**, 63–110 (2009).

[18] C. Heunen, N.P. Landsman, B. Spitters. The Bohrification of operator algebras and quantum logic. *Synthese*, in press (2011).

[19] C. Heunen, N.P. Landsman, B. Spitters. Bohrification. In *Deep Beauty*, ed. H. Halvorson, 271–313, Cambridge University Press, New York (2011).

[20] C.J. Isham. Topos methods in the foundations of physics. In *Deep Beauty*, ed. Hans Halvorson, 187–205, Cambridge University Press, New York (2011).

[21] C.J. Isham. Topos theory and consistent histories: The internal logic of the set of all consistent sets. *Int. J. Theor. Phys.*, **36**, 785–814 (1997).

[22] C.J. Isham and J. Butterfield. A topos perspective on the Kochen-Specker theorem: I. Quantum states as generalised valuations. *Int. J. Theor. Phys.* **37**, 2669–2733 (1998).

[23] C.J. Isham and J. Butterfield. A topos perspective on the Kochen-Specker theorem: II. Conceptual aspects, and classical analogues. *Int. J. Theor. Phys.* **38**, 827–859 (1999).

[24] C.J. Isham, J. Hamilton and J. Butterfield. A topos perspective on the Kochen-Specker theorem: III. Von Neumann algebras as the base category. *Int. J. Theor. Phys.* **39**, 1413-1436 (2000).

[25] C.J. Isham and J. Butterfield. Some possible roles for topos theory in quantum theory and quantum gravity. *Found. Phys.* **30**, 1707–1735 (2000).

[26] C.J. Isham and J. Butterfield. A topos perspective on the Kochen-Specker theorem: IV. Interval valuations. *Int. J. Theor. Phys* **41**, 613–639 (2002).

[27] K. Martin, P. Panangaden. A domain of spacetime intervals in general relativity. *Commun. Math. Phys.* **267**, 563-586 (2006).

[28] D.S. Scott. Outline of a mathematical theory of computation. In Proceedings of *4th Annual Princeton Conference on Information Sciences and Systems*, 169–176 (1970).

[29] D.S. Scott. Continuous lattices. In *Toposes, Algebraic Geometry and Logic*, Lecture Notes in Mathematics **274**, Springer, 93–136 (1972).

[30] D.S. Scott. Formal semantics of programming languages. In *Lattice theory, data types and semantics*, Englewood Cliffs, Prentice-Hall, 66–106 (1972).

[31] S. Willard. *General Topology.* Addison-Wesley Series in Mathematics, Addison Wesley (1970).

[32] L. Ying-Ming, L. Ji-Hua. Solutions to two problems of J.D. Lawson and M. Mislove. *Topology and its Applications* **69**, 153–164 (1996).

Andreas Döring and Rui Soares Barbosa
Quantum Group
Department of Computer Science
University of Oxford
Wolfson Building
Parks Road
Oxford OX1 3QD
UK
e-mail: `andreas.doering@cs.ox.ac.uk`
`rui.soaresbarbosa@wolfson.ox.ac.uk`

Causal Boundary of Spacetimes: Revision and Applications to AdS/CFT Correspondence

José Luis Flores, Jónatan Herrera and Miguel Sánchez

Abstract. The aim of this work is to explain the status and role of the so-called *causal boundary* of a spacetime in Mathematical Physics and Differential Geometry. This includes: (a) the consistency of its latest redefinition, (b) its role as an intrinsic conformally invariant boundary in the AdS/CFT correspondence, (c) its relation with the Gromov and Cauchy boundaries for a Riemannian manifold, and (d) its further relation with boundaries in Finsler Geometry.

Mathematics Subject Classification (2010). Primary 83C75; Secondary 53C50, 83E30, 83C35, 53C60.

Keywords. Causal boundary, conformal boundary, AdS/CFT correspondence, plane wave, static spacetime, Gromov compactification, Cauchy completion, Busemann function, stationary spacetime, Finsler metric, Randers metric, pp-wave, arrival functional, Sturm-Liouville theory.

1. Introduction

Since the seminal paper by Geroch, Kronheimer and Penrose in 1972 (see [17]), the definition and computation of a *causal boundary* (c-boundary, for short) for reasonably well-behaved spacetimes have been a long-standing issue in Mathematical Relativity. This boundary would help to understand better the global causal structure of the spacetime, and would represent a "concrete place" where asymptotic conditions could be posed. Nevertheless, the lack of a satisfactory definition, as well as the difficulty of the computation of elements associated to it, has favored the usage of the so-called *conformal boundary*. This boundary is not intrinsic to the spacetime, and it has severe

The authors are partially supported by the Spanish MICINN Grant MTM2001-60731 and Regional J. Andalucía Grant P09-FQM-4496, both with FEDER funds. Also, JH is supported by Spanish MEC Grant AP2006-02237.

problems of existence and uniqueness. However, sometimes it is easy to compute and it remains as a non-general construction which is useful in many simple particular cases. Thus it is not strange that when developments in String Theory required a *holographic boundary* for the (bulk) spacetime, the conformal boundary was chosen. However, the limitations of this boundary for both fields, Mathematical Relativity and String Theory, have become apparent. Fortunately, there has been huge recent progress in the understanding of the c-boundary, which will be briefly reviewed now.

In [11], [12] the authors have developed carefully a new notion of the c-boundary. This notion was suggested by one of the authors in [31], by selecting elements already developed by many other authors: GKP [17], Budic & Sachs [5], Rácz [30], Szabados [34], Harris [20], Marolf & Ross [25], Low [23], Flores [9], etc. In Sections 2, 3 and 4 we review briefly some intuitive motivations for the c-boundary, our formal definition of this boundary and the arguments in favor of our choice, respectively. These arguments are of two different types. First, we state some formal hypotheses on the minimum properties which any admissible c-boundary must satisfy *a priori*. Then we argue that the alternatives to our choice are very few. Second, the nice properties satisfied *a posteriori* by our chosen c-boundary are stressed. In particular, the consistency with the conformal boundary (in natural cases) is explained. This becomes essential, as many previous redefinitions of the c-boundary have been disregarded because the c-boundary did not coincide with the (obvious and natural) conformal boundary in some concrete examples. Moreover, the c-boundary allows to formalize rigorously assertions which are commonly used for the conformal boundary in a loose way (see for example Corollary 4.5). For a complete development of the material in this part, the reader is referred to [31] (historical development and AdS/CFT correspondence on plane waves) and [11] (redefinition and analysis of the c-boundary).

In Sections 5, 6 and 7 the c-boundary is computed explicitly for three families of spacetimes (where a natural conformal embedding which would define the conformal boundary is not expected in general): static, stationary and pp-wave type spacetimes, respectively. However, our aim goes much further. In fact, the c-boundary of a static spacetime suggests a compactification for any (possibly incomplete) Riemannian manifold (as first studied in [10]), which can be compared with the classical Gromov compactification. This has its own interest for Riemannian Geometry (see Remarks 5.1, 5.4), and can be extended to reversible Finsler Geometry. Moreover, the c-boundary for a stationary spacetime suggests a compactification for any non-reversible Finsler manifold. We have developed not only this compactification but also Gromov's and the Cauchy completion for a non-symmetric distance. The natural relations between these boundaries demonstrate the power of the c-boundary approach. Finally, the c-boundary for pp-wave type spacetimes suggests to associate a boundary to certain Lagrangians, an idea which might be developed further. For the complete development of the material of this part, we refer to [12] (c-boundary for static and stationary spacetimes, and the relation to the Gromov and Cauchy boundaries), [6] (connection between causality of stationary spacetimes and Finsler Geometry) and [13] (boundary of pp-wave type spacetimes).

Summing up, it seems that now we have a consistent notion of c-boundary, based on "first principles". From the practical viewpoint, this boundary agrees with the conformal one in natural cases, it is closely connected to other boundaries in Riemannian and Finslerian Geometry (Cauchy, Gromov), and specific tools for its computation have been developed. Moreover, it allows a better understanding of the AdS/CFT correspondence: the c-boundary should be used in this correspondence because its construction is intrinsic, systematic and completely general; moreover, a concrete application for plane waves has been obtained. In conclusion, the c-boundary is available as a tool for both, Mathematical Relativity and String Theory.

2. Boundaries in Mathematical Relativity and String Theory

2.1. Boundaries for spacetimes

In Differential Geometry, there are different boundaries which can be assigned to a space, depending on the problem which is being studied. For example, Cauchy's and Gromov's boundaries are two natural ones in the Riemannian setting, which will be considered later.

For a spacetime (i.e. a connected time-oriented Lorentzian $(-,+,\dots,+)$ manifold), the most-used boundary in Mathematical Relativity is the so-called *(Penrose) conformal boundary.* In principle, the rough idea is simple. First one tries to find an open conformal embedding $i : V \hookrightarrow V_0$ of the original spacetime V in some ("aphysical") spacetime V_0 with (say, compact) closure $\overline{i(V)}$. Then the topological boundary $\partial(i(V)) \subset V_0$ is taken as the conformal boundary $\partial_i V$ of the spacetime.

However, such a procedure has obvious problems of both, existence (no general procedure to find the embedding i is prescribed) and uniqueness (different choices of i may yield conformal boundaries with different properties, recall the examples in [11, Sect. 4.5]). The problem of uniqueness led Ashtekar and Hansen [1] to formulate intrinsic conditions which define *asymptotic flatness* (including the metric level, not only the conformal one). García-Parrado and Senovilla introduced *isocausal extensions* [15], because their flexibility helps to ensure existence. The *abstract boundary* [33] by Scott and Szekeres is formulated to attach a unique point-set boundary to any spacetime by using embeddings. But even though all these approaches are interesting in several cases, the general question of existence and uniqueness for the conformal boundary remains unsolved. The present status can be summarized in the recent results by Chrusciel [7, Ths. 4.5, 3.1], roughly: (a) one can ensure that a maximal extension exists assuming that a conformal extension i already exists and some technical properties between any pair of lightlike geodesics hold, and (b) uniqueness will hold under technical assumptions, which, a priori, are *not* satisfied in simple cases, for instance when the boundary changes from timelike to lightlike around some point. In spite of these problems, the success of the conformal boundary comes from the existence of very natural conformal embeddings in many physically relevant spacetimes. So, in many cases, one can define *Penrose diagrams* which allow to visualize in a simple way the conformal structure of the spacetime. As we will see later, one remarkable property of the c-boundary is that it is expected to coincide with the conformal boundary in such favorable cases.

Finally, let us point out that Schmidt's *bundle boundary* [32] and Geroch's *geodesic boundary* [16] are other boundaries (non-conformally invariant) applicable to any Lorentzian manifold. They are not used as widely in Mathematical Relativity as the conformal one, and it is known that in some simple cases these boundaries may have undesirable properties (say, a point in the boundary may be non-T_1-separated from a point of the manifold, see [18]). But in any case, they are very appealing mathematical constructions which deserve to be studied further.

2.2. Intrinsic boundary for AdS/CFT correspondence

The AdS/CFT correspondence, or Maldacena duality, is a conjectured equivalence between a string theory on a *bulk* space (typically the product of anti-de Sitter AdS_n by a round sphere S^m, or by another compact manifold) and a quantum field theory without gravity on the *boundary* of the initial space, which behaves as a hologram of lower dimension.

The pertinent question here is which boundary must be chosen as the holographic one. Anti-de Sitter spacetime has a natural open conformal embedding in Lorentz-Minkowski space which yields a simple (and non-compact) conformal boundary. Such a boundary was chosen in the AdS/CFT correspondence as holographic boundary, but the situation changes when the holography on plane waves (put forward by Berenstein, Maldacena and Nastase [2]) is considered. In fact, Berenstein and Nastase [3] found a remarkable 1-dimensional behavior for the boundary of the plane wave obtained by Blau, Figueroa-O'Farrill, Hull and Papadopoulos [4] (a Penrose limit of a lightlike geodesic in $\mathrm{AdS}_5 \times S^5$ which rotates on S^5). While studying to what extent this behavior occurs in other cases, Marolf and Ross [24] realized that, among other limitations, the conformal boundary is not available for non-conformally flat plane waves. Under some minimum hypotheses on the notion of c-boundary, they proved that the c-boundary also is 1-dimensional, and this behavior holds for other plane waves as well. Therefore the usage of the c-boundary as the natural boundary for the AdS/CFT correspondence was proposed.

3. Formal definition of the c-boundary

3.1. Intuitive ideas

In what follows V will be a *strongly causal* spacetime, i.e., the topology of V is generated by the intersections between the chronological futures and pasts,

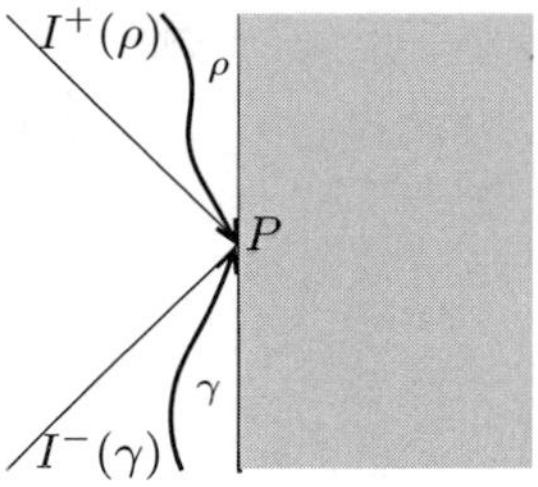

FIGURE 1. In $\mathbb{L}^2\backslash\{(t,x) : x \geq 0\}$, the boundary point P is represented by both, $I^-(\gamma)$ and $I^+(\rho)$.

$I^+(p) \cap I^-(q)$ for $p, q \in V$ (we use standard notation as in [21, 28, 26]; in particular, the chronological relation is denoted by $\ll$, i.e., $p \ll q$ iff there exists a future-directed timelike curve from p to q). Strong causality will turn out to be essential to ensure the compatibility of the topologies on V and its c-completion $\overline{V}$, among other desirable properties.

The purpose of the c-boundary ∂V is to attach a boundary endpoint $P \in \partial V$ to any inextensible future-directed (γ) or past-directed (ρ) timelike curve. The basic idea is that the boundary points are represented by the chronological past $I^-(\gamma)$ or future $I^+(\rho)$ of the curve. It is also clear intuitively that some boundary points must be represented by both, some $I^-(\gamma)$ and some $I^+(\rho)$ (see Fig. 1). So, ∂V would be the disjoint union of three parts: the *future infinity* $\partial_{+\infty} V$, reached by future-directed timelike curves but not by past-directed ones; the *past infinity* $\partial_{-\infty} V$, dual to the former; and the *timelike boundary* $\partial_0 V$, containing the points which are reached by both, future and past-directed timelike curves. The latter represents intrinsically "losses of global hyperbolicity" (the causal intersections $J^+(p) \cap J^-(q)$ are not always compact), i.e. *naked singularities.* From the viewpoint of hyperbolic partial differential equations, boundary conditions would be posed on the timelike boundary, while initial conditions would be posed on some appropriate acausal hypersurface of the spacetime, or even at future or past infinity.

Recall that the essential structure involved in the c-boundary construction is the conformal structure (causality) of the spacetime. In principle, the c-boundary does not yield directly information on singularities, except if they are conformally invariant; for example, this is the case for naked singularities. The name "singularity" may seem inappropriate, as in some cases the metric may be extensible through it (as in Fig. 1); however, such an extension will not be possible for all the representatives of the conformal class.

Next, our aim will be to give precise definitions of the c-completion $\overline{V} := V \cup \partial V$ at all the natural levels; i.e. point set, chronological and topological level.

3.2. Point set level and chronological level

As a preliminary step, we define the future and past (pre)completions $\hat{V}$, $\check{V}$, resp., of V. A subset $P \subset V$ is called a *past set* if $I^-(P) = P$. In this case, P is *decomposable* when $P = P_1 \cup P_2$ with P_i past sets such that $P_1 \neq P \neq P_2$, and is an *indecomposable past set (IP)* otherwise. The *future (pre)completion* $\hat{V}$ of V is defined as the set of all IPs. One can prove that if $P \in \hat{V}$ then either P is a *proper past set (PIP)*, i.e. $P = I^-(p)$ for some $p \in V$, or P is a *terminal past set (TIP)*, and then $P = I^-(\gamma)$ for some inextensible future-directed timelike curve γ. The *future (pre)boundary* $\hat{\partial}V$ of V is defined as the set of all TIPs. Therefore, the spacetime V, which is identifiable with the set of all PIPs, coincides with $\hat{V} \backslash \hat{\partial} V$. Analogously, one defines the *indecomposable future sets (IF)*, which constitute the *past (pre)completion* $\check{V}$. Each IF is either *terminal (TIF)* or *proper (PIF)*. The set of TIFs is the *past (pre)boundary* $\check{\partial}V$. The total precompletion is then $V^\sharp = (\hat{V} \cup \check{V})/\!\sim$, where $I^+(p) \sim I^-(p)$ $\forall p \in V$. Note, however, that this identification is insufficient, as naturally some TIPs and TIFs should represent the same boundary point (Fig. 1). To solve this problem, a new viewpoint on the boundary points is introduced.

Recall that for a point $p \in V$ one has $I^+(p) = \uparrow I^-(p)$ and $I^-(p) = \downarrow I^+(p)$, where the arrows $\uparrow, \downarrow$ denote the common future and past (e.g. $\uparrow P = I^+(\{q \in V : p \ll q \quad \forall p \in P\})$). Let $\hat{V}_\emptyset := \hat{V} \cup \{\emptyset\}$, $\check{V}_\emptyset := \check{V} \cup \{\emptyset\}$, and choose any $(P, F) \in (\hat{V}_\emptyset \times \check{V}_\emptyset) \backslash \{(\emptyset, \emptyset)\}$. We say that P is *S-related* to F if F is included and maximal in $\uparrow P$ and P is included and maximal in $\downarrow F$. By *maximal* we mean that no other $F' \in \check{V}_\emptyset$ (resp. $P' \in \hat{V}_\emptyset$) satisfies the stated property and includes strictly F (resp. P); notice that such a *maximal* set may be non-maximum (Fig. 3). In addition, if P (resp. F) does not satisfy the previous condition for any F (resp. P), we also say that P is S-related to $\emptyset$ (resp. $\emptyset$ is S-related to F); notice that $\emptyset$ is never S-related to itself. Now we define the *c-completion* $\overline{V}$ of V as:

$$\overline{V} := \{(P, F) \in \hat{V}_\emptyset \times \check{V}_\emptyset : P \sim_S F\}. \tag{3.1}$$

In particular, for $p \in V$ we have $I^-(p) \sim_S I^+(p)$; so V is regarded as the subset $\{(I^-(p), I^+(p)) : p \in V\} \subset \overline{V}$.

Recall that the c-completion has been defined only as a point set. However, it can be endowed with the following *chronological relation* (i.e. a binary relation which is transitive, anti-reflexive, without isolates and chronologically separable, see [11, Def. 2.1]) which extends naturally the chronological relation $\ll$ on V:

$$(P, F) \,\overline{\ll}\, (P', F') \Leftrightarrow F \cap P' \neq \emptyset$$

for all $(P, F), (P', F') \in \overline{V}$.

3.3. Topological level

In order to define the topology on $\overline{V}$, let us wonder how the convergence of sequences in V can be characterized in terms of chronological futures and pasts. More precisely, it is not difficult to realize that the following operator

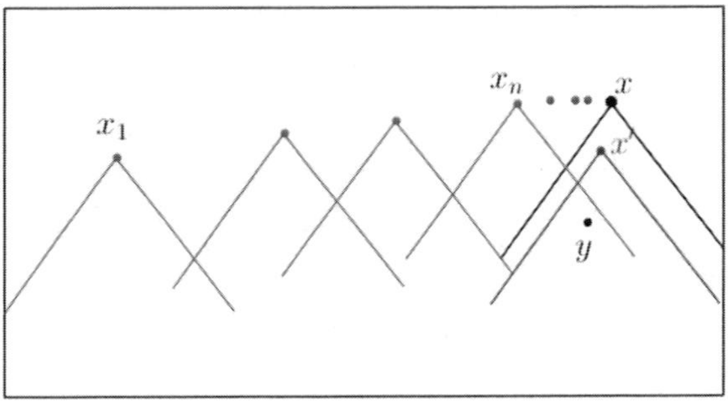

FIGURE 2. The sequence $\sigma = \{x_n\}$ converges to x. However, σ does not converge to x' as $P' = I^-(x)$ violates the second condition of the points in $\hat{L}(\sigma)$.

$\hat{L}$ determines the limit x (or the possible limits if the Hausdorffness of V were not assumed a priori) of any sequence $\sigma = \{x_n\}_n \subset V$ as follows:

$$x \in \hat{L}(\sigma) \Leftrightarrow \begin{cases} y \ll x \Rightarrow y \ll x_n, \\ I^-(x) \subset P'(\in \hat{V}),\ I^-(x) \neq P' \Rightarrow \exists z \in P' : z \not\ll x_n \end{cases} \quad \text{for large } n$$

(see Fig. 2). This operator suggests the natural notion of convergence for the future causal completion $\hat{V}$. The operator $\hat{L}$ determines the possible limits of any sequence $\sigma = \{P_n\}_n \subset \hat{V}$ as:

$$P \in \hat{L}(\sigma) \Leftrightarrow \begin{cases} P \subset LI(\{P_n\}) \\ P \text{ is a maximal IP in } LS(\{P_n\}) \end{cases}, \tag{3.2}$$

where LI and LS denote the set-theoretic lim inf and lim sup, respectively.

Now, one can check (see [11, Sect. 3.6]) that a topology is defined on $\hat{V}$ as follows: C is closed iff $\hat{L}(\sigma) \in C$ for any sequence $\sigma \subset C$. For this topology, $P \in \hat{L}(\sigma)$ only if the sequence σ converges to P. A dual operator $\check{L}$ will define a topology on $\check{V}$ analogously.

Finally, the topology on the completion $\overline{V}$ or *chronological topology* is introduced by means of a new operator L on sequences which defines the closed subsets and limits as above. Let $\sigma = \{(P_n, F_n)\} \subset \overline{V}$ $(\subset \hat{V}_\emptyset \times \check{V}_\emptyset)$, then:

$$(P, F) \in L(\sigma) \Leftrightarrow \begin{cases} \text{when } P \neq \emptyset,\ P \in \hat{L}(\{P_n\}) \\ \text{when } F \neq \emptyset,\ F \in \check{L}(\{F_n\}) \end{cases}. \tag{3.3}$$

This topology is sequential, but it may be non-Hausdorff (Fig. 3).

4. Plausibility of the boundary

4.1. Admissible redefinitions of the c-completion

Due to the long list of redefinitions of the c-boundary, the authors discussed extensively in [11] the plausible conditions to be fulfilled by such a definition

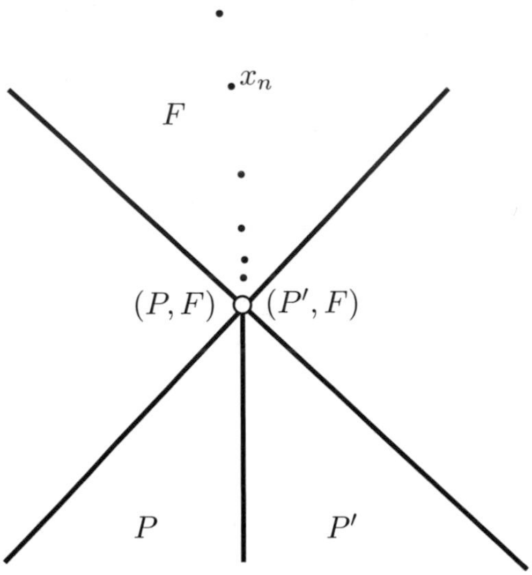

FIGURE 3. In $\mathbb{L}^2 \backslash \{(t, 0) : t \leq 0\}$, $P \sim_S F$ and $P' \sim_S F$. The sequence $\{(I^-(x_n), I^+(x_n))\}$ converges to both, (P, F) and (P', F).

in order to obtain a boundary which would be deemed satisfactory. We sketch such conditions and refer to the original article for more precise discussions.

4.1.1. Admissibility as a point set. The minimum conditions are:

- *Ambient*: the causal completion $\overline{V}$ must be a subset of $\hat{V}_\emptyset \times \check{V}_\emptyset$, and V must be recovered from the pairs $(I^+(p), I^-(p))$, $p \in V$.

  This states just a general framework which allows many different viewpoints. The possibility to consider a bigger ambient space (the product of the set of all the past sets by the set of all the future ones) was discussed, and disregarded, in [9].
- *Completeness*: every TIP and TIF in V is the component of some pair in the boundary of the completion.

  This collects the main motivation of the c-boundary: every inextensible timelike curve must determine a boundary point.
- *S-relation*: if $(P, F) \in \overline{V}$ then P and F must be S-related.

  This condition generalizes naturally what happens in the spacetime. One could weaken it by relaxing the definition of S-relation when some component is the empty set. About this, it is worth pointing out that the S-relation appears naturally under a minimality assumption for the c-boundary, among all the possible definitions satisfying that any future or past-directed timelike curve has some endpoint [9]. If one imposed the same condition for causal curves (not only timelike ones) then more pairings would be admitted. However, under very simple hypotheses, the existence of endpoints for timelike curves ensures the existence for causal curves too [11, Th. 3.35].

Among all the completions which satisfy these hypotheses, our choice (i.e. take all the S-related pairs) yields the biggest boundary. This choice is made because there exists no unique minimal choice; nevertheless, our choice is unique.

4.1.2. Admissibility as a chronological set. Let $\overline{V}$ be an admissible completion as a point set. A chronology $\tilde{\ll}$ on $\overline{V}$ is called:

- an *extended chronology* if
$$\begin{aligned} p \in P \quad &\Rightarrow \quad (I^-(p), I^+(p)) \,\tilde{\ll}\, (P, F) \\ q \in F \quad &\Rightarrow \quad (P, F) \,\tilde{\ll}\, (I^-(q), I^+(q)) \end{aligned}$$
for all $p, q \in V$ and $(P, F) \in \overline{V}$;
- an *admissible chronology* if $I^{\pm}((P, F)) \subset \overline{V}$ computed with $\tilde{\ll}$ satisfies:
$$I^-((P, F)) \cap V = P \;\text{ and }\; I^+((P, F)) \cap V = F \quad \forall\, (P, F) \in \overline{V}.$$

With these definitions, it is proved that our choice $\overline{\ll}$ for the chronology of $\overline{V}$ is the minimum extended chronology, as well as the unique admissible one compatible with a minimally well-behaved topology.

4.1.3. Admissibility as a topological space. In general the chronological futures and pasts of a spacetime generate the *Alexandrov topology* on the manifold, and this topology agrees with the manifold topology on strongly causal spacetimes. However, it is easy to realize that, for the completion, the analogous topology, called the *coarsely extended Alexandrov topology* (CEAT), is not enough (see [11, Figs. 1 (A), (B)]). Now, the conditions for the admissibility of a topology τ on $\overline{V}$ are:

- *Compatibility with Alexandrov*: τ is finer than CEAT. This only means that the chronological future and past of any $(P, F) \in \overline{V}$ must be open.
- *Compatibility of the limits with the empty set*: if $\{(P_n, F_n)\} \to (P, \emptyset)$ (resp. $(\emptyset, F)$) and $P \subset P' \subset \mathrm{LI}(\{P_n\})$ (resp. $F \subset F' \subset \mathrm{LI}(\{F_n\})$) for some $(P', F') \in \overline{V}$, then $(P', F') = (P, \emptyset)$ (resp. $(P', F') = (\emptyset, F)$). This property is natural; cf. also the extensive discussion in [11].
- *Minimality*: among all the topologies satisfying the previous two conditions, ordered by the relation "is finer than", τ is a minimal element. This requirement avoids the possibility to introduce arbitrarily new open subsets—for example, the discrete topology would not fulfill it.

Our choice of the chronological topology in (3.3) is justified because it is admissible at least when L is of first order (i.e., if $\{(P_n, F_n)\} \to (P, F)$ then $(P, F) \in L(P_n, F_n)$, which happens in all the known cases), and is then unique in very general circumstances [11, Th. 3.22]:

Theorem 4.1. *The chronological topology is the unique admissible sequential topology when L is of first order.*

4.1.4. Properties a posteriori. As a justification *a posteriori* of our choice of c-completion, recall that none of the previous choices in the literature satisfied the following simple properties simultaneously.

Theorem 4.2. *Under our definition of c-completion $\overline{V}$, one has:*

(a) *(consistence with the original motivation:) each future (resp. past) timelike curve γ converges to some pair (P, F) with $P = I^-(\gamma)$ (resp. $F = I^+(\gamma)$),*
(b) *the c-boundary ∂V is a closed subset of the completion $\overline{V}$,*
(c) *$\overline{V}$ is T_1,*
(d) *$\overline{V}$ is a sequential space,*
(e) *$I^{\pm}((P, F))$ (computed with $\overline{\ll}$) is open in $\overline{V}$.*

In the remainder of the present article, we will study other properties which also justify a posteriori our definition.

4.2. Relation to the conformal boundary

Consider an *envelopment* or open conformal embedding $i : V \hookrightarrow V_0$, as explained in Section 2.1. Let us see under which conditions the conformal completion $\overline{V}_i := \overline{i(V)}$ and boundary $\partial_i V$ can be identified with the causal ones.

4.2.1. Focusing the problem. First we have to define a chronological relation in $\overline{V}_i$. There are two natural options, which we will denote $p \ll_i q$ and $p \ll_i^S q$, for $p, q \in \overline{V}_i$. In the first option, $\ll_i$, one assumes that there exists some continuous curve $\gamma : [a, b] \to \overline{V}_i$ with $\gamma(a) = p$, $\gamma(b) = q$ and such that $\gamma|_{(a,b)}$ is future-directed smooth timelike and contained in $i(V)$. In the second option, $\ll_i^S$, one assumes additionally that γ is also smooth and timelike at the endpoints as a curve in V_0 (recall that $\partial_i V$ may be non-smooth). The second option is stronger than (and non-equivalent to) the first one, and it is not intrinsic to $\overline{V}_i$. Therefore we will choose the first one, $\ll_i$.

Notice that in order to relate the conformal and causal boundaries, one has to focus only on the *accessible (conformal) boundary* $\partial_i^* V$, that is, on those points of $\partial_i V$ which are endpoints of some timelike curve or, more precisely, which are $\ll_i$-chronologically related with some point in $i(V)$. Accordingly, one focuses on the natural *accessible completion* $\overline{V}_i^*$. The following is easy to prove:

Proposition 4.3. *If $\overline{V}_i$ is an embedded manifold with C^1 boundary then all the points of $\partial_i V$ are accessible (i.e. $\partial_i^* V = \partial_i V$).*

However, a point like the spacelike infinity i^0 in the standard conformal embedding of Lorentz-Minkowski space into the Einstein Static Universe $\mathbb{L}^n \hookrightarrow \mathrm{ESU}^n$ is not accessible. Obviously, such a point cannot correspond to any point of the c-boundary.

Analogously, we will require the envelopment i to be *chronologically complete*, that is: any inextensible (future- or past-directed) timelike curve in V has an endpoint in the conformal boundary. Obviously, otherwise the corresponding point in the causal boundary would not correspond to any point of the conformal one. If, for example, $\overline{V}_i^*$ is compact, this property is fulfilled. It is also fulfilled by $\mathbb{L}^n \hookrightarrow \mathrm{ESU}^n$ in spite of the fact that $\overline{V}_i^*$ is not compact in this case (as i^0 has been removed).

Summing up, if $\partial_i V^*$ is chronologically complete, then the natural projections

$$\hat{\pi} : \hat{\partial} V \to \partial_i^* V, \quad P = I^-(\gamma) \mapsto \text{the limit of } \gamma$$

$$\check{\pi} : \check{\partial} V \to \partial_i^* V, \quad F = I^+(\rho) \mapsto \text{the limit of } \rho$$

are well-defined. We say that the causal and (accessible) conformal completions are *identifiable*, and write $\overline{V} \equiv \overline{V}_i^*$, when the map

$$\pi : \partial V \to \partial_i^* V, \qquad \pi((P,F)) = \begin{cases} \hat{\pi}(P) & \text{if} \quad P \neq \emptyset \\ \check{\pi}(F) & \text{if} \quad F \neq \emptyset \end{cases}$$

satisfies the following conditions: (i) it is well-defined (i.e., $(P,F) \in \partial V$ and $P \neq \emptyset \neq F$ imply $\hat{\pi}(P) = \check{\pi}(F)$); (ii) π is bijective; and (iii) its natural extension to the completions $\overline{\pi} : \overline{V} \to \overline{V}_i^*$ is both, a homeomorphism and a chronological isomorphism.

4.2.2. Sufficient conditions for the identification $\overline{V} \equiv \overline{V}_i^*$. Even though there are simple examples where both boundaries differ, there are also simple conditions which ensure that they can be identified. They are studied systematically in [11] and, essentially, are fulfilled when the points of the conformal boundary are *regularly accessible*. We will not study this notion here, we will only state some consequences.

As explained above, given an envelopment there are two natural notions of chronological relation, $\ll_i$ and $\ll_i^S$. Analogously, there are two natural notions of causal relation, $\leq_i$ and $\leq_i^S$, defined as follows. We say $p \leq_i q$ if either $p = q$ or there exists some continuous curve $\gamma : [a,b] \to \overline{V}_i$ with $\gamma(a) = p$, $\gamma(b) = q$ such that γ is future-directed and *continuously causal* when regarded as a curve in V_0 (this means that $t_1 < t_2$ implies $\gamma(t_1) < \gamma(t_2)$ in V_0; it is natural to allow that γ is non-differentiable, as so may be $\partial_i V$). On the other hand, we say $p \leq_i^S q$ if there exists some curve $\gamma : [a,b] \to \overline{V}_i$ which is smooth and future-directed causal in V_0 with $\gamma(a) = p$, $\gamma(b) = q$. We will say that the envelopment is *chronologically* (resp. *causally*) *tame* if $\ll_i$ and $\ll_i^S$ are equal (resp. $\leq_i$ and $\leq_i^S$ are equal). Then we have:

Theorem 4.4. *Assume that an envelopment is chronologically complete, and satisfies that $\overline{V}_i$ is a C^1 manifold with boundary (in particular, $\overline{V}_i = \overline{V}_i^*$). Then the conformal and causal completions are identifiable ($\overline{V}_i^* \equiv \overline{V}$) if some of the following two conditions holds:*

1. *the envelopment is chronologically and causally tame,*
2. *the boundary $\partial_i V$ has no timelike points, that is, the tangent hyperplane to each point of the boundary is either spacelike or lightlike.*

In particular, this allows to prove rigourously the following frequently claimed assertion:

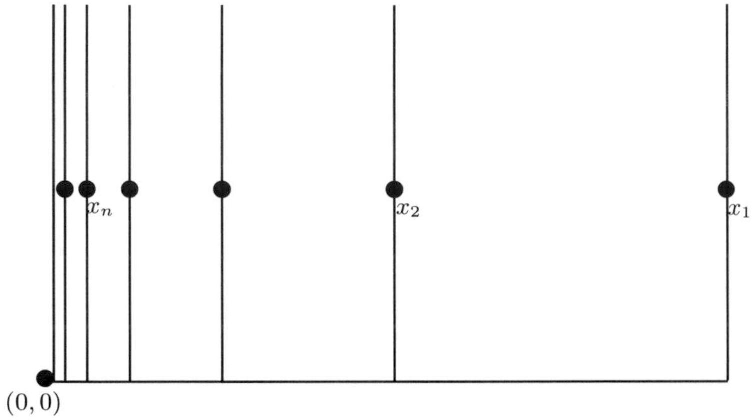

FIGURE 4. M_C and M_G are equal as point sets, and they include $(0,0)$ as a boundary point. However, their topologies differ: M_C is not locally compact, since sequences like $\{x_n\}$ violate the compactness of the balls centered at $(0,0)$.

Corollary 4.5. *Let V be a spacetime which admits a chronologically complete envelopment such that $\overline{V}_i$ is a C^1 manifold with boundary. Then V is globally hyperbolic iff $\partial_i V$ does not have any timelike point.*

5. Static spacetimes and boundaries on a Riemannian manifold

5.1. Classical Cauchy and Gromov boundaries on a Riemannian manifold

Any (connected, positive definite) Riemannian manifold (M, g), regarded as a metric space with distance d, admits the classical *Cauchy completion* M_C (say, by using classes of Cauchy sequences). Recall that such a completion is not a compactification and, even more, that $\partial_C M$ may be non-locally compact (see, for example, Fig. 4).

Gromov developed a *compactification* M_G for any complete Riemannian manifold. The idea is as follows. Let $\mathcal{L}_1(M,d)$ be the space of all the Lipschitz functions on (M,g) with Lipschitz constant equal to 1, and endow it with the topology of pointwise convergence. Consider the equivalence relation in $\mathcal{L}_1(M,d)$ which relates any two functions which only differ by an additive constant, and let $\mathcal{L}_1(M,d)/\mathbb{R}$ be the corresponding quotient topological space. Each point $x \in M$ determines the function $d_x \in \mathcal{L}_1(M,d)$ defined as $d_x(y) = d(x,y)$ for all $y \in M$. The class $[d_x] \in \mathcal{L}_1(M,d)/\mathbb{R}$ characterizes univocally x. Moreover, the map

$$M \hookrightarrow \mathcal{L}_1(M,d)/\mathbb{R}, \qquad x \mapsto [-d_x]$$

is a topological embedding, so we can regard M as a subset of $\mathcal{L}_1(M,d)/\mathbb{R}$. Then Gromov's completion M_G is just the closure of M in $\mathcal{L}_1(M,d)/\mathbb{R}$.

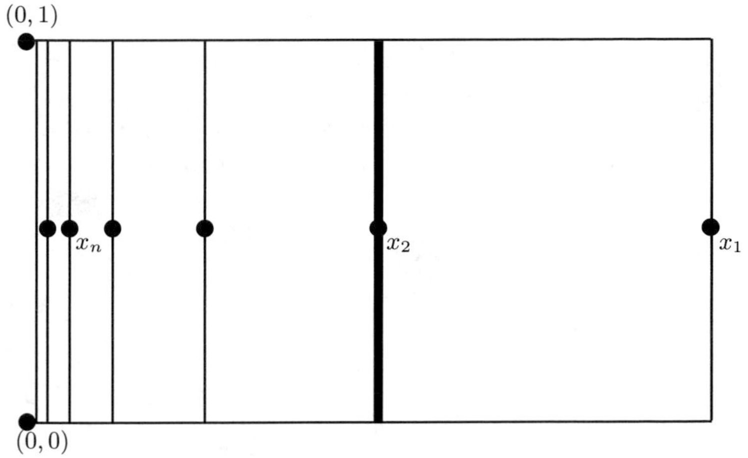

FIGURE 5. $\partial_C M$ includes only $(0,0)$ and $(0,1)$, but $\partial_G M$ includes the "segment" connecting both points. Observe that no point in that segment (but $(0,0)$ and $(0,1)$) is an endpoint of a curve in M.

Gromov's completion can be extended also to any (not necessarily complete) Riemannian manifold, and one has again a natural inclusion of the Cauchy completion M_C into M_G. However, this inclusion is only continuous in general (Fig. 4). In fact, *the inclusion is an embedding iff M_C is locally compact.*

Remark 5.1. M_G satisfies good topological properties, as being second countable, Hausdorff and compact. However, the following possibilities may occur:

(1) Even though all the points in Gromov's boundary $\partial_G M$ are limits of sequences in the manifold M, some of them *may not be the limit of any curve* (see Fig. 5).

(2) Define the *Cauchy-Gromov boundary* $\partial_{CG} M$ ($\subset \partial_G M$) as the set of boundary points which are limits of some *bounded* sequence $\sigma \subset M$ (i.e., σ is included in some ball), and the *proper Gromov boundary* as $\partial_{\mathcal{G}} M = \partial_G M \setminus \partial_{CG} M$. Clearly, the Cauchy boundary $\partial_C M$ is included in $\partial_{CG} M$, and *this inclusion may be strict* (Fig. 5)—apart from the fact that it may be not an embedding.

Summing up, we have:

Theorem 5.2. *Let (M, g) be a (possibly incomplete) Riemannian manifold. Then Gromov's completion M_G is a compact, Hausdorff and second countable space, and M is included in M_G as an open dense subset. Moreover, Gromov's boundary $\partial_G M$ is the disjoint union of the Cauchy-Gromov boundary $\partial_{CG} M$ and the proper Gromov boundary $\partial_{\mathcal{G}} M$. The former includes (perhaps strictly) the Cauchy boundary $\partial_C M$, and the inclusion $M_C \hookrightarrow M_G$*

(i) *is continuous;*
(ii) *is an embedding iff M_C is locally compact.*

Finally, we emphasize that Gromov's compactification coincides with Eberlein and O'Neill's compactification [8] for Hadamard manifolds (i.e. simply connected complete Riemannian manifolds of non-positive curvature). The latter can be described in terms of *Busemann functions* associated to *rays*. More generally, given any curve $c : [\alpha, \Omega) \to M$ with $|\dot{c}| \leq 1$ in any Riemannian manifold, we define the *Busemann function associated to c* as:

$$b_c(x) = \lim_{t \to \Omega} (t - d(x, c(t))) \qquad \forall x \in M. \tag{5.1}$$

One can check that this limit always exists and, if it is ∞ at some point, then it is constantly equal to ∞, $b_c \equiv \infty$. Let $B(M)$ be the set of all finite-valued Busemann functions. For a Hadamard manifold, $B(M)$ can be obtained by taking the curves c as rays, i.e. unit distance-minimizing complete half-geodesics (with $[\alpha, \Omega) = [0, \infty)$). The points of Eberlein and O'Neill's boundary can be described as the quotient $B(M)/\mathbb{R}$, where f, f' in $B(M)$ are identified iff $f - f'$ is a constant. Such a quotient is endowed with the *cone topology*, defined by using angles between rays.

5.2. Static spacetimes induce a new (Riemannian) Busemann boundary

Next we consider the following product spacetime:

$$V = (\mathbb{R} \times M, g_L = -dt^2 + \pi^* g), \tag{5.2}$$

where g is a Riemannian metric on M with distance d, and $\pi : \mathbb{R} \times M \to M$ the natural projection. Recall that any standard static spacetime is conformal to such a product and thus can be expressed as in (5.2) for the study of the c-boundary.

The chronological relation $\ll$ is easily characterized as follows:

$$(t_0, x_0) \ll (t_1, x_1) \Leftrightarrow d(x_0, x_1) < t_1 - t_0. \tag{5.3}$$

In order to compute an IP, notice that if $P = I^-(\gamma)$ for the future-directed timelike curve γ, this curve can be reparametrized with the t coordinate and thus we can write: $\gamma(t) = (t, c(t))$, $t \in [\alpha, \Omega)$, $|\dot{c}| < 1$. Then:

$$\begin{aligned} P &= \{(t_0, x_0) \in V : (t_0, x_0) \ll \gamma(t) \text{ for some } t \in [\alpha, \Omega)\} \\ &= \{(t_0, x_0) \in V : t_0 < t - d(x_0, c(t)) \text{ for some } t \in [\alpha, \Omega)\} \\ &= \{(t_0, x_0) \in V : t_0 < \lim_{t \to \Omega} (t - d(x_0, c(t)))\} \\ &= \{(t_0, x_0) \in V : t_0 < b_c(x_0)\}, \end{aligned} \tag{5.4}$$

where b_c is the Busemann function defined in (5.1). That is, there exists a natural identification between the future causal completion and the set of all Busemann functions

$$\hat{V} \equiv B(M) \cup \{f \equiv \infty\}. \tag{5.5}$$

Recall that $f \equiv \infty$ corresponds to the TIP $P = V$ obtained from the curve $\gamma(t) = (t, x_0)$, $t \in [\alpha, \infty)$, for any $x_0 \in M$. Moreover, the PIPs are associated to Busemann functions b_c with converging c (thus $\Omega < \infty$), and the TIPs with non-converging c.

Up to now, we have regarded $B(M) \cup \{f \equiv \infty\}$ as a point set. It is easy to check that all the functions in $B(M)$ are 1-Lipschitz, but we are not going to regard $B(M)$ as a topological subspace of $\mathcal{L}_1(M,d)$. On the contrary, as $\hat{V}$ was endowed with a topology (recall (3.2)), we can define a topology on $B(M) \cup \{f \equiv \infty\}$ so that the identification (5.5) becomes a homeomorphism (of course, this topology can be also defined directly on $B(M) \cup \{f \equiv \infty\}$, with no mention of the spacetime, see[1] [12, Sect. 5.2.2]). Now, we can consider also the quotient topological space $M_B = B(M)/\mathbb{R}$ obtained by identifying Busemann functions up to a constant. Again M can be regarded as a topological subspace of M_B. So, M_B is called the *Busemann completion* of M and $\partial_B M := M_B \backslash M$ the *Busemann boundary*.

Theorem 5.3. *Let M_B be the Busemann completion of a Riemannian manifold (M, g). Then:*

(1) *M_B is sequentially compact, and M is naturally embedded as an open dense subset in M_B.*
(2) *All the points in $\partial_B M$ can be reached as limits of curves in M.*
(3) *M_C is naturally included in M_B, and the inclusion is continuous (the Cauchy topology on M_C is finer).*
(4) *M_B is the disjoint union of M, the Cauchy boundary $\partial_C M$ and the* asymptotic Busemann boundary *$\partial_{\mathcal{B}} M$, where:*
  - *$M_C = M \cup \partial_C M$ corresponds to the subset of $B(M)$ constructed from curves c with $\Omega < \infty$ (identified with PIPs or TIPs in $\hat{V}$ which are S-related to some TIF).*
  - *$\partial_{\mathcal{B}} M$ corresponds to the subset of $B(M)$ constructed from curves c with $\Omega = \infty$ (unpaired TIPs different from $P = M$).*
(5) *M_B is T_1, and any non-T_2-related points must lie in $\partial_{\mathcal{B}} M$.*
(6) *$M_B \hookrightarrow M_G$ in a natural way, and the inverse of the inclusion is continuous (the topology on M_G is finer).*
(7) *M_B is homeomorphic to M_G iff M_B is Hausdorff. In particular, M_B coincides with Eberlein and O'Neill's compactification for Hadamard manifolds.*

Remark 5.4. The possible non-Hausdorff character of $\partial_B M$, which alerts on the discrepancy between M_B and M_G (by the assertion (7)), also points out a possible non-nice behavior of M_G. Recall that neither property (2) nor an analog to property (4) hold for M_G in general (the latter, as well as the Cauchy-Gromov boundary, may include strictly $\partial_C M$), see Remark 5.1.

In fact, the discrepancy comes from the fact that, in some sense, M_B is a compactification of M by (finite or asymptotic) directions, and M_G is a more analytical one.

[1]Therefore the Busemann boundary will be extensible to much more general spaces. In particular, it can be extended directly to *reversible* Finsler metrics. However, the Finsler case (including the non-reversible one) will be considered later.

5.3. Structure of the c-boundary for static spacetimes

Recall that, in a static spacetime, the map $(t, x) \mapsto (-t, x)$ is an isometry. So, the behavior of the past causal completion $\check{V}$ is completely analogous to the one studied for the future causal completion, and an identification as in (5.5) holds for $\check{V}$. Moreover, the S-relation, which defines the pairings in $\overline{V}$ (see above (3.1)), can be computed easily. In fact, the unique pair (P, F) with $P \neq \emptyset \neq F$ appears when both, P and F, regarded as elements of $B(M)$, project on points of the Cauchy boundary $\partial_C M \subset M_B$.

Remark 5.5. We say that a causal completion $\overline{V}$ is *simple as a point set* if any non-empty component of each pair $(P, F) \in \overline{V}$ determines univocally the pair (i.e. $(P, F), (P, F') \in \overline{V}$ with $P \neq \emptyset$ implies $F = F'$, and $(P, F), (P', F) \in \overline{V}$ with $F \neq \emptyset$ implies $P = P'$). In this case, and when L is of first order, we say that $\overline{V}$ is *simple* if, for any sequence $\{(P_n, F_n)\} \subset \overline{V}$, the convergence of the first component $\{P_n\}$ or the second one $\{F_n\}$ to a (non-empty) IP or IF implies the convergence of the sequence itself. Then the boundary of any static spacetime is always simple as a point set. If the Cauchy boundary $\partial_C M$ is locally compact, then one can prove that it is also simple.

Summing up (Fig. 6; see also [12, Cor. 6.28] for a more precise statement and definitions):

Theorem 5.6. *Let $V = \mathbb{R} \times M$ be a standard static spacetime. If M_C is locally compact and M_B Hausdorff, then the c-boundary ∂V coincides with the quotient topological space $(\hat{\partial} V \cup \check{\partial} V)/\sim_S$, where $\hat{\partial} V$ and $\check{\partial} V$ have the structures of cones with base $\partial_B M$ and apexes i^+, i^-, respectively.*

As a consequence, each point of $\partial_B M \backslash \partial_C M$ yields two lines in ∂V (which are horismotic *in a natural sense), one starting at i^- and the other one at i^+, and each point in $\partial_C M$ yields a single timelike line from i^- to i^+.*

6. Stationary spacetimes and boundaries of a Finsler manifold

6.1. Cauchy and Gromov boundaries of a Finsler manifold

6.1.1. Finsler elements. Recall that, essentially, a Finsler metric F on a manifold M assigns smoothly *a positively homogeneous tangent space norm* to each $p \in M$, where positive homogeneity means that the equality $F(\lambda v) = |\lambda| F(v)$ for $v \in TM$, $\lambda \in \mathbb{R}$ (of the usual norms) is only ensured now when $\lambda > 0$. So, given such an F, one can define the *reverse Finsler metric*: $F^{\mathrm{rev}}(v) := F(-v)$. Any Finsler metric induces a map $d : M \times M \to \mathbb{R}$, where $d(x, y)$ is the infimum of the lengths $l(c)$ of the curves $c : [a, b] \to M$ which start at x and end at y (by definition, $l(c) = \int_a^b F(\dot{c}(s))ds$). Such a d is a *generalized distance*, that is, it satisfies all the axioms of a distance except symmetry (i.e., d is a *quasidistance*) and, additionally, d satisfies the following condition: a sequence $\{x_n\} \subset M$ satisfies $d(x, x_n) \to 0$ iff $d(x_n, x) \to 0$. So, centered at any $x_0 \in M$, one can define the forward and backward closed balls of radius $r \geq 0$ as: $B^+(x_0, r) = \{x \in M : d(x_0, x) < r\}$ and $B^-(x_0, r) = \{x \in M : d(x, x_0) < r\}$, respectively. Both types of balls generate the topology of M. Recall that the backward balls for d are the forward ones for the distance d^{rev}, defined as $d^{\text{rev}}(x,y) = d(y,x)$ for all x, y. One can define also the *symmetrized distance* $d^s = (d+d^{\text{rev}})/2$, which is a (true) distance even though it cannot be obtained as a length space (i.e. as the infimum of lengths of connecting curves).

6.1.2. Cauchy completions. For any generalized distance d one can define Cauchy sequences and completions. Namely, $\{x_n\} \subset M$ is *(forward) Cauchy* if for all $\epsilon > 0$ there exists $n_0 \in \mathbb{N}$ such that $d(x_n, x_m) < \epsilon$ whenever $n_0 \leq n \leq m$; a *backward* Cauchy sequence is a forward one for d^{rev}. Accordingly, one has two types of completeness (forward, backward), of Cauchy boundaries ($\partial_C^+ M$, $\partial_C^- M$) and of Cauchy completions (M_C^+, M_C^-). Additionally, the boundary associated to the symmetrized distance d^s becomes $\partial_C^s M = \partial_C^- M \cap \partial_C^+ M$.

Remark 6.1. In the Riemannian case, the natural distance is extensible to a distance on the Cauchy completion. A generalized distance d can be also extended to M_C^+ in a natural way; however, this extension d_Q of d is *not* a generalized distance but only a quasidistance. As a consequence, d_Q generates two (in general different) topologies on M_C^+: one by using the forward balls and the other one by using the backward balls. The natural one (so that forward Cauchy sequences converge to the point represented by its class in M_C^+) is the one generated by the backward balls. This topology may be non-Hausdorff (in general, M_C^+ is only a T_0 space).

6.1.3. Finslerian Gromov completions. In order to define the Gromov completion for a (possibly incomplete) Finsler manifold, one has to start by defining an appropriate (non-symmetric) notion of *Lipschitzian function*, namely: a function f on M is *(forward) Lipschitzian* if it satisfies $f(y) - f(x) \leq d(x,y)$ for all $x, y \in M$. Then the *(forward) Gromov completion* is defined by following the same steps as in the Riemannian case. Let $\mathcal{L}_1^+(M,d)$ be the set of all Lipschitzian functions endowed with the pointwise convergence topology, and let $\mathcal{L}_1^+(M,d)/\mathbb{R}$ be its quotient by additive constants. In a natural way, M is included in this quotient, and M_G^+ is its closure. Replacing d by d^{rev}, one has the *backward Gromov completion* M_G^-. In a natural way, M_C^+ is included in M_G^+, and one can prove [12]:

Theorem 6.2. *Let (M,F) be a Finsler manifold. Gromov's completion M_G^+ satisfies all the properties stated for the Riemannian case in Theorem* 5.2 *except for (i) and (ii), which must be replaced by the following ones:*

(i') *The inclusion $M_C^+ \hookrightarrow M_G^+$ is continuous iff the natural topology generated in M_C^+ by the d_Q-backward balls is finer than the one generated by the forward balls (see Remark* 6.1*). In particular, this happens if d_Q is a generalized distance.*

(ii') *The inclusion above is an embedding if d_Q is a generalized distance and M_C^+ is locally compact.*

6.2. Stationary spacetimes induce a new (Finslerian) Busemann boundary

Consider the spacetime

$$V = (\mathbb{R} \times M, g_L = -dt^2 + \pi^*\omega \otimes dt + dt \otimes \pi^*\omega + \pi^* g), \tag{6.1}$$

where ω is a 1-form and g is a Riemannian metric, both on M, and $\pi : \mathbb{R} \times M \to M$ the natural projection. As in the static case, the conformal invariance of the c-boundary allows to consider expression (6.1) as the one of any standard stationary spacetime, without loss of generality.

The elements in (6.1) allow to construct the following Finsler metrics (of Randers type) on M:

$$F^{\pm}(v) = \sqrt{g(v,v) + \omega(v)^2} \pm \omega(v) \tag{6.2}$$

(recall that F^- is the reverse metric of F^+). It is not difficult to check that a curve $\gamma(t) = (t, c(t))$ (resp. $\gamma(t) = (-t, c(t))$), $t \in [\alpha, \Omega)$ is timelike and future (resp. past) directed iff $F^{\pm}(\dot{c}) < 1$. So the characterization of the chronological relation in (5.3) and the IPs in (5.4) hold now with d replaced by the generalized distance d^+ associated to F^+ (and with the arguments of the distances written in the same order as before, which is now essential). Thus the IPs are expressed in terms of Finslerian Busemann functions (defined as in (5.1) but with d^+), and the future causal completion $\hat{V}$ is identifiable to the set $B^+(M) \cup \{f \equiv \infty\}$ of all the F^+-Busemann functions as in (5.5). Moreover, this set is topologized to make it homeomorphic to $\hat{V}$, and the quotient topological space $M_B^+ = B^+(M)/\mathbb{R}$ is called the *(forward) Busemann completion* of (M, F^+). On the other hand, the construction can be carried out for any Finsler manifold (M, F) (where F is not necessarily a Randers metric as in (6.2)), just by defining the topology directly in terms of the Finslerian elements, with no reference to spacetimes (see [12, Sect. 5.2.2] for details). Summing up, one finds:

Theorem 6.3. *Let (M, F) be a Finsler manifold. The (forward) Busemann completion M_B^+ satisfies all the properties stated in Theorem* 5.3 *except property (3), which must be replaced by:*

(3') *M_C^+ is naturally included as a subset in M_B^+, and the inclusion is continuous if the natural topology generated in M_C^+ by the d_Q-backward balls is finer than the one generated by the forward balls (see Remark* 6.1*). In particular, this happens if d_Q is a generalized distance.*

Remark 6.4. Obviously, given a Finsler manifold (M, F), one can define also a Busemann completion M_B^- by using the reverse metric. For the Randers metrics $F^{\pm}$ in (6.2), the differences between both completions reflect asymmetries between the future and past causal boundaries, which did not appear in the static case. Moreover, the S-relation (defined for spacetimes) suggests a general relation between some points in M_C^+ with M_C^- (and M_B^+ with M_B^-) [12, Sects. 3.4, 6.3].

6.3. Structure of the c-boundary for stationary spacetimes

With all the previously obtained elements at hand, one can give a precise description of the c-boundary for any stationary spacetime. However, a notable complication appears with respect to the static case. In that case, the S-relation was trivial, and the c-boundary was always simple as a point set. In the stationary case, however, the S-relation may be highly non-trivial. In fact, it is worth emphasizing that all the classical difficulties for the definition of the S-relation and, then, for obtaining a consistent c-boundary, can be found in standard stationary spacetimes. So we give here the c-boundary of a stationary spacetime under some additional natural hypotheses, which imply that the c-boundary is *simple*, according to Remark 5.5 (see Fig. 6).

Theorem 6.5. *Let V be a standard stationary spacetime (as in* (6.1)*), and assume that the associated Finsler manifolds $(M, F^{\pm})$ satisfy: (a) the quasi-distances $d_Q^{\pm}$ are generalized distances, (b) the Cauchy completion M_C^s for the symmetrized distance is locally compact, and (c) the Busemann completions $M_B^{\pm}$ are Hausdorff.*

Then ∂V coincides with the quotient topological space $(\hat{\partial} V \cup \check{\partial} V)/\sim_S$, where $\hat{\partial} V$ and $\check{\partial} V$ have the structures of cones with bases $\partial_B^+ M$, $\partial_B^- M$ and apexes i^+, i^-, respectively.

As a consequence, the description of the c-boundary for the static case in Theorem 5.6 *holds with the following differences: (a) $\partial_B M \backslash \partial_C M$ must be replaced by $\partial_B^+ M \backslash \partial_C^s M$ (resp. $\partial_B^- M \backslash \partial_C^s M$) for the horismotic lines starting at i^+ (resp. i^-), and (b) the timelike lines correspond to points in $\partial_C^s M$ (instead of $\partial_C M$).*

In the general (non-simple) case, the points in $\partial_C^+ M \backslash \partial_C^s M$ may yield locally horismotic curves starting at i^+, which may be eventually identified with the corresponding ones starting at i^- for $\partial_C^- M \backslash \partial_C^s M$. We refer to [12] for a full description.

7. Boundary of pp-wave type spacetimes

The c-boundary of pp-wave type spacetimes was studied systematically in [13] and was explained further in the context of the AdS/CFT correspondence in [31]. Here we would just like to point out that the computation of this c-boundary is related to the critical curves associated to a Lagrangian $\mathcal{J}$ on a Riemannian manifold (M, g). So, in the line of previous cases, one could try to assign a boundary to this situation, even dropping the reference to the spacetime.

Consider the pp-wave type spacetimes:

$$V = (M_0 \times \mathbb{R}^2,\ g_L = \pi_0^* g_0 - F(x,u)\, du^2 - 2\, du\, dv),$$

where (M_0, g_0) is a Riemannian manifold, $\pi_0 : V \to M_0$ is the natural projection, (v, u) are the natural coordinates of $\mathbb{R}^2$ and $F : M_0 \times \mathbb{R} \to \mathbb{R}$ is a smooth function. These spacetimes include all the pp-waves (i.e. $(M_0, g_0) = \mathbb{R}^{n_0}$).

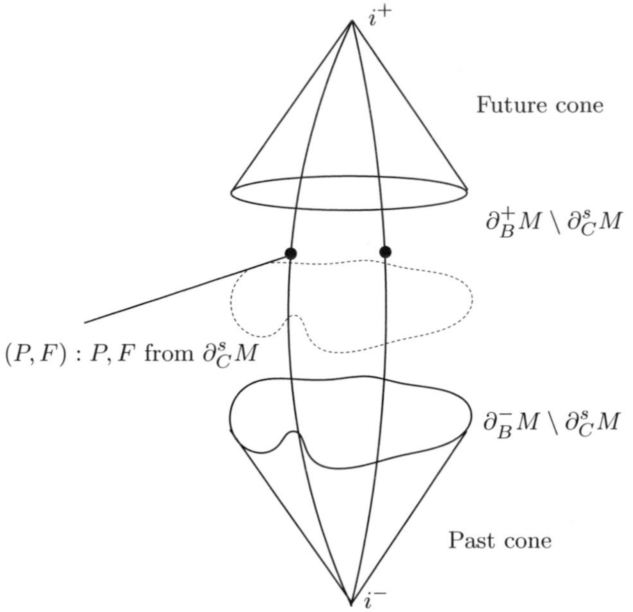

FIGURE 6. c-boundary for a standard stationary spacetime under the assumptions of Theorem 6.5. In the static case, one has the simplifications $\partial_B^{\pm} M \equiv \partial_B M$ and $\partial_C^+ M = \partial_C^- M = \partial_C^s M \equiv \partial_C M$.

Among pp-waves, the plane waves are characterized by F being quadratic in $x \in \mathbb{R}^{n_0}$.

In order to compute the TIPs and TIFs, one must take into account that the chronological relation $z_0 \ll z_1$, for $z_i = (x_i, v_i, u_i) \in V$, and thus $I^+(z_0)$ and $I^-(z_0)$, are controlled by the infimum of the "arrival functional":

$$\mathcal{J}_{u_0}^{\Delta u} : \mathcal{C} \to \mathbb{R}, \qquad \mathcal{J}_{u_0}^{\Delta u}(y) = \frac{1}{2} \int_0^{|\Delta u|} (|\dot{y}(s)|^2 - F(y(s), u_\nu(s)))ds,$$

where $\mathcal{C}$ is the space of curves in M_0 joining x_0 with x_1, $\Delta u = u_1 - u_0$, $u_\nu(s) = u_0 + \nu(\Delta u)s$, and $\nu = +1$ for $I^+(z_0)$ ($\nu = -1$ for $I^-(z_0)$). The careful study of this situation yields a precise computation of the boundary. A goal in this computation is to characterize when the TIPs "collapse" to a 1-dimensional future boundary, in agreement with the Berenstein & Nastase and Marolf & Ross results. That was achieved by studying a critical Sturm-Liouville problem. This depends on abstract technical conditions for $\mathcal{J}_{u_0}^{\Delta u}$, which turn out to depend on certain asymptotic properties of F (see Table 1).

We would like to emphasize that $\mathcal{J}_{u_0}^{\Delta u}$ is a Lagrangian which depends on the "classical time" u. This opens the possibility of further study, as remarked above.

Qualitative F	Boundary ∂V	Examples
F superquad. $-F$ at most quad.	no boundary	pp-waves yielding sine-Gordon string and related ones
λ-asymp. quad. $\lambda > 1/2$	1-dimensional, lightlike	plane waves with some eigenvalue $\mu_1 \geq \lambda^2/(1+u^2)$ for $\|u\|$ large
λ-asymp. quad. $\lambda \leq 1/2$	critical	pp-wave with $F(x,u) = \lambda^2 x^2/(1+u)^2$ (for $u > 0$)
subquadratic	no identif. in $\hat{\partial}V, \check{\partial}V$ expected higher dim.	(1) $\mathbb{L}^n$ and static type Mp-waves (2) plane waves with $-F$ quadratic

TABLE 1. Rough properties of the c-boundary for a pp-wave type spacetime depending on the qualitative behavior of F.

References

[1] A. Ashtekar, R. O. Hansen, A unified treatment of null and spatial infinity in general relativity. I. Universal structure, asymptotic symmetries, and conserved quantities at spatial infinity, *J. Math. Phys.* **19** (1978), no. 7, 1542–1566.

[2] D. Berenstein, J. M. Maldacena and H. Nastase, Strings in flat space and pp waves from N = 4 super Yang Mills, *J. High Energy Phys.* **0204** (2002) 013, 30 pp.

[3] D. Berenstein and H. Nastase, *On lightcone string field theory from super Yang-Mills and holography.* Available at arXiv:hep-th/0205048.

[4] M. Blau, J. Figueroa-O'Farrill, C. Hull, G. Papadopoulos, Penrose limits and maximal supersymmetry, *Class. Quant. Grav.* **19** (2002) L87–L95.

[5] R. Budic, R. K. Sachs, Causal boundaries for general relativistic spacetimes, *J. Math. Phys.* **15** (1974) 1302–1309.

[6] E. Caponio, M. A. Javaloyes, M. Sánchez, On the interplay between Lorentzian causality and Finsler metrics of Randers type, *Rev. Matem. Iberoam.* **27** (2011) 919–952.

[7] P. T. Chrusciel, Conformal boundary extensions of Lorentzian manifolds, *J. Diff. Geom.* **84** (2010) 19–44.

[8] P. Eberlein, B. O'Neill, Visibility manifolds, *Pacific J. Math.* **46** (1973) 45–109.

[9] J. L. Flores, The causal boundary of spacetimes revisited, *Commun. Math. Phys.* **276** (2007) 611–643.

[10] J. L. Flores, S. G. Harris, Topology of the causal boundary for standard static spacetimes, *Class. Quant. Grav.* **24** (2007), no. 5, 1211–1260.

[11] J. L. Flores, J. Herrera, M. Sánchez, On the final definition of the causal boundary and its relation with the conformal boundary, *Adv. Theor. Math. Phys.* **15** (2011), to appear. Available at arXiv:1001.3270v3.

[12] J. L. Flores, J. Herrera, M. Sánchez, *Gromov, Cauchy and causal boundaries for Riemannian, Finslerian and Lorentzian manifolds*, preprint 2010. Available at arXiv:1011.1154.

[13] J. L. Flores, M. Sánchez, The causal boundary of wave-type spacetimes, *J. High Energy Phys.* (2008), no. 3, 036, 43 pp.

[14] C. Frances, The conformal boundary of anti-de Sitter space-times, in: *AdS/CFT correspondence: Einstein metrics and their conformal boundaries*, 205–216, ed. O. Biquard, Euro. Math. Soc., Zürich, 2005.

[15] A. García-Parrado, J. M. M. Senovilla, Causal relationship: a new tool for the causal characterization of Lorentzian manifolds, *Class. Quantum Grav.* **20** (2003) 625–64.

[16] R. P. Geroch, Local characterization of singularities in general relativity, *J. Math. Phys.* **9** (1968) 450–465.

[17] R. P. Geroch, E. H. Kronheimer and R. Penrose, Ideal points in spacetime, *Proc. Roy. Soc. Lond. A* **237** (1972) 545–67.

[18] R. P. Geroch, C. B. Liang, R. M. Wald, Singular boundaries of space-times, *J. Math. Phys.* **23** (1982), no. 3, 432–435.

[19] S. G. Harris, Universality of the future chronological boundary, *J. Math. Phys.* **39** (1998), no. 10, 5427–5445.

[20] S. G. Harris, Topology of the future chronological boundary: universality for spacelike boundaries, *Classical Quantum Gravity* **17** (2000), no. 3, 551–603.

[21] S. W. Hawking, G. F. R. Ellis, *The large scale structure of space-time*, Cambridge University, Cambridge, 1973.

[22] Z. Q. Kuang, J. Z. Li, C. B. Liang, c-boundary of Taub's plane-symmetric static vacuum spacetime, *Phys. Rev. D* **33** (1986) 1533–1537.

[23] R. Low, The space of null geodesics (and a new causal boundary), in: *Analytical and numerical approaches to mathematical relativity*, 35-50, *Lecture Notes in Phys.* **692**, Springer, Berlin, 2006.

[24] D. Marolf, S. Ross, Plane waves: to infinity and beyond! *Class. Quant. Grav.* **19** (2002) 6289–6302.

[25] D. Marolf, S. R. Ross, A new recipe for causal completions, *Class. Quant. Grav.* **20** (2003) 4085–4117.

[26] E. Minguzzi, M. Sánchez, The causal hierarchy of spacetimes, in: *Recent developments in pseudo-Riemannian geometry* (2008) 359–418. ESI Lect. in Math. Phys., European Mathematical Society Publishing House. Available at gr-qc/0609119.

[27] O. Müller, M. Sánchez, Lorentzian manifolds isometrically embeddable in $\mathbb{L}^N$, *Trans. Amer. Math. Soc.* **363** (2011) 5367–5379.

[28] B. O'Neill, *Semi-Riemannian geometry. With applications to relativity*, Academic Press, 1983.

[29] S. Scott, P. Szekeres, The abstract boundary—a new approach to singularities of manifolds, *J. Geom. Phys.* **13** (1994), no. 3, 223–253.

[30] I. Rácz, Causal boundary of space-times, *Phys. Rev. D* **36** (1987) 1673–1675. Causal boundary for stably causal space-times, *Gen. Relat. Grav.* **20** (1988) 893–904.

[31] M. Sánchez, Causal boundaries and holography on wave type spacetimes, *Nonlinear Anal.*, **71** (2009) e1744–e1764.

[32] B. G. Schmidt, A new definition of singular points in general relativity. *Gen. Relat. Grav.* **1** (1970/71), no. 3, 269–280.

[33] S. Scott, P. Szekeres, The abstract boundary–a new approach to singularities of manifolds, *J. Geom. Phys.* **13** (1994) 223–253.

[34] L. B. Szabados, Causal boundary for strongly causal spaces, *Class. Quant. Grav.* **5** (1988) 121–34. Causal boundary for strongly causal spacetimes: II, *Class. Quant. Grav.* **6** (1989) 77–91.

José Luis Flores and Jónatan Herrera
Departamento de Álgebra, Geometría y Topología
Fac. Ciencias, Campus Teatinos s/n,
Universidad de Málaga
E 29071 Málaga
Spain
e-mail: `floresj@agt.cie.uma.es`
`jherrera@uma.es`

Miguel Sánchez
Departamento de Geometría y Topología
Fac. Ciencias, Campus de Fuentenueva s/n,
Universidad de Granada
E 18071 Granada
Spain
e-mail: `sanchezm@ugr.es`

Some Mathematical Aspects of the Hawking Effect for Rotating Black Holes

Dietrich Häfner

Abstract. The aim of this work is to give a mathematically rigorous description of the Hawking effect for fermions in the setting of the collapse of a rotating charged star.

Mathematics Subject Classification (2010). 35P25, 35Q75, 58J45, 83C47, 83C57.

Keywords. General relativity, Kerr-Newman metric, quantum field theory, Hawking effect, Dirac equation, scattering theory.

1. Introduction

It was in 1975 that S. W. Hawking published his famous paper about the creation of particles by black holes (see [10]). Later this effect was analyzed by other authors in more detail (see e.g. [13]), and we can say that the effect was well-understood from a physical point of view at the end of the 1970s. From a mathematical point of view, however, fundamental questions linked to the Hawking radiation, such as scattering theory for field equations on black hole space-times, were not addressed at that time.

Scattering theory for field equations on the Schwarzschild metric has been studied from a mathematical point of view since the 1980s, see e.g. [7]. In 1999 A. Bachelot [2] gave a mathematically rigorous description of the Hawking effect in the spherically symmetric case. The methods used by Dimock, Kay and Bachelot rely in an essential way on the spherical symmetry of the problem and can't be generalized to the rotating case.

The aim of the present work is to give a mathematically precise description of the Hawking effect for spin-1/2 fields in the setting of the collapse of a rotating charged star, see [9] for a detailed exposition. We show that an observer who is located far away from the black hole and at rest with respect to the Boyer-Lindquist coordinates observes the emergence of a thermal state when his proper time t goes to infinity. In the proof we use the results of [8] as well as their generalizations to the Kerr-Newman case in [4].

Let us give an idea of the theorem describing the effect. Let r_* be the Regge-Wheeler coordinate. We suppose that the boundary of the star is described by $r_* = z(t,\theta)$. The space-time is then given by

$$\mathcal{M}_{col} = \bigcup_t \Sigma_t^{col}, \qquad \Sigma_t^{col} = \big\{(t,r_*,\omega) \in \mathbb{R}_t \times \mathbb{R}_{r_*} \times S^2 \; ; \; r_* \geq z(t,\theta)\big\}.$$

The typical asymptotic behavior of $z(t,\theta)$ is ($A(\theta) > 0$, $\kappa_+ > 0$):

$$z(t,\theta) = -t - A(\theta)e^{-2\kappa_+ t} + B(\theta) + \mathcal{O}(e^{-4\kappa_+ t}), \quad t \to \infty.$$

Here κ_+ is the surface gravity of the outer horizon. Let

$$\mathcal{H}_t = L^2\big((\Sigma_t^{col}, \mathrm{dVol}); \mathbb{C}^4\big).$$

The Dirac equation can be written as

$$\partial_t \Psi = i\not{D}_t \Psi \quad + \quad \text{boundary condition.} \tag{1.1}$$

We will put an MIT boundary condition on the surface of the star. This condition makes the boundary totally reflecting, we refer to [9, Section 4.5] for details. The evolution of the Dirac field is then described by an isometric propagator $U(s,t) : \mathcal{H}_s \to \mathcal{H}_t$. The Dirac equation on the whole exterior Kerr-Newman space-time $\mathcal{M}_{BH}$ will be written as

$$\partial_t \Psi = i\not{D}\Psi.$$

Here $\not{D}$ is a selfadjoint operator on $\mathcal{H} = L^2((\mathbb{R}_{r_*} \times S^2, dr_* d\omega); \mathbb{C}^4)$. There exists an asymptotic velocity operator $P^{\pm}$ such that for all continuous functions J with $\lim_{|x|\to\infty} J(x) = 0$ we have

$$J(P^{\pm}) = \operatorname*{s-lim}_{t\to\pm\infty} e^{-it\not{D}} J\left(\frac{r_*}{t}\right) e^{it\not{D}}.$$

Let $\mathcal{U}_{col}(\mathcal{M}_{col})$ (resp. $\mathcal{U}_{BH}(\mathcal{M}_{BH})$) be the algebras of observables outside the collapsing body (resp. on the space-time describing the eternal black hole) generated by $\Psi_{col}^*(\Phi_1)\Psi_{col}(\Phi_2)$ (resp. $\Psi_{BH}^*(\Phi_1)\Psi_{BH}(\Phi_2)$). Here $\Psi_{col}(\Phi)$ (resp. $\Psi_{BH}(\Phi)$) are the quantum spin fields on $\mathcal{M}_{col}$ (resp. $\mathcal{M}_{BH}$). Let ω_{col} be a vacuum state on $\mathcal{U}_{col}(\mathcal{M}_{col})$; ω_{vac} a vacuum state on $\mathcal{U}_{BH}(\mathcal{M}_{BH})$ and $\omega_{Haw}^{\eta,\sigma}$ be a KMS-state on $\mathcal{U}_{BH}(\mathcal{M}_{BH})$ with inverse temperature $\sigma > 0$ and chemical potential $\mu = e^{\sigma\eta}$ (see Section 5 for details). For a function $\Phi \in C_0^\infty(\mathcal{M}_{BH})$ we define:

$$\Phi^T(t,r_*,\omega) = \Phi(t-T,r_*,\omega).$$

The theorem about the Hawking effect is the following:

Theorem 1.1 (Hawking effect). *Let*

$$\Phi_j \in \big(C_0^\infty(\mathcal{M}_{col})\big)^4, \quad j = 1,2.$$

Then we have

$$\begin{aligned}\lim_{T\to\infty} \omega_{col}\big(\Psi_{col}^*(\Phi_1^T)\Psi_{col}(\Phi_2^T)\big) &= \omega_{Haw}^{\eta,\sigma}\big(\Psi_{BH}^*(\mathbf{1}_{\mathbb{R}^+}(P^-)\Phi_1)\Psi_{BH}(\mathbf{1}_{\mathbb{R}^+}(P^-)\Phi_2)\big)\\ &\quad + \omega_{vac}\big(\Psi_{BH}^*(\mathbf{1}_{\mathbb{R}^-}(P^-)\Phi_1)\Psi_{BH}(\mathbf{1}_{\mathbb{R}^-}(P^-)\Phi_2)\big),\end{aligned} \tag{1.2}$$

$$T_{Haw} = 1/\sigma = \kappa_+/2\pi, \qquad \mu = e^{\sigma\eta}, \qquad \eta = \frac{qQr_+}{r_+^2 + a^2} + \frac{aD_\varphi}{r_+^2 + a^2}.$$

Here q is the charge of the field, Q the charge of the black hole, a is the angular momentum per unit mass of the black hole, $r_+ = M + \sqrt{M^2 - (a^2 + Q^2)}$ defines the outer event horizon, and κ_+ is the surface gravity of this horizon. The interpretation of (1.2) is the following. We start with a vacuum state which we evolve in the proper time of an observer at rest with respect to the Boyer-Lindquist coordinates. The limit as the proper time of this observer goes to infinity is a thermal state coming from the event horizon in formation and a vacuum state coming from infinity as expressed on the R.H.S. of (1.2). The Hawking effect comes from an infinite Doppler effect and the mixing of positive and negative frequencies. To explain this a little bit more, we describe the analytic problem behind the effect. Let $f(r_*, \omega) \in C_0^\infty(\mathbb{R} \times S^2)$. The key result about the Hawking effect is:

$$\lim_{T\to\infty} \left\| \mathbf{1}_{[0,\infty)}(\not{D}_0) U(0,T) f \right\|_0^2 = \left\langle \mathbf{1}_{\mathbb{R}^+}(P^-) f,\, \mu e^{\sigma \not{D}} (1 + \mu e^{\sigma \not{D}})^{-1} \mathbf{1}_{\mathbb{R}^+}(P^-) f \right\rangle + \left\| \mathbf{1}_{[0,\infty)}(\not{D}) \mathbf{1}_{\mathbb{R}^-}(P^-) f \right\|^2, \tag{1.3}$$

where μ, η, σ are as in the above theorem. Here $\langle .,. \rangle$ and $\|.\|$ (resp. $\|.\|_0$) are the standard inner product and norm on $\mathcal{H}$ (resp. $\mathcal{H}_0$). Equation (1.3) implies (1.2).

The term on the L.H.S. comes from the vacuum state we consider. We have to project onto the positive frequency solutions (see Section 5 for details). Note that in (1.3) we consider the time-reversed evolution. This comes from the quantization procedure. When time becomes large, the solution hits the surface of the star at a point closer and closer to the future event horizon. Figure 1 shows the situation for an asymptotic comparison dynamics, which satisfies Huygens' principle. For this asymptotic comparison dynamics the support of the solution concentrates more and more when time becomes large, which means that the frequency increases. The consequence of the change in frequency is that the system does not stay in the vacuum state.

2. The analytic problem

Let us consider a model where the eternal black hole is described by a static space-time (although the Kerr-Newman space-time is not even stationary, the problem will be essentially reduced to this kind of situation). Then the problem can be described as follows. Consider a Riemannian manifold Σ_0 with one asymptotically euclidean end and a boundary. The boundary will move when t becomes large asymptotically with the speed of light. The manifold at time t is denoted Σ_t. The "limit" manifold Σ is a manifold with two ends, one asymptotically euclidean and the other asymptotically hyperbolic (see Figure 2). The problem consists in evaluating the limit

$$\lim_{T\to\infty} \left\| \mathbf{1}_{[0,\infty)}(\not{D}_0) U(0,T) f \right\|_0,$$

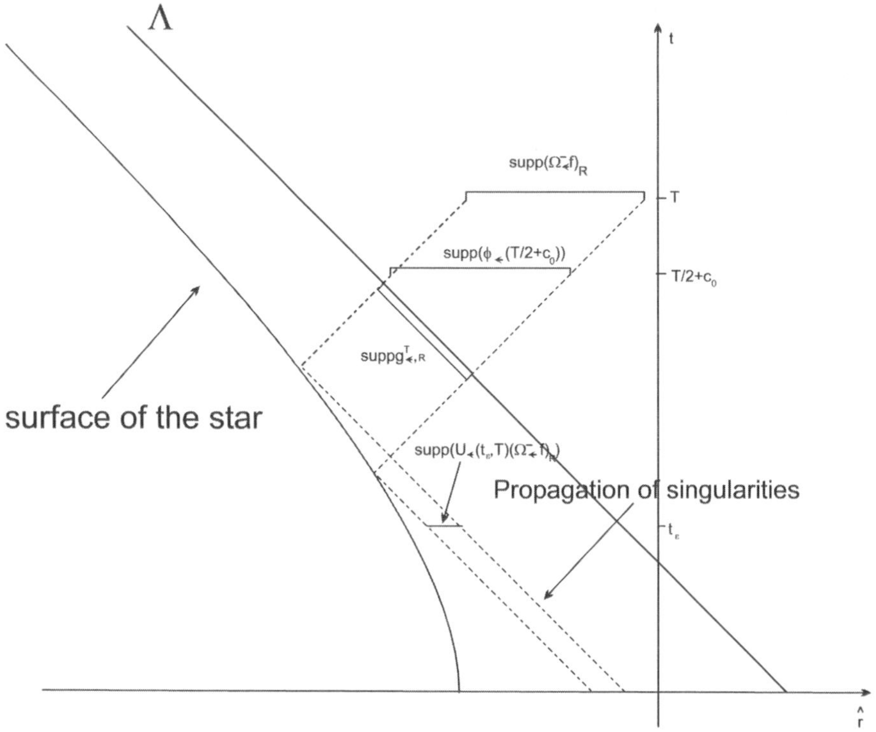

FIGURE 1. The collapse of the star

where $U(0, T)$ is the isometric propagator for the Dirac equation on the manifold with moving boundary and suitable boundary conditions and $\not{D}_0$ is the Dirac Hamiltonian at time $t = 0$. It is worth noting that the underlying scattering theory is not the scattering theory for the problem with moving boundary but the scattering theory on the "limit" manifold. It is shown in [9] that the result does not depend on the chiral angle in the MIT boundary condition. Note also that the boundary viewed in $\bigcup_t \{t\} \times \Sigma_t$ is only weakly timelike, a problem that has been rarely considered (but see [1]).

One of the problems for the description of the Hawking effect is to derive a reasonable model for the collapse of the star. We will suppose that the metric outside the collapsing star is always given by the Kerr-Newman metric. Whereas this is a genuine assumption in the rotational case, in the spherically symmetric case Birkhoff's theorem assures that the metric outside the star is the Reissner-Nordström metric. We will suppose that a point on the surface of the star will move along a curve which behaves asymptotically like a timelike geodesic with $L = \mathcal{Q} = \tilde{E} = 0$, where L is the angular momentum, $\tilde{E}$ the rotational energy and $\mathcal{Q}$ the Carter constant. The choice of geodesics is justified by the fact that the collapse creates the space-time, i.e., angular momenta and rotational energy should be zero with respect to the space-time. We will need an additional asymptotic condition on the collapse. It turns out that there is a natural coordinate system $(t, \hat{r}, \omega)$ associated to the collapse. In this coordinate system the surface of the star is described by $\hat{r} = \hat{z}(t, \theta)$. We need to assume the existence of a constant C such that

$$|\hat{z}(t,\theta) + t + C| \to 0, \ \ t \to \infty. \tag{2.1}$$

It can be checked that this asymptotic condition is fulfilled if we use the above geodesics for some appropriate initial condition. We think that it is more natural to impose a (symmetric) asymptotic condition than an initial condition. If we would allow in (2.1) a function $C(\theta)$ rather than a constant, the problem would become more difficult. Indeed one of the problems for treating the Hawking radiation in the rotational case is the high frequencies of the solution. In contrast with the spherically symmetric case, the difference between the Dirac operator and an operator with constant coefficients is near the horizon always a differential operator of order one[1]. This explains that in the high-energy regime we are interested in, the Dirac operator is not close to a constant-coefficient operator. Our method for proving (1.3) is to use scattering arguments to reduce the problem to a problem with a constant-coefficient operator, for which we can compute the radiation explicitly. If we do not impose a condition of type (2.1), then in all coordinate systems the solution has high frequencies, in the radial as well as in the angular directions. With condition (2.1) these high frequencies only occur in the radial direction. Our asymptotic comparison dynamics will differ from the real dynamics only by derivatives in angular directions and by potentials.

Let us now give some ideas of the proof of (1.3). We want to reduce the problem to the evaluation of a limit that can be explicitly computed. To do so, we use the asymptotic completeness results obtained in [8] and [4]. There exists a constant-coefficient operator $\not{D}_{\leftarrow}$ such that the following limits exist:

$$\begin{aligned} W_{\leftarrow}^{\pm} &:= \operatorname*{s-lim}_{t\to\pm\infty} e^{-it\not{D}} e^{it\not{D}_{\leftarrow}} \mathbf{1}_{\mathbb{R}^{\mp}}(P_{\leftarrow}^{\pm}), \\ \Omega_{\leftarrow}^{\pm} &:= \operatorname*{s-lim}_{t\to\pm\infty} e^{-it\not{D}_{\leftarrow}} e^{it\not{D}} \mathbf{1}_{\mathbb{R}^{\mp}}(P^{\pm}). \end{aligned}$$

Here $P_{\leftarrow}^{\pm}$ is the asymptotic velocity operator associated to the dynamics $e^{it\not{D}_{\leftarrow}}$. Then the R.H.S. of (1.3) equals:

$$\left\|\mathbf{1}_{[0,\infty)}(\not{D})\mathbf{1}_{\mathbb{R}^-}(P^-)f\right\|^2 + \langle \Omega_{\leftarrow}^- f,\ \mu e^{\sigma \not{D}_{\leftarrow}}(1+\mu e^{\sigma\not{D}_{\leftarrow}})^{-1}\Omega_{\leftarrow}^- f\rangle.$$

The aim is to show that the incoming part is:

$$\lim_{T\to\infty} \left\|\mathbf{1}_{[0,\infty)}(D_{\leftarrow,0})U_{\leftarrow}(0,T)\Omega_{\leftarrow}^- f\right\|_0^2 = \langle \Omega_{\leftarrow}^- f,\ \mu e^{\sigma\not{D}_{\leftarrow}}(1+\mu e^{\sigma\not{D}_{\leftarrow}})^{-1}\Omega_{\leftarrow}^- f\rangle,$$

where the equality can be shown by explicit calculation. Here $\not{D}_{\leftarrow,t}$ and $U_{\leftarrow}(s,t)$ are the asymptotic operator with boundary condition and the associated propagator. The outgoing part is easy to treat.

[1]In the spherically symmetric case we can diagonalize the operator. After diagonalization the difference is just a potential.

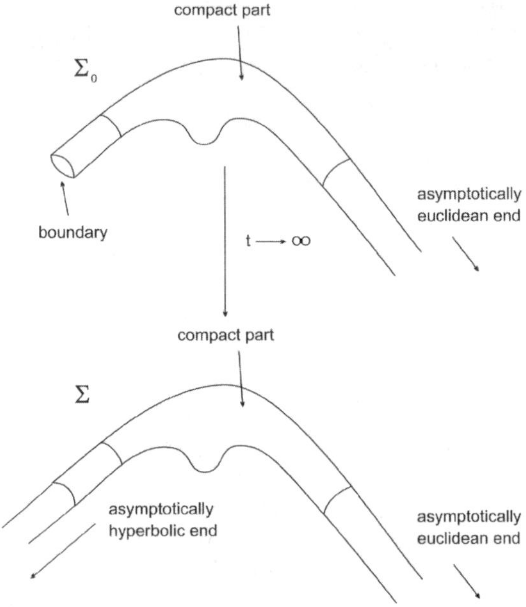

FIGURE 2. The manifold at time $t = 0$ Σ_0 and the limit manifold Σ.

Note that (1.3) is of course independent of the choice of the coordinate system and the tetrad, i.e., both sides of (1.3) are independent of these choices.

The proofs of all the results stated in this work can be found in [9]. The work is organized as follows:

- In Section 3 we present the model of the collapsing star. We first analyze geodesics in the Kerr-Newman space-time and explain how the Carter constant can be understood in terms of the Hamiltonian flow. We construct a coordinate system which is well adapted to the collapse.
- In Section 4 we describe classical Dirac fields. The form of the Dirac equation with an adequate choice of the Newman-Penrose tetrad is given. Scattering results are stated.
- Dirac quantum fields are discussed in Section 5. The theorem about the Hawking effect is formulated and discussed in Subsection 5.2.
- In Section 6 we give the main ideas of the proof.

3. The model of the collapsing star

The purpose of this section is to describe the model of the collapsing star. We will suppose that the metric outside the star is given by the Kerr-Newman metric. Geodesics are discussed in Subsection 3.2. We give a description of the Carter constant in terms of the associated Hamiltonian flow. A new position variable is introduced. In Subsection 3.3 we give the asymptotic behavior of the boundary of the star using this new position variable. We require that a point on the surface behaves asymptotically like incoming timelike geodesics with $L = Q = \tilde{E} = 0$, which are studied in Subsection 3.3.1.

3.1. The Kerr-Newman metric

We give a brief description of the Kerr-Newman metric, which describes an eternal rotating charged black hole. In Boyer-Lindquist coordinates, a Kerr-Newman black hole is described by a smooth 4-dimensional Lorentzian manifold $\mathcal{M}_{BH} = \mathbb{R}_t \times \mathbb{R}_r \times S^2_\omega$, whose space-time metric g and electromagnetic vector potential Φ_a are given by:

$$\begin{aligned}
g &= \left(1 + \frac{Q^2 - 2Mr}{\rho^2}\right) \mathrm{d}t^2 + \frac{2a \sin^2\theta\,(2Mr - Q^2)}{\rho^2}\,\mathrm{d}t\,\mathrm{d}\varphi \\
&\quad - \frac{\rho^2}{\Delta}\mathrm{d}r^2 - \rho^2\,\mathrm{d}\theta^2 - \frac{\sigma^2}{\rho^2}\sin^2\theta\,\mathrm{d}\varphi^2, \\
\rho^2 &= r^2 + a^2\cos^2\theta, \\
\Delta &= r^2 - 2Mr + a^2 + Q^2, \\
\sigma^2 &= (r^2 + a^2)^2 - a^2\Delta\sin^2\theta, \\
\Phi_a\,\mathrm{d}x^a &= -\frac{Qr}{\rho^2}(\mathrm{d}t - a\sin^2\theta\,\mathrm{d}\varphi).
\end{aligned} \tag{3.1}$$

Here M is the mass of the black hole, a its angular momentum per unit mass, and Q the charge of the black hole. If $Q = 0$, g reduces to the Kerr metric, and if $Q = a = 0$ we recover the Schwarzschild metric. The expression (3.1) of the Kerr metric has two types of singularities. While the set of points $\{\rho^2 = 0\}$ (the equatorial ring $\{r = 0,\ \theta = \pi/2\}$ of the $\{r = 0\}$ sphere) is a true curvature singularity, the spheres where Δ vanishes, called horizons, are mere coordinate singularities. We will consider in this work subextremal Kerr-Newman space-times, that is, we suppose $Q^2 + a^2 < M^2$. In this case Δ has two real roots:

$$r_\pm = M \pm \sqrt{M^2 - (a^2 + Q^2)}. \tag{3.2}$$

The spheres $\{r = r_-\}$ and $\{r = r_+\}$ are called event horizons. The two horizons separate $\mathcal{M}_{BH}$ into three connected components called Boyer-Lindquist blocks: B_I, B_{II}, B_{III} $(r_+ < r,\ r_- < r < r_+,\ r < r_-)$. No Boyer-Lindquist block is stationary, that is to say there exists no globally defined timelike Killing vector field on any given block. In the following $\mathcal{M}_{BH}$ will denote only block I of the Kerr-Newman space-time.

3.2. Some remarks about geodesics in the Kerr-Newman space-time

It is one of the most remarkable facts about the Kerr-Newman metric that there exist four first integrals for the geodesic equations. If γ is a geodesic in the Kerr-Newman space-time, then $p := \langle \gamma', \gamma' \rangle$ is conserved. The two Killing vector fields ∂_t, ∂_φ give two first integrals, the energy $E := \langle \gamma', \partial_t \rangle$

and the angular momentum $L := -\langle \gamma', \partial_\varphi \rangle$. There exists a fourth constant of motion, the so-called Carter constant $\mathcal{K}$ (see e.g. [3]). We will also use the Carter constant $\mathcal{Q} = \mathcal{K} - (L - aE)^2$, which has a somewhat more geometrical meaning, but gives in general more complicated formulas. Let

$$\mathbf{P} := (r^2 + a^2)E - aL, \qquad \mathbf{D} := L - aE\sin^2\theta. \tag{3.3}$$

Let $\Box_g$ be the d'Alembertian associated to the Kerr-Newman metric. We will consider the Hamiltonian flow of the principal symbol of $\frac{1}{2}\Box_g$ and then use the fact that a geodesic can be understood as the projection of the Hamiltonian flow on $\mathcal{M}_{BH}$. The principal symbol of $\frac{1}{2}\Box_g$ is:

$$P := \frac{1}{2\rho^2}\left(\frac{\sigma^2}{\Delta}\tau^2 - \frac{2a(Q^2 - 2Mr)}{\Delta}q_\varphi\tau - \frac{\Delta - a^2\sin^2\theta}{\Delta\sin^2\theta}q_\varphi^2 - \Delta|\xi|^2 - q_\theta^2\right). \tag{3.4}$$

Let

$$\mathcal{C}_p := \left\{(t, r, \theta, \varphi; \tau, \xi, q_\theta, q_\varphi) \ ; \ P(t, r, \theta, \varphi; \tau, \xi, q_\theta, q_\varphi) = \frac{1}{2}p\right\}.$$

Here $(\tau, \xi, q_\theta, q_\varphi)$ is dual to (t, r, θ, φ). We have the following:

Theorem 3.1. *Let $x_0 = (t_0, r_0, \varphi_0, \theta_0, \tau_0, \xi_0, q_{\theta_0}, q_{\varphi_0}) \in \mathcal{C}_p$, and let $x(s) = (t(s), r(s), \theta(s), \varphi(s); \tau(s), \xi(s), q_\theta(s), q_\varphi(s))$ be the associated Hamiltonian flow line. Then we have the following constants of motion:*

$$\begin{aligned} &p = 2P, \qquad E = \tau, \qquad L = -q_\varphi, \\ &\mathcal{K} = q_\theta^2 + \frac{\mathbf{D}^2}{\sin^2\theta} + pa^2\cos^2\theta = \frac{\mathbf{P}^2}{\Delta} - \Delta|\xi|^2 - pr^2, \end{aligned} \tag{3.5}$$

where $\mathbf{D}$, $\mathbf{P}$ *are defined in* (3.3).

The case $L = \mathcal{Q} = 0$ is of particular interest. Let γ be a null geodesic with energy $E > 0$, Carter constant $\mathcal{Q} = 0$, angular momentum $L = 0$ and given signs of ξ_0 and q_{θ_0}. We can associate a Hamiltonian flow line using (3.5) to define the initial data $\tau_0, \xi_0, q_{\theta_0}, q_{\varphi_0}$ given $t_0, r_0, \theta_0, \varphi_0$. From (3.5) we infer that ξ, q_θ do not change their signs. Note that γ is either in the equatorial plane or it does not cross it. Under the above conditions ξ (resp. q_θ) can be understood as a function of r (resp. θ) alone. In this case let k and l such that

$$\frac{dk(r)}{dr} = \frac{\xi(r)}{E}, \qquad l'(\theta) = \frac{q_\theta(\theta)}{E}, \qquad \hat{r} := k(r) + l(\theta). \tag{3.6}$$

It is easy to check that $(t, \hat{r}, \omega)$ is a coordinate system on block I.

Lemma 3.2. *We have:*

$$\frac{\partial \hat{r}}{\partial t} = -1 \quad \text{along } \gamma, \tag{3.7}$$

where t is the Boyer-Lindquist time.

We will suppose from now on that our construction is based on incoming null geodesics. From the above lemma follows that for given sign of q_{θ_0} the surfaces $\mathcal{C}^{c,\pm} = \{(t, r_*, \theta, \varphi) \ ; \ \pm t = \hat{r}(r_*, \theta) + c\}$ are characteristic.

Remark 3.3. The variable $\hat{r}$ is a Bondi-Sachs type coordinate. This coordinate system is discussed in some detail in [12]. As in [12], we will call the null geodesics with $L = \mathcal{Q} = 0$ *simple null geodesics (SNGs)*.

3.3. The model of the collapsing star

Let $\mathcal{S}_0$ be the surface of the star at time $t = 0$. We suppose that elements $x_0 \in \mathcal{S}_0$ will move along curves which behave asymptotically like certain incoming timelike geodesics γ_p. All these geodesics should have the same energy E, angular momentum L, Carter constant $\mathcal{K}$ (resp. $\mathcal{Q} = \mathcal{K} - (L - aE)^2$) and "mass" $p := \langle \gamma_p', \gamma_p' \rangle$. We will suppose:

(A) The angular momentum L vanishes: $L = 0$.
(B) The rotational energy vanishes: $\tilde{E} = a^2(E^2 - p) = 0$.
(C) The total angular momentum about the axis of symmetry vanishes: $\mathcal{Q} = 0$.

The conditions (A)–(C) are imposed by the fact that the collapse itself creates the space-time, so that momenta and rotational energy should be zero with respect to the space-time.

3.3.1. Timelike geodesics with $L = \mathcal{Q} = \tilde{E} = 0$. Next, we will study the above family of geodesics. The starting point of the geodesic is denoted $(0, r_0, \theta_0, \varphi_0)$. Given a point in the space-time, the conditions (A)–(C) define a unique cotangent vector provided one adds the condition that the corresponding tangent vector is incoming. The choice of p is irrelevant because it just corresponds to a normalization of the proper time.

Lemma 3.4. *Let γ_p be a geodesic as above. Along this geodesic we have:*

$$\frac{\partial \theta}{\partial t} = 0, \qquad \frac{\partial \varphi}{\partial t} = \frac{a(2Mr - Q^2)}{\sigma^2}, \tag{3.8}$$

where t is the Boyer-Lindquist time.

The function $\frac{\partial \varphi}{\partial t} = \frac{a(2Mr-Q^2)}{\sigma^2}$ is usually called the *local angular velocity* of the space-time. Our next aim is to adapt our coordinate system to the collapse of the star. The most natural way of doing this is to choose an incoming null geodesic γ with $L = Q = 0$ and then use the Bondi-Sachs type coordinate as in the previous subsection. In addition we want that $k(r_*)$ behaves like r_* when $r_* \to -\infty$. We therefore put:

$$k(r_*) = r_* + \int_{-\infty}^{r_*} \left(\sqrt{1 - \frac{a^2 \Delta(s)}{(r(s)^2 + a^2)^2}} - 1 \right) \mathrm{d}s, \qquad l(\theta) = a \sin \theta. \tag{3.9}$$

The choice of the sign of l' is not important, the opposite sign would have been possible. Recall that $\cos \theta$ does not change its sign along a null geodesic with $L = \mathcal{Q} = 0$. We put $\hat{r} = k(r_*) + l(\theta)$, and by Lemma 3.2 we have $\frac{\partial \hat{r}}{\partial t} = -1$ along γ.

In order to describe the model of the collapsing star we have to evaluate $\frac{\partial \hat{r}}{\partial t}$ along γ_p. Note that $\theta(t) = \theta_0 = \text{const}$ along γ_p.

Lemma 3.5. *There exist smooth functions $\hat{A}(\theta, r_0) > 0$, $\hat{B}(\theta, r_0)$ such that along γ_p we have uniformly in θ, $r_0 \in [r_1, r_2] \subset (r_+, \infty)$:*

$$\hat{r} = -t - \hat{A}(\theta, r_0)e^{-2\kappa_+ t} + \hat{B}(\theta, r_0) + \mathcal{O}(e^{-4\kappa_+ t}), \quad t \to \infty, \tag{3.10}$$

where $\kappa_+ = \frac{r_+ - r_-}{2(r_+^2 + a^2)}$ is the surface gravity of the outer horizon.

3.3.2. Assumptions. We will suppose that the surface at time $t = 0$ is given in the $(t, \hat{r}, \theta, \varphi)$ coordinate system by $\mathcal{S}_0 = \{(\hat{r}_0(\theta_0), \theta_0, \varphi_0) \; ; \; (\theta_0, \varphi_0) \in S^2\}$, where $\hat{r}_0(\theta_0)$ is a smooth function. As $\hat{r}_0$ does not depend on φ_0, we will suppose that $\hat{z}(t, \theta_0, \varphi_0)$ will be independent of $\varphi_0 : \hat{z}(t, \theta_0, \varphi_0) = \hat{z}(t, \theta_0) = \hat{z}(t, \theta)$ as this is the case for $\hat{r}(t)$ along timelike geodesics with $L = \mathcal{Q} = 0$. We suppose that $\hat{z}(t, \theta)$ satisfies the asymptotics (3.10) with $\hat{B}(\theta, r_0)$ independent of θ, see [9] for the precise assumptions. Thus the surface of the star is given by:

$$\mathcal{S} = \{(t, \hat{z}(t, \theta), \omega) \; ; \; t \in \mathbb{R}, \, \omega \in S^2\}. \tag{3.11}$$

The space-time of the collapsing star is given by:

$$\mathcal{M}_{col} = \{(t, \hat{r}, \theta, \varphi) \; ; \; \hat{r} \geq \hat{z}(t, \theta)\}.$$

We will also note:

$$\Sigma_t^{col} = \{(\hat{r}, \theta, \varphi) \; ; \; \hat{r} \geq \hat{z}(t, \theta)\}, \quad \text{thus } \mathcal{M}_{col} = \bigcup_t \Sigma_t^{col}.$$

The asymptotic form (3.10) with $\hat{B}(\theta, r_0)$ can be achieved by incoming timelike geodesics with $L = \mathcal{Q} = \tilde{E} = 0$, see [9, Lemma 3.5].

4. Classical Dirac fields

In this section we will state some results on classical Dirac fields and explain in particular how to overcome the difficulty linked to the high-frequency problem. The key point is the appropriate choice of a Newman-Penrose tetrad. Let $(\mathcal{M}, g)$ be a general globally hyperbolic space-time. Using the Newman-Penrose formalism, the Dirac equation can be expressed as a system of partial differential equations with respect to a coordinate basis. This formalism is based on the choice of a null tetrad, i.e. a set of four vector fields l^a, n^a, m^a and $\bar{m}^a$, the first two being real and future oriented, $\bar{m}^a$ being the complex conjugate of m^a, such that all four vector fields are null and m^a is orthogonal to l^a and n^a, that is to say, $l_a l^a = n_a n^a = m_a m^a = l_a m^a = n_a m^a = 0$. The tetrad is said to be normalized if in addition $l_a n^a = 1$, $m_a \bar{m}^a = -1$. The vectors l^a and n^a usually describe "dynamic" or scattering directions, i.e. directions along which light rays may escape towards infinity (or more generally asymptotic regions corresponding to scattering channels). The vector m^a tends to have, at least spatially, bounded integral curves; typically m^a and $\bar{m}^a$ generate rotations. To this Newman-Penrose tetrad is associated a spin frame. The Dirac equation is then a system of partial differential equations for the components of the spinor in this spin frame. For the Weyl equation ($m = 0$) we obtain:

$$\begin{cases} n^a\partial_a\phi_0 - m^a\partial_a\phi_1 + (\mu-\gamma)\phi_0 + (\tau-\beta)\phi_1 = 0, \\ l^a\partial_a\phi_1 - \bar{m}^a\partial_a\phi_1 + (\alpha-\pi)\phi_0 + (\epsilon-\tilde{\rho})\phi_1 = 0. \end{cases}$$

The μ, γ etc. are the so-called spin coefficients, for example

$$\mu = -\bar{m}^a\delta n_a, \qquad \delta = m^a\nabla_a.$$

For the formulas of the spin coefficients and details about the Newman-Penrose formalism see e.g. [11].

Our first result is that there exists a tetrad well-adapted to the high-frequency problem. Let $\mathcal{H} = L^2((\mathbb{R}_{\hat{r}} \times S^2, \mathrm{d}\hat{r}\,\mathrm{d}\omega); \mathbb{C}^4)$, $\Gamma^1 = Diag(1,-1,-1,1)$.

Proposition 4.1. *There exists a Newman-Penrose tetrad such that the Dirac equation in the Kerr-Newman space-time can be written as*

$$\partial_t\psi = iH\psi; \qquad H = \Gamma^1 D_{\hat{r}} + P_\omega + W,$$

where W is a real potential and P_ω is a differential operator of order one with derivatives only in the angular directions. The operator H is selfadjoint with domain $D(H) = \{v \in \mathcal{H}\ ;\ Hv \in \mathcal{H}\}$.

Remark 4.2.

(i) l^a, n^a are chosen to be generators of the simple null geodesics.

(ii) Note that the local velocity in $\hat{r}$ direction is ± 1:

$$v = [\hat{r}, H] = \Gamma^1.$$

This comes from the fact that $\frac{\partial\hat{r}}{\partial t} = \pm 1$ along simple null geodesics ($\pm$ depending on whether the geodesic is incoming or outgoing).

(iii) Whereas the above tetrad is well-adapted to the high-frequency analysis, it is not the good choice for the proof of asymptotic completeness results. See [8] for an adequate choice.

(iv) ψ are the components of the spinor which is multiplied by some density.

Let

$$\begin{aligned} H_{\leftarrow} &= \Gamma^1 D_{\hat{r}} - \frac{a}{r_+^2 + a^2} D_\varphi - \frac{qQr_+}{r_+^2 + a^2}, \\ \mathcal{H}^+ &= \big\{v = (v_1, v_2, v_3, v_4) \in \mathcal{H}\ ;\ v_1 = v_4 = 0\big\}, \\ \mathcal{H}^- &= \big\{v = (v_1, v_2, v_3, v_4) \in \mathcal{H}\ ;\ v_2 = v_3 = 0\big\}. \end{aligned}$$

The operator $H_{\leftarrow}$ is selfadjoint on $\mathcal{H}$ with domain

$$D(H_{\leftarrow}) = \big\{v \in \mathcal{H}\ ;\ H_{\leftarrow}v \in \mathcal{H}\big\}.$$

Remark 4.3. The above operator is our comparison dynamics. Note that the difference between the full dynamics and the comparison dynamics is a differential operator with derivatives only in the angular directions. The high frequencies will only be present in the $\hat{r}$ directions; this solves the high-frequency problem.

Proposition 4.4. *There exist selfadjoint operators $P^\pm$ such that for all $g \in C(\mathbb{R})$ with $\lim_{|x|\to\infty} g(x) = 0$, we have:*

$$g(P^\pm) = \operatorname*{s-lim}_{t\to\pm\infty} e^{-itH} g\left(\frac{\hat{r}}{t}\right) e^{itH}. \tag{4.1}$$

Let $P_{\mathcal{H}^\mp}$ be the projections from $\mathcal{H}$ to $\mathcal{H}^\mp$.

Theorem 4.5. *The wave operators*

$$\begin{aligned} W_{\leftarrow}^{\pm} &= \operatorname*{s-lim}_{t\to\pm\infty} e^{-itH} e^{itH_\leftarrow} P_{\mathcal{H}^\mp}, \\ \Omega_{\leftarrow}^{\pm} &= \operatorname*{s-lim}_{t\to\pm\infty} e^{-itH_\leftarrow} e^{itH}\, \mathbf{1}_{\mathbb{R}^\mp}(P^\pm) \end{aligned}$$

exist.

Using the above tetrad, the Dirac equation with MIT boundary condition on the surface of the star (chiral angle ν) can be written in the following form:

$$\left.\begin{aligned} \partial_t \Psi &= iH\Psi, \qquad \hat{z}(t,\theta) < \hat{r}, \\ \big(\textstyle\sum_{\hat{\mu}\in\{t,\hat{r},\theta,\varphi\}} \mathcal{N}_{\hat{\mu}} \hat{\gamma}^{\hat{\mu}}\big)\Psi\big(t, \hat{z}(t,\theta),\omega\big) &= -ie^{-i\nu\gamma^5}\Psi\big(t,\hat{z}(t,\theta),\omega\big), \\ \Psi(t=s,.) &= \Psi_s(.). \end{aligned}\right\} \tag{4.2}$$

Here $\mathcal{N}_{\hat{\mu}}$ are the coordinates of the conormal, $\hat{\gamma}^{\hat{\mu}}$ are some appropriate Dirac matrices and $\gamma^5 = Diag(1,1,-1,-1)$. Let

$$\mathcal{H}_t = L^2\big(\big(\{(\hat{r},\omega) \in \mathbb{R}\times S^2\ ;\ \hat{r} \geq \hat{z}(t,\theta)\}, \mathrm{d}\hat{r}\,\mathrm{d}\omega\big);\ \mathbb{C}^4\big).$$

Proposition 4.6. *The equation* (4.2) *can be solved by a unitary propagator* $U(t,s) : \mathcal{H}_s \to \mathcal{H}_t$.

5. Dirac quantum fields

We adopt the approach of Dirac quantum fields in the spirit of [5] and [6]. This approach is explained in Section 5.1. In Section 5.2 we present the theorem about the Hawking effect.

5.1. Quantization in a globally hyperbolic space-time

Following J. Dimock [6] we construct the local algebra of observables in the space-time outside the collapsing star. This construction does not depend on the choice of the representation of the CARs, or on the spin structure of the Dirac field, or on the choice of the hypersurface. In particular we can consider the Fermi-Dirac-Fock representation and the following foliation of our space-time (see Subsection 3.3):

$$\mathcal{M}_{col} = \bigcup_{t\in\mathbb{R}} \Sigma_t^{col}, \qquad \Sigma_t^{col} = \big\{(t,\hat{r},\theta,\varphi)\ ;\ \hat{r} \geq \hat{z}(t,\theta)\big\}.$$

We construct the Dirac field Ψ_0 and the C^*-algebra $\mathcal{U}(\mathcal{H}_0)$ in the usual way. We define the operator:

$$S_{col}:\ \Phi \in \big(C_0^\infty(\mathcal{M}_{col})\big)^4 \ \mapsto\ S_{col}\Phi := \int_{\mathbb{R}} U(0,t)\Phi(t)\mathrm{d}t \in \mathcal{H}_0, \tag{5.1}$$

where $U(0,t)$ is the propagator defined in Proposition 4.6. The quantum spin field is defined by:

$$\Psi_{col}\,:\Phi\in\left(C_0^\infty(\mathcal{M}_{col})\right)^4\ \mapsto\ \Psi_{col}(\Phi):=\Psi_0(S_{col}\Phi)\in\mathcal{L}(\mathcal{F}(\mathcal{H}_0)).$$

Here $\mathcal{F}(\mathcal{H}_0)$ is the Dirac-Fermi-Fock space associated to $\mathcal{H}_0$. For an arbitrary set $\mathcal{O}\subset\mathcal{M}_{col}$, we introduce $\mathcal{U}_{col}(\mathcal{O})$, the C^*-algebra generated by $\psi^*_{col}(\Phi_1)\Psi_{col}(\Phi_2)$, $\operatorname{supp}\Phi_j\subset\mathcal{O}$, $j=1,2$. Eventually, we have:

$$\mathcal{U}_{col}(\mathcal{M}_{col})=\overline{\bigcup_{\mathcal{O}\subset\mathcal{M}_{col}}\mathcal{U}_{col}(\mathcal{O})}.$$

Then we define the fundamental state on $\mathcal{U}_{col}(\mathcal{M}_{col})$ as follows:

$$\begin{aligned}\omega_{col}\big(\Psi^*_{col}(\Phi_1)\Psi_{col}(\Phi_2)\big)&:=\omega_{vac}\big(\Psi_0^*(S_{col}\Phi_1)\Psi_0(S_{col}\Phi_2)\big)\\&=\big\langle\mathbf{1}_{[0,\infty)}(H_0)S_{col}\Phi_1,\,S_{col}\Phi_2\big\rangle.\end{aligned}$$

Let us now consider the future black hole. The algebra $\mathcal{U}_{BH}(\mathcal{M}_{BH})$ and the vacuum state ω_{vac} are constructed as before working now with the group e^{itH} rather than the evolution system $U(t,s)$. We also define the thermal Hawking state (S is analogous to S_{col}, Ψ_{BH} to Ψ_{col}, and Ψ to Ψ_0):

$$\begin{aligned}\omega^{\eta,\sigma}_{Haw}\big(\Psi^*_{BH}(\Phi_1)\Psi_{BH}(\Phi_2)\big)&=\big\langle\mu e^{\sigma H}(1+\mu e^{\sigma H})^{-1}S\Phi_1,\,S\Phi_2\big\rangle_{\mathcal{H}}\\&=:\omega^{\eta,\sigma}_{KMS}\big(\Psi^*(S\Phi_1)\Psi(S\Phi_2)\big)\end{aligned}$$

with

$$T_{Haw}=\sigma^{-1},\qquad \mu=e^{\sigma\eta},\qquad \sigma>0,$$

where T_{Haw} is the Hawking temperature and μ is the chemical potential.

5.2. The Hawking effect

In this subsection we formulate the main result of this work.
Let $\Phi\in(C_0^\infty(\mathcal{M}_{col}))^4$. We put

$$\Phi^T(t,\hat{r},\omega)=\Phi(t-T,\hat{r},\omega). \tag{5.2}$$

Theorem 5.1 (Hawking effect). *Let*

$$\Phi_j\in(C_0^\infty(\mathcal{M}_{col}))^4,\ \ j=1,2.$$

Then we have

$$\begin{aligned}\lim_{T\to\infty}\omega_{col}\big(\Psi^*_{col}(\Phi_1^T)\Psi_{col}(\Phi_2^T)\big)&=\omega^{\eta,\sigma}_{Haw}\big(\Psi^*_{BH}(\mathbf{1}_{\mathbb{R}^+}(P^-)\Phi_1)\Psi_{BH}(\mathbf{1}_{\mathbb{R}^+}(P^-)\Phi_2)\big)\\&\quad+\omega_{vac}\big(\Psi^*_{BH}(\mathbf{1}_{\mathbb{R}^-}(P^-)\Phi_1)\Psi_{BH}(\mathbf{1}_{\mathbb{R}^-}(P^-)\Phi_2)\big),\end{aligned} \tag{5.3}$$

$$T_{Haw}=1/\sigma=\kappa_+/2\pi,\qquad \mu=e^{\sigma\eta},\qquad \eta=\frac{qQr_+}{r_+^2+a^2}+\frac{aD_\varphi}{r_+^2+a^2}.$$

In the above theorem $P^\pm$ is the asymptotic velocity introduced in Section 4. The projections $\mathbf{1}_{\mathbb{R}^\pm}(P^\pm)$ separate outgoing and incoming solutions.

Remark 5.2. The result is independent of the choices of coordinate system, tetrad and chiral angle in the boundary condition.

6. Strategy of the proof

The radiation can be explicitly computed for the asymptotic dynamics near the horizon. For $f = (0, f_2, f_3, 0)$ and T large, the time-reversed solution of the mixed problem for the asymptotic dynamics is well approximated by the so called geometric optics approximation:

$$F^T(\hat{r},\omega) := \frac{1}{\sqrt{-\kappa_+\hat{r}}}(f_3, 0, 0, -f_2)\left(T + \frac{1}{\kappa_+}\ln(-\hat{r}) - \frac{1}{\kappa_+}\ln\hat{A}(\theta),\ \omega\right).$$

For this approximation the radiation can be computed explicitly:

Lemma 6.1. *We have:*

$$\lim_{T\to\infty}\left\|\mathbf{1}_{[0,\infty)}(H_{\leftarrow})F^T\right\|^2 = \left\langle f,\ e^{\frac{2\pi}{\kappa_+}H_{\leftarrow}}\left(1 + e^{\frac{2\pi}{\kappa_+}H_{\leftarrow}}\right)^{-1} f\right\rangle.$$

The strategy of the proof is now the following:

i) We decouple the problem at infinity from the problem near the horizon by cut-off functions. The problem at infinity is easy to treat.

ii) We consider $U(t,T)f$ on a characteristic hypersurface Λ. The resulting characteristic data is denoted g^T. We will approximate $\Omega_{\leftarrow}^- f$ by a function $(\Omega_{\leftarrow}^- f)_R$ with compact support and higher regularity in the angular derivatives. Let $U_{\leftarrow}(s,t)$ be the isometric propagator associated to the asymptotic Hamiltonian $H_{\leftarrow}$ with MIT boundary conditions. We also consider $U_{\leftarrow}(t,T)(\Omega_{\leftarrow}^- f)_R$ on Λ. The resulting characteristic data is denoted $g^T_{\leftarrow,R}$. The situation for the asymptotic comparison dynamics is shown in Figure 1.

iii) We solve a characteristic Cauchy problem for the Dirac equation with data $g^T_{\leftarrow,R}$. The solution at time zero can be written in a region near the boundary as

$$G(g^T_{\leftarrow,R}) = U(0, T/2 + c_0)\,\Phi(T/2 + c_0), \tag{6.1}$$

where Φ is the solution of a characteristic Cauchy problem in the whole space (without the star). The solutions of the characteristic problems for the asymptotic Hamiltonian are written in a similar way and denoted $G_{\leftarrow}(g^T_{\leftarrow,R})$ and $\Phi_{\leftarrow}$, respectively.

iv) Using the asymptotic completeness result we show that $g^T - g^T_{\leftarrow,R} \to 0$ when $T, R \to \infty$. By continuous dependence on the characteristic data we see that:

$$G(g^T) - G(g^T_{\leftarrow,R}) \to 0,\quad T, R \to \infty.$$

v) We write

$$G(g^T_{\leftarrow,R}) - G_{\leftarrow}(g^T_{\leftarrow,R}) = U(0,T/2+c_0)\big(\Phi(T/2+c_0) - \Phi_{\leftarrow}(T/2+c_0)\big) + \big(U(0,T/2+c_0) - U_{\leftarrow}(0,T/2+c_0)\big)\Phi_{\leftarrow}(T/2+c_0).$$

The first term becomes small near the boundary when T becomes large. We then note that for all $\epsilon > 0$ there exists $t_\epsilon > 0$ such that

$$\left\|\big(U(t_\epsilon, T/2+c_0) - U_{\leftarrow}(t_\epsilon, T/2+c_0)\big)\Phi_{\leftarrow}(T/2+c_0)\right\| < \epsilon$$

uniformly in T when T is large. We fix the angular momentum $D_\varphi = n$. The function $U_{\leftarrow}(t_\epsilon, T/2 + c_0)\,\Phi_{\leftarrow}(T/2 + c_0)$ will be replaced by a geometric optics approximation $F_{t_\epsilon}^T$ which has the following properties:

$$\operatorname{supp} F_{t_\epsilon}^T \subset \big(-t_\epsilon - |\mathcal{O}(e^{-\kappa_+ T})|,\ -t_\epsilon\big), \tag{6.2}$$

$$F_{t_\epsilon}^T \rightharpoonup 0, \quad T \to \infty, \tag{6.3}$$

$$\forall \lambda > 0 \quad Op\big(\chi(\langle\xi\rangle \le \lambda\langle q\rangle)\big) F_{t_\epsilon}^T \to 0, \quad T \to \infty. \tag{6.4}$$

Here ξ and q are the dual coordinates to $\hat{r}$ and θ, respectively. $Op(a)$ is the pseudo-differential operator associated to the symbol a (Weyl calculus). The notation $\chi(\langle\xi\rangle \le \lambda\langle q\rangle)$ means that the symbol is supported in $\langle\xi\rangle \le \lambda\langle q\rangle$.

vi) We show that for λ sufficiently large possible singularities of

$$Op\big(\chi(\langle\xi\rangle \ge \lambda\langle q\rangle)\big)\, F_{t_\epsilon}^T$$

are transported by the group $e^{-it_\epsilon H}$ in such a way that they always stay away from the surface of the star.

vii) The points *i)* to *v)* imply:

$$\lim_{T\to\infty} \big\| \mathbf{1}_{[0,\infty)}(H_0)\, j_- U(0,T) f \big\|_0^2 = \lim_{T\to\infty} \big\| \mathbf{1}_{[0,\infty)}(H_0)\, U(0,t_\epsilon)\, F_{t_\epsilon}^T \big\|_0^2,$$

where j_- is a smooth cut-off which equals 1 near the boundary and 0 at infinity. Let ϕ_δ be a cut-off outside the surface of the star at time 0. If $\phi_\delta = 1$ sufficiently close to the surface of the star at time 0, we see by the previous point that

$$(1 - \phi_\delta) e^{-it_\epsilon H} F_{t_\epsilon}^T \to 0, \quad T \to \infty. \tag{6.5}$$

Using (6.5) we show that (modulo a small error term):

$$\big(U(0,t_\epsilon) - \phi_\delta e^{-it_\epsilon H}\big) F_{t_\epsilon}^T \to 0, \quad T \to \infty.$$

Therefore it remains to consider:

$$\lim_{T\to\infty} \big\| \mathbf{1}_{[0,\infty)}(H_0) \phi_\delta\, e^{-it_\epsilon H} F_{t_\epsilon}^T \big\|_0.$$

viii) We show that we can replace $\mathbf{1}_{[0,\infty)}(H_0)$ by $\mathbf{1}_{[0,\infty)}(H)$. This will essentially allow us to commute the energy cut-off and the group. We then show that we can replace the energy cut-off by $\mathbf{1}_{[0,\infty)}(H_{\leftarrow})$. We end up with:

$$\lim_{T\to\infty} \big\| \mathbf{1}_{[0,\infty)}(H_{\leftarrow})\, e^{-it_\epsilon H_{\leftarrow}} F_{t_\epsilon}^T \big\|, \tag{6.6}$$

which we have already computed.

References

[1] A. Bachelot, *Scattering of scalar fields by spherical gravitational collapse*, J. Math. Pures Appl. (9) **76**, 155–210 (1997).

[2] A. Bachelot, *The Hawking effect*, Ann. Inst. H. Poincaré Phys. Théor. **70**, 41–99 (1999).

[3] B. Carter, *Black hole equilibrium states*, Black holes/Les astres occlus (École d'Été Phys. Théor., Les Houches, 1972), 57–214, Gordon and Breach, New York, 1973.

[4] T. Daudé, *Sur la théorie de la diffusion pour des champs de Dirac dans divers espaces-temps de la relativité générale*, PhD thesis Université Bordeaux 1 (2004).

[5] J. Dimock, *Algebras of local observables on a manifold*, Comm. Math. Phys. **77**, 219–228 (1980).

[6] J. Dimock, *Dirac quantum fields on a manifold*, Trans. Amer. Math. Soc. **269**, 133–147 (1982).

[7] J. Dimock, B. S. Kay, *Classical and quantum scattering theory for linear scalar fields on the Schwarzschild metric. I*, Ann. Physics **175**, 366–426 (1987).

[8] D. Häfner and J.-P. Nicolas, *Scattering of massless Dirac fields by a Kerr black hole*, Rev. Math. Phys. **16**, 29–123 (2004).

[9] D. Häfner, *Creation of fermions by rotating charged black holes*, Mémoires de la SMF **117** (2009), 158 pp.

[10] S. W. Hawking, *Particle creation by black holes*, Comm. Math. Phys. **43**, 199–220 (1975).

[11] R. Penrose, W. Rindler, *Spinors and space-time*, Vol. I, Cambridge monographs on mathematical physics, Cambridge University Press 1984.

[12] S. J. Fletcher and A. W. C. Lun, *The Kerr spacetime in generalized Bondi-Sachs coordinates*, Classical Quantum Gravity **20**, 4153–4167 (2003).

[13] R. Wald, *On particle creation by black holes*, Comm. Math. Phys. **45**, 9–34 (1974).

Dietrich Häfner
Université de Grenoble 1
Institut Fourier-UMR CNRS 5582
100, rue des Maths
BP 74
38402 Saint Martin d'Hères Cedex
France
e-mail: `Dietrich.Hafner@ujf-grenoble.fr`

Observables in the General Boundary Formulation

Robert Oeckl

Abstract. We develop a notion of quantum observable for the general boundary formulation of quantum theory. This notion is adapted to spacetime regions rather than to hypersurfaces and naturally fits into the topological-quantum-field-theory-like axiomatic structure of the general boundary formulation. We also provide a proposal for a generalized concept of expectation value adapted to this type of observable. We show how the standard notion of quantum observable arises as a special case together with the usual expectation values. We proceed to introduce various quantization schemes to obtain such quantum observables including path integral quantization (yielding the time-ordered product), Berezin-Toeplitz (antinormal-ordered) quantization and normal-ordered quantization, and discuss some of their properties.

Mathematics Subject Classification (2010). Primary 81P15; Secondary 81P16, 81T70, 53D50, 81S40, 81R30.

Keywords. Quantum field theory, general boundary formulation, observable, expectation value, quantization, coherent state.

1. Motivation: Commutation relations and quantum field theory

In standard quantum theory one is used to think of observables as encoded in operators on the Hilbert space of states. The algebra formed by these is then seen as encoding fundamental structure of the quantum theory. Moreover, this algebra often constitutes the primary target of quantization schemes that aim to produce a quantum theory from a classical theory. Commutation relations in this algebra then provide a key ingredient of correspondence principles between a classical theory and its quantization. We shall argue in the following that while this point of view is natural in a non-relativistic setting, it is less compelling in a special-relativistic setting and becomes questionable in a general-relativistic setting.[1]

In non-relativistic quantum mechanics (certain) operators correspond to measurements that can be applied at any given time, meaning that the measurement is performed at that time. Let us say we consider the measurement of two quantities, one associated with the operator A and another associated with the operator B. In particular, we can then also say which operator is associated with the *consecutive* measurement of both quantities. If we first measure A and then B the operator is the product BA, and if we first measure B and then A the operator is the product AB.[2] Hence, the operator product has the operational meaning of describing the *temporal composition* of measurements. One of the key features of quantum mechanics is of course the fact that in general $AB \neq BA$, i.e., a different temporal ordering of measurements leads to a different outcome.

The treatment of operators representing observables is different in quantum field theory. Here, such operators are *labeled* with the time at which they are applied. For example, we write $\phi(t,x)$ for a field operator at time t. Hence, if we want to combine the measurement processes associated with operators $\phi(t,x)$ and $\phi(t',x')$ say, there is only one operationally meaningful way to do so. The operator associated with the combined process is the *time-ordered* product of the two operators, $\mathbf{T}\phi(t,x)\phi(t',x')$. Of course, this time-ordered product is commutative since the information about the temporal ordering of the processes associated with the operators is already contained in their labels. Nevertheless, in traditional treatments of quantum field theory one first constructs a non-commutative algebra of field operators starting with equal-time commutation relations. Since the concept of equal-time hypersurface is not Poincaré-invariant, one then goes on to generalize these commutation relations to field operators at different times. In particular, one finds that for two localized operators $A(t,x)$ and $B(t',x')$, the commutator obeys

$$[A(t,x),B(t',x')] = 0 \qquad \text{if } (t,x) \text{ and } (t',x') \text{ are spacelike separated,} \quad (1)$$

which is indeed a Poincaré-invariant condition. The time-ordered product is usually treated as a concept that is *derived* from the non-commutative operator product. From this point of view, condition (1) serves to make sure that it is well-defined and does not depend on the inertial frame. Nevertheless, it is the former and not the latter that has a direct operational meaning. Indeed, essentially all the predictive power of quantum field theory derives from the amplitudes and the S-matrix which are defined entirely in terms

[1]By "general-relativistic setting" we shall understand here a context where the metric of spacetime is a dynamical object, but which is not necessarily limited to Einstein's theory of General Relativity.

[2]The notion of composition of measurements considered here is one where the output value generated by the composite measurement can be understood as the product of output values of the individual measurements, rather than one where one would obtain separate output values for both measurements.

of time-ordered products. On the other hand, the non-commutative operator product can be recovered from the time-ordered product. Equal-time commutation relations can be obtained as the limit

$$[A(t,x), B(t,x')] = \lim_{\epsilon\to+0} \mathbf{T}A(t+\epsilon,x)B(t-\epsilon,x') - \mathbf{T}A(t-\epsilon,x)B(t+\epsilon,x').$$

The property (1) can then be seen as arising from the transformation properties of this limit and its non-equal time generalization.

We conclude that in a special-relativistic setting, there are good reasons to regard the time-ordered product of observables as more fundamental than the non-commutative operator product. This suggests to try to formulate the theory of observables in terms of the former rather than the latter. In a (quantum) general-relativistic setting with no predefined background metric a condition such as (1) makes no longer sense, making the postulation of a non-commutative algebra structure for observables even more questionable.

In this paper we shall consider a proposal for a concept of quantum observable that takes these concerns into account. The wider framework in which we embed this is the general boundary formulation of quantum theory (GBF) [1]. We start in Section 2 with a short review of the relevant ingredients of the GBF. In Section 3 we introduce a concept of quantum observable in an axiomatic way, provide a suitably general notion of expectation value and show how standard concepts of quantum observable and expectation values arise as special cases. In Section 4 we consider different quantization prescriptions of classical observables that produce such quantum observables, mainly in a field-theoretic context.

2. Short review of the general boundary formulation

2.1. Core axioms

The basic data of a general-boundary quantum field theory consists of two types: geometric objects that encode a basic structure of spacetime and algebraic objects that encode notions of quantum states and amplitudes. The algebraic objects are assigned to the geometric objects in such a way that the core axioms of the general boundary formulation are satisfied. These may be viewed as a special variant of the axioms of a topological quantum field theory [2]. They have been elaborated, with increasing level of precision, in [3, 1, 4, 5]. In order for this article to be reasonably self-contained, we repeat below the version from [5].

The geometric objects are of two kinds:

Regions. These are (certain) oriented manifolds of dimension d (the spacetime dimension), usually with boundary.

Hypersurfaces. These are (certain) oriented manifolds of dimension $d-1$, here assumed without boundary.[3]

[3]The setting may be generalized to allow for hypersurfaces with boundaries along the lines of [4]. However, as the required modifications are of little relevance in the context of the present paper, we restrict to the simpler setting.

Depending on the theory to be modeled, the manifolds may carry additional structure such as that of a Lorentzian metric in the case of quantum field theory. For more details see the references mentioned above. The core axioms may be stated as follows:

(T1) Associated to each hypersurface Σ is a complex separable Hilbert space $\mathcal{H}_\Sigma$, called the *state space* of Σ. We denote its inner product by $\langle\cdot,\cdot\rangle_\Sigma$.

(T1b) Associated to each hypersurface Σ is a conjugate-linear isometry $\iota_\Sigma : \mathcal{H}_\Sigma \to \mathcal{H}_{\bar\Sigma}$. This map is an involution in the sense that $\iota_{\bar\Sigma} \circ \iota_\Sigma$ is the identity on $\mathcal{H}_\Sigma$.

(T2) Suppose the hypersurface Σ decomposes into a disjoint union of hypersurfaces $\Sigma = \Sigma_1 \cup \cdots \cup \Sigma_n$. Then, there is an isometric isomorphism of Hilbert spaces $\tau_{\Sigma_1,\ldots,\Sigma_n;\Sigma} : \mathcal{H}_{\Sigma_1} \hat{\otimes} \cdots \hat{\otimes} \mathcal{H}_{\Sigma_n} \to \mathcal{H}_\Sigma$. The composition of the maps τ associated with two consecutive decompositions is identical to the map τ associated to the resulting decomposition.

(T2b) The involution ι is compatible with the above decomposition. That is, $\tau_{\bar\Sigma_1,\ldots,\bar\Sigma_n;\bar\Sigma} \circ (\iota_{\Sigma_1} \hat{\otimes} \cdots \hat{\otimes} \iota_{\Sigma_n}) = \iota_\Sigma \circ \tau_{\Sigma_1,\ldots,\Sigma_n;\Sigma}$.

(T4) Associated with each region M is a linear map from a dense subspace $\mathcal{H}^\circ_{\partial M}$ of the state space $\mathcal{H}_{\partial M}$ of its boundary ∂M (which carries the induced orientation) to the complex numbers, $\rho_M : \mathcal{H}^\circ_{\partial M} \to \mathbb{C}$. This is called the *amplitude map*.

(T3x) Let Σ be a hypersurface. The boundary $\partial\hat\Sigma$ of the associated empty region $\hat\Sigma$ decomposes into the disjoint union $\partial\hat\Sigma = \bar\Sigma \cup \Sigma'$, where Σ' denotes a second copy of Σ. Then, $\tau_{\bar\Sigma,\Sigma';\partial\hat\Sigma}(\mathcal{H}_{\bar\Sigma} \otimes \mathcal{H}_{\Sigma'}) \subseteq \mathcal{H}^\circ_{\partial\hat\Sigma}$. Moreover, $\rho_{\hat\Sigma} \circ \tau_{\bar\Sigma,\Sigma';\partial\hat\Sigma}$ restricts to a bilinear pairing $(\cdot,\cdot)_\Sigma : \mathcal{H}_{\bar\Sigma} \times \mathcal{H}_{\Sigma'} \to \mathbb{C}$ such that $\langle\cdot,\cdot\rangle_\Sigma = (\iota_\Sigma(\cdot),\cdot)_\Sigma$.

(T5a) Let M_1 and M_2 be regions and $M := M_1 \cup M_2$ be their disjoint union. Then $\partial M = \partial M_1 \cup \partial M_2$ is also a disjoint union and $\tau_{\partial M_1,\partial M_2;\partial M}(\mathcal{H}^\circ_{\partial M_1} \otimes \mathcal{H}^\circ_{\partial M_2}) \subseteq \mathcal{H}^\circ_{\partial M}$. Then, for all $\psi_1 \in \mathcal{H}^\circ_{\partial M_1}$ and $\psi_2 \in \mathcal{H}^\circ_{\partial M_2}$,

$$\rho_M \circ \tau_{\partial M_1,\partial M_2;\partial M}(\psi_1 \otimes \psi_2) = \rho_{M_1}(\psi_1)\rho_{M_2}(\psi_2). \tag{2}$$

(T5b) Let M be a region with its boundary decomposing as a disjoint union $\partial M = \Sigma_1 \cup \Sigma \cup \overline{\Sigma'}$, where Σ' is a copy of Σ. Let M_1 denote the gluing of M with itself along $\Sigma, \overline{\Sigma'}$ and suppose that M_1 is a region. Note that $\partial M_1 = \Sigma_1$. Then, $\tau_{\Sigma_1,\Sigma,\overline{\Sigma'};\partial M}(\psi \otimes \xi \otimes \iota_\Sigma(\xi)) \in \mathcal{H}^\circ_{\partial M}$ for all $\psi \in \mathcal{H}^\circ_{\partial M_1}$ and $\xi \in \mathcal{H}_\Sigma$. Moreover, for any orthonormal basis $\{\xi_i\}_{i\in I}$ of $\mathcal{H}_\Sigma$, we have for all $\psi \in \mathcal{H}^\circ_{\partial M_1}$:

$$\rho_{M_1}(\psi) \cdot c(M;\Sigma,\overline{\Sigma'}) = \sum_{i\in I} \rho_M \circ \tau_{\Sigma_1,\Sigma,\overline{\Sigma'};\partial M}(\psi \otimes \xi_i \otimes \iota_\Sigma(\xi_i)), \tag{3}$$

where $c(M;\Sigma,\overline{\Sigma'}) \in \mathbb{C}\backslash\{0\}$ is called the *gluing-anomaly factor* and depends only on the geometric data.

As in [5] we omit in the following the explicit mention of the maps τ.

2.2. Amplitudes and probabilities

In standard quantum theory transition amplitudes can be used to encode measurements. The setup, in its simplest form, involves an initial state ψ and a final state η. The initial state encodes a *preparation* of or *knowledge* about the measurement, while the final state encodes a *question* about or *observation* of the system. The modulus square $|\langle\eta, U\psi\rangle|^2$, where U is the time-evolution operator between initial and final time, is then the probability for the answer to the question to be affirmative. (We assume states to be normalized.) This is a *conditional probability* $P(\eta|\psi)$, namely the probability to observe η given that ψ was prepared.

In the GBF this type of measurement setup generalizes considerably.[4] Given a spacetime region M, a *preparation* of or *knowledge* about the measurement is encoded through a closed subspace $\mathcal{S}$ of the boundary Hilbert space $\mathcal{H}_{\partial M}$. Similarly, the *question* or *observation* is encoded in another closed subspace $\mathcal{A}$ of $\mathcal{H}_{\partial M}$. The *conditional probability* for observing $\mathcal{A}$ given that $\mathcal{S}$ is prepared (or known to be the case) is given by the following formula [1, 6]:

$$P(\mathcal{A}|\mathcal{S}) = \frac{\|\rho_M \circ P_\mathcal{S} \circ P_\mathcal{A}\|^2}{\|\rho_M \circ P_\mathcal{S}\|^2}. \tag{4}$$

Here $P_\mathcal{S}$ and $P_\mathcal{A}$ are the orthogonal projectors onto the subspaces $\mathcal{S}$ and $\mathcal{A}$ respectively. $\rho_M \circ P_\mathcal{S}$ is the linear map $\mathcal{H}_{\partial M} \to \mathbb{C}$ given by the composition of the amplitude map ρ_M with the projector $P_\mathcal{S}$. A requirement for (4) to make sense is that this composed map is continuous but does not vanish. (The amplitude map ρ_M is generically not continuous.) That is, $\mathcal{S}$ must be neither "too large" nor "too small". Physically this means that $\mathcal{S}$ must on the one hand be sufficiently restrictive while on the other hand not imposing an impossibility. The continuity of $\rho_M \circ P_\mathcal{S}$ means that it is an element in the dual Hilbert space $\mathcal{H}^*_{\partial M}$. The norm in $\mathcal{H}^*_{\partial M}$ is denoted in formula (4) by $\|\cdot\|$. With an analogous explanation for the numerator the mathematical meaning of (4) is thus clear.

In [1], where this probability interpretation of the GBF was originally proposed, the additional assumption $\mathcal{A} \subseteq \mathcal{S}$ was made, and with good reason. Physically speaking, this condition enforces that we only ask questions in a way that takes into account fully what we already know. Since it is of relevance in the following, we remark that formula (4) might be rewritten in this case as follows:

$$P(\mathcal{A}|\mathcal{S}) = \frac{\langle \rho_M \circ P_\mathcal{S}, \rho_M \circ P_\mathcal{A}\rangle}{\|\rho_M \circ P_\mathcal{S}\|^2}. \tag{5}$$

Here the inner product $\langle\cdot,\cdot\rangle$ is the inner product of the dual Hilbert space $\mathcal{H}^*_{\partial M}$. Indeed, whenever $P_\mathcal{S}$ and $P_\mathcal{A}$ commute, (5) coincides with (4).

---

[4]Even in standard quantum theory, generalizations are possible which involve subspaces of the Hilbert space instead of states. A broader analysis of this situation shows that formula (4) is a much milder generalization of standard probability rules than might seem at first sight, see [1].

2.3. Recovery of standard transition amplitudes and probabilities

We briefly recall in the following how standard transition amplitudes are recovered from amplitude functions. Similarly, we recall how standard transition probabilities arise as special cases of the formula (4). Say t_1 is some initial time and $t_2 > t_1$ some final time, and we consider the spacetime region $M = [t_1, t_2] \times \mathbb{R}^3$ in Minkowski space. ∂M is the disjoint union $\Sigma_1 \cup \bar{\Sigma}_2$ of hypersurfaces of constant $t = t_1$ and $t = t_2$ respectively. We have chosen the orientation of Σ_2 here to be opposite to that induced by M, but equal (under time-translation) to that of Σ_1. Due to axioms (T2) and (T1b), we can identify the Hilbert space $\mathcal{H}_{\partial M}$ with the tensor product $\mathcal{H}_{\Sigma_1} \hat{\otimes} \mathcal{H}^*_{\Sigma_2}$. The amplitude map ρ_M associated with M can thus be viewed as a linear map $\mathcal{H}_{\Sigma_1} \hat{\otimes} \mathcal{H}^*_{\Sigma_2} \to \mathbb{C}$.

In the standard formalism, we have on the other hand a single Hilbert space H of states and a unitary time-evolution map $U(t_1, t_2) : H \to H$. To relate the two settings we should think of H, $\mathcal{H}_{\Sigma_1}$ and $\mathcal{H}_{\Sigma_2}$ as really identical (due to time-translation being an isometry). Then, for any $\psi, \eta \in H$, the amplitude map ρ_M and the operator U are related as

$$\rho_M(\psi \otimes \iota(\eta)) = \langle \eta, U(t_1, t_2)\psi \rangle. \tag{6}$$

Consider now a measurement in the same spacetime region, where an initial state ψ is prepared at time t_1 and a final state η is tested at time t_2. The standard formalism tells us that the probability for this is (assuming normalized states):

$$P(\eta|\psi) = |\langle \eta, U(t_1, t_2)\psi \rangle|^2. \tag{7}$$

In the GBF, the preparation of ψ and observation of η are encoded in the following subspaces of $\mathcal{H}_{\Sigma_1} \hat{\otimes} \mathcal{H}^*_{\Sigma_2}$:

$$\mathcal{S} = \{\psi \otimes \xi : \xi \in \mathcal{H}_{\Sigma_2}\} \quad \text{and} \quad \mathcal{A} = \{\lambda\psi \otimes \iota(\eta) : \lambda \in \mathbb{C}\}. \tag{8}$$

Using (6) one can easily show that then $P(\mathcal{A}|\mathcal{S}) = P(\eta|\psi)$, i.e., the expressions (4) and (7) coincide. This remains true if, alternatively, we define $\mathcal{A}$ disregarding the knowledge encoded in $\mathcal{S}$, i.e., as

$$\mathcal{A} = \{\xi \otimes \iota(\eta) : \xi \in \mathcal{H}_{\Sigma_1}\}. \tag{9}$$

3. A conceptual framework for observables

3.1. Axiomatics

Taking account of the fact that realistic measurements are extended both in space as well as in time, it appears sensible to locate also the mathematical objects that represent observables in spacetime regions. This is familiar for example from algebraic quantum field theory, while being in contrast to idealizing measurements as happening at instants of time as in the standard formulation of quantum theory.

Mathematically, we model an observable associated with a given spacetime region M as a replacement of the corresponding amplitude map ρ_M. That is, an observable in M is a linear map $\mathcal{H}^\circ_{\partial M} \to \mathbb{C}$, where $\mathcal{H}^\circ_{\partial M}$ is the dense subspace of $\mathcal{H}_{\partial M}$ appearing in core axiom (T4). Not any such map needs to be an observable, though. Which map exactly qualifies as an observable may generally depend on the theory under consideration.

(O1) Associated to each spacetime region M is a real vector space $\mathcal{O}_M$ of linear maps $\mathcal{H}^\circ_{\partial M} \to \mathbb{C}$, called *observable maps*. In particular, $\rho_M \in \mathcal{O}_M$.

The most important operation that can be performed with observables is that of *composition*. This composition is performed exactly in the same way as prescribed for amplitude maps in core axioms (T5a) and (T5b). This leads to an additional condition on the spaces $\mathcal{O}_M$ of observables, namely that they be closed under composition.

(O2a) Let M_1 and M_2 be regions as in (T5a), and let $O_1 \in \mathcal{O}_{M_1}$, $O_2 \in \mathcal{O}_{M_2}$. Then there is $O_3 \in \mathcal{O}_{M_1 \cup M_2}$ such that for all $\psi_1 \in \mathcal{H}^\circ_{\partial M_1}$ and $\psi_2 \in \mathcal{H}^\circ_{\partial M_2}$,

$$O_3(\psi_1 \otimes \psi_2) = \rho_{M_1}(\psi_1)\rho_{M_2}(\psi_2). \tag{10}$$

(O2b) Let M be a region with its boundary decomposing as a disjoint union $\partial M = \Sigma_1 \cup \Sigma \cup \overline{\Sigma'}$ and M_1 given as in (T5b) and $O \in \mathcal{O}_M$. Then, there exists $O_1 \in \mathcal{O}_{M_1}$ such that for any orthonormal basis $\{\xi_i\}_{i\in I}$ of $\mathcal{H}_\Sigma$ and for all $\psi \in \mathcal{H}^\circ_{\partial M_1}$,

$$O_1(\psi) \cdot c(M;\Sigma,\overline{\Sigma'}) = \sum_{i\in I} O(\psi \otimes \xi_i \otimes \iota_\Sigma(\xi_i)). \tag{11}$$

We generally refer to the gluing operations of observables of the types described in (O2a) and (O2b) as well as their iterations and combinations as *compositions of observables*. Physically, the composition is meant to represent the combination of measurements. Combination is here to be understood as in classical physics, when the product of observables is taken.

3.2. Expectation values

As in the standard formulation of quantum theory, the *expectation value* of an observable depends on a preparation of or knowledge about a system. As recalled in Section 2.2, this is encoded for a region M in a closed subspace $\mathcal{S}$ of the boundary Hilbert space $\mathcal{H}_{\partial M}$. Given an observable $O \in \mathcal{O}_M$ and a closed subspace $\mathcal{S} \subseteq \mathcal{H}_{\partial M}$, the expectation value of O with respect to $\mathcal{S}$ is defined as

$$\langle O \rangle_{\mathcal{S}} := \frac{\langle \rho_M \circ P_{\mathcal{S}}, O\rangle}{\|\rho_M \circ P_{\mathcal{S}}\|^2}. \tag{12}$$

We use notation here from Section 2.2. Also, as there we need $\rho_M \circ P_{\mathcal{S}}$ to be continuous and different from zero for the expectation value to make sense.

We proceed to make some remarks about the motivation for postulating the expression (12). Clearly, the expectation value must be linear in the observable. Another important requirement is that we would like probabilities in the sense of Section 2.2 to arise as a special case of expectation values. Indeed, given a closed subspace $\mathcal{A}$ of $\mathcal{H}_{\partial M}$ and setting $O = \rho_M \circ P_{\mathcal{A}}$, we see that expression (12) reproduces exactly expression (5). At least in the case where the condition $\mathcal{A} \subseteq \mathcal{S}$ is met, this coincides with expression (4) and represents the conditional probability to observe $\mathcal{A}$ given $\mathcal{S}$.

3.3. Recovery of standard observables and expectation values

Of course it is essential that the present proposal for implementing observables in the GBF can reproduce observables and their expectation values as occurring in the standard formulation of quantum theory. There observables are associated to instants of time, i.e., equal-time hypersurfaces. To model these we use "infinitesimally thin" regions, also called *empty regions*, which geometrically speaking are really hypersurfaces, but are treated as regions.

Concretely, consider the equal-time hypersurface at time t in Minkowski space, i.e., $\Sigma = \{t\} \times \mathbb{R}^3$. We denote the empty region defined by the hypersurface Σ as $\hat{\Sigma}$. The relation between an observable map $O \in \mathcal{O}_M$ and the corresponding operator $\tilde{O}$ is then analogous to the relation between the amplitude map and the time-evolution operator as expressed in equation (6). By definition, $\partial\hat{\Sigma}$ is equal to the disjoint union $\Sigma \cup \bar{\Sigma}$, so that $\mathcal{H}_{\partial\hat{\Sigma}} = \mathcal{H}_\Sigma \hat{\otimes} \mathcal{H}_\Sigma^*$. The Hilbert space $\mathcal{H}_\Sigma$ is identified with the conventional Hilbert space H, and for $\psi, \eta \in H$ we require

$$O(\psi \otimes \iota(\eta)) = \langle \eta, \tilde{O}\psi \rangle_\Sigma. \tag{13}$$

Note that we can glue two copies of $\hat{\Sigma}$ together, yielding again a copy of $\hat{\Sigma}$. The induced composition of observable maps then translates via (13) precisely to the composition of the corresponding operators. In this way we recover the usual operator product for observables of the standard formulation.

Consider now a normalized state $\psi \in H = \mathcal{H}_\Sigma$ encoding a preparation. This translates in the GBF language to the subspace $\mathcal{S} = \{\psi \otimes \xi : \xi \in \mathcal{H}_\Sigma^*\}$ of $\mathcal{H}_{\partial\hat{\Sigma}}$ as reviewed in Section 2.3. The amplitude map $\rho_{\hat{\Sigma}}$ can be identified with the inner product of $H = \mathcal{H}_\Sigma$ due to core axiom (T3x). Thus, $\rho_{\hat{\Sigma}} \circ P_\mathcal{S}(\xi \otimes \iota(\eta)) = \langle \eta, P_\psi \xi \rangle_\Sigma$, where P_ψ is the orthogonal projector in $\mathcal{H}_\Sigma$ onto the subspace spanned by ψ. This makes it straightforward to evaluate the denominator of (12). Let $\{\xi_i\}_{i\in\mathbb{N}}$ be an orthonormal basis of $\mathcal{H}_\Sigma$, which, moreover, we choose for convenience such that $\xi_1 = \psi$. Then

$$\|\rho_M \circ P_\mathcal{S}\|^2 = \sum_{i,j=1}^{\infty} |\rho_M \circ P_\mathcal{S}(\xi_i \otimes \iota(\xi_j))|^2 = \sum_{i,j=1}^{\infty} |\langle \xi_j, P_\psi \xi_i \rangle_\Sigma|^2 = 1. \tag{14}$$

For the numerator of (12) we observe

$$\begin{aligned}\langle \rho_M \circ P_\mathcal{S}, O \rangle &= \sum_{i,j=1}^{\infty} \overline{\rho_M \circ P_\mathcal{S}(\xi_i \otimes \iota(\xi_j))}\, O(\xi_i \otimes \iota(\xi_j)) \\ &= \sum_{i,j=1}^{\infty} \langle P_\psi \xi_i, \xi_j \rangle_\Sigma \, \langle \xi_j, \tilde{O}\xi_i \rangle_\Sigma = \langle \psi, \tilde{O}\psi \rangle_\Sigma. \end{aligned} \tag{15}$$

Hence, the GBF formula (12) recovers in this case the conventional expectation value of $\tilde{O}$ with respect to the state ψ.

4. Quantization

We turn in this section to the problem of the *quantization* of classical observables. On the one hand, we consider the question of how specific quantization schemes that produce Hilbert spaces and amplitude functions satisfying the core axioms can be extended to produce observables. On the other hand, we discuss general features of quantization schemes for observables and the relation to conventional schemes.

4.1. Schrödinger-Feynman quantization

Combining the Schrödinger representation with the Feynman path integral yields a quantization scheme that produces Hilbert spaces for hypersurfaces and amplitude maps for regions in a way that "obviously" satisfies the core axioms [3, 7, 8]. We shall see that it is quite straightforward to include observables into this scheme. Moreover, the resulting quantization can be seen to be in complete agreement with the results of standard approaches to quantum field theory.

We recall that in this scheme states on a hypersurface Σ arise as wave functions on the space space K_Σ of field configurations on Σ. These form a Hilbert space $\mathcal{H}_\Sigma$ of square-integrable functions with respect to a (fictitious) translation-invariant measure μ_Σ:

$$\langle \psi', \psi \rangle_\Sigma := \int_{K_\Sigma} \overline{\psi'(\varphi)} \psi(\varphi) \, \mathrm{d}\mu_\Sigma(\varphi). \tag{16}$$

The amplitude map for a region M arises as the Feynman path integral,

$$\rho_M(\psi) := \int_{K_M} \psi\left(\phi|_\Sigma\right) e^{\mathrm{i}S_M(\phi)} \, \mathrm{d}\mu_M(\phi), \tag{17}$$

where S_M is the action evaluated in M, and K_M is the space of field configurations in M.

The Feynman path integral is of course famous for resisting a rigorous definition, and it is a highly non-trivial task to make sense of expressions (17) or even (16) in general. Nevertheless, much of text-book quantum field theory relies on the Feynman path integral and can be carried over to the present context relatively easily for equal-time hypersurfaces in Minkowski space and regions bounded by such. Moreover, for other special types of regions and hypersurfaces this quantization program has also been successfully carried through for linear or perturbative quantum field theories. Notably, this includes timelike hypersurfaces [7, 8] and has led to a widening of the concept of an asymptotic S-matrix [9, 10].

We proceed to incorporate observables into the quantization scheme. To this end, a classical observable F in a region M is modeled as a real- (or complex-) valued function on K_M. According to Section 3.1 the quantization of F, which we denote here by ρ_M^F, must be a linear map $\mathcal{H}_{\partial M}^\circ \to \mathbb{C}$. We define it as

$$\rho_M^F(\psi) := \int_{K_M} \psi\left(\phi|_\Sigma\right) F(\phi) \, e^{\mathrm{i}S_M(\phi)} \, \mathrm{d}\mu_M(\phi). \tag{18}$$

Before we proceed to interpret this formula in terms of text-book quantum field theory language, we emphasize a key property of this quantization prescription. Suppose we have disjoint but adjacent spacetime regions M_1 and M_2 supporting classical observables $F_1 : K_{M_1} \to \mathbb{R}$ and $F_2 : K_{M_2} \to \mathbb{R}$, respectively. Applying first the operation of (O2a) and then that of (O2b), we can compose the corresponding quantum observables $\rho_{M_1}^{F_1}$ and $\rho_{M_2}^{F_2}$ to a new observable, which we shall denote $\rho_{M_1}^{F_1} \diamond \rho_{M_2}^{F_2}$, supported on the spacetime region $M := M_1 \cup M_2$. On the other hand, the classical observables F_1 and F_2 can be extended trivially to the spacetime region M and there be multiplied to a classical observable $F_1 \cdot F_2 : K_M \to \mathbb{R}$. The composition property of the Feynman path integral now implies the identity

$$\rho_M^{F_1 \cdot F_2} = \rho_{M_1}^{F_1} \diamond \rho_{M_2}^{F_2}. \tag{19}$$

That is, there is a direct correspondence between the product of classical observables and the spacetime composition of quantum observables. This *composition correspondence*, as we shall call it, is not to be confused with what is usually meant with the term "correspondence principle" such as a relation between the commutator of operators and the Poisson bracket of classical observables that these are representing. Indeed, at a careless glance these concepts might even seem to be in contradiction.

Consider now in Minkowski space a region $M = [t_1, t_2] \times \mathbb{R}^3$, where $t_1 < t_2$. Then $\mathcal{H}_{\partial M} = \mathcal{H}_{\Sigma_1} \hat{\otimes} \mathcal{H}_{\Sigma_2}^*$, with notation as in Section 2.3. Consider a classical observable $F_{x_1,\ldots,x_n} : K_M \to \mathbb{R}$ that encodes an n-point function,[5]

$$F_{x_1,\ldots,x_n} : \phi \mapsto \phi(x_1) \cdots \phi(x_n), \tag{20}$$

where $x_1, \ldots, x_n \in M$. Given an initial state $\psi \in \mathcal{H}_{\Sigma_1}$ at time t_1 and a final state $\eta \in \mathcal{H}_{\Sigma_2}$ at time t_2, the quantization of $F_{x_1,\ldots,x_n}$ according to formula (18) can be written in the more familiar form

$$\begin{aligned}\rho_M^{F_{x_1,\ldots,x_n}}(\psi \otimes \iota(\eta)) &= \int_{K_M} \psi(\phi|_{\Sigma_1}) \overline{\eta(\phi|_{\Sigma_2})}\, \phi(x_1) \cdots \phi(x_n)\, e^{\mathrm{i} S_M(\phi)}\, \mathrm{d}\mu_M(\phi) \\ &= \left\langle \eta,\, \mathbf{T} \tilde{\phi}(x_1) \cdots \tilde{\phi}(x_n) e^{-\mathrm{i} \int_{t_1}^{t_2} \tilde{H}(t)\, \mathrm{d}t} \psi \right\rangle,\end{aligned} \tag{21}$$

where $\tilde{\phi}(x_i)$ are the usual quantizations of the classical observables $\phi \mapsto \phi(x_i)$, $\tilde{H}(t)$ is the usual quantization of the Hamiltonian operator at time t, and $\mathbf{T}$ signifies time-ordering. Thus, in familiar situations the prescription (18) really is the "usual" quantization performed in quantum field theory, but with time-ordering of operators. From formula (21) the correspondence property (19) is also clear, although in the more limited context of temporal composition. We realize thus the goal, mentioned in the introduction, of implementing the time-ordered product as more fundamental than the non-commutative operator product.

For a linear field theory, it turns out that the quantization prescription encoded in (18) exhibits an interesting factorization property with respect

[5] For simplicity we use notation here that suggests a real scalar field.

to coherent states. We consider the simple setting of a massive free scalar field theory in Minkowski space with equal-time hypersurfaces. Recall ([10], equation (26)) that a coherent state in the Schrödinger representation at time t can be written as

$$\psi_{t,\eta}(\varphi) := C_{t,\eta} \exp\left(\int \frac{\mathrm{d}^3x\,\mathrm{d}^3k}{(2\pi)^3}\,\eta(k)\,e^{-\mathrm{i}(Et-kx)}\,\varphi(x)\right)\psi_0(\varphi), \tag{22}$$

where η is a complex function on momentum space encoding a solution of the Klein-Gordon equation. ψ_0 is the vacuum wave function and $C_{t,\eta}$ is a normalization constant. Consider as above an initial time t_1, a final time $t_2 > t_1$ and the region $M := [t_1, t_2]\times\mathbb{R}^3$ in Minkowski space. Let $F : K_M \to \mathbb{C}$ represent a classical observable. Evaluating the quantized observable map ρ_M^F on an initial coherent state encoded by η_1 and a final coherent state encoded by η_2 yields:

$$\begin{aligned}
&\rho_M^F\left(\psi_{t_1,\eta_1}\otimes\overline{\psi_{t_2,\eta_2}}\right)\\
&= C_{t_1,\eta_1}\overline{C_{t_2,\eta_2}}\int_{K_M}\psi_0(\varphi_1)\,\overline{\psi_0(\varphi_2)}\\
&\qquad\cdot\exp\left(\int\frac{\mathrm{d}^3x\,\mathrm{d}^3k}{(2\pi)^3}\left(\eta_1(k)\,e^{-\mathrm{i}(Et_1-kx)}\,\varphi_1(x)+\overline{\eta_2(k)}\,e^{\mathrm{i}(Et_2-kx)}\,\varphi_2(x)\right)\right)\\
&\qquad\cdot F(\phi)\,e^{\mathrm{i}S_M(\phi)}\,\mathrm{d}\mu_M(\phi)\\
&= \rho_M\left(\psi_{t_1,\eta_1}\otimes\overline{\psi_{t_2,\eta_2}}\right)\int_{K_M}\psi_0(\varphi_1)\,\overline{\psi_0(\varphi_2)}\,F(\phi+\hat{\eta})\,e^{\mathrm{i}S_M(\phi)}\,\mathrm{d}\mu_M(\phi).
\end{aligned} \tag{23}$$

Here φ_i denote the restrictions of the configuration ϕ to time t_i. To obtain the second equality we have shifted the integration variable ϕ by

$$\hat{\eta}(t,x) := \int\frac{\mathrm{d}^3k}{(2\pi)^3 2E}\left(\eta_1(k)e^{-\mathrm{i}(Et-kx)}+\overline{\eta_2(k)}e^{\mathrm{i}(Et-kx)}\right) \tag{24}$$

and used the conventions of [10]. Note that $\hat{\eta}$ is a *complexified* classical solution in M determined by η_1 and η_2. We have supposed that F naturally extends to a function on the *complexified* configuration space $K_M^{\mathbb{C}}$. Viewing the function $\phi\to F(\phi+\hat{\eta})$ as a new observable $F^{\hat{\eta}}$, the remaining integral in (23) can be interpreted in terms of (18), and we obtain the factorization identity

$$\rho_M^F\left(\psi_{t_1,\eta_1}\otimes\overline{\psi_{t_2,\eta_2}}\right) = \rho_M\left(\psi_{t_1,\eta_1}\otimes\overline{\psi_{t_2,\eta_2}}\right)\rho_M^{F^{\hat{\eta}}}\left(\psi_0\otimes\psi_0\right). \tag{25}$$

That is, the quantum observable map evaluated on a pair of coherent states factorizes into the plain amplitude for the same pair of states and the quantum observable map for a shifted observable evaluated on the vacuum. Note that the second term on the right-hand side here is a vacuum expectation value.

It turns out that factorization identities analogous to (25) are generic rather than special to the types of hypersurfaces and regions considered here. We will come back to this issue in the next section, where also the role of the complex classical solution $\hat{\eta}$ will become clearer from the point of view of holomorphic quantization. For the moment let us consider the particularly simple case where F is a linear observable. In this case $F^{\hat{\eta}}(\phi) = F(\phi)+F(\hat{\eta})$, and the second term on the right-hand side of (25) decomposes into a sum of two terms:

$$\rho_M^{F^{\hat{\eta}}}(\psi_0 \otimes \psi_0) = \rho_M^F(\psi_0 \otimes \psi_0) + F(\hat{\eta})\rho_M(\psi_0 \otimes \psi_0). \tag{26}$$

The first term on the right-hand side is a one-point function which vanishes in the present case of a linear field theory. (F is antisymmetric under exchange of ϕ and $-\phi$, while the other expressions in (18) are symmetric.) The second factor in the second term is the amplitude of the vacuum and hence equal to unity. Thus, in the case of a linear observable, (25) simplifies to

$$\rho_M^F\left(\psi_{t_1,\eta_1} \otimes \overline{\psi_{t_2,\eta_2}}\right) = F(\hat{\eta})\rho_M\left(\psi_{t_1,\eta_1} \otimes \overline{\psi_{t_2,\eta_2}}\right). \tag{27}$$

4.2. Holomorphic quantization

A more rigorous quantization scheme that produces a GBF from a classical field theory is the holomorphic quantization scheme introduced in [5]. It is based on ideas from geometric quantization, and its Hilbert spaces are versions of "Fock representations". An advantage of this scheme is that taking an axiomatically described classical field theory as input, it produces a GBF as output that can be rigorously proved to satisfy the core axioms of Section 2.1. A shortcoming so far is that only the case of linear field theory has been worked out.

Concretely, the classical field theory is to be provided in the form of a real vector space L_Σ of (germs of) solutions near each hypersurface Σ. Moreover, for each region M there is to be given a subspace $L_{\tilde{M}}$ of the space $L_{\partial M}$ of solutions on the boundary of M. This space $L_{\tilde{M}}$ has the interpretation of being the space of solutions in the interior of M (restricted to the boundary). Also, the spaces L_Σ carry non-degenerate symplectic structures ω_Σ as well as complex structures J_Σ. Moreover, for each hypersurface Σ, the symplectic and complex structures combine to a complete real inner product $g_\Sigma(\cdot,\cdot) = 2\omega_\Sigma(\cdot, J_\Sigma\cdot)$ and to a complete complex inner product $\{\cdot,\cdot\}_\Sigma = g_\Sigma(\cdot,\cdot)+2\mathrm{i}\omega_\Sigma(\cdot,\cdot)$. Another important condition is that the subspace $L_{\tilde{M}} \subseteq L_{\partial M}$ is Lagrangian with respect to the symplectic structure $\omega_{\partial M}$.

The Hilbert space $\mathcal{H}_\Sigma$ associated with a hypersurface Σ is the space of *holomorphic* square-integrable functions on $\hat{L}_\Sigma$ with respect to a Gaussian measure ν_Σ.[6] That is, the inner product in $\mathcal{H}_\Sigma$ is

$$\langle\psi',\psi\rangle_\Sigma := \int_{\hat{L}_\Sigma} \psi(\phi)\overline{\psi'(\phi)}\,\mathrm{d}\nu_\Sigma(\phi). \tag{28}$$

[6]The space $\hat{L}_\Sigma$ is a certain extension of the space L_Σ, namely the algebraic dual of its topological dual. Nevertheless, due to Theorem 3.18 of [5] it is justified to think of wave functions ψ as functions merely on L_Σ rather than on $\hat{L}_\Sigma$, and to essentially ignore the distinction between L_Σ and $\hat{L}_\Sigma$.

Heuristically, the measure ν_Σ can be understood as

$$\mathrm{d}\nu_\Sigma(\phi) \approx \exp\left(-\frac{1}{2} g_{\partial M}(\phi,\phi)\right) \mathrm{d}\mu_\Sigma(\phi), \tag{29}$$

where μ_Σ is a fictitious translation-invariant measure on $\hat{L}_\Sigma$. The space $\mathcal{H}_\Sigma$ is essentially the Fock space constructed from L_Σ viewed as a 1-particle space with the inner product $\{\cdot,\cdot\}_\Sigma$.

The amplitude map $\rho_M : \mathcal{H}_{\partial M} \to \mathbb{C}$ associated with a region M is given by the integral formula

$$\rho_M(\psi) := \int_{\hat{L}_{\tilde{M}}} \psi(\phi)\,\mathrm{d}\nu_{\tilde{M}}(\phi). \tag{30}$$

The integration here is over the space $\hat{L}_{\tilde{M}} \subseteq \hat{L}_{\partial M}$ of solutions in M with the measure $\nu_{\tilde{M}}$, which heuristically can be understood as

$$\mathrm{d}\nu_{\tilde{M}}(\phi) \approx \exp\left(-\frac{1}{4} g_{\partial M}(\phi,\phi)\right) \mathrm{d}\mu_{\tilde{M}}(\phi), \tag{31}$$

where again $\mu_{\tilde{M}}$ is a fictitious translation-invariant measure on $\hat{L}_{\tilde{M}}$.

Particularly useful in the holomorphic quantization scheme turn out to be the coherent states that are associated to classical solutions near the corresponding hypersurface. On a hypersurface Σ the coherent state $K_\xi \in \mathcal{H}_\Sigma$ associated to $\xi \in L_\Sigma$ is given by the wave function

$$K_\xi(\phi) := \exp\left(\frac{1}{2}\{\xi,\phi\}_\Sigma\right). \tag{32}$$

The natural vacuum, which we denote by $\mathbf{1}$, is the constant wave function of unit value. Note that $\mathbf{1} = K_0$.

4.2.1. Creation and annihilation operators. One-particle states on a hypersurface Σ are represented by non-zero continuous complex-linear maps $p : L_\Sigma \to \mathbb{C}$, where complex-linearity here implies $p(J\xi) = \mathrm{i}p(\xi)$. By the Riesz Representation Theorem such maps are thus in one-to-one correspondence with non-zero elements of L_Σ. Concretely, for a non-zero element $\xi \in L_\Sigma$ the corresponding one-particle state is represented by the wave function $p_\xi \in \mathcal{H}_\Sigma$ given by

$$p_\xi(\phi) = \frac{1}{\sqrt{2}}\{\xi,\phi\}_\Sigma. \tag{33}$$

The normalization is chosen here such that $\|p_\xi\| = \|\xi\|$. Physically distinct one-particle states thus correspond to the distinct rays in L_Σ, viewed as a complex Hilbert space. An n-particle state is represented by a (possibly infinite) linear combination of the product of n wave functions of this type. The creation operator $a^\dagger_\xi$ for a particle state corresponding to $\xi \in L_\Sigma$ is given by multiplication,

$$(a^\dagger_\xi \psi)(\phi) = p_\xi(\phi)\psi(\phi) = \frac{1}{\sqrt{2}}\{\xi,\phi\}_\Sigma \psi(\phi). \tag{34}$$

The corresponding annihilation operator is the adjoint. Using the reproducing property of the coherent states $K_\phi \in \mathcal{H}_\Sigma$, we can write it as

$$(a_\xi \psi)(\phi) = \langle K_\phi, a_\xi \psi \rangle_\Sigma = \langle a_\xi^\dagger K_\phi, \psi \rangle_\Sigma. \tag{35}$$

Note in particular, that the action of an annihilation operator on a coherent state is by multiplication,

$$a_\xi K_\phi = \frac{1}{\sqrt{2}} \{\phi, \xi\}_\Sigma K_\phi. \tag{36}$$

For $\xi, \eta \in L_\Sigma$ the commutation relations are, as usual,

$$[a_\xi, a_\eta^\dagger] = \{\eta, \xi\}_\Sigma, \qquad [a_\xi, a_\eta] = 0, \qquad [a_\xi^\dagger, a_\eta^\dagger] = 0. \tag{37}$$

4.2.2. Berezin-Toeplitz quantization. A natural way to include observables into this quantization scheme seems to be the following. We model a classical observable F on a spacetime region M as a map $L_{\tilde{M}} \to \mathbb{C}$ (or $L_{\tilde{M}} \to \mathbb{R}$) and define the associated quantized observable map via

$$\rho_M^{\blacktriangleleft F \blacktriangleright}(\psi) := \int_{\hat{L}_{\tilde{M}}} \psi(\phi) F(\phi)\, \mathrm{d}\nu_{\tilde{M}}(\phi). \tag{38}$$

To bring this into a more familiar form, we consider, as in Section 3.3, the special case of an empty region $\hat{\Sigma}$, given geometrically by a hypersurface Σ. Then, for $\psi_1, \psi_1 \in \mathcal{H}_\Sigma$ encoding "initial" and "final" state, we have

$$\rho_{\hat{\Sigma}}^{\blacktriangleleft F \blacktriangleright}(\psi_1 \otimes \iota(\psi_2)) = \int_{\hat{L}_\Sigma} \psi_1(\phi) \overline{\psi_2(\phi)} F(\phi)\, \mathrm{d}\nu_\Sigma(\phi). \tag{39}$$

We can interpret this formula as follows: The wave function ψ_1 is multiplied by the function F. The resulting function is an element of the Hilbert space $\mathrm{L}^2(\hat{L}_\Sigma, \nu_\Sigma)$ (supposing F to be essentially bounded), but not of the subspace $\mathcal{H}_\Sigma$ of holomorphic functions. We thus project back onto this subspace and finally take the inner product with the state ψ_2. This is precisely accomplished by the integral. We may recognize this as a version of Berezin-Toeplitz quantization, where in the language of Berezin [11] the function F is the contravariant symbol of the operator $\tilde{F}$ that is related to $\rho_{\hat{\Sigma}}^{\blacktriangleleft F \blacktriangleright}$ by formula (13). That is,

$$\rho_{\hat{\Sigma}}^{\blacktriangleleft F \blacktriangleright}(\psi_1 \otimes \iota(\psi_2)) = \langle \psi_2, \tilde{F} \psi_1 \rangle_\Sigma. \tag{40}$$

In the following we shall refer to the prescription encoded in (38) simply as Berezin-Toeplitz quantization.

Note that any complex-valued continuous real-linear observable $F : L_\Sigma \to \mathbb{C}$ can be decomposed into its holomorphic (complex-linear) and anti-holomorphic (complex-conjugate-linear) part

$$F(\phi) = F^+(\phi) + F^-(\phi), \quad \text{where} \quad F^\pm(\phi) = \frac{1}{2}\big(F(\phi) \mp \mathrm{i} F(J_\Sigma \phi)\big). \tag{41}$$

If we consider real-valued observables only, we can parametrize them by elements of L_Σ due to the Riesz Representation Theorem. (In the complex-valued case the parametrization is by elements of $L_\Sigma^\mathbb{C}$, the complexification of

L_Σ, instead.) If we associate to $\xi \in L_\Sigma$ the real-linear observable F_ξ given by

$$F_\xi(\phi) := \sqrt{2}\, g_\Sigma(\xi, \phi), \tag{42}$$

then

$$F_\xi^+(\phi) = \frac{1}{\sqrt{2}}\{\xi, \phi\}_\Sigma, \qquad F_\xi^-(\phi) = \frac{1}{\sqrt{2}}\{\phi, \xi\}_\Sigma. \tag{43}$$

Using the results of Section 4.2.1, we see that $F_\xi^\pm$ quantized according to the prescription (39) and expressed in terms of operators $\tilde{F}_\xi^\pm$ as in (40) yields

$$\tilde{F}_\xi^+ = a_\xi^\dagger, \quad \tilde{F}_\xi^- = a_\xi, \quad \text{and for the sum,} \quad \tilde{F}_\xi = a_\xi^\dagger + a_\xi. \tag{44}$$

Consider now n real-linear observables $F_1, \ldots, F_n : L_\Sigma \to \mathbb{C}$. We shall be interested in the *antinormal-ordered product* of the corresponding operators $\tilde{F}_1, \ldots, \tilde{F}_n$, which we denote by $\blacktriangleleft \tilde{F}_1 \cdots \tilde{F}_n \blacktriangleright$. To evaluate matrix elements of this antinormal-ordered product, we decompose the observables F_i according to (41) into holomorphic and anti-holomorphic parts, corresponding to creation operators and annihilation operators, respectively. The creation operators $\tilde{F}_i^+$ then act on wave functions by multiplication with F_i^+ according to (34). Converting the annihilation operators into creation operators by moving them to the left-hand side of the inner product, we see that these correspondingly contribute factors F_i^- in the inner product (28). We obtain:

$$\begin{aligned}\left\langle \psi_2, \blacktriangleleft \tilde{F}_1 \cdots \tilde{F}_n \blacktriangleright \psi_1 \right\rangle_\Sigma &= \left\langle \psi_2, \blacktriangleleft \prod_{i=1}^n (\tilde{F}_i^+ + \tilde{F}_i^-) \blacktriangleright \psi_1 \right\rangle_\Sigma \\ &= \int_{\hat{L}_\Sigma} \overline{\psi_2(\phi)} \left(\prod_{i=1}^n (F_i^+(\phi) + F_i^-(\phi)) \right) \psi_1(\phi)\, \mathrm{d}\nu_\Sigma(\phi) \\ &= \int_{\hat{L}_\Sigma} \overline{\psi_2(\phi)} F_1(\phi) \cdots F_n(\phi) \psi_1(\phi)\, \mathrm{d}\nu_\Sigma(\phi).\end{aligned}$$

Setting $F := F_1 \cdots F_n$, this coincides precisely with the right-hand side of (39). Thus in the case of a hypersurface (empty region) the Berezin-Toeplitz quantization precisely implements antinormal ordering.

Remarkably, the Berezin-Toeplitz quantization shares with the Schrödinger-Feynman quantization the factorization property exhibited in equation (25). In fact, it is in the present context of holomorphic quantization that this property attains a strikingly simple form. In order to state it rigorously, we need a bit of technical language. For a map $F : L_{\tilde{M}} \to \mathbb{C}$ and an element $\xi \in L_{\tilde{M}}$, we denote by $F^\xi : L_{\tilde{M}} \to \mathbb{C}$ the translated map $\phi \mapsto F(\phi + \xi)$. We say that $F : L_{\tilde{M}} \to \mathbb{C}$ is *analytic* iff for each pair $\phi, \xi \in L_{\tilde{M}}$ the map $z \mapsto F(\phi + z\xi)$ is real analytic. We denote the induced extension $L_{\tilde{M}}^{\mathbb{C}} \to \mathbb{C}$ also by F, where $L_{\tilde{M}}^{\mathbb{C}}$ is the complexification of $L_{\tilde{M}}$. We say that $F : L_{\tilde{M}} \to \mathbb{C}$ is analytic and *sufficiently integrable* iff for any $\eta \in L_{\tilde{M}}^{\mathbb{C}}$ the map F^η is integrable in $(\hat{L}_{\tilde{M}}, \nu_{\tilde{M}})$. We recall (Lemma 4.1 of [5]) that elements ξ of $L_{\partial M}$ decompose uniquely as $\xi = \xi^{\mathrm{R}} + J_\Sigma \xi^{\mathrm{I}}$, where $\xi^{\mathrm{R}}, \xi^{\mathrm{I}}$ are elements of $L_{\tilde{M}}$.

Proposition 4.1 (Coherent factorization property). *Let $F : L_M \to \mathbb{C}$ be analytic and sufficiently integrable. Then for any $\xi \in L_{\partial M}$, we have*

$$\rho_M^{\blacktriangleleft F \blacktriangleright}(K_\xi) = \rho_M(K_\xi)\, \rho_M^{\blacktriangleleft F^{\hat{\xi}} \blacktriangleright}(\mathbf{1}), \tag{45}$$

where $\hat{\xi} \in L^{\mathbb{C}}_{\tilde{M}}$ is given by $\hat{\xi} = \xi^{\mathrm{R}} - \mathrm{i}\xi^{\mathrm{I}}$.

Proof. Recall that for $\phi \in \hat{L}_{\tilde{M}}$ we can rewrite the wave function of the coherent state K_ξ as follows:

$$K_\xi(\phi) = \exp\left(\frac{1}{2} g_{\partial M}(\xi^{\mathrm{R}}, \phi) - \frac{\mathrm{i}}{2} g_{\partial M}(\xi^{\mathrm{I}}, \phi)\right). \tag{46}$$

We restrict first to the special case $\xi \in L_{\tilde{M}}$, i.e., $\xi^{\mathrm{I}} = 0$. Translating the integrand by ξ (using Proposition 3.11 of [5]), we find:

$$\begin{aligned}
&\int_{\hat{L}_{\tilde{M}}} F(\phi) \exp\left(\frac{1}{2} g_{\partial M}(\xi, \phi)\right) \mathrm{d}\nu(\phi) \\
&= \int_{\hat{L}_{\tilde{M}}} F(\phi + \xi) \exp\left(\frac{1}{2} g_{\partial M}(\xi, \phi + \xi) - \frac{1}{4} g_{\partial M}(2\phi + \xi, \xi)\right) \mathrm{d}\nu(\phi) \\
&= \exp\left(\frac{1}{4} g_{\partial M}(\xi, \xi)\right) \int_{\hat{L}_{\tilde{M}}} F(\phi + \xi)\, \mathrm{d}\nu(\phi) \\
&= \rho_M(K_\xi)\, \rho_M^{\blacktriangleleft F^{\xi} \blacktriangleright}(\mathbf{1}).
\end{aligned}$$

In order to work out the general case, we follow the strategy outlined in the proof of Proposition 4.2 of [5]: We replace the i in (46) by a complex parameter and note that all relevant expressions are holomorphic in this parameter. This must also hold for the result of the integration performed above. But performing the integration is straightforward when this parameter is real, since we can then combine both terms in the exponential in (46). On the other hand, a holomorphic function is completely determined by its values on the real line. This leads to the stated result. □

It is clear at this point that equation (25) is just a special case of (the analog for ρ_M^F of) equation (45). Indeed, it turns out that with a suitable choice of complex structure (see [5]) the complexified classical solution $\hat{\eta}$ given by (24) decomposes precisely as $\hat{\eta} = \eta^{\mathrm{R}} - \mathrm{i}\eta^{\mathrm{I}}$.[7] From here onwards we shall say that a quantization scheme satisfying equation (45) has the *coherent factorization property*.

The coherent factorization property may also be interpreted as suggesting an intrinsic definition of a *real* observable in the quantum setting. It is clear that quantum observable maps must take values in the complex numbers and not merely in the real numbers since for example the amplitude map is a special kind of quantum observable map.[8] On the other hand, we have in axiom (O1) deliberately only required that the observable maps in a

---

[7] We differ here slightly from the conventions in [5] to obtain exact agreement.

[8] Proposition 4.2 of [5] implies that amplitude maps generically take complex and not merely real values.

region M form a real vector space $\mathcal{O}_M$, to allow for a restriction to "real" observables, analogous to hermitian operators in the standard formulation and to real-valued maps in the classical theory. Of course, given a quantization prescription such as (18) or (38), we can simply restrict the quantization to real classical observables. However, equation (45) suggests a more intrinsic definition in case of availability of coherent states. Namely, we could say that a quantum observable map is *real* iff its evaluation on a coherent state K_ξ associated to any element ξ in the subspace $L_{\tilde{M}} \subseteq L_{\partial M}$ yields a real multiple of the amplitude map evaluated on the same coherent state. Note that this characterization is closed under real-linear combinations. Also, if a quantization scheme satisfies the coherent factorization property, this characterization coincides with the condition for the classical observable to be real-valued, as is easily deduced using the completeness of the coherent states.

Let us briefly return to the special case of linear observables. Suppose that $F : L_{\tilde{M}} \to \mathbb{R}$ is linear (hence analytic) and sufficiently integrable. We evaluate the Berezin-Toeplitz quantum observable map $\rho_M^{\blacktriangleleft F \blacktriangleright}$ on the coherent state K_ξ associated to $\xi \in L_{\partial M}$. As above we define $\hat{\xi} \in L_{\tilde{M}}^{\mathbb{C}}$ as $\hat{\xi} = \xi^{\mathrm{R}} - \mathrm{i}\xi^{\mathrm{I}}$. Using the coherent factorization property (45) as well as linearity of F, we obtain

$$\rho_M^{\blacktriangleleft F \blacktriangleright}(K_\xi) = \rho_M(K_\xi)\rho_M^{\blacktriangleleft F^{\hat{\xi}} \blacktriangleright}(\mathbf{1}) = \rho_M(K_\xi)\left(\rho_M^{\blacktriangleleft F \blacktriangleright}(\mathbf{1}) + F(\hat{\xi})\rho_M(\mathbf{1})\right). \quad (47)$$

The first term in brackets vanishes by inspection of (38) due to anti-symmetry of F under exchange of ϕ and $-\phi$, while $\rho_M(\mathbf{1}) = 1$. Thus, analogously to (27), we obtain

$$\rho_M^{\blacktriangleleft F \blacktriangleright}(K_\xi) = F(\hat{\xi})\rho_M(K_\xi). \quad (48)$$

Supposing that the amplitudes of coherent states coincide between the Schrödinger-Feynman scheme and the holomorphic scheme (that is, if the complex structure of the holomorphic scheme and the vacuum of the Schrödinger-Feynman scheme are mutually adapted), also the quantization of linear observables according to (18) coincides with that of (38). Nevertheless, the quantization of non-linear observables is necessarily different. For one, classical observables in the Schrödinger-Feynman scheme are defined on configuration spaces rather than on spaces of solutions. Indeed, the quantization of observables that coincide when viewed merely as functions on solutions differs in general. However, it is also clear that the Berezin-Toeplitz quantization cannot satisfy the composition correspondence property (19) that is satisfied by the Schrödinger-Feynman scheme. Indeed, consider adjacent regions M_1 and M_2 that can be glued to a joint region M. Then the classical observables in the disjoint region induce classical observables in the glued region, but not the other way round. While the former are functions on $L_{\tilde{M}_1} \times L_{\tilde{M}_2}$, the latter are functions on the subspace $L_{\tilde{M}} \subseteq L_{\tilde{M}_1} \times L_{\tilde{M}_2}$. In spite of the summation involved in axiom (O2b), one can use this to cook up a contradiction to the composition correspondence property (19). It is easy to see how this problem is avoided in the Schrödinger-Feynman scheme: There, classical observables in a region M are functions on the space of field configurations K_M, which is much larger than the space of classical solutions L_M and permits the "decoupling" of observables in adjacent regions. Indeed, the present considerations indicate that in order for a quantization scheme to satisfy the composition correspondence property, this kind of modification of the definition of what constitutes a classical observable is a necessity.

The composition correspondence property suggests also a different route to quantization of observables: We may consider a quantization scheme only for linear observables and impose the composition correspondence property to *define* a quantization scheme for more general observables. In the Berezin-Toeplitz case this would lead to a scheme equivalent to the path integral (18). However, recall that the composition of quantum observable maps is only possible between disjoint regions. On the other hand, well-known difficulties (related to renormalization) arise also for the path integral (18) when considering field observables at coincident points.

4.2.3. Normal-ordered quantization. Consider a hypersurface Σ and linear observables $F_1, \ldots, F_n : L_\Sigma \to \mathbb{C}$ in the associated empty region $\hat{\Sigma}$. Consider now the *normal-ordered product* $:\tilde{F}_1 \cdots \tilde{F}_n:$ and its matrix elements. These matrix elements turn out to be particularly simple for coherent states. To evaluate these we decompose the maps $F_1, \ldots, F_n$ into holomorphic (creation) and anti-holomorphic (annihilation) parts according to (41). The annihilation operators act on coherent states simply by multiplication according to (36). The creation operators on the other hand can be converted to annihilation operators by moving them to the left-hand side of the inner product. We find:

$$\left\langle K_\eta, :F_1 \cdots F_n: K_\xi \right\rangle_\Sigma = \prod_{i=1}^{n} \left(F_i^+(\eta) + F_i^-(\xi)\right) \langle K_\eta, K_\xi \rangle_\Sigma. \tag{49}$$

While this expression looks quite simple, it can be further simplified by taking seriously the point of view that $\hat{\Sigma}$ is an (empty) region. Hence, $K_\xi \otimes \iota(K_\eta)$ is really the coherent state $K_{(\xi,\eta)} \in \mathcal{H}_{\partial\hat{\Sigma}}$ associated to the solution $(\xi, \eta) \in L_{\partial\hat{\Sigma}}$. As above we may decompose $(\xi, \eta) = (\xi, \eta)^{\mathrm{R}} + J_{\partial\hat{\Sigma}}(\xi, \eta)^{\mathrm{I}}$, where $(\xi, \eta)^{\mathrm{R}}, (\xi, \eta)^{\mathrm{I}} \in L_{\tilde{\hat{\Sigma}}}$. Identifying $L_{\tilde{\hat{\Sigma}}}$ with L_Σ (and taking into account $J_{\partial\hat{\Sigma}} = (J_\Sigma, -J_\Sigma)$), we have

$$(\xi, \eta)^{\mathrm{R}} = \frac{1}{2}(\xi + \eta), \quad (\xi, \eta)^{\mathrm{I}} = -\frac{1}{2}(J_\Sigma \xi - J_\Sigma \eta). \tag{50}$$

But observe:

$$\begin{aligned} F_i^+(\eta) + F_i^-(\xi) &= \frac{1}{2}\left(F_i(\eta + \xi) - \mathrm{i} F_i(J_\Sigma(\eta - \xi))\right) \\ &= F_i\left((\xi, \eta)^{\mathrm{R}}\right) - \mathrm{i} F_i\left((\xi, \eta)^{\mathrm{I}}\right) = F_i\left((\xi, \eta)^{\mathrm{R}} - \mathrm{i}(\xi, \eta)^{\mathrm{I}}\right), \end{aligned} \tag{51}$$

where in the last step we have extended the domain of F_i from $L_{\tilde{\hat{\Sigma}}}$ to its complexification $L^{\mathbb{C}}_{\tilde{\hat{\Sigma}}}$.

Defining a quantum observable map encoding the normal-ordered product

$$\rho_{\hat{\Sigma}}^{:F_1 \cdots F_n:}(\psi_1 \otimes \iota(\psi_2)) := \left\langle \psi_2, :\tilde{F}_1 \cdots \tilde{F}_n: \psi_1 \right\rangle_\Sigma, \tag{52}$$

the identity (49) becomes thus

$$\rho_{\hat{\Sigma}}^{:F_1\cdots F_n:}(K_{(\xi,\eta)}) = \prod_{i=1}^{n} F_i\left((\xi,\eta)^{\mathrm{R}} - \mathrm{i}(\xi,\eta)^{\mathrm{I}}\right)\rho_{\hat{\Sigma}}(K_{(\xi,\eta)}). \tag{53}$$

Note that in the above expression the fact that we consider an empty region rather than a generic region is no longer essential. Rather, we may consider a region M and replace (ξ,η) by some solution $\phi \in L_{\tilde{M}}$. Also there is no longer a necessity to write the observable explicitly as a product of linear observables. A generic observable $F : L_{\tilde{M}} \to \mathbb{C}$ that has the analyticity property will do. We obtain:

$$\rho_M^{:F:}(K_\phi) := F(\hat{\phi})\rho_M(K_\phi), \tag{54}$$

where $\hat{\phi} := \phi^{\mathrm{R}} - \mathrm{i}\phi^{\mathrm{I}}$. We may take this now as the *definition* of a quantization prescription that we shall call *normal-ordered quantization.* It coincides with — and provides an extremely concise expression for — the usual concept of normal ordering in the case when M is the empty region associated to a hypersurface.

Interestingly, expression (54) also coincides with expression (48) and with expression (27). However, the latter two expressions were only valid in the case where F is linear. So, unsurprisingly, we obtain agreement of normal-ordered quantization with Berezin-Toeplitz quantization and with Schrödinger-Feynman quantization in the case of linear observables, while they differ in general. Remarkably, however, normal-ordered quantization shares with these other quantization prescriptions the coherent factorization property (45). To see this, note that (using (54)):

$$\rho_M^{:F^{\hat{\phi}}:}(\mathbf{1}) = F^{\hat{\phi}}(0)\rho_M(\mathbf{1}) = F(\hat{\phi}). \tag{55}$$

4.2.4. Geometric quantization. Since the holomorphic quantization scheme draws on certain ingredients of geometric quantization, it is natural to also consider what geometric quantization has to say about the quantization of observables [12]. For hypersurfaces Σ (empty regions $\hat{\Sigma}$) the geometric quantization of a classical observable $F : L_\Sigma \to \mathbb{C}$ is given by an operator $\check{F} : \mathcal{H}_\Sigma \to \mathcal{H}_\Sigma$. If the observable F preserves the polarization (which is the case for example for linear observables), then $\check{F}$ is given by the formula

$$\check{F}\psi = -\mathrm{i}\,\mathrm{d}\psi(\mathbf{F}) - \theta(\mathbf{F})\psi + F\psi. \tag{56}$$

Here $\mathbf{F}$ denotes the Hamiltonian vector field generated by the function F, θ is the symplectic potential given here by $\theta_\eta(\Phi) = -\frac{\mathrm{i}}{2}\{\eta,\Phi_\eta\}_\Sigma$, and $\mathrm{d}\psi$ is the exterior derivative of ψ.

Consider a real-linear observable $F : L_\Sigma \to \mathbb{C}$. Without explaining the details we remark that for the holomorphic part F^+ (recall (41)) we obtain $\mathrm{d}\psi(\mathbf{F}^+) = 0$ as well as $\theta(\mathbf{F}^+) = 0$. On the other hand, for the anti-holomorphic part F^- we have $\theta(\mathbf{F}^-) = F^-$. This simplifies (56) to

$$\check{F}\psi = -\mathrm{i}\,\mathrm{d}\psi(\mathbf{F}^-) + F^+\psi. \tag{57}$$

Setting F equal to F_ξ given by (42) for $\xi \in L_\Sigma$ results in $F^+\psi = a^\dagger_\xi \psi$ and $-\mathrm{i}\, \mathrm{d}\psi(\mathbf{F}^-) = a_\xi \psi$. That, is the operator $\check{F}$ coincides with the operator $\tilde{F}$ obtained by quantizing F with any of the other prescriptions discussed. On the other hand, the quantization of non-linear observables will in general differ from any of the other prescriptions. Indeed, it is at present not even clear whether or how the prescription (56) can be generalized to non-empty regions.

Acknowledgments

I would like to thank the organizers of the *Regensburg Conference 2010: Quantum Field Theory and Gravity* for inviting me to present aspects of this work as well as for providing a stimulating and well-organized conference. This work was supported in part by CONACyT grant 49093.

References

[1] R. Oeckl, *General boundary quantum field theory: Foundations and probability interpretation*, Adv. Theor. Math. Phys. **12** (2008), 319–352, hep-th/0509122.

[2] M. Atiyah, *Topological quantum field theories*, Inst. Hautes Études Sci. Publ. Math. (1989), no. 68, 175–186.

[3] R. Oeckl, *A "general boundary" formulation for quantum mechanics and quantum gravity*, Phys. Lett. **B 575** (2003), 318–324, hep-th/0306025.

[4] R. Oeckl, *Two-dimensional quantum Yang-Mills theory with corners*, J. Phys. A **41** (2008), 135401, hep-th/0608218.

[5] R. Oeckl, *Holomorphic quantization of linear field theory in the general boundary formulation*, preprint arXiv:1009.5615.

[6] R. Oeckl, *Probabilities in the general boundary formulation*, J. Phys.: Conf. Ser. **67** (2007), 012049, hep-th/0612076.

[7] R. Oeckl, *States on timelike hypersurfaces in quantum field theory*, Phys. Lett. **B 622** (2005), 172–177, hep-th/0505267.

[8] R. Oeckl, *General boundary quantum field theory: Timelike hypersurfaces in Klein-Gordon theory*, Phys. Rev. **D 73** (2006), 065017, hep-th/0509123.

[9] D. Colosi and R. Oeckl, *S-matrix at spatial infinity*, Phys. Lett. **B 665** (2008), 310–313, arXiv:0710.5203.

[10] D. Colosi and R. Oeckl, *Spatially asymptotic S-matrix from general boundary formulation*, Phys. Rev. **D 78** (2008), 025020, arXiv:0802.2274.

[11] F. A. Berezin, *Covariant and contravariant symbols of operators*, Math. USSR Izvestija **6** (1972), 1117–1151.

[12] N. M. J. Woodhouse, *Geometric quantization*, 2nd ed., Oxford University Press, Oxford, 1991.

Robert Oeckl
Instituto de Matemáticas
Universidad Nacional Autónoma de México, Campus Morelia
C.P. 58190, Morelia, Michoacán
Mexico
e-mail: `robert@matmor.unam.mx`

Causal Fermion Systems: A Quantum Space-Time Emerging From an Action Principle

Felix Finster, Andreas Grotz and Daniela Schiefeneder

Abstract. Causal fermion systems are introduced as a general mathematical framework for formulating relativistic quantum theory. By specializing, we recover earlier notions like fermion systems in discrete space-time, the fermionic projector and causal variational principles. We review how an effect of spontaneous structure formation gives rise to a topology and a causal structure in space-time. Moreover, we outline how to construct a spin connection and curvature, leading to a proposal for a "quantum geometry" in the Lorentzian setting. We review recent numerical and analytical results on the support of minimizers of causal variational principles which reveal a "quantization effect" resulting in a discreteness of space-time. A brief survey is given on the correspondence to quantum field theory and gauge theories.

Mathematics Subject Classification (2010). Primary 51P05, 81T20; Secondary 49Q20, 83C45, 47B50, 47B07.

Keywords. Quantum geometry, causal fermion systems, fermionic projector approach.

1. The general framework of causal fermion systems

Causal fermion systems provide a general mathematical framework for the formulation of relativistic quantum theory. They arise by generalizing and harmonizing earlier notions like the "fermionic projector," "fermion systems in discrete space-time" and "causal variational principles." After a brief motivation of the basic objects (Section 1.1), we introduce the general framework, trying to work out the mathematical essence from an abstract point of view (Sections 1.2 and 1.3). By specializing, we then recover the earlier notions (Section 1.4). Our presentation is intended as a mathematical introduction, which can clearly be supplemented by the more physical introductions in the survey articles [9, 12, 13].

Supported in part by the Deutsche Forschungsgemeinschaft.

1.1. Motivation of the basic objects

In order to put the general objects into a simple and concrete context, we begin with the free Dirac equation in Minkowski space. Thus we let $(M, \langle .,. \rangle)$ be Minkowski space (with the signature convention $(+---)$) and $d\mu$ the standard volume measure (thus $d\mu = d^4x$ in a reference frame $x = (x^0, \ldots, x^3)$). We consider a subspace I of the solution space of the Dirac equation $(i\gamma^j \partial_j - m)\psi = 0$ (I may be finite- or infinite-dimensional). On I we introduce a scalar product $\langle .|. \rangle_{\mathcal{H}}$. The most natural choice is to take the scalar product associated to the probability integral,

$$\langle \psi | \phi \rangle_{\mathcal{H}} = 2\pi \int_{t=\text{const}} (\overline{\psi} \gamma^0 \phi)(t, \vec{x})\, d\vec{x} \tag{1}$$

(where $\overline{\psi} = \psi^\dagger \gamma^0$ is the usual adjoint spinor; note that due to current conservation, the value of the integral is independent of t), but other choices are also possible. In order not to distract from the main ideas, in this motivation we disregard technical issues by implicitly assuming that the scalar product $\langle .|. \rangle_{\mathcal{H}}$ is well-defined on I and by ignoring the fact that mappings on I may be defined only on a dense subspace (for details on how to make the following consideration rigorous see [14, Section 4]). Forming the completion of I, we obtain a Hilbert space $(\mathcal{H}, \langle .|. \rangle_{\mathcal{H}})$.

Next, for any $x \in M$ we introduce the sesquilinear form

$$b \,:\, \mathcal{H} \times \mathcal{H} \to \mathbb{C} \,:\, (\psi, \phi) \mapsto -(\overline{\psi}\phi)(x)\,. \tag{2}$$

As the inner product $\overline{\psi}\phi$ on the Dirac spinors is indefinite of signature $(2, 2)$, the sesquilinear form b has signature (p, q) with $p, q \leq 2$. Thus we may uniquely represent it as

$$b(\psi, \phi) = \langle \psi | F\phi \rangle_{\mathcal{H}} \tag{3}$$

with a self-adjoint operator $F \in \mathrm{L}(\mathcal{H})$ of finite rank, which (counting with multiplicities) has at most two positive and at most two negative eigenvalues. Introducing this operator for every $x \in M$, we obtain a mapping

$$F \,:\, M \to \mathcal{F}\,, \tag{4}$$

where $\mathcal{F} \subset \mathrm{L}(\mathcal{H})$ denotes the set of all self-adjoint operators of finite rank with at most two positive and at most two negative eigenvalues, equipped with the topology induced by the Banach space $\mathrm{L}(\mathcal{H})$.

It is convenient to simplify this setting in the following way. In most physical applications, the mapping F will be injective with a closed image. Then we can identify M with the subset $F(M) \subset \mathcal{F}$. Likewise, we can identify the measure μ with the push-forward measure $\rho = F_* \mu$ on $F(M)$ (defined by $\rho(\Omega) = \mu(F^{-1}(\Omega))$). The measure ρ is defined even on all of $\mathcal{F}$, and the image of F coincides with the support of ρ. Thus setting $M = \operatorname{supp} \rho$, we can reconstruct space-time from ρ. This construction allows us to describe the physical system by a single object: the measure ρ on $\mathcal{F}$. Moreover, we can extend our notion of space-time simply by allowing ρ to be a more general measure (i.e. a measure which can no longer be realized as the push-forward $\rho = F_* \mu$ of the volume measure in Minkowski space by a continuous function F).

We have the situation in mind when I is composed of all the occupied fermionic states of a physical system, including the states of the Dirac sea (for a physical discussion see [13]). In this situation, the causal structure is encoded in the spectrum of the operator product $F(x) \cdot F(y)$. In the remainder of this section, we explain how this works for the vacuum. Thus let us assume that I is the subspace of all negative-energy solutions of the Dirac equation. We first compute F.

Lemma 1.1. *Let ψ, ϕ be two smooth negative-energy solutions of the free Dirac equation. We set*

$$\big(F(y)\,\phi\big)(x) = P(x,y)\,\phi(y)\,,$$

where $P(x,y)$ is the distribution

$$P(x,y) = \int \frac{d^4k}{(2\pi)^4}\,(\not{k}+m)\,\delta(k^2-m^2)\,\Theta(-k^0)\,e^{-ik(x-y)}\,. \tag{5}$$

Then the equation

$$\langle \psi | F(y)\,\phi \rangle_{\mathcal{H}} = -(\overline{\psi}\phi)(y)$$

holds, where all integrals are to be understood in the distributional sense.

Proof. We can clearly assume that ψ is a plane-wave solution, which for convenience we write as

$$\psi(x) = (\not{q}+m)\,\chi\,e^{-iqx}\,, \tag{6}$$

where $q = (q^0, \vec{q})$ with $\vec{q} \in \mathbb{R}^3$ and $q^0 = -\sqrt{|\vec{q}|^2+m^2}$ is a momentum on the lower mass shell and χ is a constant spinor. A straightforward calculation yields

$$\begin{aligned}
\langle \psi | F(y)\,\phi \rangle_{\mathcal{H}} &\overset{(1)}{=} 2\pi \int_{\mathbb{R}^3} d\vec{x}\,\overline{\chi}\,(\not{q}+m)\,e^{iqx}\,\gamma^0\,P(x,y)\,\phi(y) \\
&\overset{(5)}{=} \int d^4k\,\delta^3(\vec{k}-\vec{q})\,\overline{\chi}\,(\not{q}+m)\gamma^0(\not{k}+m)\,\delta(k^2-m^2)\,\Theta(-k^0)\,e^{iky}\phi(y) \\
&= \frac{1}{2|q^0|}\,\overline{\chi}\,(\not{q}+m)\gamma^0(\not{q}+m)\,e^{iqy}\phi(y) \\
&\overset{(*)}{=} -\overline{\chi}\,(\not{q}+m)\,e^{iqy}\,\phi(y) = -(\overline{\psi}\phi)(y)\,,
\end{aligned}$$

where in $(*)$ we used the anti-commutation relations of the Dirac matrices. □

This lemma gives an explicit solution to (3) and (2). The fact that $F(y)\phi$ is merely a distribution shows that an ultraviolet regularization is needed in order for $F(y)$ to be a well-defined operator on $\mathcal{H}$. We will come back to this technical point after (41) and refer to [14, Section 4] for details. For clarity, we now proceed simply by computing the eigenvalues of the operator product $F(y) \cdot F(x)$ formally (indeed, the following calculation is mathematically rigorous except if y lies on the boundary of the light cone centered at x, in which case the expressions become singular). First of all, as the operators $F(y)$ and $F(x)$ have rank at most four, we know that their product $F(y) \cdot F(x)$ also has at most four non-trivial eigenvalues, which counting with algebraic multiplicities we denote by $\lambda_1 \ldots, \lambda_4$. Since this operator product is self-adjoint only if the factors $F(x)$ and $F(y)$ commute, the eigenvalues $\lambda_1, \ldots, \lambda_4$ will in general be complex. By iterating Lemma 1.1, we find that for any $n \geq 0$,

$$\Big(F(x)\,\big(F(y)\,F(x)\big)^n \phi\Big)(z) = P(z,x)\,\Big(P(x,y)\,P(y,x)\Big)^n \phi(x)\,.$$

Forming a characteristic polynomial, one sees that the non-trivial eigenvalues of $F(y){\cdot}F(x)$ coincide precisely with the eigenvalues of the (4×4)-matrix A_{xy} defined by

$$A_{xy} = P(x,y)\,P(y,x)\,.$$

The qualitative properties of the eigenvalues of A_{xy} can easily be determined even without computing the Fourier integral (5): From Lorentz symmetry, we know that for all x and y for which the Fourier integral exists, $P(x,y)$ can be written as

$$P(x,y) \;=\; \alpha\,(y-x)_j \gamma^j + \beta\,\mathbb{1} \tag{7}$$

with two complex coefficients α and β. Taking the conjugate, we see that

$$P(y,x) \;=\; \overline{\alpha}\,(y-x)_j \gamma^j + \overline{\beta}\,\mathbb{1}\,.$$

As a consequence,

$$A_{xy} \;=\; P(x,y)\,P(y,x) \;=\; a\,(y-x)_j \gamma^j + b\,\mathbb{1} \tag{8}$$

with two real parameters a and b given by

$$a \;=\; \alpha\overline{\beta} + \beta\overline{\alpha}\,, \qquad b \;=\; |\alpha|^2\,(y-x)^2 + |\beta|^2\,. \tag{9}$$

Applying the formula $(A_{xy} - b\mathbb{1})^2 = a^2\,(y-x)^2\,\mathbb{1}$, we find that the roots of the characteristic polynomial of A_{xy} are given by

$$b \pm \sqrt{a^2\,(y-x)^2}\,.$$

Thus if the vector $(y-x)$ is timelike, the term $(y-x)^2$ is positive, so that the λ_j are all real. If conversely the vector $(y-x)$ is spacelike, the term $(y-x)^2$ is negative, and the λ_j form a complex conjugate pair. We conclude that the the causal structure of Minkowski space has the following spectral correspondence:

The non-trivial eigenvalues of $F(x){\cdot}F(y)$

$$\left\{ \begin{array}{c} \text{are real} \\ \text{form a complex conjugate pair} \end{array} \right\} \text{ if } x \text{ and } y \text{ are } \left\{ \begin{array}{c} \text{timelike} \\ \text{spacelike} \end{array} \right\} \text{ separated.} \tag{10}$$

1.2. Causal fermion systems

Causal fermion systems have two formulations, referred to as the particle and the space-time representation. We now introduce both formulations and explain their relation. After that, we introduce the setting of the fermionic projector as a special case.

1.2.1. From the particle to the space-time representation.

Definition 1.2. Given a complex Hilbert space $(\mathcal{H}, \langle .|. \rangle_{\mathcal{H}})$ (the *particle space*) and a parameter $n \in \mathbb{N}$ (the *spin dimension*), we let $\mathcal{F} \subset \mathrm{L}(\mathcal{H})$ be the set of all self-adjoint operators on $\mathcal{H}$ of finite rank which (counting with multiplicities) have at most n positive and at most n negative eigenvalues. On $\mathcal{F}$ we are given a positive measure ρ (defined on a σ-algebra of subsets of $\mathcal{F}$), the so-called *universal measure*. We refer to $(\mathcal{H}, \mathcal{F}, \rho)$ as a *causal fermion system in the particle representation*.

Vectors in the particle space have the interpretation as the occupied fermionic states of our system. The name "universal measure" is motivated by the fact that ρ describes the distribution of the fermions in a space-time "universe", with causal relations defined as follows.

Definition 1.3. (causal structure) For any $x, y \in \mathcal{F}$, the product xy is an operator of rank at most $2n$. We denote its non-trivial eigenvalues (counting with algebraic multiplicities) by $\lambda_1^{xy}, \ldots, \lambda_{2n}^{xy}$. The points x and y are called *timelike separated* if the λ_j^{xy} are all real. They are said to be *spacelike separated* if all the λ_j^{xy} are complex and have the same absolute value. In all other cases, the points x and y are said to be *lightlike separated.*

Since the operators xy and yx are isospectral (this follows from the matrix identity $\det(BC - \lambda \mathbb{1}) = \det(CB - \lambda \mathbb{1})$; see for example [7, Section 3]), this definition is symmetric in x and y.

We now construct additional objects, leading us to the more familiar space-time representation. First, on $\mathcal{F}$ we consider the topology induced by the operator norm $\|A\| := \sup\{\|Au\|_{\mathcal{H}} \text{ with } \|u\|_{\mathcal{H}} = 1\}$. For every $x \in \mathcal{F}$ we define the *spin space* S_x by $S_x = x(\mathcal{H})$; it is a subspace of $\mathcal{H}$ of dimension at most $2n$. On S_x we introduce the *spin scalar product* $\prec .|. \succ_x$ by

$$\prec u|v \succ_x = -\langle u|xu \rangle_{\mathcal{H}} \qquad (\text{for all } u, v \in S_x)\,; \tag{11}$$

it is an indefinite inner product of signature (p, q) with $p, q \leq n$. We define *space-time* M as the support of the universal measure, $M = \operatorname{supp} \rho$. It is a closed subset of $\mathcal{F}$, and by restricting the causal structure of $\mathcal{F}$ to M, we get causal relations in space-time. A *wave function* ψ is defined as a function which to every $x \in M$ associates a vector of the corresponding spin space,

$$\psi \,:\, M \to \mathcal{H} \qquad \text{with} \qquad \psi(x) \in S_x \quad \text{for all } x \in M\,. \tag{12}$$

On the wave functions we introduce the indefinite inner product

$$<\psi|\phi> = \int_M \prec \psi(x)|\phi(x) \succ_x \, d\rho(x)\,. \tag{13}$$

In order to ensure that the last integral converges, we also introduce the norm $||| \,.\, |||$ by

$$||| \,\psi\, |||^2 = \int_M \langle \psi(x)|\, |x|\, \psi(x) \rangle_{\mathcal{H}} \, d\rho(x) \tag{14}$$

(where $|x|$ is the absolute value of the operator x on $\mathcal{H}$). The *one-particle space* $\mathcal{K}$ is defined as the space of wave functions for which the norm $||| \,.\, |||$

is finite, with the topology induced by this norm, and endowed with the inner product $<.|.>$. Then $(\mathcal{K}, <.|.>)$ is a Krein space (see [2]). Next, for any $x, y \in M$ we define the kernel of the fermionic operator $P(x,y)$ by

$$P(x,y) = \pi_x \, y \,:\, S_y \to S_x \,, \tag{15}$$

where π_x is the orthogonal projection onto the subspace $S_x \subset \mathcal{H}$. The *closed chain* is defined as the product

$$A_{xy} = P(x,y)\, P(y,x) \,:\, S_x \to S_x \,.$$

As it is an endomorphism of S_x, we can compute its eigenvalues. The calculation $A_{xy} = (\pi_x y)(\pi_y x) = \pi_x \, yx$ shows that these eigenvalues coincide precisely with the non-trivial eigenvalues $\lambda_1^{xy}, \ldots, \lambda_{2n}^{xy}$ of the operator xy as considered in Definition 1.3. In this way, the kernel of the fermionic operator encodes the causal structure of M. Choosing a suitable dense domain of definition[1] $\mathcal{D}(P)$, we can regard $P(x,y)$ as the integral kernel of a corresponding operator P,

$$P \,:\, \mathcal{D}(P) \subset \mathcal{K} \to \mathcal{K} \,, \qquad (P\psi)(x) = \int_M P(x,y)\, \psi(y)\, d\rho(y) \,, \tag{16}$$

referred to as the *fermionic operator*. We collect two properties of the fermionic operator:

(A) P is *symmetric* in the sense that $<P\psi|\phi> = <\psi|P\phi>$ for all $\psi, \phi \in \mathcal{D}(P)$:

According to the definitions (15) and (11),

$$\begin{aligned} \prec P(x,y)\, \psi(y) \,|\, \psi(x) \succ_x &= -\langle (\pi_x \, y\, \psi(y)) \,|\, x\, \phi(x) \rangle_{\mathcal{H}} \\ &= -\langle \psi(y) \,|\, yx\, \phi(x) \rangle_{\mathcal{H}} = \prec \psi(y) \,|\, P(y,x)\, \psi(x) \succ_y \,. \end{aligned}$$

We now integrate over x and y and apply (16) and (13).

(B) $(-P)$ is *positive* in the sense that $<\psi|(-P)\psi> \geq 0$ for all $\psi \in \mathcal{D}(P)$:
This follows immediately from the calculation

$$\begin{aligned} <\psi|(-P)\psi> &= -\iint_{M\times M} \prec \psi(x) \,|\, P(x,y)\, \psi(y) \succ_x \, d\rho(x)\, d\rho(y) \\ &= \iint_{M\times M} \langle \psi(x) \,|\, x\, \pi_x \, y\, \psi(y) \rangle_{\mathcal{H}} \, d\rho(x)\, d\rho(y) = \langle \phi | \phi \rangle_{\mathcal{H}} \geq 0 \,, \end{aligned}$$

where we again used (13) and (15) and set

$$\phi = \int_M x\, \psi(x)\, d\rho(x) \,.$$

[1]For example, one may choose $\mathcal{D}(P)$ as the set of all vectors $\psi \in \mathcal{K}$ satisfying the conditions

$$\phi := \int_M x\, \psi(x)\, d\rho(x) \,\in\, \mathcal{H} \qquad \text{and} \qquad |\!|\!|\, \phi \,|\!|\!| < \infty \,.$$

The *space-time representation* of the causal fermion system consists of the Krein space $(\mathcal{K}, <.|.>)$, whose vectors are represented as functions on M (see (12), (13)), together with the fermionic operator P in the integral representation (16) with the above properties (A) and (B).

Before going on, it is instructive to consider the symmetries of our framework. First of all, unitary transformations

$$\psi \to U\psi \qquad \text{with } U \in \mathrm{L}(\mathcal{H}) \text{ unitary}$$

give rise to isomorphic systems. This symmetry corresponds to the fact that the fermions are *indistinguishable particles* (for details see [5, §3.2] and [11, Section 3]). Another symmetry becomes apparent if we choose basis representations of the spin spaces and write the wave functions in components. Denoting the signature of $(S_x, \prec.|.\succ_x)$ by $(p(x), q(x))$, we choose a pseudo-orthonormal basis $(\mathfrak{e}_\alpha(x))_{\alpha=1,\ldots,p+q}$ of S_x,

$$\prec \mathfrak{e}_\alpha | \mathfrak{e}_\beta \succ = s_\alpha \, \delta_{\alpha\beta} \qquad \text{with} \qquad s_1, \ldots, s_p = 1\,, \quad s_{p+1}, \ldots, s_{p+q} = -1\,.$$

Then a wave function $\psi \in \mathcal{K}$ can be represented as

$$\psi(x) = \sum_{\alpha=1}^{p+q} \psi^\alpha(x)\, \mathfrak{e}_\alpha(x)$$

with component functions $\psi^1, \ldots, \psi^{p+q}$. The freedom in choosing the basis $(\mathfrak{e}_\alpha)$ is described by the group $\mathrm{U}(p,q)$ of unitary transformations with respect to an inner product of signature (p,q),

$$\mathfrak{e}_\alpha \to \sum_{\beta=1}^{p+q} (U^{-1})^\beta_\alpha \, \mathfrak{e}_\beta \qquad \text{with } U \in \mathrm{U}(p,q)\,. \tag{17}$$

As the basis $(\mathfrak{e}_\alpha)$ can be chosen independently at each space-time point, this gives rise to local unitary transformations of the wave functions,

$$\psi^\alpha(x) \to \sum_{\beta=1}^{p+q} U(x)^\alpha_\beta \, \psi^\beta(x)\,. \tag{18}$$

These transformations can be interpreted as *local gauge transformations* (see also Section 5). Thus in our framework, the gauge group is the isometry group of the spin scalar product; it is a non-compact group whenever the spin scalar product is indefinite. *Gauge invariance* is incorporated in our framework simply because the basic definitions are basis-independent.

The fact that we have a distinguished representation of the wave functions as functions on M can be expressed by the *space-time projectors*, defined as the operators of multiplication by a characteristic function. Thus for any measurable $\Omega \subset M$, we define the space-time projector E_Ω by

$$E_\Omega \,:\, \mathcal{K} \to \mathcal{K}\,, \qquad (E_\Omega \psi)(x) = \chi_\Omega(x)\, \psi(x)\,.$$

Obviously, the space-time projectors satisfy the relations

$$E_U E_V = E_{U \cap V}\,, \quad E_U + E_V = E_{U \cup V} + E_{U \cap V}\,, \qquad E_M = \mathbb{1}_\mathcal{K}\,, \tag{19}$$

which are familiar in functional analysis as the relations which characterize spectral projectors. We can now take the measure space (M, ρ) and the Krein space $(\mathcal{K}, <.|.>)$ together with the fermionic operator and the space-time projectors as the abstract starting point.

1.2.2. From the space-time to the particle representation.

Definition 1.4. Let (M, ρ) be a measure space (*"space-time"*) and $(\mathcal{K}, <.|.>)$ a Krein space (the *"one-particle space"*). Furthermore, we let $P : \mathcal{D}(P) \subset \mathcal{K} \to \mathcal{K}$ be an operator with dense domain of definition $\mathcal{D}(P)$ (the *"fermionic operator"*), such that P is symmetric and $(-P)$ is positive (see (A) and (B) on page 162). Moreover, to every ρ-measurable set $\Omega \subset M$ we associate a projector E_Ω onto a closed subspace $E_\Omega(\mathcal{K}) \subset \mathcal{K}$, such that the resulting family of operators (E_Ω) (the *"space-time projectors"*) satisfies the relations (19). We refer to (M, ρ) together with $(\mathcal{K}, <.|.>, E_\Omega, P)$ as a *causal fermion system in the space-time representation.*

This definition is more general than the previous setting because it does not involve a notion of spin dimension. Before one can introduce this notion, one needs to "localize" the vectors in $\mathcal{K}$ with the help of the space-time projectors to obtain wave functions on M. If ρ were a discrete measure, this localization could be obtained by considering the vectors $E_x\psi$ with $x \in \operatorname{supp}\rho$. If ψ could be expected to be a continuous function, we could consider the vectors $E_{\Omega_n}\psi$ for Ω_n a decreasing sequence of neighborhoods of a single point. In the general setting of Definition 1.4, however, we must use a functional analytic construction, which in the easier Hilbert space setting was worked out in [4]. We now sketch how the essential parts of the construction can be carried over to Krein spaces. First, we need some technical assumptions.

Definition 1.5. A causal fermion system in the space-time representation has *spin dimension at most n* if there are vectors $\psi_1, \ldots, \psi_{2n} \in \mathcal{K}$ with the following properties:

(i) For every measurable set Ω, the matrix S with components $S_{ij} = <\psi_i|E_\Omega\psi_j>$ has at most n positive and at most n negative eigenvalues.

(ii) The set
$$\{E_\Omega\psi_k \text{ with } \Omega \text{ measurable and } k = 1, \ldots, 2n\} \tag{20}$$
generates a dense subset of $\mathcal{K}$.

(iii) For all $j, k \in \{1, \ldots, 2n\}$, the mapping
$$\mu_{jk} : \Omega \to <\psi_j|E_\Omega\psi_k>$$
defines a complex measure on M which is absolutely continuous with respect to ρ.

This definition allows us to use the following construction. In order to introduce the spin scalar product between the vectors $\psi_1, \ldots \psi_{2n}$, we use property (iii) to form the Radon-Nikodym decomposition

$$<\psi_j|E_\Omega\,\psi_k> = \int_\Omega \prec\psi_j|\psi_k\succ_x \,d\rho(x) \quad \text{with} \quad \prec\psi_j|\psi_k\succ \in L^1(M,d\rho)\,,$$

valid for any measurable set $\Omega \subset M$. Setting

$$\prec E_U\psi_j|E_V\psi_k\succ_x = \chi_U(x)\,\chi_V(x)\;\prec\psi_j|\psi_k\succ_x\,,$$

we can extend the spin scalar product to the sets (20). Property (ii) allows us to extend the spin scalar product by approximation to all of $\mathcal{K}$. Property (i) ensures that the spin scalar product has the signature (p,q) with $p,q \leq n$.

Having introduced the spin scalar product, we can now get a simple connection to the particle representation: The range of the fermionic operator $I := P(\mathcal{D}(P))$ is a (not necessarily closed) subspace of $\mathcal{K}$. By

$$\langle P(\phi)\,|\,P(\phi')\rangle := <\phi|(-P)\phi'>$$

we introduce on I an inner product $\langle .|. \rangle$, which by the positivity property (B) is positive semi-definite. Thus its abstract completion $\mathcal{H} := \overline{I}$ is a Hilbert space $(\mathcal{H}, \langle .|. \rangle_\mathcal{H})$. We again let $\mathcal{F} \subset \mathrm{L}(\mathcal{H})$ be the set of all self-adjoint operators of finite rank which have at most n positive and at most n negative eigenvalues. For any $x \in M$, the conditions

$$\langle\psi|F\phi\rangle_\mathcal{H} = -\prec\psi|\phi\succ_x \quad \text{for all } \psi,\phi \in I \tag{21}$$

uniquely define a self-adjoint operator F on I, which has finite rank and at most n positive and at most n negative eigenvalues. By continuity, this operator uniquely extends to an operator $F \in \mathcal{F}$, referred to as the *local correlation operator* at x. We thus obtain a mapping $F : M \to \mathcal{F}$. Identifying points of M which have the same image (see the discussion below), we can also consider the subset $\mathcal{F}(M)$ of $\mathcal{F}$ as our space-time. Replacing M by $F(M)$ and ρ by the push-forward measure $F_*\rho$ on $\mathcal{F}$, we get back to the setting of Definition 1.2.

We point out that, despite the fact that the particle and space-time representations can be constructed from each other, the two representations are *not* equivalent. The reason is that the construction of the particle representation involves an identification of points of M which have the same local correlation operators. Thus it is possible that two causal fermion systems in the space-time representation which are not gauge-equivalent may have the same particle representation[2]. In this case, the two systems have identical causal structures and give rise to exactly the same densities and correlation functions. In other words, the two systems are indistinguishable by any measurements, and thus one can take the point of view that they are equivalent descriptions of the same physical system. Moreover, the particle

[2]As a simple example consider the case $M = \{0,1\}$ with ρ the counting measure, $\mathcal{K} = \mathbb{C}^4$ with $<\psi|\phi> = \langle\psi, S\phi\rangle_{\mathbb{C}^4}$ and the signature matrix $S = \mathrm{diag}(1,-1,1,-1)$. Moreover, we choose the space-time projectors as $E_1 = \mathrm{diag}(1,1,0,0)$, $E_2 = \mathrm{diag}(0,0,1,1)$ and consider a one-particle fermionic operator $P = -|\psi><\psi|$. Then the systems obtained by choosing $\psi = (0,1,0,0)$ and $\psi = (0,1,1,1)$ are not gauge-equivalent, although they give rise to the same particle representation.

representation gives a cleaner framework, without the need for technical assumptions as in Definition 1.5. For these reasons, it seems preferable to take the point of view that the particle representation is more fundamental, and to always deduce the space-time representation by the constructions given after Definition 1.2.

1.2.3. The setting of the fermionic projector. A particularly appealing special case is the setting of the *fermionic projector*, which we now review. Beginning in the particle representation, we impose the additional constraint

$$\int_M x\, d\rho(x) = \mathbb{1}_{\mathcal{H}}\,, \tag{22}$$

where the integral is assumed to converge in the strong sense, meaning that

$$\int_M \|x\,\psi\|\, d\rho(x) < \infty \quad \text{for all } \psi \in \mathcal{H}$$

(where $\|\psi\| = \sqrt{\langle\psi|\psi\rangle_{\mathcal{H}}}$ is the norm on $\mathcal{H}$). Under these assumptions, it is straightforward to verify from (14) that the mapping

$$\iota\, :\, \mathcal{H} \to \mathcal{K}\,, \qquad (\iota\psi)(x) = \pi_x\psi$$

is well-defined. Moreover, the calculation

$$<\!\iota\psi|\iota\psi\!> = \int_M \prec\!\pi_x\psi\,|\,\pi_x\phi\!\succ_x\, d\rho(x) \overset{(11)}{=} -\int_M \langle\psi\,|\,x\phi\rangle_{\mathcal{H}}\, d\rho(x) \overset{(22)}{=} -\langle\psi|\phi\rangle_{\mathcal{H}}$$

shows that ι is, up to a minus sign, an isometric embedding of $\mathcal{H}$ into $\mathcal{K}$. Thus we may identify $\mathcal{H}$ with the subspace $\iota(\mathcal{H}) \subset \mathcal{K}$, and on this closed subspace the inner products $\prec.|.\succ_{\mathcal{H}}$ and $<.|.>|_{\mathcal{H}\times\mathcal{H}}$ coincide up to a sign. Moreover, the calculation

$$(P\iota\psi)(x) = \int_M \pi_x\, y\, \pi_y\, \psi\, d\rho(y) = \int_M \pi_x\, y\, \psi\, d\rho(y) = \pi_x\psi = (\iota\psi)(x)$$

yields that P restricted to $\mathcal{H}$ is the identity. Next, for every $\psi \in \mathcal{D}(P)$, the estimate

$$\begin{aligned}\Big\| \int_M y\,\psi(y)\, d\rho(y)\Big\|^2 &\overset{(22)}{=} \int_M d\rho(x) \iint_{M\times M} d\rho(y)\, d\rho(z)\, \langle y\,\psi(y)\,|\,xz\,\psi(z)\rangle_{\mathcal{H}} \\ &= -<\!P\psi\,|\,P\psi\!> < \infty\end{aligned}$$

shows (after a straightforward approximation argument) that

$$\phi := \int_M y\,\psi(y)\, d\rho(y) \;\in\; \mathcal{H}\,.$$

On the other hand, we know from (16) and (15) that $P\psi = \iota\phi$. This shows that the image of P is contained in $\mathcal{H}$. We conclude that P is a *projection operator* in $\mathcal{K}$ onto the negative definite, closed subspace $\mathcal{H} \subset \mathcal{K}$.

1.3. An action principle

We now return to the general setting of Definitions 1.2 and 1.3. For two points $x, y \in \mathcal{F}$ we define the *spectral weight* $|.|$ of the operator products xy and $(xy)^2$ by

$$|xy| = \sum_{i=1}^{2n} |\lambda_i^{xy}| \qquad \text{and} \qquad \big|(xy)^2\big| = \sum_{i=1}^{2n} |\lambda_i^{xy}|^2 \,.$$

We also introduce the

$$\textit{Lagrangian} \qquad \mathcal{L}(x,y) = \big|(xy)^2\big| - \frac{1}{2n}\, |xy|^2 \,. \tag{23}$$

For a given universal measure ρ on $\mathcal{F}$, we define the non-negative functionals

$$\textit{action} \qquad \mathcal{S}[\rho] = \iint_{\mathcal{F}\times\mathcal{F}} \mathcal{L}(x,y)\, d\rho(x)\, d\rho(y) \tag{24}$$

$$\textit{constraint} \qquad \mathcal{T}[\rho] = \iint_{\mathcal{F}\times\mathcal{F}} |xy|^2\, d\rho(x)\, d\rho(y) \,. \tag{25}$$

Our action principle is to

$$\text{minimize } \mathcal{S} \text{ for fixed } \mathcal{T} \,, \tag{26}$$

under variations of the universal measure. These variations should keep the total volume unchanged, which means that a variation $(\rho(\tau))_{\tau\in(-\varepsilon,\varepsilon)}$ should for all $\tau, \tau' \in (-\varepsilon, \varepsilon)$ satisfy the conditions

$$\big|\rho(\tau) - \rho(\tau')\big|(\mathcal{F}) < \infty \qquad \text{and} \qquad \big(\rho(\tau) - \rho(\tau')\big)(\mathcal{F}) = 0$$

(where $|.|$ denotes the total variation of a measure; see [16, §28]). Depending on the application, one may impose additional constraints. For example, in the setting of the fermionic projector, the variations should obey the condition (22). Moreover, one may prescribe properties of the universal measure by choosing a measure space $(\hat{M}, \hat{\mu})$ and restricting attention to universal measures which can be represented as the push-forward of $\hat{\mu}$,

$$\rho = F_* \hat{\mu} \qquad \text{with} \qquad F \,:\, \hat{M} \to \mathcal{F} \text{ measurable} \,. \tag{27}$$

One then minimizes the action under variations of the mapping F.

The Lagrangian (23) is compatible with our notion of causality in the following sense. Suppose that two points $x, y \in \mathcal{F}$ are spacelike separated (see Definition 1.3). Then the eigenvalues λ_i^{xy} all have the same absolute value, so that the Lagrangian (23) vanishes. Thus pairs of points with spacelike separation do not enter the action. This can be seen in analogy to the usual notion of causality where points with spacelike separation cannot influence each other.

1.4. Special cases

We now discuss modifications and special cases of the above setting as considered earlier. First of all, in all previous papers except for [14] it was assumed that the Hilbert space $\mathcal{H}$ is finite-dimensional and that the measure ρ is finite. Then by rescaling, one can normalize ρ such that $\rho(M) = 1$. Moreover, the Hilbert space $(\mathcal{H}, \langle .|. \rangle_{\mathcal{H}})$ can be replaced by $\mathbb{C}^f$ with the canonical scalar product (the parameter $f \in \mathbb{N}$ has the interpretation as the number of particles of the system). These two simplifications lead to the setting of *causal variational principles* introduced in [10]. More precisely, the particle and space-time representations are considered in [10, Section 1 and 2] and [10, Section 3], respectively. The connection between the two representations is established in [10, Section 3] by considering the relation (21) in a matrix representation. In this context, F is referred to as the *local correlation matrix* at x. Moreover, in [10] the universal measure is mainly represented as in (27) as the push-forward by a mapping F. This procedure is of advantage when analyzing the variational principle, because by varying F while keeping $(\hat{M}, \hat{\mu})$ fixed, one can prescribe properties of the measure ρ. For example, if $\hat{\mu}$ is a counting measure, then one varies ρ in the restricted class of measures whose support consists of at most $\#\hat{M}$ points with integer weights. More generally, if $\hat{\mu}$ is a discrete measure, then ρ is also discrete. However, if $\hat{\mu}$ is a continuous measure (or in more technical terms a so-called non-atomic measure), then we do not get any constraints for ρ, so that varying F is equivalent to varying ρ in the class of positive normalized regular Borel measures.

Another setting is to begin in the space-time representation (see Definition 1.4), but assuming that ρ is a finite counting measure. Then the relations (19) become

$$E_x E_y = \delta_{xy} E_x \qquad \text{and} \qquad \sum_{x \in M} E_x = \mathbb{1}_{\mathcal{K}} \,,$$

whereas the "localization" discussed after (19) reduces to multiplication with the space-time projectors,

$$\psi(x) = E_x \psi \,, \qquad P(x,y) = E_x P E_y \,, \qquad \prec \psi(x) \,|\, \phi(x) \succ_x \,=\, <\psi \,|\, E_x\, \phi>.$$

This is the setting of *fermion systems in discrete space-time* as considered in [5] and [7, 6]. We point out that all the work before 2006 deals with the space-time representation, simply because the particle representation had not yet been found.

We finally remark that, in contrast to the settings considered previously, here the dimension and signature of the spin space S_x may depend on x. We only know that it is finite-dimensional, and that its positive and negative signatures are at most n. In order to get into the setting of constant spin dimension, one can isometrically embed every S_x into an indefinite inner product space of signature (n, n) (for details see [10, Section 3.3]).

2. Spontaneous structure formation

For a given measure ρ, the structures of $\mathcal{F}$ induce corresponding structures on space-time $M = \operatorname{supp} \rho \subset \mathcal{F}$. Two of these structures are obvious: First, on M we have the relative *topology* inherited from $\mathcal{F}$. Second, the causal structure on $\mathcal{F}$ (see Definition 1.3) also induces a *causal structure* on M. Additional structures like a spin connection and curvature are less evident; their construction will be outlined in Section 3 below.

The appearance of the above structures in space-time can also be understood as an effect of *spontaneous structure formation* driven by our action principle. We now explain this effect in the particle representation (for a discussion in the space-time representation see [12]). For clarity, we consider the situation (27) where the universal measure is represented as the push-forward of a given measure $\hat{\mu}$ on $\hat{M}$ (this is no loss of generality because choosing a non-atomic measure space $(\hat{M}, \hat{\mu})$, any measure ρ on $\mathcal{F}$ can be represented in this way; see [10, Lemma 1.4]). Thus our starting point is a measure space $(\hat{M}, \hat{\mu})$, without any additional structures. The symmetries are described by the group of mappings T of the form

$$T \,:\, \hat{M} \to \hat{M} \quad \text{is bijective and preserves the measure } \hat{\mu}\,. \tag{28}$$

We now consider measurable mappings $F : \hat{M} \to \mathcal{F}$ and minimize $\mathcal{S}$ under variations of F. The resulting minimizer gives rise to a measure $\rho = F_* \hat{\mu}$ on $\mathcal{F}$. On $M := \operatorname{supp} \rho$ we then have the above structures inherited from $\mathcal{F}$. Taking the pull-back by F, we get corresponding structures on $\hat{M}$. The symmetry group reduces to the mappings T which in addition to (28) preserve these structures. In this way, minimizing our action principle triggers an effect of spontaneous symmetry breaking, leading to additional structures in space-time.

3. A Lorentzian quantum geometry

We now outline constructions from [14] which give general notions of a connection and curvature (see Theorem 3.9, Definition 3.11 and Definition 3.12). We also explain how these notions correspond to the usual objects of differential geometry in Minkowski space (Theorem 3.15) and on a globally hyperbolic Lorentzian manifold (Theorem 3.16).

3.1. Construction of the spin connection

Having Dirac spinors in a four-dimensional space-time in mind, we consider as in Section 1.1 a causal fermion system of spin dimension two. Moreover, we only consider space-time points $x \in M$ which are *regular* in the sense that the corresponding spin spaces S_x have the maximal dimension four.

An important structure from spin geometry missing so far is Clifford multiplication. To this end, we need a Clifford algebra represented by symmetric operators on S_x. For convenience, we first consider Clifford algebras with the maximal number of five generators; later we reduce to four space-time dimensions (see Definition 3.14 below). We denote the set of symmetric linear endomorphisms of S_x by $\text{Symm}(S_x)$; it is a 16-dimensional real vector space.

Definition 3.1. A five-dimensional subspace $K \subset \text{Symm}(S_x)$ is called a *Clifford subspace* if the following conditions hold:

(i) For any $u, v \in K$, the anti-commutator $\{u, v\} \equiv uv + vu$ is a multiple of the identity on S_x.
(ii) The bilinear form $\langle .,. \rangle$ on K defined by

$$\frac{1}{2} \{u, v\} = \langle u, v \rangle \, \mathbb{1} \qquad \text{for all } u, v \in K \tag{29}$$

is non-degenerate and has signature $(1, 4)$.

In view of the situation in spin geometry, we would like to distinguish a specific Clifford subspace. In order to partially fix the freedom in choosing Clifford subspaces, it is useful to impose that K should contain a given so-called sign operator.

Definition 3.2. An operator $v \in \text{Symm}(S_x)$ is called a *sign operator* if $v^2 = \mathbb{1}$ and if the inner product $\prec .|v\, .\succ \,:\, S_x \times S_x \to \mathbb{C}$ is positive definite.

Definition 3.3. For a given sign operator v, the set of *Clifford extensions* $\mathcal{T}^v$ is defined as the set of all Clifford subspaces containing v,

$$\mathcal{T}^v = \{K \text{ Clifford subspace with } v \in K\} .$$

Considering x as an operator on S_x, this operator has by definition of the spin dimension two positive and two negative eigenvalues. Moreover, the calculation

$$\prec u|(-x)\, u\succ_x \overset{(11)}{=} \langle u|x^2 u\rangle_{\mathcal{H}} > 0 \quad \text{for all } u \in S_x \setminus \{0\}$$

shows that the operator $(-x)$ is positive definite on S_x. Thus we can introduce a unique sign operator s_x by demanding that the eigenspaces of s_x corresponding to the eigenvalues ± 1 are precisely the positive and negative spectral subspaces of the operator $(-x)$. This sign operator is referred to as the *Euclidean sign operator.*

A straightforward calculation shows that for two Clifford extensions $K, \tilde{K} \in \mathcal{T}^v$, there is a unitary transformation $U \in e^{i\mathbb{R}v}$ such that $\tilde{K} = UKU^{-1}$ (for details see [14, Section 3]). By dividing out this group action, we obtain a five-dimensional vector space, endowed with the inner product $\langle ., \rangle$. Taking for v the Euclidean signature operator, we regard this vector space as a generalization of the usual tangent space.

Definition 3.4. The *tangent space* T_x is defined by

$$T_x = \mathcal{T}_x^{s_x} / \exp(i\mathbb{R}s_x) .$$

It is endowed with an inner product $\langle .,. \rangle$ of signature $(1, 4)$.

We next consider two space-time points, for which we need to make the following assumption.

Definition 3.5. Two points $x, y \in M$ are said to be *properly time-like separated* if the closed chain A_{xy} has a strictly positive spectrum and if the corresponding eigenspaces are definite subspaces of S_x.

This definition clearly implies that x and y are time-like separated (see Definition 1.3). Moreover, the eigenspaces of A_{xy} are definite if and only if those of A_{yx} are, showing that Definition 3.5 is again symmetric in x and y. As a consequence, the spin space can be decomposed uniquely into an orthogonal direct sum $S_x = I^+ \oplus I^-$ of a positive definite subspace I^+ and a negative definite subspace I^- of A_{xy}. This allows us to introduce a unique sign operator v_{xy} by demanding that its eigenspaces corresponding to the eigenvalues ± 1 are the subspaces $I^\pm$. This sign operator is referred to as the *directional sign operator* of A_{xy}. Having two sign operators s_x and v_{xy} at our disposal, we can distinguish unique corresponding Clifford extensions, provided that the two sign operators satisfy the following generic condition.

Definition 3.6. Two sign operators $v, \tilde{v}$ are said to be *generically separated* if their commutator $[v, \tilde{v}]$ has rank four.

Lemma 3.7. *Assume that the sign operators s_x and v_{xy} are generically separated. Then there are unique Clifford extensions $K_x^{(y)} \in \mathcal{T}^{s_x}$ and $K_{xy} \in \mathcal{T}^{v_{xy}}$ and a unique operator $\rho \in K_x^{(y)} \cap K_{xy}$ with the following properties:*

(i) *The relations $\{s_x, \rho\} = 0 = \{v_{xy}, \rho\}$ hold.*

(ii) *The operator $U_{xy} := e^{i\rho}$ transforms one Clifford extension to the other,*

$$K_{xy} = U_{xy}\, K_x^{(y)}\, U_{xy}^{-1}\,. \tag{30}$$

(iii) *If $\{s_x, v_{xy}\}$ is a multiple of the identity, then $\rho = 0$.*

The operator ρ depends continuously on s_x and v_{xy}.

We refer to U_{xy} as the *synchronization map*. Exchanging the roles of x and y, we also have two sign operators s_y and v_{yx} at the point y. Assuming that these sign operators are again generically separated, we also obtain a unique Clifford extension $K_{yx} \in \mathcal{T}^{v_{yx}}$.

After these preparations, we can now explain the construction of the spin connection D (for details see [14, Section 3]). For two space-time points $x, y \in M$ with the above properties, we want to introduce an operator

$$D_{x,y} \;:\; S_y \to S_x \tag{31}$$

(generally speaking, by the subscript $_{xy}$ we always denote an object at the point x, whereas the additional comma $_{x,y}$ denotes an operator which maps an object at y to an object at x). It is natural to demand that $D_{x,y}$ is unitary, that $D_{y,x}$ is its inverse, and that these operators map the directional sign operators at x and y to each other,

$$D_{x,y} = (D_{y,x})^* = (D_{y,x})^{-1} \tag{32}$$

$$v_{xy} = D_{x,y}\, v_{yx}\, D_{y,x}\,. \tag{33}$$

The obvious idea for constructing an operator with these properties is to take a polar decomposition of $P(x,y)$; this amounts to setting

$$D_{x,y} = A_{xy}^{-\frac{1}{2}}\, P(x,y)\,. \tag{34}$$

This definition has the shortcoming that it is not compatible with the chosen Clifford extensions. In particular, it does not give rise to a connection on the corresponding tangent spaces. In order to resolve this problem, we modify (34) by the ansatz

$$D_{x,y} = e^{i\varphi_{xy}\, v_{xy}}\, A_{xy}^{-\frac{1}{2}}\, P(x,y) \tag{35}$$

with a free real parameter φ_{xy}. In order to comply with (32), we need to demand that

$$\varphi_{xy} = -\varphi_{yx} \mod 2\pi\,; \tag{36}$$

then (33) is again satisfied. We can now use the freedom in choosing φ_{xy} to arrange that the distinguished Clifford subspaces K_{xy} and K_{yx} are mapped onto each other,

$$K_{xy} = D_{x,y}\, K_{yx}\, D_{y,x}\,. \tag{37}$$

It turns out that this condition determines φ_{xy} up to multiples of $\frac{\pi}{2}$. In order to fix φ_{xy} uniquely in agreement with (36), we need to assume that φ_{xy} is not a multiple of $\frac{\pi}{4}$. This leads us to the following definition.

Definition 3.8. Two points $x, y \in M$ are called *spin-connectable* if the following conditions hold:

(a) The points x and y are properly timelike separated (note that this already implies that x and y are regular as defined at the beginning of Section 3).
(b) The Euclidean sign operators s_x and s_y are generically separated from the directional sign operators v_{xy} and v_{yx}, respectively.
(c) Employing the ansatz (35), the phases φ_{xy} which satisfy condition (37) are not multiples of $\frac{\pi}{4}$.

We denote the set of points which are spin-connectable to x by $\mathcal{I}(x)$. It is straightforward to verify that $\mathcal{I}(x)$ is an open subset of M.

Under these assumptions, we can fix φ_{xy} uniquely by imposing that

$$\varphi_{xy} \in \Big(-\frac{\pi}{2}, -\frac{\pi}{4}\Big) \cup \Big(\frac{\pi}{4}, \frac{\pi}{2}\Big)\,, \tag{38}$$

giving the following result (for the proofs see [14, Section 3.3]).

Theorem 3.9. *Assume that two points $x, y \in M$ are spin-connectable. Then there is a unique* **spin connection** $D_{x,y} : S_y \to S_x$ *of the form* (35) *having the properties* (32), (33), (37) *and* (38).

3.2. A time direction, the metric connection and curvature

We now outline a few further constructions from [14, Section 3]. First, for spin-connectable points we can distinguish a direction of time.

Definition 3.10. Assume that the points $x, y \in M$ are spin-connectable. We say that y lies in the *future* of x if the phase φ_{xy} as defined by (35) and (38) is positive. Otherwise, y is said to lie in the *past* of x.

According to (36), y lies in the future of x if and only if x lies in the past of y. By distinguishing a direction of time, we get a structure similar to a causal set (see for example [3]). However, in contrast to a causal set, our notion of "lies in the future of" is not necessarily transitive.

The spin connection induces a connection on the corresponding tangent spaces, as we now explain. Suppose that $u_y \in T_y$. Then, according to Definition 3.4 and Lemma 3.7, we can consider u_y as a vector of the representative $K_y^{(x)} \in \mathcal{T}^{s_y}$. By applying the synchronization map, we obtain a vector in K_{yx},

$$u_{yx} := U_{yx}\, u_y\, U_{yx}^{-1} \in K_{yx} \,.$$

According to (37), we can now "parallel-transport" the vector to the Clifford subspace K_{xy},

$$u_{xy} := D_{x,y}\, u_{yx}\, D_{y,x} \in K_{xy} \,.$$

Finally, we apply the inverse of the synchronization map to obtain the vector

$$u_x := U_{xy}^{-1}\, u_{xy}\, U_{xy} \in K_x^{(y)} \,.$$

As $K_x^{(y)}$ is a representative of the tangent space T_x and all transformations were unitary, we obtain an isometry from T_y to T_x.

Definition 3.11. The isometry between the tangent spaces defined by

$$\nabla_{x,y} \;:\; T_y \to T_x \;:\; u_y \mapsto u_x$$

is referred to as the *metric connection* corresponding to the spin connection D.

We next introduce a notion of curvature.

Definition 3.12. Suppose that three points $x, y, z \in M$ are pairwise spin-connectable. Then the associated *metric curvature* R is defined by

$$R(x,y,z) = \nabla_{x,y}\, \nabla_{y,z}\, \nabla_{z,x} \;:\; T_x \to T_x \,. \qquad (39)$$

The metric curvature $R(x, y, z)$ can be thought of as a discrete analog of the holonomy of the Levi-Civita connection on a manifold, where a tangent vector is parallel-transported along a loop starting and ending at x. On a manifold, the curvature at x is immediately obtained from the holonomy by considering the loops in a small neighborhood of x. With this in mind, Definition 3.12 indeed generalizes the usual notion of curvature to causal fermion systems.

The following construction relates directional sign operators to vectors of the tangent space. Suppose that y is spin-connectable to x. By synchronizing the directional sign operator v_{xy}, we obtain the vector

$$\hat{y}_x := U_{xy}^{-1} \, v_{xy} \, U_{xy} \in K_x^{(y)} \,. \tag{40}$$

As $K_x^{(y)} \in \mathcal{T}^{s_x}$ is a representative of the tangent space, we can regard $\hat{y}_x$ as a tangent vector. We thus obtain a mapping

$$\mathcal{I}(x) \to T_x \;:\; y \mapsto \hat{y}_x \,.$$

We refer to $\hat{y}_x$ as the *directional tangent vector* of y in T_x. As v_{xy} is a sign operator and the transformations in (40) are unitary, the directional tangent vector is a timelike unit vector with the additional property that the inner product $\prec . | \hat{y}_x . \succ_x$ is positive definite.

We finally explain how to reduce the dimension of the tangent space to four, with the desired Lorentzian signature $(1,3)$.

Definition 3.13. The fermion system is called *chirally symmetric* if to every $x \in M$ we can associate a spacelike vector $u(x) \in T_x$ which is orthogonal to all directional tangent vectors,

$$\langle u(x), \hat{y}_x \rangle = 0 \qquad \text{for all } y \in \mathcal{I}(x) \,,$$

and is parallel with respect to the metric connection, i.e.

$$u(x) = \nabla_{x,y} \, u(y) \, \nabla_{y,x} \qquad \text{for all } y \in \mathcal{I}(x) \,.$$

Definition 3.14. For a chirally symmetric fermion system, we introduce the *reduced tangent space* T_x^{red} by

$$T_x^{\text{red}} = \langle u_x \rangle^\perp \subset T_x \,.$$

Clearly, the reduced tangent space has dimension four and signature $(1,3)$. Moreover, the operator $\nabla_{x,y}$ maps the reduced tangent spaces isometrically to each other. The local operator $\gamma^5 := -iu/\sqrt{-u^2}$ takes the role of the *pseudoscalar matrix*.

3.3. The correspondence to Lorentzian geometry

We now explain how the above spin connection is related to the usual spin connection used in spin geometry (see for example [17, 1]). To this end, let (M, g) be a time-oriented Lorentzian spin manifold with spinor bundle SM (thus $S_x M$ is a 4-dimensional complex vector space endowed with an inner product $\prec . | . \succ_x$ of signature $(2,2)$). Assume that $\gamma(t)$ is a smooth, future-directed and timelike curve, for simplicity parametrized by the arc length, defined on the interval $[0, T]$ with $\gamma(0) = y$ and $\gamma(T) = x$. Then the parallel transport of tangent vectors along γ with respect to the Levi-Civita connection ∇^{LC} gives rise to the isometry

$$\nabla^{\text{LC}}_{x,y} \;:\; T_y \to T_x \,.$$

In order to compare with the metric connection ∇ of Definition 3.11, we subdivide γ (for simplicity with equal spacing, although a non-uniform spacing would work just as well). Thus for any given N, we define the points $x_0, \ldots, x_N$ by

$$x_n = \gamma(t_n) \qquad \text{with} \qquad t_n = \frac{nT}{N} .$$

We define the parallel transport $\nabla^N_{x,y}$ by successively composing the parallel transport between neighboring points,

$$\nabla^N_{x,y} := \nabla_{x_N,x_{N-1}} \nabla_{x_{N-1},x_{N-2}} \cdots \nabla_{x_1,x_0} \;:\; T_y \to T_x .$$

Our first theorem gives a connection to the Minkowski vacuum. For any $\varepsilon > 0$ we regularize on the scale $\varepsilon > 0$ by inserting a convergence-generating factor into the integrand in (5),

$$P^\varepsilon(x,y) = \int \frac{d^4k}{(2\pi)^4}\, (\not{k} + m)\, \delta(k^2 - m^2)\, \Theta(-k^0)\, e^{\varepsilon k^0}\, e^{-ik(x-y)} . \tag{41}$$

This function can indeed be realized as the kernel of the fermionic operator (15) corresponding to a causal fermion system $(\mathcal{H}, \mathcal{F}, \rho^\varepsilon)$. Here the measure ρ^ε is the push-forward of the volume measure in Minkowski space by an operator F^ε, being an ultraviolet regularization of the operator F in (2)-(4) (for details see [14, Section 4]).

Theorem 3.15. *For given γ, we consider the family of regularized fermionic projectors of the vacuum $(P^\varepsilon)_{\varepsilon>0}$ as given by (41). Then for a generic curve γ and for every $N \in \mathbb{N}$, there is ε_0 such that for all $\varepsilon \in (0, \varepsilon_0]$ and all $n = 1, \ldots, N$, the points x_n and x_{n-1} are spin-connectable, and x_{n+1} lies in the future of x_n (according to Definition 3.10). Moreover,*

$$\nabla^{\text{LC}}_{x,y} = \lim_{N\to\infty} \lim_{\varepsilon \searrow 0} \nabla^N_{x,y} .$$

By a *generic curve* we mean that the admissible curves are dense in the C^∞-topology (i.e., for any smooth γ and every $K \in \mathbb{N}$, there is a sequence γ_ℓ of admissible curves such that $D^k\gamma_\ell \to D^k\gamma$ uniformly for all $k = 0, \ldots, K$). The restriction to generic curves is needed in order to ensure that the Euclidean and directional sign operators are generically separated (see Definition 3.8(b)). The proof of the above theorem is given in [14, Section 4].

Clearly, in this theorem the connection $\nabla^{\text{LC}}_{x,y}$ is trivial. In order to show that our connection also coincides with the Levi-Civita connection in the case with curvature, in [14, Section 5] a globally hyperbolic Lorentzian manifold is considered. For technical simplicity, we assume that the manifold is flat Minkowski space in the past of a given Cauchy hypersurface.

Theorem 3.16. *Let (M, g) be a globally hyperbolic manifold which is isometric to Minkowski space in the past of a given Cauchy hypersurface $\mathcal{N}$. For given γ, we consider the family of regularized fermionic projectors $(P^\varepsilon)_{\varepsilon>0}$ such that $P^\varepsilon(x,y)$ coincides with the distribution (41) if x and y lie in the past of $\mathcal{N}$. Then for a generic curve γ and for every sufficiently large N, there is ε_0 such that for all $\varepsilon \in (0, \varepsilon_0]$ and all $n = 1, \ldots, N$, the points x_n and x_{n-1} are* *spin-connectable, and x_{n+1} lies in the future of x_n (according to Definition 3.10). Moreover,*

$$\lim_{N\to\infty} \lim_{\varepsilon\searrow 0} \nabla^N_{x,y} - \nabla^{\text{LC}}_{x,y} = \mathcal{O}\Big(L(\gamma)\, \frac{\nabla R}{m^2} \Big) \Big(1 + \mathcal{O}\Big(\frac{\text{scal}}{m^2} \Big) \Big) , \tag{42}$$

where R denotes the Riemann curvature tensor, scal *is scalar curvature, and $L(\gamma)$ is the length of the curve γ.*

Thus the metric connection of Definition 3.11 indeed coincides with the Levi-Civita connection, up to higher-order curvature corrections. For detailed explanations and the proof we refer to [14, Section 5].

At first sight, one might conjecture that Theorem 3.16 should also apply to the spin connection in the sense that

$$D^{\text{LC}}_{x,y} = \lim_{N\to\infty} \lim_{\varepsilon\searrow 0} D^N_{x,y} , \tag{43}$$

where D^{LC} is the spin connection on SM induced by the Levi-Civita connection and

$$D^N_{x,y} := D_{x_N, x_{N-1}} D_{x_{N-1}, x_{N-2}} \cdots D_{x_1, x_0} \;:\; S_y \to S_x \tag{44}$$

(and D is the spin connection of Theorem 3.9). It turns out that this conjecture is false. But the conjecture becomes true if we replace (44) by the operator product

$$D^N_{(x,y)} := D_{x_N, x_{N-1}} U^{(x_N|x_{N-2})}_{x_{N-1}} D_{x_{N-1}, x_{N-2}} U^{(x_{N-1}|x_{N-3})}_{x_{N-2}} \cdots U^{(x_2|x_0)}_{x_1} D_{x_1, x_0} .$$

Here the intermediate factors $U^{(.|.)}_{.}$ are the so-called *splice maps* given by

$$U^{(z|y)}_x = U_{xz}\, V\, U^{-1}_{xy} ,$$

where U_{xz} and U_{xy} are synchronization maps, and $V \in \exp(i\mathbb{R}s_x)$ is an operator which identifies the representatives $K_{xy}, K_{xz} \in T_x$ (for details see [14, Section 3.7 and Section 5]). The splice maps also enter the *spin curvature* $\mathfrak{R}$, which is defined in analogy to the metric curvature (39) by

$$\mathfrak{R}(x,y,z) = U^{(z|y)}_x\, D_{x,y}\, U^{(x|z)}_y\, D_{y,z}\, U^{(y|x)}_z\, D_{z,x} \;:\; S_x \to S_x .$$

4. A "quantization effect" for the support of minimizers

The recent paper [15] contains a first numerical and analytical study of the minimizers of the action principle (26). We now explain a few results and discuss their potential physical significance. We return to the setting of causal variational principles (see Section 1.4). In order to simplify the problem as much as possible, we only consider the case of spin dimension $n = 1$ and two particles $f = 2$ (although many results in [15] apply similarly to a general finite number of particles). Thus we identify the particle space $(\mathcal{H}, \langle .|. \rangle_{\mathcal{H}})$ with $\mathbb{C}^2$. Every point $F \in \mathcal{F}$ is a Hermitian (2×2)-matrix with at most one positive and at most one negative eigenvalue. We represent it in terms of the Pauli matrices as $F = \alpha\, \mathbb{1} + \vec{u}\vec{\sigma}$ with $|\vec{u}| \geq |\alpha|$. In order to further simplify the problem, we prescribe the eigenvalues in the support of the universal measure

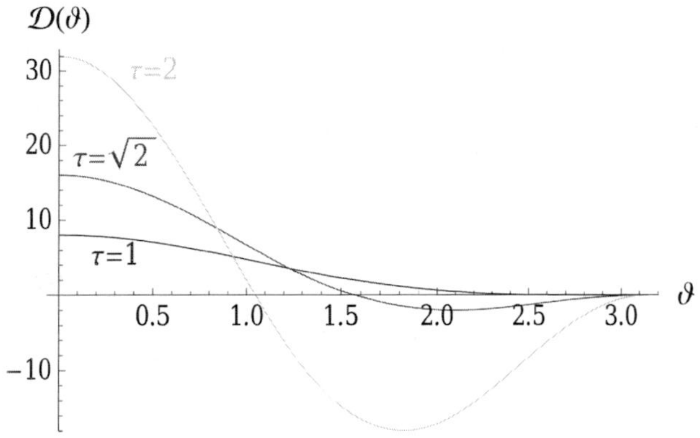

FIGURE 1. The function $\mathcal{D}$.

to be $1+\tau$ and $1-\tau$, where $\tau \geq 1$ is a given parameter. Then F can be represented as

$$F = \tau\, x \cdot \sigma + \mathbb{1} \qquad \text{with} \qquad x \in S^2 \subset \mathbb{R}^3 .$$

Thus we may identify $\mathcal{F}$ with the unit sphere S^2. The Lagrangian $\mathcal{L}(x,y)$ given in (23) simplifies to a function only of the angle ϑ between the vectors x and y. More precisely,

$$\begin{aligned} \mathcal{L}(x,y) &= \max\big(0, \mathcal{D}(\langle x, y\rangle)\big) \\ \mathcal{D}(\cos\vartheta) &= 2\tau^2\,(1+\cos\vartheta)\,\big(2-\tau^2\,(1-\cos\vartheta)\big) . \end{aligned}$$

As shown in Figure 1 in typical examples, the function $\mathcal{D}$ is positive for small ϑ and becomes negative if ϑ exceeds a certain value $\vartheta_{\max}(\tau)$. Following Definition 1.3, two points x and y are timelike separated if $\vartheta < \vartheta_{\max}$ and spacelike separated if $\vartheta > \vartheta_{\max}$.

Our action principle is to minimize the action (24) by varying the measure ρ in the family of normalized Borel measures on the sphere. In order to solve this problem numerically, we approximate the minimizing measure by a weighted counting measure. Thus for any given integer m, we choose points $x_1, \ldots, x_m \in S^2$ together with corresponding weights $\rho_1, \ldots, \rho_m$ with

$$\rho_i \geq 0 \qquad \text{and} \qquad \sum_{i=1}^m \rho_i = 1$$

and introduce the measure ρ by

$$\rho = \sum_{i=1}^m \rho_i\, \delta_{x_i} , \tag{45}$$

where δ_x denotes the Dirac measure. Fixing different values of m and seeking for numerical minimizers by varying both the points x_i and the weights ρ_i, we obtain the plots shown in Figure 2. It is surprising that for each fixed τ, the obtained minimal action no longer changes if m is increased beyond a certain value $m_0(\tau)$. The numerics shows that if $m > m_0$, some of the points

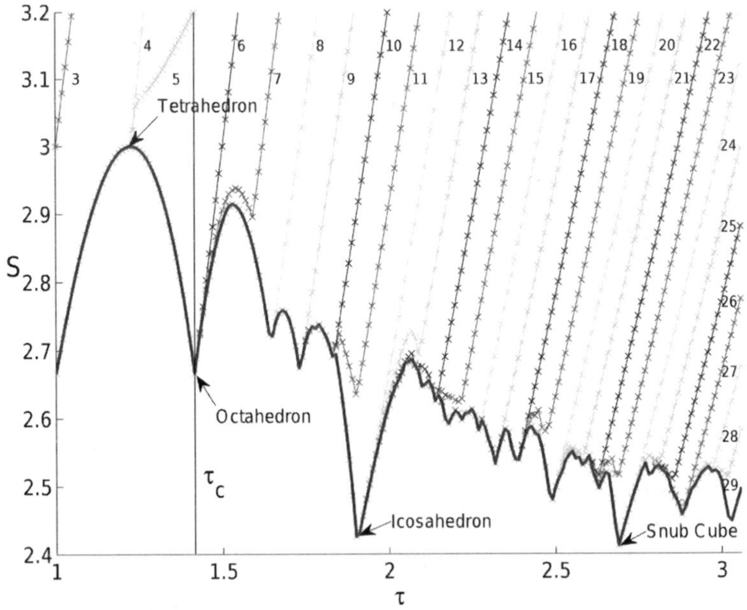

FIGURE 2. Numerical minima of the action on the two-dimensional sphere.

x_i coincide, so that the support of the minimizing measure never consists of more than m_0 points. Since this property remains true in the limit $m \to \infty$, our numerical analysis shows that the *minimizing measure is discrete* in the sense that its support consists of a finite number of m_0 points. Another interesting effect is that the action seems to favor symmetric configurations on the sphere. Namely, the most distinct local minima in Figure 2 correspond to configurations where the points x_i lie on the vertices of Platonic solids. The analysis in [15] gives an explanation for this "discreteness" of the minimizing measure, as is made precise in the following theorem (for more general results in the same spirit see [15, Theorems 4.15 and 4.17]).

Theorem 4.1. *If $\tau > \tau_c := \sqrt{2}$, the support of every minimizing measure on the two-dimensional sphere is singular in the sense that it has empty interior.*

Extrapolating to the general situation, this result indicates that our variational principle favors discrete over continuous configurations. Again interpreting $M := \operatorname{supp} \rho$ as our space-time, this might correspond to a mechanism driven by our action principle which makes space-time discrete. Using a more graphic language, one could say that space-time "discretizes itself" on the Planck scale, thus avoiding the ultraviolet divergences of quantum field theory.

Another possible interpretation gives a connection to field quantization: Our model on the two-sphere involves one continuous parameter τ. If we allow τ to be varied while minimizing the action (as is indeed possible if we drop the constraint of prescribed eigenvalues), then the local minima of the action attained at discrete values of τ (like the configurations of the Platonic solids) are favored. Regarding τ as the amplitude of a "classical field", our action principle gives rise to a "quantization" of this field, in the sense that the amplitude only takes discrete values.

The observed "discreteness" might also account for effects related to the wave-particle duality and the collapse of the wave function in the measurement process (for details see [12]).

5. The correspondence to quantum field theory and gauge theories

The correspondence to Minkowski space mentioned in Section 3.3 can also be used to analyze our action principle for interacting systems in the so-called *continuum limit*. We now outline a few ideas and constructions (for details see [5, Chapter 4], [8] and the survey article [13]). We first observe that the vacuum fermionic projector (5) is a solution of the Dirac equation $(i\gamma^j\partial_j - m)P^{\text{sea}}(x,y) = 0$. To introduce the interaction, we replace the free Dirac operator by a more general Dirac operator, which may in particular involve gauge potentials or a gravitational field. For simplicity, we here only consider an electromagnetic potential A,

$$\left(i\gamma^j(\partial_j - ieA_j) - m\right) P(x,y) = 0\,. \tag{46}$$

Next, we introduce particles and anti-particles by occupying (suitably normalized) positive-energy states and removing states of the sea,

$$P(x,y) = P^{\text{sea}}(x,y) - \frac{1}{2\pi}\sum_{k=1}^{n_f} |\psi_k(x)\succ \prec\psi_k(y)| + \frac{1}{2\pi}\sum_{l=1}^{n_a} |\phi_l(x)\succ \prec\phi_l(y)|\,. \tag{47}$$

Using the so-called causal perturbation expansion and light-cone expansion, the fermionic projector can be introduced uniquely from (46) and (47).

It is important that our setting so far does not involve the field equations; in particular, the electromagnetic potential in the Dirac equation (46) does not need to satisfy the Maxwell equations. Instead, the field equations should be derived from our action principle (26). Indeed, analyzing the corresponding Euler-Lagrange equations, one finds that they are satisfied only if the potentials in the Dirac equation satisfy certain constraints. Some of these constraints are partial differential equations involving the potentials as well as the wave functions of the particles and anti-particles in (47). In [8], such field equations are analyzed in detail for a system involving an axial field. In order to keep the setting as simple as possible, we here consider the analogous field equation for the electromagnetic field:

$$\partial_{jk}A^k - \Box A_j = e\sum_{k=1}^{n_f} \prec\psi_k|\gamma_j\psi_k\succ - e\sum_{l=1}^{n_a} \prec\phi_l|\gamma_j\phi_l\succ\,. \tag{48}$$

With (46) and (48), the interaction as described by the action principle (26) reduces in the continuum limit to the *coupled Dirac-Maxwell equations.* The many-fermion state is again described by the fermionic projector, which is built up of *one-particle wave functions.* The electromagnetic field merely is a *classical bosonic field.* Nevertheless, regarding (46) and (48) as a nonlinear hyperbolic system of partial differential equations and treating it perturbatively, one obtains all the Feynman diagrams which do not involve fermion loops. Taking into account that by exciting sea states we can describe pair creation and annihilation processes, we also get all diagrams involving fermion loops. In this way, we obtain agreement with perturbative quantum field theory (for details see [8, §8.4] and the references therein).

We finally remark that in the continuum limit, the freedom in choosing the spinor basis (17) can be described in the language of standard gauge theories. Namely, introducing a gauge-covariant derivative $D_j = \partial_j - iC_j$ with gauge potentials C_j (see for example [18]), the transformation (18) gives rise to the local gauge transformations

$$\begin{aligned} \psi(x) \to U(x)\,\psi(x)\,, \qquad\qquad & D_j \to U D_j U^{-1} \\ C_j(x) \to U(x)C(x)U(x)^{-1} &+ iU(x)\,(\partial_j U(x)^{-1}) \end{aligned} \tag{49}$$

with $U(x) \in \mathrm{U}(p,q)$. The difference to standard gauge theories is that the gauge group cannot be chosen arbitrarily, but it is determined to be the isometry group of the spin space. In the case of spin dimension two, the corresponding gauge group $\mathrm{U}(2,2)$ allows for a unified description of electrodynamics and general relativity (see [5, Section 5.1]). By choosing a higher spin dimension (see [5, Section 5.1]), one gets a larger gauge group. Our mathematical framework ensures that our action principle and thus also the continuum limit is *gauge-symmetric* in the sense that the transformations (49) with $U(x) \in \mathrm{U}(p,q)$ map solutions of the equations of the continuum limit to each other. However, our action is *not* invariant under local transformations of the form (49) if $U(x) \notin \mathrm{U}(p,q)$ is not unitary. An important example of such non-unitary transformations are chiral gauge transformations like

$$U(x) = \chi_L\, U_L(x) + \chi_R\, U_R(x) \quad \text{with} \quad U_{L/R} \in \mathrm{U}(1)\,, \;\; U_L \not\equiv U_R\,.$$

Thus chiral gauge transformations do not describe a gauge symmetry in the above sense. In the continuum limit, this leads to a mechanism which gives chiral gauge fields a rest mass (see [8, Section 8.5] and [13, Section 7]). Moreover, in systems of higher spin dimension, the presence of chiral gauge fields gives rise to a spontaneous breaking of the gauge symmetry, resulting in a smaller "effective" gauge group. As shown in [5, Chapters 6-8], these mechanisms make it possible to realize the gauge groups and couplings of the standard model.

Acknowledgment

We thank the referee for helpful suggestions on the manuscript.

References

[1] H. Baum, *Spinor structures and Dirac operators on pseudo-Riemannian manifolds*, Bull. Polish Acad. Sci. Math. **33** (1985), no. 3-4, 165–171.

[2] J. Bognár, *Indefinite inner product spaces*, Springer-Verlag, New York, 1974, Ergebnisse der Mathematik und ihrer Grenzgebiete, Band 78.

[3] L. Bombelli, J. Lee, D. Meyer, and R.D. Sorkin, *Space-time as a causal set*, Phys. Rev. Lett. **59** (1987), no. 5, 521–524.

[4] F. Finster, *Derivation of local gauge freedom from a measurement principle*, arXiv:funct-an/9701002, Photon and Poincare Group (V. Dvoeglazov, ed.), Nova Science Publishers, 1999, pp. 315–325.

[5] ———, *The principle of the fermionic projector*, hep-th/0001048, hep-th/0202059, hep-th/0210121, AMS/IP Studies in Advanced Mathematics, vol. 35, American Mathematical Society, Providence, RI, 2006.

[6] ———, *Fermion systems in discrete space-time—outer symmetries and spontaneous symmetry breaking*, arXiv:math-ph/0601039, Adv. Theor. Math. Phys. **11** (2007), no. 1, 91–146.

[7] ———, *A variational principle in discrete space-time: Existence of minimizers*, arXiv:math-ph/0503069, Calc. Var. Partial Differential Equations **29** (2007), no. 4, 431–453.

[8] ———, *An action principle for an interacting fermion system and its analysis in the continuum limit*, arXiv:0908.1542 [math-ph] (2009).

[9] ———, *From discrete space-time to Minkowski space: Basic mechanisms, methods and perspectives*, arXiv:0712.0685 [math-ph], Quantum Field Theory (B. Fauser, J. Tolksdorf, and E. Zeidler, eds.), Birkhäuser Verlag, 2009, pp. 235–259.

[10] ———, *Causal variational principles on measure spaces*, arXiv:0811.2666 [math-ph], J. Reine Angew. Math. **646** (2010), 141–194.

[11] ———, *Entanglement and second quantization in the framework of the fermionic projector*, arXiv:0911.0076 [math-ph], J. Phys. A: Math. Theor. **43** (2010), 395302.

[12] ———, *The fermionic projector, entanglement, and the collapse of the wave function*, arXiv:1011.2162 [quant-ph], to appear in the Proceedings of DICE2010 (2011).

[13] ———, *A formulation of quantum field theory realizing a sea of interacting Dirac particles*, arXiv:0911.2102 [hep-th], to appear in Lett. Math. Phys (2011).

[14] F. Finster and A. Grotz, *A Lorentzian quantum geometry*, arXiv:1107.2026 [math-ph] (2011).

[15] F. Finster and D. Schiefeneder, *On the support of minimizers of causal variational principles*, arXiv:1012.1589 [math-ph] (2010).

[16] P.R. Halmos, *Measure theory*, Springer, New York, 1974.

[17] H.B. Lawson, Jr. and M.-L. Michelsohn, *Spin geometry*, Princeton Mathematical Series, vol. 38, Princeton University Press, Princeton, NJ, 1989.

[18] S. Pokorski, *Gauge field theories*, second ed., Cambridge Monographs on Mathematical Physics, Cambridge University Press, Cambridge, 2000.

Felix Finster, Andreas Grotz and Daniela Schiefeneder
Fakultät für Mathematik
Universität Regensburg
D-93040 Regensburg, Germany
e-mail: Felix.Finster@mathematik.uni-regensburg.de
Andreas.Grotz@mathematik.uni-regensburg.de
Daniela.Schiefeneder@mathematik.uni-regensburg.de

CCR- versus CAR-Quantization on Curved Spacetimes

Christian Bär and Nicolas Ginoux

Abstract. We provide a systematic construction of bosonic and fermionic locally covariant quantum field theories on curved backgrounds for large classes of free fields. It turns out that bosonic quantization is possible under much more general assumptions than fermionic quantization.

Mathematics Subject Classification (2010). 58J45, 35Lxx, 81T20.

Keywords. Wave operator, Dirac-type operator, globally hyperbolic spacetime, Green's operator, CCR-algebra, CAR-algebra, locally covariant quantum field theory.

1. Introduction

Classical fields on spacetime are mathematically modeled by sections of a vector bundle over a Lorentzian manifold. The field equations are usually partial differential equations. We introduce a class of differential operators, called Green-hyperbolic operators, which have good analytical solubility properties. This class includes wave operators as well as Dirac-type operators but also the Proca and the Rarita-Schwinger operator.

In order to quantize such a classical field theory on a curved background, we need local algebras of observables. They come in two flavors, bosonic algebras encoding the canonical commutation relations and fermionic algebras encoding the canonical anti-commutation relations. We show how such algebras can be associated to manifolds equipped with suitable Green-hyperbolic operators. We prove that we obtain locally covariant quantum field theories in the sense of [12]. There is a large literature where such constructions are carried out for particular examples of fields, see e.g. [15, 16, 17, 22, 30]. In all these papers the well-posedness of the Cauchy problem plays an important role. We avoid using the Cauchy problem altogether and only make use of Green's operators. In this respect, our approach is similar to the one in [31]. This allows us to deal with larger classes of fields, see Section 3.7, and to treat them systematically. Much of the work on particular examples can be subsumed under this general approach.

It turns out that bosonic algebras can be obtained in much more general situations than fermionic algebras. For instance, for the classical Dirac field both constructions are possible. Hence, on the level of observable algebras, there is no spin-statistics theorem.

This is a condensed version of our paper [4] where full details are given. Here we confine ourselves to the results and the main arguments while we leave aside all technicalities. Moreover, [4] contains a discussion of states and the induced quantum fields.

Acknowledgments. It is a pleasure to thank Alexander Strohmaier for very valuable discussion and the anonymous referee for his remarks. The authors would like to thank SPP 1154 "Globale Differentialgeometrie" and SFB 647 "Raum-Zeit-Materie", both funded by Deutsche Forschungsgemeinschaft, for financial support. Last but not the least, the authors thank the organizers of the international conference "Quantum field theory and gravity".

2. Algebras of canonical (anti-) commutation relations

We start with algebraic preparations and collect the necessary algebraic facts about CAR- and CCR-algebras.

2.1. CAR-algebras

The symbol "CAR" stands for "canonical anti-commutation relations". These algebras are related to pre-Hilbert spaces. We always assume the Hermitian inner product $(\cdot,\cdot)$ to be linear in the first argument and anti-linear in the second.

Definition 2.1. A CAR-*representation* of a complex pre-Hilbert space $(V,(\cdot,\cdot))$ is a pair $(\mathbf{a}, A)$, where A is a unital C*-algebra and $\mathbf{a}: V \to A$ is an anti-linear map satisfying:

(i) $A = C^*(\mathbf{a}(V))$, that is, A is the C*-algebra generated by A,
(ii) $\{\mathbf{a}(v_1), \mathbf{a}(v_2)\} = 0$ and
(iii) $\{\mathbf{a}(v_1)^*, \mathbf{a}(v_2)\} = (v_1, v_2) \cdot 1$,

for all $v_1, v_2 \in V$.

As an example, for any complex pre-Hilbert vector space $(V,(\cdot,\cdot))$, the C*-completion $\mathrm{Cl}(V_{\mathbb{C}}, q_{\mathbb{C}})$ of the algebraic Clifford algebra of the complexification $(V_{\mathbb{C}}, q_{\mathbb{C}})$ of $(V,(\cdot,\cdot))$ is a CAR-representation of $(V,(\cdot,\cdot))$. See [4, App. A.1] for the details, in particular for the construction of the map $\mathbf{a}: V \to \mathrm{Cl}(V_{\mathbb{C}}, q_{\mathbb{C}})$.

Theorem 2.2. *Let $(V,(\cdot,\cdot))$ be an arbitrary complex pre-Hilbert space. Let $\widehat{A}$ be any unital C*-algebra and $\widehat{\mathbf{a}}: V \to \widehat{A}$ be any anti-linear map satisfying Axioms* (ii) *and* (iii) *of Definition* 2.1. *Then there exists a unique C*-morphism*

$\widetilde{\alpha} : \mathrm{Cl}(V_{\mathbb{C}}, q_{\mathbb{C}}) \to \widehat{A}$ *such that*

$$\begin{array}{ccc} V & \xrightarrow{\ \widehat{\mathbf{a}}\ } & \widehat{A} \\ {\scriptstyle \mathbf{a}}\big\downarrow & \nearrow_{\widetilde{\alpha}} & \\ \mathrm{Cl}(V_{\mathbb{C}}, q_{\mathbb{C}}) & & \end{array}$$

commutes. Furthermore, $\widetilde{\alpha}$ *is injective.*
In particular, $(V, (\cdot,\cdot))$ *has, up to* C^**-isomorphism, a unique* CAR*-representation.*

For an alternative description of the CAR-representation in terms of creation and annihilation operators on the fermionic Fock space we refer to [10, Prop. 5.2.2].

From now on, given a complex pre-Hilbert space $(V, (\cdot,\cdot))$, we denote the C*-algebra $\mathrm{Cl}(V_{\mathbb{C}}, q_{\mathbb{C}})$ associated with the CAR-representation $(\mathbf{a}, \mathrm{Cl}(V_{\mathbb{C}}, q_{\mathbb{C}}))$ of $(V, (\cdot,\cdot))$ by $\mathrm{CAR}(V, (\cdot,\cdot))$. We list the properties of CAR-representations which are relevant for quantization, see also [10, Vol. II, Thm. 5.2.5, p. 15].

Proposition 2.3. *Let* $(\mathbf{a}, \mathrm{CAR}(V, (\cdot,\cdot)))$ *be the* CAR*-representation of a complex pre-Hilbert space* $(V, (\cdot,\cdot))$.

(i) *For every* $v \in V$ *one has* $\|\mathbf{a}(v)\| = |v| = (v, v)^{\frac{1}{2}}$, *where* $\|\cdot\|$ *denotes the* C^**-norm on* $\mathrm{CAR}(V, (\cdot,\cdot))$.
(ii) *The* C^**-algebra* $\mathrm{CAR}(V, (\cdot,\cdot))$ *is simple, i.e., it has no closed two-sided* $*$*-ideals other than* $\{0\}$ *and the algebra itself.*
(iii) *The algebra* $\mathrm{CAR}(V, (\cdot,\cdot))$ *is* $\mathbb{Z}_2$*-graded,*

$$\mathrm{CAR}(V, (\cdot,\cdot)) = \mathrm{CAR}^{\mathrm{even}}(V, (\cdot,\cdot)) \oplus \mathrm{CAR}^{\mathrm{odd}}(V, (\cdot,\cdot)),$$

and $\mathbf{a}(V) \subset \mathrm{CAR}^{\mathrm{odd}}(V, (\cdot,\cdot))$.
(iv) *Let* $f : V \to V'$ *be an isometric linear embedding, where* $(V', (\cdot,\cdot)')$ *is another complex pre-Hilbert space. Then there exists a unique injective* C^**-morphism* $\mathrm{CAR}(f) : \mathrm{CAR}(V, (\cdot,\cdot)) \to \mathrm{CAR}(V', (\cdot,\cdot)')$ *such that*

$$\begin{array}{ccc} V & \xrightarrow{\quad f \quad} & V' \\ {\scriptstyle \mathbf{a}}\big\downarrow & & \big\downarrow{\scriptstyle \mathbf{a}'} \\ \mathrm{CAR}(V, (\cdot,\cdot)) & \xrightarrow{\ \mathrm{CAR}(f)\ } & \mathrm{CAR}(V', (\cdot,\cdot)') \end{array}$$

commutes.

One easily sees that $\mathrm{CAR}(\mathrm{id}) = \mathrm{id}$ and that $\mathrm{CAR}(f' \circ f) = \mathrm{CAR}(f') \circ \mathrm{CAR}(f)$ for all isometric linear embeddings $V \xrightarrow{f} V' \xrightarrow{f'} V''$. Therefore we have constructed a covariant functor

$$\mathrm{CAR} : \mathsf{HILB} \longrightarrow \mathsf{C^*Alg},$$

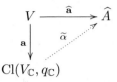

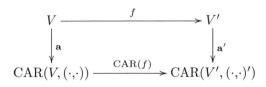

where HILB denotes the category whose objects are the complex pre-Hilbert spaces and whose morphisms are the isometric linear embeddings and $\mathsf{C^*Alg}$ is the category whose objects are the unital C*-algebras and whose morphisms are the injective unit-preserving C*-morphisms.

For *real* pre-Hilbert spaces there is the concept of *self-dual* CAR-representations.

Definition 2.4. A *self-dual* CAR-*representation* of a real pre-Hilbert space $(V,(\cdot,\cdot))$ is a pair $(\mathbf{b}, A)$, where A is a unital C*-algebra and $\mathbf{b} : V \to A$ is an $\mathbb{R}$-linear map satisfying:

(i) $A = C^*(\mathbf{b}(V))$,
(ii) $\mathbf{b}(v) = \mathbf{b}(v)^*$ and
(iii) $\{\mathbf{b}(v_1), \mathbf{b}(v_2)\} = (v_1, v_2) \cdot 1$,

for all $v, v_1, v_2 \in V$.

Note that a self-dual CAR-representation is not a CAR-representation in the sense of Definition 2.1. Given a self-dual CAR-representation, one can extend $\mathbf{b}$ to a $\mathbb{C}$-linear map from the complexification $V_{\mathbb{C}}$ to A. This extension $\mathbf{b} : V_{\mathbb{C}} \to A$ then satisfies $\mathbf{b}(\bar{v}) = \mathbf{b}(v)^*$ and $\{\mathbf{b}(v_1), \mathbf{b}(v_2)\} = (v_1, \bar{v}_2) \cdot 1$ for all $v, v_1, v_2 \in V_{\mathbb{C}}$. These are the axioms of a self-dual CAR-representation as in [1, p. 386].

Theorem 2.5. *For every real pre-Hilbert space $(V,(\cdot,\cdot))$, the C*-Clifford algebra* $\mathrm{Cl}(V_{\mathbb{C}}, q_{\mathbb{C}})$ *provides a self-dual* CAR-*representation of* $(V,(\cdot,\cdot))$ *via* $\mathbf{b}(v) = \frac{i}{\sqrt{2}}v$.

Moreover, self-dual CAR-*representations have the following universal property: Let $\widehat{A}$ be any unital C*-algebra and $\widehat{\mathbf{b}} : V \to \widehat{A}$ be any $\mathbb{R}$-linear map satisfying Axioms* (ii) *and* (iii) *of Definition* 2.4. *Then there exists a unique C*-morphism $\widetilde{\beta} : \mathrm{Cl}(V_{\mathbb{C}}, q_{\mathbb{C}}) \to \widehat{A}$ such that*

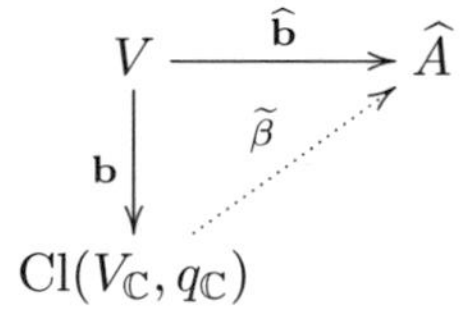

commutes. Furthermore, $\widetilde{\beta}$ is injective.

In particular, $(V,(\cdot,\cdot))$ has, up to C-isomorphism, a unique self-dual* CAR-*representation.*

From now on, given a real pre-Hilbert space $(V,(\cdot,\cdot))$, we denote the C*-algebra $\mathrm{Cl}(V_{\mathbb{C}}, q_{\mathbb{C}})$ associated with the self-dual CAR-representation $(\mathbf{b}, \mathrm{Cl}(V_{\mathbb{C}}, q_{\mathbb{C}}))$ of $(V,(\cdot,\cdot))$ by $\mathrm{CAR}_{\mathrm{sd}}(V,(\cdot,\cdot))$.

Proposition 2.6. *Let $(\mathbf{b}, \mathrm{CAR}_{\mathrm{sd}}(V,(\cdot,\cdot)))$ be the self-dual* CAR-*representation of a real pre-Hilbert space $(V,(\cdot,\cdot))$.*

(i) *For every $v \in V$ one has $\|\mathbf{b}(v)\| = \frac{1}{\sqrt{2}}|v|$, where $\|\cdot\|$ denotes the C*-norm on $\mathrm{CAR}_{\mathrm{sd}}(V,(\cdot,\cdot))$.*
(ii) *The C*-algebra $\mathrm{CAR}_{\mathrm{sd}}(V,(\cdot,\cdot))$ is simple.*

(iii) *The algebra* $\mathrm{CAR}_{\mathrm{sd}}(V,(\cdot,\cdot))$ *is* $\mathbb{Z}_2$*-graded,*

$$\mathrm{CAR}_{\mathrm{sd}}(V,(\cdot,\cdot)) = \mathrm{CAR}_{\mathrm{sd}}^{\mathrm{even}}(V,(\cdot,\cdot)) \oplus \mathrm{CAR}_{\mathrm{sd}}^{\mathrm{odd}}(V,(\cdot,\cdot)),$$

and $\mathbf{b}(V) \subset \mathrm{CAR}_{\mathrm{sd}}^{\mathrm{odd}}(V,(\cdot,\cdot))$.

(iv) *Let* $f : V \to V'$ *be an isometric linear embedding, where* $(V',(\cdot,\cdot)')$ *is another real pre-Hilbert space. Then there exists a unique injective* C^**-morphism* $\mathrm{CAR}_{\mathrm{sd}}(f) : \mathrm{CAR}_{\mathrm{sd}}(V,(\cdot,\cdot)) \to \mathrm{CAR}_{\mathrm{sd}}(V',(\cdot,\cdot)')$ *such that*

$$\begin{array}{ccc} V & \xrightarrow{\quad f \quad} & V' \\ \Big\downarrow{\scriptstyle \mathbf{b}} & & \Big\downarrow{\scriptstyle \mathbf{b}'} \\ \mathrm{CAR}_{\mathrm{sd}}(V,(\cdot,\cdot)) & \xrightarrow{\mathrm{CAR}_{\mathrm{sd}}(f)} & \mathrm{CAR}_{\mathrm{sd}}(V',(\cdot,\cdot)') \end{array}$$

commutes.

The proofs are similar to the ones for CAR-representations of complex pre-Hilbert spaces as given in [4, App. A]. We have constructed a functor

$$\mathrm{CAR}_{\mathrm{sd}} : \mathsf{HILB}_{\mathbb{R}} \longrightarrow \mathsf{C^*Alg},$$

where $\mathsf{HILB}_{\mathbb{R}}$ denotes the category whose objects are the real pre-Hilbert spaces and whose morphisms are the isometric linear embeddings.

Remark 2.7. Let $(V,(\cdot,\cdot))$ be a complex pre-Hilbert space. If we consider V as a real vector space, then we have the real pre-Hilbert space $(V,\mathfrak{Re}(\cdot,\cdot))$. For the corresponding CAR-representations we have

$$\mathrm{CAR}(V,(\cdot,\cdot)) = \mathrm{CAR}_{\mathrm{sd}}(V,\mathfrak{Re}(\cdot,\cdot)) = \mathrm{Cl}(V_{\mathbb{C}},q_{\mathbb{C}})$$

and

$$\mathbf{b}(v) = \frac{i}{\sqrt{2}}(\mathbf{a}(v) - \mathbf{a}(v)^*).$$

2.2. CCR-algebras

In this section, we recall the construction of the representation of any (real) symplectic vector space by the so-called canonical commutation relations (CCR). Proofs can be found in [5, Sec. 4.2].

Definition 2.8. A CCR-*representation* of a symplectic vector space (V,ω) is a pair (w,A), where A is a unital C*-algebra and w is a map $V \to A$ satisfying:

(i) $A = C^*(w(V))$,
(ii) $w(0) = 1$,
(iii) $w(-\varphi) = w(\varphi)^*$,
(iv) $w(\varphi + \psi) = e^{i\omega(\varphi,\psi)/2} w(\varphi) \cdot w(\psi)$,

for all $\varphi, \psi \in V$.

The map w is in general neither linear, nor any kind of group homomorphism, nor continuous as soon as V carries a topology which is different from the discrete one [5, Prop. 4.2.3].

Example 2.9. Given any symplectic vector space (V, ω), consider the Hilbert space $H := L^2(V, \mathbb{C})$, where V is endowed with the counting measure. Define the map w from V into the space $\mathcal{L}(H)$ of bounded endomorphisms of H by

$$(w(\varphi)F)(\psi) := e^{i\omega(\varphi,\psi)/2}F(\varphi + \psi),$$

for all $\varphi, \psi \in V$ and $F \in H$. It is well-known that $\mathcal{L}(H)$ is a C*-algebra with the operator norm as C*-norm, and that the map w satisfies the Axioms (ii)-(iv) from Definition 2.8, see e.g. [5, Ex. 4.2.2]. Hence setting $A := C^*(w(V))$, the pair (w, A) provides a CCR-representation of (V, ω).

This is essentially the only CCR-representation:

Theorem 2.10. *Let (V, ω) be a symplectic vector space and $(\hat{w}, \widehat{A})$ be a pair satisfying the Axioms* (ii)-(iv) *of Definition* 2.8. *Then there exists a unique C*-morphism $\Phi : A \to \widehat{A}$ such that $\Phi \circ w = \hat{w}$, where (w, A) is the* CCR-*representation from Example* 2.9. *Moreover, Φ is injective.*

In particular, (V, ω) has a CCR-*representation, unique up to C*-isomorphism.*

We denote the C*-algebra associated to the CCR-representation of (V, ω) from Example 2.9 by $\mathrm{CCR}(V, \omega)$. As a consequence of Theorem 2.10, we obtain the following important corollary.

Corollary 2.11. *Let (V, ω) be a symplectic vector space and $(w, \mathrm{CCR}(V, \omega))$ its* CCR-*representation.*

(i) *The C*-algebra $\mathrm{CCR}(V, \omega)$ is simple, i.e., it has no closed two-sided $*$-ideals other than $\{0\}$ and the algebra itself.*
(ii) *Let (V', ω') be another symplectic vector space and $f : V \to V'$ a symplectic linear map. Then there exists a unique injective C*-morphism $\mathrm{CCR}(f) : \mathrm{CCR}(V, \omega) \to \mathrm{CCR}(V', \omega')$ such that*

$$\begin{array}{ccc} V & \xrightarrow{\quad f \quad} & V' \\ \downarrow{\scriptstyle w} & & \downarrow{\scriptstyle w'} \\ \mathrm{CCR}(V, \omega) & \xrightarrow{\mathrm{CCR}(f)} & \mathrm{CCR}(V', \omega') \end{array}$$

commutes.

Obviously $\mathrm{CCR}(\mathrm{id}) = \mathrm{id}$ and $\mathrm{CCR}(f' \circ f) = \mathrm{CCR}(f') \circ \mathrm{CCR}(f)$ for all symplectic linear maps $V \xrightarrow{f} V' \xrightarrow{f'} V''$, so that we have constructed a covariant functor

$$\mathrm{CCR} : \mathsf{Sympl} \longrightarrow \mathsf{C^*Alg}.$$

3. Field equations on Lorentzian manifolds

3.1. Globally hyperbolic manifolds

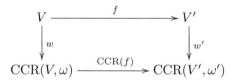

We begin by fixing notation and recalling general facts about Lorentzian manifolds, see e.g. [26] or [5] for more details. Unless mentioned otherwise, the pair (M, g) will stand for a smooth m-dimensional manifold M equipped with a smooth Lorentzian metric g, where our convention for Lorentzian signature is $(-+\cdots+)$. The associated volume element will be denoted by dV. We shall also assume our Lorentzian manifold (M, g) to be time-orientable, i.e., that there exists a smooth timelike vector field on M. Time-oriented Lorentzian manifolds will be also referred to as *spacetimes*. Note that in contrast to conventions found elsewhere, we do not assume that a spacetime be connected nor that its dimension be $m = 4$.

For every subset A of a spacetime M we denote the causal future and past of A in M by $J_+(A)$ and $J_-(A)$, respectively. If we want to emphasize the ambient space M in which the causal future or past of A is considered, we write $J_\pm^M(A)$ instead of $J_\pm(A)$. Causal curves will always be implicitly assumed (future or past) oriented.

Definition 3.1. A *Cauchy hypersurface* in a spacetime (M, g) is a subset of M which is met exactly once by every inextensible timelike curve.

Cauchy hypersurfaces are always topological hypersurfaces but need not be smooth. All Cauchy hypersurfaces of a spacetime are homeomorphic.

Definition 3.2. A spacetime (M, g) is called *globally hyperbolic* if and only if it contains a Cauchy hypersurface.

A classical result of R. Geroch [18] says that a globally hyperbolic spacetime can be foliated by Cauchy hypersurfaces. It is a rather recent and very important result that this also holds in the smooth category: any globally hyperbolic spacetime is of the form $(\mathbb{R} \times \Sigma, -\beta dt^2 \oplus g_t)$, where each $\{t\} \times \Sigma$ is a smooth spacelike Cauchy hypersurface, β a smooth positive function and $(g_t)_t$ a smooth one-parameter family of Riemannian metrics on Σ [7, Thm. 1.1]. The hypersurface Σ can be even chosen such that $\{0\} \times \Sigma$ coincides with a given smooth spacelike Cauchy hypersurface [8, Thm. 1.2]. Moreover, any compact acausal smooth spacelike submanifold with boundary in a globally hyperbolic spacetime is contained in a smooth spacelike Cauchy hypersurface [8, Thm. 1.1].

Definition 3.3. A closed subset $A \subset M$ is called

- *spacelike compact* if there exists a compact subset $K \subset M$ such that $A \subset J^M(K) := J_-^M(K) \cup J_+^M(K)$,
- *future-compact* if $A \cap J_+(x)$ is compact for any $x \in M$,
- *past-compact* if $A \cap J_-(x)$ is compact for any $x \in M$.

A spacelike compact subset is in general not compact, but its intersection with any Cauchy hypersurface is compact, see e.g. [5, Cor. A.5.4].

Definition 3.4. A subset Ω of a spacetime M is called *causally compatible* if and only if $J_\pm^\Omega(x) = J_\pm^M(x) \cap \Omega$ for every $x \in \Omega$.

This means that every causal curve joining two points in Ω must be contained entirely in Ω.

3.2. Differential operators and Green's functions

A *differential operator* of order (at most) k on a vector bundle $S \to M$ over $\mathbb{K} = \mathbb{R}$ or $\mathbb{K} = \mathbb{C}$ is a linear map $P : C^\infty(M, S) \to C^\infty(M, S)$ which in local coordinates $x = (x^1, \ldots, x^m)$ of M and with respect to a local trivialization looks like

$$P = \sum_{|\alpha| \leq k} A_\alpha(x) \frac{\partial^\alpha}{\partial x^\alpha}.$$

Here $C^\infty(M, S)$ denotes the space of smooth sections of $S \to M$, $\alpha = (\alpha_1, \ldots, \alpha_m) \in \mathbb{N}_0 \times \cdots \times \mathbb{N}_0$ runs over multi-indices, $|\alpha| = \sum_{j=1}^m \alpha_j$ and $\frac{\partial^\alpha}{\partial x^\alpha} = \frac{\partial^{|\alpha|}}{\partial(x^1)^{\alpha_1} \cdots \partial(x^m)^{\alpha_m}}$. The *principal symbol* σ_P of P associates to each covector $\xi \in T_x^*M$ a linear map $\sigma_P(\xi) : S_x \to S_x$. Locally, it is given by

$$\sigma_P(\xi) = \sum_{|\alpha| = k} A_\alpha(x) \xi^\alpha,$$

where $\xi^\alpha = \xi_1^{\alpha_1} \cdots \xi_m^{\alpha_m}$ and $\xi = \sum_j \xi_j dx^j$. If P and Q are two differential operators of order k and ℓ respectively, then $Q \circ P$ is a differential operator of order $k + \ell$ and

$$\sigma_{Q \circ P}(\xi) = \sigma_Q(\xi) \circ \sigma_P(\xi).$$

For any linear differential operator $P : C^\infty(M, S) \to C^\infty(M, S)$ there is a unique formally dual operator $P^* : C^\infty(M, S^*) \to C^\infty(M, S^*)$ of the same order characterized by

$$\int_M \langle \varphi, P\psi \rangle \, \mathrm{dV} = \int_M \langle P^*\varphi, \psi \rangle \, \mathrm{dV}$$

for all $\psi \in C^\infty(M, S)$ and $\varphi \in C^\infty(M, S^*)$ with $\mathrm{supp}(\varphi) \cap \mathrm{supp}(\psi)$ compact. Here $\langle \cdot, \cdot \rangle : S^* \otimes S \to \mathbb{K}$ denotes the canonical pairing, i.e., the evaluation of a linear form in S_x^* on an element of S_x, where $x \in M$. We have $\sigma_{P^*}(\xi) = (-1)^k \sigma_P(\xi)^*$ where k is the order of P.

Definition 3.5. Let a vector bundle $S \to M$ be endowed with a non-degenerate inner product $\langle \cdot\,, \cdot \rangle$. A linear differential operator P on S is called *formally self-adjoint* if and only if

$$\int_M \langle P\varphi, \psi \rangle \, \mathrm{dV} = \int_M \langle \varphi, P\psi \rangle \, \mathrm{dV}$$

holds for all $\varphi, \psi \in C^\infty(M, S)$ with $\mathrm{supp}(\varphi) \cap \mathrm{supp}(\psi)$ compact.

Similarly, we call P *formally skew-adjoint* if instead

$$\int_M \langle P\varphi, \psi \rangle \, \mathrm{dV} = -\int_M \langle \varphi, P\psi \rangle \, \mathrm{dV}\,.$$

We recall the definition of advanced and retarded Green's operators for a linear differential operator.

Definition 3.6. Let P be a linear differential operator acting on the sections of a vector bundle S over a Lorentzian manifold M. An *advanced Green's operator* for P on M is a linear map

$$G_+ : C_c^\infty(M, S) \to C^\infty(M, S)$$

satisfying:

(G_1) $P \circ G_+ = \mathrm{id}_{C_c^\infty(M,S)}$;

(G_2) $G_+ \circ P|_{C_c^\infty(M,S)} = \mathrm{id}_{C_c^\infty(M,S)}$;

(G_3^+) $\mathrm{supp}(G_+\varphi) \subset J_+^M(\mathrm{supp}(\varphi))$ for any $\varphi \in C_c^\infty(M,S)$.

A *retarded Green's operator* for P on M is a linear map $G_- : C_c^\infty(M,S) \to C^\infty(M,S)$ satisfying (G_1), (G_2), and

(G_3^-) $\mathrm{supp}(G_-\varphi) \subset J_-^M(\mathrm{supp}(\varphi))$ for any $\varphi \in C_c^\infty(M,S)$.

Here we denote by $C_c^\infty(M,S)$ the space of compactly supported smooth sections of S.

Definition 3.7. Let $P : C^\infty(M,S) \to C^\infty(M,S)$ be a linear differential operator. We call P *Green-hyperbolic* if the restriction of P to any globally hyperbolic subregion of M has advanced and retarded Green's operators.

The Green's operators for a given Green-hyperbolic operator P provide solutions φ of $P\varphi = 0$. More precisely, denoting by $C_{sc}^\infty(M,S)$ the set of smooth sections in S with spacelike compact support, we have the following

Theorem 3.8. *Let M be a Lorentzian manifold, let $S \to M$ be a vector bundle, and let P be a Green-hyperbolic operator acting on sections of S. Let $G_\pm$ be advanced and retarded Green's operators for P, respectively. Put*

$$G := G_+ - G_- : C_c^\infty(M,S) \to C_{sc}^\infty(M,S).$$

Then the following linear maps form a complex:

$$\{0\} \to C_c^\infty(M,S) \xrightarrow{P} C_c^\infty(M,S) \xrightarrow{G} C_{sc}^\infty(M,S) \xrightarrow{P} C_{sc}^\infty(M,S). \qquad (1)$$

This complex is always exact at the first $C_c^\infty(M,S)$. If M is globally hyperbolic, then the complex is exact everywhere.

We refer to [4, Theorem 3.5] for the proof. Note that exactness at the first $C_c^\infty(M,S)$ in sequence (1) says that there are no non-trivial smooth solutions of $P\varphi = 0$ with compact support. Indeed, if M is globally hyperbolic, more is true. Namely, if $\varphi \in C^\infty(M,S)$ solves $P\varphi = 0$ and $\mathrm{supp}(\varphi)$ is future or past-compact, then $\varphi = 0$ (see e.g. [4, Remark 3.6] for a proof). As a straightforward consequence, the Green's operators for a Green-hyperbolic operator on a globally hyperbolic spacetime are unique [4, Remark 3.7].

3.3. Wave operators

The most prominent class of Green-hyperbolic operators are wave operators, sometimes also called normally hyperbolic operators.

Definition 3.9. A linear differential operator of second order $P : C^\infty(M,S) \to C^\infty(M,S)$ is called a *wave operator* if its principal symbol is given by the Lorentzian metric, i.e., for all $\xi \in T^*M$ we have

$$\sigma_P(\xi) = -\langle \xi, \xi \rangle \cdot \mathrm{id}.$$

In other words, if we choose local coordinates $x^1, \dots, x^m$ on M and a local trivialization of S, then

$$P = -\sum_{i,j=1}^{m} g^{ij}(x)\frac{\partial^2}{\partial x^i \partial x^j} + \sum_{j=1}^{m} A_j(x)\frac{\partial}{\partial x^j} + B(x),$$

where A_j and B are matrix-valued coefficients depending smoothly on x and (g^{ij}) is the inverse matrix of (g_{ij}) with $g_{ij} = \langle \frac{\partial}{\partial x^i}, \frac{\partial}{\partial x^j}\rangle$. If P is a wave operator, then so is its dual operator P^*. In [5, Cor. 3.4.3] it has been shown that wave operators are Green-hyperbolic.

Example 3.10 (d'Alembert operator). Let S be the trivial line bundle so that sections of S are just functions. The d'Alembert operator $P = \Box = -\mathrm{div} \circ \mathrm{grad}$ is a formally self-adjoint wave operator, see e.g. [5, p. 26].

Example 3.11 (connection-d'Alembert operator). More generally, let S be a vector bundle and let ∇ be a connection on S. This connection and the Levi-Civita connection on T^*M induce a connection on $T^*M \otimes S$, again denoted ∇. We define the connection-d'Alembert operator $\Box^\nabla$ to be the composition of the following three maps

$$C^\infty(M,S) \xrightarrow{\nabla} C^\infty(M,T^*M \otimes S) \xrightarrow{\nabla} C^\infty(M,T^*M \otimes T^*M \otimes S) \xrightarrow{-\mathrm{tr}\otimes \mathrm{id}_S} C^\infty(M,S),$$

where $\mathrm{tr} : T^*M \otimes T^*M \to \mathbb{R}$ denotes the metric trace, $\mathrm{tr}(\xi \otimes \eta) = \langle \xi, \eta \rangle$. We compute the principal symbol,

$$\sigma_{\Box^\nabla}(\xi)\varphi = -(\mathrm{tr}\otimes \mathrm{id}_S)\circ\sigma_\nabla(\xi)\circ\sigma_\nabla(\xi)(\varphi) = -(\mathrm{tr}\otimes \mathrm{id}_S)(\xi\otimes\xi\otimes\varphi) = -\langle \xi, \xi\rangle\, \varphi.$$

Hence $\Box^\nabla$ is a wave operator.

Example 3.12 (Hodge-d'Alembert operator). For $S = \Lambda^k T^*M$, exterior differentiation $d : C^\infty(M, \Lambda^k T^*M) \to C^\infty(M, \Lambda^{k+1} T^*M)$ increases the degree by one, the codifferential $\delta = d^* : C^\infty(M, \Lambda^k T^*M) \to C^\infty(M, \Lambda^{k-1} T^*M)$ decreases the degree by one. While d is independent of the metric, the codifferential δ does depend on the Lorentzian metric. The operator $P = -d\delta - \delta d$ is a formally self-adjoint wave operator.

3.4. The Proca equation

The Proca operator is an example of a Green-hyperbolic operator of second order which is not a wave operator.

Example 3.13 (Proca operator). The discussion of this example follows [31, p. 116f]. The Proca equation describes massive vector bosons. We take $S = T^*M$ and let $m_0 > 0$. The Proca equation is

$$P\varphi := \delta d\varphi + m_0^2\varphi = 0\ , \tag{2}$$

where $\varphi \in C^\infty(M, S)$. Applying δ to (2) we obtain, using $\delta^2 = 0$ and $m_0 \neq 0$,

$$\delta\varphi = 0 \tag{3}$$

and hence

$$(d\delta + \delta d)\varphi + m_0^2\varphi = 0. \tag{4}$$

Conversely, (3) and (4) clearly imply (2).

Since $\tilde{P} := d\delta + \delta d + m_0^2$ is minus a wave operator, it has Green's operators $\tilde{G}_\pm$. We define $G_\pm : C_c^\infty(M,S) \to C_{sc}^\infty(M,S)$ by

$$G_\pm := (m_0^{-2} d\delta + \mathrm{id}) \circ \tilde{G}_\pm = \tilde{G}_\pm \circ (m_0^{-2} d\delta + \mathrm{id}).$$

The last equality holds because d and δ commute with $\tilde{P}$, see [4, Lemma 2.16]. For $\varphi \in C_c^\infty(M,S)$ we compute

$$G_\pm P\varphi = \tilde{G}_\pm(m_0^{-2} d\delta + \mathrm{id})(\delta d + m_0^2)\varphi = \tilde{G}_\pm \tilde{P}\varphi = \varphi$$

and similarly $PG_\pm\varphi = \varphi$. Since the differential operator $m_0^{-2} d\delta + \mathrm{id}$ does not increase supports, the third axiom in the definition of advanced and retarded Green's operators holds as well.

This shows that G_+ and G_- are advanced and retarded Green's operators for P, respectively. Thus P is not a wave operator but Green-hyperbolic.

3.5. Dirac-type operators

The most important Green-hyperbolic operators of first order are the so-called Dirac-type operators.

Definition 3.14. A linear differential operator $D : C^\infty(M,S) \to C^\infty(M,S)$ of first order is called *of Dirac type* if $-D^2$ is a wave operator.

Remark 3.15. If D is of Dirac type, then i times its principal symbol satisfies the Clifford relations

$$(i\sigma_D(\xi))^2 = -\sigma_{D^2}(\xi) = -\langle \xi, \xi \rangle \cdot \mathrm{id},$$

hence by polarization

$$(i\sigma_D(\xi))(i\sigma_D(\eta)) + (i\sigma_D(\eta))(i\sigma_D(\xi)) = -2\langle \xi, \eta \rangle \cdot \mathrm{id}.$$

The bundle S thus becomes a module over the bundle of Clifford algebras $\mathrm{Cl}(TM)$ associated with $(TM, \langle \cdot\,, \cdot \rangle)$. See [6, Sec. 1.1] or [23, Ch. I] for the definition and properties of the Clifford algebra $\mathrm{Cl}(V)$ associated with a vector space V with inner product.

Remark 3.16. If D is of Dirac type, then so is its dual operator D^*. On a globally hyperbolic region let G_+ be the advanced Green's operator for D^2, which exists since $-D^2$ is a wave operator. Then it is not hard to check that $D \circ G_+$ is an advanced Green's operator for D, see [25, Thm. 3.2]. The same discussion applies to the retarded Green's operator. Hence any Dirac-type operator is Green-hyperbolic.

Example 3.17 (classical Dirac operator). If the spacetime M carries a spin structure, then one can define the spinor bundle $S = \Sigma M$ and the classical Dirac operator

$$D : C^\infty(M, \Sigma M) \to C^\infty(M, \Sigma M), \quad D\varphi := i \sum_{j=1}^{m} \varepsilon_j e_j \cdot \nabla_{e_j} \varphi.$$

Here $(e_j)_{1\leq j\leq m}$ is a local orthonormal basis of the tangent bundle, $\varepsilon_j = \langle e_j, e_j\rangle = \pm 1$ and "$\cdot$" denotes the Clifford multiplication, see e.g. [6] or [3, Sec. 2]. The principal symbol of D is given by

$$\sigma_D(\xi)\psi = i\xi^\sharp \cdot \psi.$$

Here $\xi^\sharp$ denotes the tangent vector dual to the 1-form ξ via the Lorentzian metric, i.e., $\langle \xi^\sharp, Y\rangle = \xi(Y)$ for all tangent vectors Y over the same point of the manifold. Hence

$$\sigma_{D^2}(\xi)\psi = \sigma_D(\xi)\sigma_D(\xi)\psi = -\xi^\sharp \cdot \xi^\sharp \cdot \psi = \langle \xi, \xi\rangle\, \psi.$$

Thus $P = -D^2$ is a wave operator. Moreover, D is formally self-adjoint, see e.g. [3, p. 552].

Example 3.18 (twisted Dirac operators). More generally, let $E \to M$ be a complex vector bundle equipped with a non-degenerate Hermitian inner product and a metric connection ∇^E over a spin spacetime M. In the notation of Example 3.17, one may define the Dirac operator of M twisted with E by

$$D^E := i\sum_{j=1}^m \varepsilon_j e_j \cdot \nabla^{\Sigma M\otimes E}_{e_j} : C^\infty(M, \Sigma M \otimes E) \to C^\infty(M, \Sigma M \otimes E),$$

where $\nabla^{\Sigma M\otimes E}$ is the tensor product connection on $\Sigma M \otimes E$. Again, D^E is a formally self-adjoint Dirac-type operator.

Example 3.19 (Euler operator). In Example 3.12, replacing $\Lambda^k T^*M$ by $S := \Lambda T^*M \otimes \mathbb{C} = \oplus_{k=0}^m \Lambda^k T^*M \otimes \mathbb{C}$, the Euler operator $D = i(d - \delta)$ defines a formally self-adjoint Dirac-type operator. In case M is spin, the Euler operator coincides with the Dirac operator of M twisted with ΣM if m is even and twisted with $\Sigma M \oplus \Sigma M$ if m is odd.

Example 3.20 (Buchdahl operators). On a 4-dimensional spin spacetime M, consider the standard orthogonal and parallel splitting $\Sigma M = \Sigma_+ M \oplus \Sigma_- M$ of the complex spinor bundle of M into spinors of positive and negative chirality. The finite dimensional irreducible representations of the simply-connected Lie group $\mathrm{Spin}^0(3,1)$ are given by $\Sigma_+^{(k/2)} \otimes \Sigma_-^{(\ell/2)}$ where $k, \ell \in \mathbb{N}$. Here $\Sigma_+^{(k/2)} = \Sigma_+^{\odot k}$ is the k-th symmetric tensor product of the positive half-spinor representation Σ_+ and similarly for $\Sigma_-^{(\ell/2)}$. Let the associated vector bundles $\Sigma_\pm^{(k/2)} M$ carry the induced inner product and connection.

For $s \in \mathbb{N}$, $s \geq 1$, consider the twisted Dirac operator $D^{(s)}$ acting on sections of $\Sigma M \otimes \Sigma_+^{((s-1)/2)} M$. In the induced splitting

$$\Sigma M \otimes \Sigma_+^{((s-1)/2)} M = \Sigma_+ M \otimes \Sigma_+^{((s-1)/2)} M \oplus \Sigma_- M \otimes \Sigma_+^{((s-1)/2)} M$$

the operator $D^{(s)}$ is of the form

$$\begin{pmatrix} 0 & D_-^{(s)} \\ D_+^{(s)} & 0 \end{pmatrix}$$

because Clifford multiplication by vectors exchanges the chiralities. The Clebsch-Gordan formulas [11, Prop. II.5.5] tell us that the representation $\Sigma_+ \otimes \Sigma_+^{(\frac{s-1}{2})}$ splits as

$$\Sigma_+ \otimes \Sigma_+^{(\frac{s-1}{2})} = \Sigma_+^{(\frac{s}{2})} \oplus \Sigma_+^{(\frac{s}{2}-1)}.$$

Hence we have the corresponding parallel orthogonal projections

$$\pi_s : \Sigma_+ M \otimes \Sigma_+^{(\frac{s-1}{2})} M \to \Sigma_+^{(\frac{s}{2})} M \quad \text{and} \quad \pi_s' : \Sigma_+ M \otimes \Sigma_+^{(\frac{s-1}{2})} M \to \Sigma_+^{(\frac{s}{2}-1)} M.$$

On the other hand, the representation $\Sigma_- \otimes \Sigma_+^{(\frac{s-1}{2})}$ is irreducible. Now *Buchdahl operators* are the operators of the form

$$B^{(s)}_{\mu_1,\mu_2,\mu_3} := \begin{pmatrix} \mu_1 \cdot \pi_s + \mu_2 \cdot \pi_s' & D_-^{(s)} \\ D_+^{(s)} & \mu_3 \cdot \mathrm{id} \end{pmatrix},$$

where $\mu_1, \mu_2, \mu_3 \in \mathbb{C}$ are constants. By definition, $B^{(s)}_{\mu_1,\mu_2,\mu_3}$ is of the form $D^{(s)} + b$, where b is of order zero. In particular, $B^{(s)}_{\mu_1,\mu_2,\mu_3}$ is a Dirac-type operator, hence it is Green-hyperbolic. For a definition of Buchdahl operators using indices we refer to [13, 14, 35] and to [24, Def. 8.1.4, p. 104].

3.6. The Rarita-Schwinger operator

For the Rarita-Schwinger operator on Riemannian manifolds, we refer to [34, Sec. 2], see also [9, Sec. 2]. In this section let the spacetime M be spin and consider the Clifford multiplication $\gamma : T^*M \otimes \Sigma M \to \Sigma M$, $\theta \otimes \psi \mapsto \theta^\sharp \cdot \psi$, where ΣM is the complex spinor bundle of M. Then there is the representation-theoretic splitting of $T^*M \otimes \Sigma M$ into the orthogonal and parallel sum

$$T^*M \otimes \Sigma M = \iota(\Sigma M) \oplus \Sigma^{3/2} M,$$

where $\Sigma^{3/2}M := \ker(\gamma)$ and $\iota(\psi) := -\frac{1}{m}\sum_{j=1}^m e_j^* \otimes e_j \cdot \psi$. Here again $(e_j)_{1\le j\le m}$ is a local orthonormal basis of the tangent bundle. Let $\mathcal{D}$ be the twisted Dirac operator on $T^*M \otimes \Sigma M$, that is, $\mathcal{D} := i \cdot (\mathrm{id} \otimes \gamma) \circ \nabla$, where ∇ denotes the induced covariant derivative on $T^*M \otimes \Sigma M$.

Definition 3.21. The *Rarita-Schwinger operator* on the spin spacetime M is defined by $\mathcal{Q} := (\mathrm{id} - \iota \circ \gamma) \circ \mathcal{D} : C^\infty(M, \Sigma^{3/2}M) \to C^\infty(M, \Sigma^{3/2}M)$.

By definition, the Rarita-Schwinger operator is pointwise obtained as the orthogonal projection onto $\Sigma^{3/2}M$ of the twisted Dirac operator $\mathcal{D}$ restricted to a section of $\Sigma^{3/2}M$. As for the Dirac operator, its characteristic variety coincides with the set of lightlike covectors, at least when $m \ge 3$, see [4, Lemma 2.26]. In particular, [21, Thms. 23.2.4 & 23.2.7] imply that the Cauchy problem for $\mathcal{Q}$ is well-posed in case M is globally hyperbolic. Since the well-posedness of the Cauchy problem implies the existence of advanced and retarded Green's operators (compare e.g. [4, Theorem 3.3.1 & Prop. 3.4.2] for wave operators), the operator $\mathcal{Q}$ has advanced and retarded Green's operators. Hence $\mathcal{Q}$ is not of Dirac type but is Green-hyperbolic.

Remark 3.22. The equations originally considered by Rarita and Schwinger in [28] correspond to the twisted Dirac operator $\mathcal{D}$ restricted to $\Sigma^{3/2}M$ but not projected back to $\Sigma^{3/2}M$. In other words, they considered the operator

$$\mathcal{D}|_{C^\infty(M,\Sigma^{3/2}M)} : C^\infty(M,\Sigma^{3/2}M) \to C^\infty(M, T^*M \otimes \Sigma M).$$

These equations are over-determined. Therefore it is not a surprise that non-trivial solutions restrict the geometry of the underlying manifold as observed by Gibbons [19] and that this operator has no Green's operators.

3.7. Combining given operators into a new one

Given two Green-hyperbolic operators we can form the direct sum and obtain a new operator in a trivial fashion. Namely, let $S_1, S_2 \to M$ be two vector bundles over a globally hyperbolic manifold M and let P_1 and P_2 be two Green-hyperbolic operators acting on sections of S_1 and S_2 respectively. Then

$$P_1 \oplus P_2 := \begin{pmatrix} P_1 & 0 \\ 0 & P_2 \end{pmatrix} : C^\infty(M, S_1 \oplus S_2) \to C^\infty(M, S_1 \oplus S_2)$$

is Green-hyperbolic [5, Lemma 2.27]. Note that the two operators need not have the same order. Hence Green-hyperbolic operators need not be hyperbolic in the usual sense.

4. Algebras of observables

Our next aim is to quantize the classical fields governed by Green-hyperbolic differential operators. We construct local algebras of observables and we prove that we obtain locally covariant quantum field theories in the sense of [12].

4.1. Bosonic quantization

In this section we show how a quantization process based on canonical commutation relations (CCR) can be carried out for formally self-adjoint Green-hyperbolic operators. This is a functorial procedure. We define the first category involved in the quantization process.

Definition 4.1. The category $\mathsf{GlobHypGreen}$ consists of the following objects and morphisms:

- An object in $\mathsf{GlobHypGreen}$ is a triple (M, S, P), where
  - ▸ M is a globally hyperbolic spacetime,
  - ▸ S is a real vector bundle over M endowed with a non-degenerate inner product $\langle\cdot\,,\cdot\rangle$ and
  - ▸ P is a formally self-adjoint Green-hyperbolic operator acting on sections of S.
- A morphism between objects (M_1, S_1, P_1), (M_2, S_2, P_2) of $\mathsf{GlobHypGreen}$ is a pair (f, F), where
  - ▸ f is a time-orientation preserving isometric embedding $M_1 \to M_2$ with $f(M_1)$ causally compatible and open in M_2,

► F is a fiberwise isometric vector bundle isomorphism over f such that the following diagram commutes:

$$\begin{array}{ccc} C^\infty(M_2, S_2) & \xrightarrow{P_2} & C^\infty(M_2, S_2) \\ \Big\downarrow \text{res} & & \Big\downarrow \text{res} \\ C^\infty(M_1, S_1) & \xrightarrow{P_1} & C^\infty(M_1, S_1), \end{array} \tag{5}$$

where $\text{res}(\varphi) := F^{-1} \circ \varphi \circ f$ for every $\varphi \in C^\infty(M_2, S_2)$.

Note that morphisms exist only if the manifolds have equal dimension and the vector bundles have the same rank. Note, furthermore, that the inner product $\langle \cdot , \cdot \rangle$ on S is not required to be positive or negative definite.

The causal compatibility condition, which is not automatically satisfied (see e.g. [5, Fig. 33]), ensures the commutation of the extension and restriction maps with the Green's operators. Namely, if (f, F) be a morphism between two objects (M_1, S_1, P_1) and (M_2, S_2, P_2) in the category GlobHypGreen, and if $(G_1)_\pm$ and $(G_2)_\pm$ denote the respective Green's operators for P_1 and P_2, then we have

$$\text{res} \circ (G_2)_\pm \circ \text{ext} = (G_1)_\pm .$$

Here $\text{ext}(\varphi) \in C_c^\infty(M_2, S_2)$ is the extension by 0 of $F \circ \varphi \circ f^{-1} : f(M_1) \to S_2$ to M_2, for every $\varphi \in C_c^\infty(M_1, S_1)$, see [4, Lemma 3.2].

What is most important for our purpose is that the Green's operators for a formally self-adjoint Green-hyperbolic operator provide a symplectic vector space in a canonical way. First recall how the Green's operators of an operator and of its formally dual operator are related: if M is a globally hyperbolic spacetime, G_+, G_- are the advanced and retarded Green's operators for a Green-hyperbolic operator P acting on sections of $S \to M$ and G_+^*, G_-^* denote the advanced and retarded Green's operators for P^*, then

$$\int_M \langle G_\pm^* \varphi, \psi \rangle \, \mathrm{dV} = \int_M \langle \varphi, G_\mp \psi \rangle \, \mathrm{dV} \tag{6}$$

for all $\varphi \in C_c^\infty(M, S^*)$ and $\psi \in C_c^\infty(M, S)$, see e.g. [4, Lemma 3.3]. This implies:

Proposition 4.2. *Let (M, S, P) be an object in the category* GlobHypGreen*. Set $G := G_+ - G_-$, where G_+, G_- are the advanced and retarded Green's operator for P, respectively.*

Then the pair $(\mathrm{SYMPL}(M, S, P), \omega)$ is a symplectic vector space, where

$$\mathrm{SYMPL}(M, S, P) := C_c^\infty(M, S)/\ker(G) \quad \textit{and} \quad \omega([\varphi], [\psi]) := \int_M \langle G\varphi, \psi \rangle \, \mathrm{dV}.$$

Here the square brackets $[\cdot]$ denote residue classes modulo $\ker(G)$.

Proof. The bilinear form $(\varphi, \psi) \mapsto \int_M \langle G\varphi, \psi\rangle \,\mathrm{dV}$ on $C_c^\infty(M,S)$ is skew-symmetric as a consequence of (6) because P is formally self-adjoint. Its null space is exactly $\ker(G)$. Therefore the induced bilinear form ω on the quotient space $\mathrm{SYMPL}(M,S,P)$ is non-degenerate and hence a symplectic form. □

Theorem 3.8 shows that $G(C_c^\infty(M,S))$ coincides with the space of smooth solutions of the equation $P\varphi = 0$ which have spacelike compact support. In particular, given an object (M,S,P) in GlobHypGreen, the map G induces an isomorphism

$$\mathrm{SYMPL}(M,S,P) = C_c^\infty(M,S)/\ker(G) \xrightarrow{\cong} \ker(P) \cap C_{sc}^\infty(M,S).$$

Hence we may think of $\mathrm{SYMPL}(M,S,P)$ as the space of classical solutions of the equation $P\varphi = 0$ with spacelike compact support.

Now, let (f,F) be a morphism between objects (M_1,S_1,P_1) and (M_2,S_2,P_2) in the category GlobHypGreen. Then the extension by zero induces a symplectic linear map $\mathrm{SYMPL}(f,F) : \mathrm{SYMPL}(M_1,S_1,P_1) \to \mathrm{SYMPL}(M_2,S_2,P_2)$ with

$$\mathrm{SYMPL}(\mathrm{id}_M, \mathrm{id}_S) = \mathrm{id}_{\mathrm{SYMPL}(M,S,P)} \tag{7}$$

and, for any further morphism $(f',F') : (M_2,S_2,P_2) \to (M_3,S_3,P_3)$,

$$\mathrm{SYMPL}((f',F') \circ (f,F)) = \mathrm{SYMPL}(f',F') \circ \mathrm{SYMPL}(f,F). \tag{8}$$

Remark 4.3. Under the isomorphism $\mathrm{SYMPL}(M,S,P) \to \ker(P) \cap C_{sc}^\infty(M,S)$ induced by G, the extension by zero corresponds to an extension as a smooth solution of $P\varphi = 0$ with spacelike compact support. In other words, for any morphism (f,F) from (M_1,S_1,P_1) to (M_2,S_2,P_2) in GlobHypGreen we have the following commutative diagram:

$$\begin{array}{ccc}
\mathrm{SYMPL}(M_1,S_1,P_1) & \xrightarrow{\mathrm{SYMPL}(f,F)} & \mathrm{SYMPL}(M_2,S_2,P_2) \\
\cong \downarrow & & \downarrow \cong \\
\ker(P_1) \cap C_{sc}^\infty(M_1,S_1) & \xrightarrow[\text{a solution}]{\text{extension as}} & \ker(P_2) \cap C_{sc}^\infty(M_2,S_2).
\end{array}$$

Summarizing, we have constructed a covariant functor

$$\mathrm{SYMPL} : \mathsf{GlobHypGreen} \longrightarrow \mathsf{Sympl},$$

where Sympl denotes the category of real symplectic vector spaces with symplectic linear maps as morphisms. In order to obtain an algebra-valued functor, we compose SYMPL with the functor CCR which associates to any symplectic vector space its Weyl algebra. Here "CCR" stands for "canonical commutation relations". This is a general algebraic construction which is independent of the context of Green-hyperbolic operators and which is carried out in Section 2.2. As a result, we obtain the functor

$$\mathfrak{A}_{\mathrm{bos}} := \mathrm{CCR} \circ \mathrm{SYMPL} : \mathsf{GlobHypGreen} \longrightarrow \mathsf{C^*Alg},$$

where $\mathsf{C^*Alg}$ is the category whose objects are the unital C*-algebras and whose morphisms are the injective unit-preserving C*-morphisms.

In the remainder of this section we show that the functor $\mathfrak{A}_{\mathrm{bos}}$ is a bosonic locally covariant quantum field theory. We call two subregions M_1 and M_2 of a spacetime M *causally disjoint* if and only if $J^M(M_1) \cap M_2 = \emptyset$. In other words, there are no causal curves joining M_1 and M_2.

Theorem 4.4. *The functor* $\mathfrak{A}_{\mathrm{bos}} : \mathsf{GlobHypGreen} \longrightarrow \mathsf{C^*Alg}$ *is a bosonic locally covariant quantum field theory, i.e., the following axioms hold:*

(i) **(Quantum causality)** *Let* (M_j, S_j, P_j) *be objects in* $\mathsf{GlobHypGreen}$, $j = 1, 2, 3$, *and* (f_j, F_j) *morphisms from* (M_j, S_j, P_j) *to* (M_3, S_3, P_3), $j = 1, 2$, *such that* $f_1(M_1)$ *and* $f_2(M_2)$ *are causally disjoint regions in* M_3. *Then the subalgebras* $\mathfrak{A}_{\mathrm{bos}}(f_1, F_1)(\mathfrak{A}_{\mathrm{bos}}(M_1, S_1, P_1))$ *and* $\mathfrak{A}_{\mathrm{bos}}(f_2, F_2)(\mathfrak{A}_{\mathrm{bos}}(M_2, S_2, P_2))$ *of* $\mathfrak{A}_{\mathrm{bos}}(M_3, S_3, P_3)$ *commute.*

(ii) **(Time-slice axiom)** *Let* (M_j, S_j, P_j) *be objects in* $\mathsf{GlobHypGreen}$, $j = 1, 2$, *and* (f, F) *a morphism from* (M_1, S_1, P_1) *to* (M_2, S_2, P_2) *such that there is a Cauchy hypersurface* $\Sigma \subset M_1$ *for which* $f(\Sigma)$ *is a Cauchy hypersurface of* M_2. *Then*
$$\mathfrak{A}_{\mathrm{bos}}(f, F) : \mathfrak{A}_{\mathrm{bos}}(M_1, S_1, P_1) \to \mathfrak{A}_{\mathrm{bos}}(M_2, S_2, P_2)$$
is an isomorphism.

Proof. We first show (i). For notational simplicity we assume without loss of generality that f_j and F_j are inclusions, $j = 1, 2$. Let $\varphi_j \in C_c^\infty(M_j, S_j)$. Since M_1 and M_2 are causally disjoint, the sections $G\varphi_1$ and φ_2 have disjoint support, thus
$$\omega([\varphi_1], [\varphi_2]) = \int_M \langle G\varphi_1, \varphi_2 \rangle \,\mathrm{dV} = 0.$$
Now relation (iv) in Definition 2.8 tells us
$$w([\varphi_1]) \cdot w([\varphi_2]) = w([\varphi_1] + [\varphi_2]) = w([\varphi_2]) \cdot w([\varphi_1]).$$
Since $\mathfrak{A}_{\mathrm{bos}}(f_1, F_1)(\mathfrak{A}_{\mathrm{bos}}(M_1, S_1, P_1))$ is generated by elements of the form $w([\varphi_1])$ and $\mathfrak{A}_{\mathrm{bos}}(f_2, F_2)(\mathfrak{A}_{\mathrm{bos}}(M_2, S_2, P_2))$ by elements of the form $w([\varphi_2])$, the assertion follows.

In order to prove (ii) we show that $\mathrm{SYMPL}(f, F)$ is an isomorphism of symplectic vector spaces provided f maps a Cauchy hypersurface of M_1 onto a Cauchy hypersurface of M_2. Since symplectic linear maps are always injective, we only need to show surjectivity of $\mathrm{SYMPL}(f, F)$. This is most easily seen by replacing $\mathrm{SYMPL}(M_j, S_j, P_j)$ by $\ker(P_j) \cap C_{\mathrm{sc}}^\infty(M_j, S_j)$ as in Remark 4.3. Again we assume without loss of generality that f and F are inclusions.

Let $\psi \in C_{\mathrm{sc}}^\infty(M_2, S_2)$ be a solution of $P_2\psi = 0$. Let φ be the restriction of ψ to M_1. Then φ solves $P_1\varphi = 0$ and has spacelike compact support in M_1, see [4, Lemma 3.11]. We will show that there is only one solution in M_2 with spacelike compact support extending φ. It will then follow that ψ is the image of φ under the extension map corresponding to $\mathrm{SYMPL}(f, F)$ and surjectivity will be shown.

To prove uniqueness of the extension, we may, by linearity, assume that $\varphi = 0$. Then ψ_+ defined by

$$\psi_+(x) := \begin{cases} \psi(x), & \text{if } x \in J_+^{M_2}(\Sigma), \\ 0, & \text{otherwise,} \end{cases}$$

is smooth since ψ vanishes in an open neighborhood of Σ. Now ψ_+ solves $P_2\psi_+ = 0$ and has past-compact support. As noticed just below Theorem 3.8, this implies $\psi_+ \equiv 0$, i.e., ψ vanishes on $J_+^{M_2}(\Sigma)$. One shows similarly that ψ vanishes on $J_-^{M_2}(\Sigma)$, hence $\psi = 0$. □

The quantization process described in this subsection applies in particular to formally self-adjoint wave and Dirac-type operators.

4.2. Fermionic quantization

Next we construct a fermionic quantization. For this we need a functorial construction of Hilbert spaces rather than symplectic vector spaces. As we shall see this seems to be possible only under much more restrictive assumptions. The underlying Lorentzian manifold M is assumed to be a globally hyperbolic spacetime as before. The vector bundle S is assumed to be complex with Hermitian inner product $\langle\cdot\,,\cdot\rangle$ which may be indefinite. The formally self-adjoint Green-hyperbolic operator P is assumed to be of first order.

Definition 4.5. A formally self-adjoint Green-hyperbolic operator P of first order acting on sections of a complex vector bundle S over a spacetime M is of *definite type* if and only if for any $x \in M$ and any future-directed timelike tangent vector $\mathfrak{n} \in T_xM$, the bilinear map

$$S_x \times S_x \to \mathbb{C}, \qquad (\varphi, \psi) \mapsto \langle i\sigma_P(\mathfrak{n}^\flat)\cdot\varphi, \psi\rangle,$$

yields a positive definite Hermitian scalar product on S_x.

Example 4.6. The classical Dirac operator P from Example 3.17 is, when defined with the correct sign, of definite type, see e.g. [6, Sec. 1.1.5] or [3, Sec. 2].

Example 4.7. If $E \to M$ is a semi-Riemannian or semi-Hermitian vector bundle endowed with a metric connection over a spin spacetime M, then the twisted Dirac operator from Example 3.18 is of definite type if and only if the metric on E is positive definite. This can be seen by evaluating the tensorized inner product on elements of the form $\sigma \otimes v$, where $v \in E_x$ is null.

Example 4.8. The operator $P = i(d-\delta)$ on $S = \Lambda T^*M \otimes \mathbb{C}$ is of Dirac type but not of definite type. This follows from Example 4.7 applied to Example 3.19, since the natural inner product on ΣM is not positive definite. An alternative elementary proof is the following: for any timelike tangent vector $\mathfrak{n}$ on M and the corresponding covector $\mathfrak{n}^\flat$, one has

$$\langle i\sigma_P(\mathfrak{n}^\flat)\mathfrak{n}^\flat, \mathfrak{n}^\flat\rangle = -\langle \mathfrak{n}^\flat \wedge \mathfrak{n}^\flat - \mathfrak{n} \lrcorner \mathfrak{n}^\flat, \mathfrak{n}^\flat\rangle = \langle \mathfrak{n}, \mathfrak{n}\rangle\langle 1, \mathfrak{n}^\flat\rangle = 0.$$

Example 4.9. An elementary computation shows that the Rarita-Schwinger operator defined in Section 3.6 is not of definite type if $m \geq 3$, see [4, Ex. 3.16].

We define the category GlobHypDef, whose objects are triples (M, S, P), where M is a globally hyperbolic spacetime, S is a complex vector bundle equipped with a complex inner product $\langle\cdot\,,\cdot\rangle$, and P is a formally self-adjoint Green-hyperbolic operator of definite type acting on sections of S. The morphisms are the same as in the category GlobHypGreen.

We construct a covariant functor from GlobHypDef to HILB, where HILB denotes the category whose objects are complex pre-Hilbert spaces and whose morphisms are isometric linear embeddings. As in Section 4.1, the underlying vector space is the space of classical solutions to the equation $P\varphi = 0$ with spacelike compact support. We put

$$\mathrm{SOL}(M, S, P) := \ker(P) \cap C^\infty_{\mathrm{sc}}(M, S).$$

Here "SOL" stands for classical solutions of the equation $P\varphi = 0$ with spacelike compact support. We endow $\mathrm{SOL}(M, S, P)$ with a positive definite Hermitian scalar product as follows: consider a smooth spacelike Cauchy hypersurface $\Sigma \subset M$ with its future-oriented unit normal vector field $\mathfrak{n}$ and its induced volume element dA and set

$$(\varphi, \psi) := \int_\Sigma \langle i\sigma_P(\mathfrak{n}^\flat) \cdot \varphi|_\Sigma, \psi|_\Sigma\rangle \,\mathrm{dA}, \tag{9}$$

for all $\varphi, \psi \in C^\infty_{\mathrm{sc}}(M, S)$. The Green's formula for formally self-adjoint first-order differential operators [32, p. 160, Prop. 9.1] (see also [4, Lemma 3.17]) implies that $(\cdot,\cdot)$ does not depend on the choice of Σ. Of course, it is positive definite because of the assumption that P is of definite type. In case P is not of definite type, the sesquilinear form $(\cdot,\cdot)$ is still independent of the choice of Σ but may be degenerate, see [4, Remark 3.18].

For any object (M, S, P) in GlobHypDef we equip $\mathrm{SOL}(M, S, P)$ with the Hermitian scalar product in (9) and thus turn $\mathrm{SOL}(M, S, P)$ into a pre-Hilbert space.

Given a morphism $(f, F)\colon (M_1, S_1, P_1) \to (M_2, S_2, P_2)$ in GlobHypDef, then this is also a morphism in GlobHypGreen and hence induces a homomorphism $\mathrm{SYMPL}(f, F) : \mathrm{SYMPL}(M_1, S_1, P_1) \to \mathrm{SYMPL}(M_2, S_2, P_2)$. As explained in Remark 4.3, there is a corresponding extension homomorphism $\mathrm{SOL}(f, F) : \mathrm{SOL}(M_1, S_1, P_1) \to \mathrm{SOL}(M_2, S_2, P_2)$. In other words, $\mathrm{SOL}(f, F)$ is defined such that the diagram

$$\begin{array}{ccc} \mathrm{SYMPL}(M_1, S_1, P_1) & \xrightarrow{\mathrm{SYMPL}(f,F)} & \mathrm{SYMPL}(M_2, S_2, P_2) \\ \cong\downarrow & & \downarrow\cong \\ \mathrm{SOL}(M_1, S_1, P_1) & \xrightarrow{\mathrm{SOL}(f,F)} & \mathrm{SOL}(M_2, S_2, P_2) \end{array} \tag{10}$$

commutes. The vertical arrows are the vector space isomorphisms induced be the Green's propagators G_1 and G_2, respectively.

Lemma 4.10. *The vector space homomorphism* $\mathrm{SOL}(f, F)\colon \mathrm{SOL}(M_1, S_1, P_1) \to \mathrm{SOL}(M_2, S_2, P_2)$ *preserves the scalar products, i.e., it is an isometric linear embedding of pre-Hilbert spaces.*

We refer to [4, Lemma 3.19] for a proof. The functoriality of SYMPL and diagram (10) show that SOL is a functor from GlobHypDef to HILB, the category of pre-Hilbert spaces with isometric linear embeddings. Composing with the functor CAR (see Section 2.1), we obtain the covariant functor

$$\mathfrak{A}_{\mathrm{ferm}} := \mathrm{CAR} \circ \mathrm{SOL} : \mathsf{GlobHypDef} \longrightarrow \mathsf{C^*Alg}.$$

The fermionic algebras $\mathfrak{A}_{\mathrm{ferm}}(M, S, P)$ are actually $\mathbb{Z}_2$-graded algebras, see Proposition 2.3 (iii).

Theorem 4.11. *The functor* $\mathfrak{A}_{\mathrm{ferm}} : \mathsf{GlobHypDef} \longrightarrow \mathsf{C^*Alg}$ *is a fermionic locally covariant quantum field theory, i.e., the following axioms hold:*

(i) **(Quantum causality)** *Let* (M_j, S_j, P_j) *be objects in* GlobHypDef, $j = 1, 2, 3$, *and* (f_j, F_j) *morphisms from* (M_j, S_j, P_j) *to* (M_3, S_3, P_3), $j = 1, 2$, *such that* $f_1(M_1)$ *and* $f_2(M_2)$ *are causally disjoint regions in* M_3. *Then the subalgebras* $\mathfrak{A}_{\mathrm{ferm}}(f_1, F_1)(\mathfrak{A}_{\mathrm{ferm}}(M_1, S_1, P_1))$ *and* $\mathfrak{A}_{\mathrm{ferm}}(f_2, F_2)(\mathfrak{A}_{\mathrm{ferm}}(M_2, S_2, P_2))$ *of* $\mathfrak{A}_{\mathrm{ferm}}(M_3, S_3, P_3)$ *super-commute*[1].

(ii) **(Time-slice axiom)** *Let* (M_j, S_j, P_j) *be objects in* GlobHypDef, $j = 1, 2$, *and* (f, F) *a morphism from* (M_1, S_1, P_1) *to* (M_2, S_2, P_2) *such that there is a Cauchy hypersurface* $\Sigma \subset M_1$ *for which* $f(\Sigma)$ *is a Cauchy hypersurface of* M_2. *Then*

$$\mathfrak{A}_{\mathrm{ferm}}(f, F) : \mathfrak{A}_{\mathrm{ferm}}(M_1, S_1, P_1) \to \mathfrak{A}_{\mathrm{ferm}}(M_2, S_2, P_2)$$

is an isomorphism.

Proof. To show (i), we assume without loss of generality that f_j and F_j are inclusions. Let $\varphi_1 \in \mathrm{SOL}(M_1, S_1, P_1)$ and $\psi_1 \in \mathrm{SOL}(M_2, S_2, P_2)$. Denote the extensions to M_3 by $\varphi_2 := \mathrm{SOL}(f_1, F_1)(\varphi_1)$ and $\psi_2 := \mathrm{SOL}(f_2, F_2)(\psi_1)$. Choose a compact submanifold K_1 (with boundary) in a spacelike Cauchy hypersurface Σ_1 of M_1 such that $\mathrm{supp}(\varphi_1) \cap \Sigma_1 \subset K_1$ and similarly K_2 for ψ_1. Since M_1 and M_2 are causally disjoint, $K_1 \cup K_2$ is acausal. Hence, by [8, Thm. 1.1], there exists a Cauchy hypersurface Σ_3 of M_3 containing K_1 and K_2. As in the proof of Lemma 4.10 one sees that $\mathrm{supp}(\varphi_2) \cap \Sigma_3 = \mathrm{supp}(\varphi_1) \cap \Sigma_1$ and similarly for ψ_2. Thus, when restricted to Σ_3, φ_2 and ψ_2 have disjoint support. Hence $(\varphi_2, \psi_2) = 0$. This shows that the subspaces $\mathrm{SOL}(f_1, F_1)(\mathrm{SOL}(M_1, S_1, P_1))$ and $\mathrm{SOL}(f_2, F_2)(\mathrm{SOL}(M_2, S_2, P_2))$ of $\mathrm{SOL}(M_3, S_3, P_3)$ are perpendicular. Since the even (resp. odd) part of the Clifford algebra of a vector space V with quadratic form is linearly spanned by the even (resp. odd) products of vectors in V, Definition 2.1 shows that the corresponding CAR-algebras must super-commute.

To see (ii) we recall that (f, F) is also a morphism in GlobHypGreen and that we know from Theorem 4.4 that $\mathrm{SYMPL}(f, F)$ is an isomorphism. From diagram (10) we see that $\mathrm{SOL}(f, F)$ is an isomorphism. Hence $\mathfrak{A}_{\mathrm{ferm}}(f, F)$ is also an isomorphism. □

[1]This means that the odd parts of the algebras anti-commute while the even parts commute with everything.

Remark 4.12. Since causally disjoint regions should lead to commuting observables also in the fermionic case, one usually considers only the even part $\mathfrak{A}^{\mathrm{even}}_{\mathrm{ferm}}(M,S,P)$ as the observable algebra while the full algebra $\mathfrak{A}_{\mathrm{ferm}}(M,S,P)$ is called the *field algebra.*

There is a slightly different description of the functor $\mathfrak{A}_{\mathrm{ferm}}$. Let $\mathsf{HILB}_{\mathbb{R}}$ denote the category whose objects are the real pre-Hilbert spaces and whose morphisms are the isometric linear embeddings. We have the functor REAL : $\mathsf{HILB} \to \mathsf{HILB}_{\mathbb{R}}$ which associates to each complex pre-Hilbert space $(V,(\cdot,\cdot))$ its underlying real pre-Hilbert space $(V,\mathfrak{Re}(\cdot,\cdot))$. By Remark 2.7,

$$\mathfrak{A}_{\mathrm{ferm}} = \mathrm{CAR}_{\mathrm{sd}} \circ \mathrm{REAL} \circ \mathrm{SOL}.$$

Since the self-dual CAR-algebra of a real pre-Hilbert space is the Clifford algebra of its complexification and since for any complex pre-Hilbert space V we have

$$\mathrm{REAL}(V) \otimes_{\mathbb{R}} \mathbb{C} = V \oplus V^*,$$

$\mathfrak{A}_{\mathrm{ferm}}(M,S,P)$ is also the Clifford algebra of $\mathrm{SOL}(M,S,P) \oplus \mathrm{SOL}(M,S,P)^* = \mathrm{SOL}(M, S \oplus S^*, P \oplus P^*)$. This is the way this functor is often described in the physics literature, see e.g. [31, p. 115f].

Self-dual CAR-representations are more natural for real fields. Let M be globally hyperbolic and let $S \to M$ be a *real* vector bundle equipped with a real inner product $\langle \cdot\,,\cdot \rangle$. A formally skew-adjoint[2] differential operator P acting on sections of S is called of *definite type* if and only if for any $x \in M$ and any future-directed timelike tangent vector $\mathfrak{n} \in T_xM$, the bilinear map

$$S_x \times S_x \to \mathbb{R}, \qquad (\varphi,\psi) \mapsto \langle \sigma_P(\mathfrak{n}^\flat)\cdot\varphi,\psi\rangle,$$

yields a positive definite Euclidean scalar product on S_x. An example is given by the real Dirac operator

$$D := \sum_{j=1}^{m} \varepsilon_j e_j \cdot \nabla_{e_j}$$

acting on sections of the real spinor bundle $\Sigma^{\mathbb{R}}M$.

Given a smooth spacelike Cauchy hypersurface $\Sigma \subset M$ with future-directed timelike unit normal field $\mathfrak{n}$, we define a scalar product on $\mathrm{SOL}(M,S,P) = \ker(P) \cap C^\infty_{\mathrm{sc}}(M,S,P)$ by

$$(\varphi,\psi) := \int_\Sigma \langle \sigma_P(\mathfrak{n}^\flat)\cdot \varphi|_\Sigma, \psi|_\Sigma\rangle \,\mathrm{dA}.$$

With essentially the same proofs as before, one sees that this scalar product does not depend on the choice of Cauchy hypersurface Σ and that a morphism $(f,F) : (M_1,S_1,P_1) \to (M_2,S_2,P_2)$ gives rise to an extension operator $\mathrm{SOL}(f,F) : \mathrm{SOL}(M_1,S_1,P_1) \to \mathrm{SOL}(M_2,S_2,P_2)$ preserving the scalar product. We have constructed a functor

$$\mathrm{SOL} : \mathsf{GlobHypSkewDef} \longrightarrow \mathsf{HILB}_{\mathbb{R}},$$

[2] instead of self-adjoint!

where GlobHypSkewDef denotes the category whose objects are triples (M, S, P) with M globally hyperbolic, $S \to M$ a real vector bundle with real inner product and P a formally skew-adjoint, Green-hyperbolic differential operator of definite type acting on sections of S. The morphisms are the same as before.

Now the functor

$$\mathfrak{A}_{\mathrm{ferm}}^{\mathrm{sd}} := \mathrm{CAR}_{\mathrm{sd}} \circ \mathrm{SOL} : \mathsf{GlobHypSkewDef} \longrightarrow \mathsf{C^*Alg}$$

is a locally covariant quantum field theory in the sense that Theorem 4.11 holds with $\mathfrak{A}_{\mathrm{ferm}}$ replaced by $\mathfrak{A}_{\mathrm{ferm}}^{\mathrm{sd}}$.

5. Conclusion

We have constructed three functors,

$$\begin{aligned} \mathfrak{A}_{\mathrm{bos}} &: \mathsf{GlobHypGreen} \longrightarrow \mathsf{C^*Alg}, \\ \mathfrak{A}_{\mathrm{ferm}} &: \mathsf{GlobHypDef} \longrightarrow \mathsf{C^*Alg}, \\ \mathfrak{A}_{\mathrm{ferm}}^{\mathrm{sd}} &: \mathsf{GlobHypSkewDef} \longrightarrow \mathsf{C^*Alg}. \end{aligned}$$

The first functor turns out to be a bosonic locally covariant quantum field theory while the second and third are fermionic locally covariant quantum field theories.

The category GlobHypGreen seems to contain basically all physically relevant free fields such as fields governed by wave equations, Dirac equations, the Proca equation and the Rarita-Schwinger equation. It contains operators of all orders. Bosonic quantization of Dirac fields might be considered unphysical but the discussion shows that there is no spin-statistics theorem on the level of observable algebras. In order to obtain results like Theorem 5.1 in [33] one needs more structure, namely representations of the observable algebras with good properties.

The categories GlobHypDef and GlobHypSkewDef are much smaller. They contain only operators of first order with Dirac operators as main examples. But even certain twisted Dirac operators such as the Euler operator do not belong to this class. The category GlobHypSkewDef is essentially the real analogue of GlobHypDef.

References

[1] H. Araki: *On quasifree states of* CAR *and Bogoliubov automorphisms.* Publ. Res. Inst. Math. Sci. **6** (1970/71), 385–442.

[2] C. Bär and C. Becker: *C*-algebras.* In: C. Bär and K. Fredenhagen (Eds.): *Quantum field theory on curved spacetimes.* 1–37, Lecture Notes in Phys. **786**, Springer-Verlag, Berlin, 2009.

[3] C. Bär, P. Gauduchon, and A. Moroianu: *Generalized cylinders in semi-Riemannian and spin geometry.* Math. Zeitschr. **249** (2005), 545–580.

[4] C. Bär and N. Ginoux: *Classical and quantum fields on Lorentzian manifolds.* submitted.

[5] C. Bär, N. Ginoux, and F. Pfäffle: *Wave equations on Lorentzian manifolds and quantization.* EMS, Zürich, 2007.

[6] H. Baum: *Spin-Strukturen und Dirac-Operatoren über pseudoriemannschen Mannigfaltigkeiten.* Teubner, Leipzig, 1981.

[7] A. N. Bernal and M. Sánchez: *Smoothness of time functions and the metric splitting of globally hyperbolic spacetimes.* Commun. Math. Phys. **257** (2005), 43–50.

[8] A. N. Bernal and M. Sánchez: *Further results on the smoothability of Cauchy hypersurfaces and Cauchy time functions.* Lett. Math. Phys. **77** (2006), 183–197.

[9] T. Branson and O. Hijazi: *Bochner-Weitzenböck formulas associated with the Rarita-Schwinger operator.* Internat. J. Math. **13** (2002), 137–182.

[10] O. Bratteli and D. W. Robinson: *Operator algebras and quantum statistical mechanics, I-II* (second edition). Texts and Monographs in Physics, Springer, Berlin, 1997.

[11] T. Bröcker and T. tom Dieck: *Representations of compact Lie groups.* Graduate Texts in Mathematics **98**, Springer-Verlag, New York, 1995.

[12] R. Brunetti, K. Fredenhagen and R. Verch: *The generally covariant locality principle – a new paradigm for local quantum field theory.* Commun. Math. Phys. **237** (2003), 31–68.

[13] H. A. Buchdahl: *On the compatibility of relativistic wave equations in Riemann spaces. II.* J. Phys. A **15** (1982), 1–5.

[14] H. A. Buchdahl: *On the compatibility of relativistic wave equations in Riemann spaces. III.* J. Phys. A **15** (1982), 1057–1062.

[15] J. Dimock: *Algebras of local observables on a manifold.* Commun. Math. Phys. **77** (1980), 219–228.

[16] J. Dimock: *Dirac quantum fields on a manifold.* Trans. Amer. Math. Soc. **269** (1982), 133–147.

[17] E. Furlani: *Quantization of massive vector fields in curved space-time.* J. Math. Phys. **40** (1999), 2611–2626.

[18] R.P. Geroch: *Domain of dependence.* J. Math. Phys. **11** (1970), 437–449.

[19] G. W. Gibbons: *A note on the Rarita-Schwinger equation in a gravitational background.* J. Phys. A **9** (1976), 145–148.

[20] L. Hörmander: *The analysis of linear partial differential operators. I. Distribution theory and Fourier analysis.* 2nd ed. Grundlehren der Mathematischen Wissenschaften **256**, Springer-Verlag, Berlin, 1990.

[21] L. Hörmander: *The analysis of linear partial differential operators. III. Pseudodifferential operators.* Grundlehren der Mathematischen Wissenschaften **274**, Springer-Verlag, Berlin, 1985.

[22] B. S. Kay: *Linear spin-zero quantum fields in external gravitational and scalar fields.* Commun. Math. Phys. **62** (1978), 55–70.

[23] H. B. Lawson and M.-L. Michelsohn: *Spin geometry.* Princeton University Press, Princeton, 1989.

[24] R. Mühlhoff: *Higher spin fields on curved spacetimes.* Diplomarbeit, Universität Leipzig, 2007.

[25] R. Mühlhoff: *Cauchy problem and Green's functions for first order differential operators and algebraic quantization.* J. Math. Phys. **52** (2011), 022303.

[26] B. O'Neill: *Semi-Riemannian geometry.* Academic Press, San Diego, 1983.

[27] R. J. Plymen and P.L. Robinson: *Spinors in Hilbert space.* Cambridge Tracts in Mathematics **114**, Cambridge University Press, Cambridge, 1994.

[28] W. Rarita and J. Schwinger: *On a theory of particles with half-integral spin.* Phys. Rev. **60** (1941), 61.

[29] M. Reed and B. Simon: *Methods of modern mathematical physics I: Functional analysis.* Academic Press, Orlando, 1980.

[30] K. Sanders: *The locally covariant Dirac field.* Rev. Math. Phys. 22 (2010), 381–430.

[31] A. Strohmaier: *The Reeh-Schlieder property for quantum fields on stationary spacetimes.* Commun. Math. Phys. **215** (2000), 105–118.

[32] M. E. Taylor: *Partial differential equations I - Basic theory.* Springer-Verlag, New York - Berlin - Heidelberg, 1996.

[33] R. Verch: *A spin-statistics theorem for quantum fields on curved spacetime manifolds in a generally covariant framework.* Commun. Math. Phys. **223** (2001), 261–288.

[34] McK. Y. Wang: *Preserving parallel spinors under metric deformations.* Indiana Univ. Math. J. **40** (1991), 815–844.

[35] V. Wünsch: *Cauchy's problem and Huygens' principle for relativistic higher spin wave equations in an arbitrary curved space-time.* Gen. Relativity Gravitation **17** (1985), 15–38.

Christian Bär
Universität Potsdam
Institut für Mathematik
Am Neuen Palais 10
Haus 8
14469 Potsdam
Germany
e-mail: `baer@math.uni-potsdam.de`

Nicolas Ginoux
Fakultät für Mathematik
Universität Regensburg
93040 Regensburg
Germany
e-mail: `nicolas.ginoux@mathematik.uni-regensburg.de`

On the Notion of 'the Same Physics in All Spacetimes'

Christopher J. Fewster

Abstract. Brunetti, Fredenhagen and Verch (BFV) have shown how the notion of local covariance for quantum field theories can be formulated in terms of category theory: a theory being described as a functor from a category of spacetimes to a category of $(C)^*$-algebras. We discuss whether this condition is sufficient to guarantee that a theory represents 'the same physics' in all spacetimes, giving examples to show that it does not. A new criterion, *dynamical locality*, is formulated, which requires that descriptions of local physics based on kinematical and dynamical considerations should coincide. Various applications are given, including a proof that dynamical locality for quantum fields is incompatible with the possibility of covariantly choosing a preferred state in each spacetime.

As part of this discussion we state a precise condition that should hold on any class of theories each representing the same physics in all spacetimes. This condition holds for the dynamically local theories but is violated by the full class of locally covariant theories in the BFV sense.

The majority of results described form part of forthcoming papers with Rainer Verch [16].

Mathematics Subject Classification (2010). 81T05, 81T20, 81P99.

Keywords. Locality, covariance, quantum field theory in curved spacetime.

1. Introduction

This contribution is devoted to the issue of how a physical theory should be formulated in arbitrary spacetime backgrounds in such a way that the physical content is preserved. Our motivation arises from various directions. First, it is essential for the extension of axiomatic quantum field theory to curved spacetimes. Second, one would expect that any quantization of gravity coupled to matter should, in certain regimes, resemble a common theory of matter on different fixed backgrounds. Third, as our universe appears to be well-described by a curved spacetime on many scales, one may well wonder what physical relevance should be ascribed to a theory that could only be formulated in Minkowski space.

In Lagrangian theories, there is an apparently satisfactory answer to our question: namely that the Lagrangian should transform covariantly under coordinate transformations. Actually, even here there are subtleties, as a simple example reveals. Consider the nonminimally coupled scalar field with Lagrangian density

$$\mathcal{L} = \frac{1}{2}\sqrt{-g}\left(\nabla^a\phi\nabla_a\phi - \xi R\phi^2\right).$$

In Minkowski space the equation of motion does not depend on the coupling constant ξ but theories with different coupling constant can still be distinguished by their stress-tensors, which contain terms proportional to ξ. Suppose, however, that ξR is replaced by $\zeta(R)$, where ζ is a smooth function vanishing in a neighbourhood of the origin and taking a constant value $\xi_0 \neq 0$ for all sufficiently large $|R|$. This gives a new theory that coincides with the $\xi = 0$ theory in Minkowski space in terms of the field equation, stress-energy tensor and any other quantity formed by functional differentiation of the action and then evaluated in Minkowski space. But in de Sitter space (with sufficiently large cosmological constant) the theory coincides, in the same sense, with the $\xi = \xi_0$ theory. Of course, it is unlikely that a theory of this type would be physically relevant, but the example serves to illustrate that covariance of the Lagrangian does not guarantee that the physical content of a theory will be the same in all spacetimes. Additional conditions would be required; perhaps that the (undensitized) Lagrangian should depend analytically on the metric and curvature quantities.

The situation is more acute when one attempts to generalize axiomatic quantum field theory to the curved spacetime context. After all, these approaches do not take a classical action as their starting point, but focus instead on properties that ought to hold for any reasonable quantum field theory. In Minkowski space, Poincaré covariance and the existence of a unique invariant vacuum state obeying the spectrum condition provide strong constraints which ultimately account to a large part for the successes of axiomatic QFT in both its Wightman–Gårding [30] and Araki–Haag–Kastler [20] formulations. As a generic spacetime has no symmetries, and moreover attempts to define distinguished vacuum states in general curved spacetimes lead to failure even for free fields,[1] there are severe difficulties associated with generalizing the axiomatic setting to curved spacetime.

Significant progress was made by Brunetti, Fredenhagen and Verch [5] (henceforth BFV) who formalized the idea of a locally covariant quantum field theory in the language of category theory (see [26, 1] as general references). As we will see, this paper opens up new ways to analyze physical theories; in this sense it justifies its subtitle, 'A new paradigm for local quantum physics'. At the structural level, the ideas involved have led to a number

[1]Theorem 5.2 below strengthens this to a general no-go result.

of new model-independent results for QFT in curved spacetime such as the spin-statistics connection [31], Reeh–Schlieder type results [27] and the analysis of superselection sectors [6, 7]; they were also crucial in completing the perturbative construction of interacting QFT in curved spacetime [4, 22, 23]. It should also be mentioned that antecedents of the ideas underlying the BFV formalism can be found in [17, 24, 11]. However, we stress that the BFV approach is not simply a matter of formalism: it suggests and facilitates new calculations that can lead to concrete physical predictions such as *a priori* bounds on Casimir energy densities [15, 14] and new viewpoints in cosmology [8, 9] (see also Verch's contribution to these proceedings [32]).

The focus in this paper is on the physical content of local covariance in the BFV formulation: in particular, is it sufficiently strong to enforce the same physics in all spacetimes? And, if not, what does? We confine ourselves here to the main ideas, referring to forthcoming papers [16] for the details.

2. Locally covariant physical theories

To begin, let us consider what is required for a local, causal description of physics. A minimal list might be the following:

- Experiments can be conducted over a finite timespan and spatial extent in reasonable isolation from the rest of the world.
- We may distinguish a 'before' and an 'after' for such experiments.
- The 'same' experiment could, in principle, be conducted at other locations and times with the 'same' outcomes (to within experimental accuracy, and possibly in a statistical sense).
- The theoretical account of such experiments should be independent (as far as possible) of the rest of the world.

Stated simply, we should not need to know where in the Universe our laboratory is, nor whether the Universe has any extent much beyond its walls (depending on the duration of the experiment and the degree of shielding the walls provide). Of course, these statements contain a number of undefined terms; and, when referring to the 'same' experiment, we should take into account the motion of the apparatus relative to local inertial frames (compare a Foucault pendulum in Regensburg (latitude 49°N) with one in York (54°N)). Anticipating Machian objections to the last requirement, we assume that sufficient structures are present to permit identification of (approximately) inertial frames.

If spacetime is modelled as a Lorentzian manifold, the minimal requirements can be met by restricting to the globally hyperbolic spacetimes. Recall that an orientable and time-orientable Lorentzian manifold $\boldsymbol{M}$ (the symbol $\boldsymbol{M}$ will encompass the manifold, choice of metric,[2] orientation and time-orientation) is said to be globally hyperbolic if it has no closed causal curves

[2] We adopt signature $+ - \cdots -$.

and $J^+_{\boldsymbol{M}}(p) \cap J^-_{\boldsymbol{M}}(q)$ is compact for all $p, q \in \boldsymbol{M}$ (see [3, Thm 3.2] for equivalence with the older definition [21]).[3] Compactness of these regions ensures that experiments can be conducted within bounded spacetime regions, which could in principle be shielded, and that communication within the region can be achieved by finitely many radio links of finite range. Technically, of course, the main use of global hyperbolicity is that it guarantees well-posedness of the Klein–Gordon equation and other field equations with metric principal part (as described, for example, in Prof. Bär's contribution to this workshop). Henceforth we will assume that all spacetimes are globally hyperbolic and have at most finitely many connected components.

The globally hyperbolic spacetimes constitute the objects of a category Loc, in which the morphisms are taken to be hyperbolic embeddings.

Definition 2.1. A *hyperbolic embedding* of $\boldsymbol{M}$ into $\boldsymbol{N}$ is an isometry $\psi : \boldsymbol{M} \to \boldsymbol{N}$ preserving time and space orientations and such that $\psi(\boldsymbol{M})$ is a causally convex[4] (and hence globally hyperbolic) subset of $\boldsymbol{N}$.

A hyperbolic embedding $\psi : \boldsymbol{M} \to \boldsymbol{N}$ allows us to regard $\boldsymbol{N}$ as an enlarged version of $\boldsymbol{M}$: any experiment taking place in $\boldsymbol{M}$ should have an analogue in $\psi(\boldsymbol{M})$ which should yield indistinguishable results – at least if 'physics is the same' in both spacetimes.

BFV implemented this idea by regarding a physical theory as a functor from the category Loc to a category Phys, which encodes the type of physical system under consideration: the objects representing systems and the morphisms representing embeddings of one system in a larger one. BFV were interested in quantum field theories, described in terms of their algebras of observables. Here, natural candidates for Phys are provided by Alg (resp., C*-Alg), whose objects are unital $(C)^*$-algebras with unit-preserving injective $*$-homomorphisms as the morphisms. However, the idea applies more widely. For example, in the context of linear Hamiltonian systems, a natural choice of Phys would be the category Sympl of real symplectic spaces with symplectic maps as morphisms. As a more elaborate example, take as objects of a category Sys all pairs $(\mathcal{A}, S)$ where $\mathcal{A} \in$ Alg (or C*-Alg) and S is a nonempty convex subset of the states on $\mathcal{A}$, closed under operations induced by $\mathcal{A}$.[5] Whenever $\alpha : \mathcal{A} \to \mathcal{B}$ is an Alg-morphism such that $\alpha^*(T) \subset S$, we will say that α induces a morphism $(\mathcal{A}, S) \to (\mathcal{B}, T)$ in Sys. It may be shown that Sys is a category under the composition inherited from Alg. This category provides an arena for discussing algebras of observables equipped with a distinguished class of states. The beauty of the categorical description is that it allows us to treat different types of physical theory in very similar ways.

---

[3] Here, $J^\pm_{\boldsymbol{M}}(p)$ denotes the set of points that can be reached by future (+) or past (−) directed causal curves originating from a point p in $\boldsymbol{M}$ (including p itself). If S is a subset of $\boldsymbol{M}$, we write $J^\pm(S)$ for the union $J^\pm(S) = \bigcup_{p \in S} J^\pm_{\boldsymbol{M}}(p)$, and $J_{\boldsymbol{M}}(S) = J^+_{\boldsymbol{M}}(S) \cup J^-_{\boldsymbol{M}}(S)$.

[4] A subset S of $\boldsymbol{N}$ is causally convex if every causal curve in $\boldsymbol{N}$ with endpoints in S is contained wholly in S.

[5] That is, if $\sigma \in S$ and $A \in \mathcal{A}$ has $\sigma(A^*A) = 1$, then $\sigma_A(X) := \sigma(A^*XA)$ defines $\sigma_A \in S$.

According to BFV, then, a theory should assign to each spacetime $\boldsymbol{M} \in$ Loc a mathematical object $\mathscr{A}(\boldsymbol{M})$ modelling 'the physics on $\boldsymbol{M}$' and to each hyperbolic embedding $\psi : \boldsymbol{M} \to \boldsymbol{N}$ a means of embedding the physics on $\boldsymbol{M}$ inside the physics on $\boldsymbol{N}$, expressed as a morphism $\mathscr{A}(\psi) : \mathscr{A}(\boldsymbol{M}) \to \mathscr{A}(\boldsymbol{N})$. The natural conditions that $\mathscr{A}(\mathrm{id}_{\boldsymbol{M}}) = \mathrm{id}_{\mathscr{A}(\boldsymbol{M})}$ and $\mathscr{A}(\varphi \circ \psi) = \mathscr{A}(\varphi) \circ \mathscr{A}(\psi)$ are precisely those that make $\mathscr{A}$ a (covariant) functor.

In particular, this gives an immediate definition of the 'local physics' in any nonempty open globally hyperbolic subset O of $\boldsymbol{M}$ as

$$\mathscr{A}^{\mathrm{kin}}(\boldsymbol{M}; O) := \mathscr{A}(\boldsymbol{M}|_O),$$

where $\boldsymbol{M}|_O$ denotes the region O with geometry induced from $\boldsymbol{M}$ and considered as a spacetime in its own right; this embeds in $\mathscr{A}(\boldsymbol{M})$ by

$$\mathscr{A}^{\mathrm{kin}}(\boldsymbol{M}; O) \xrightarrow{\mathscr{A}(\iota_{\boldsymbol{M};O})} \mathscr{A}(\boldsymbol{M}),$$

where $\iota_{\boldsymbol{M};O} : \boldsymbol{M}|_O \to \boldsymbol{M}$ is the obvious embedding. Anticipating future developments, we refer to this as a *kinematic* description of the local physics.

The kinematic description has many nice properties. Restricted to Minkowski space, and with $\mathsf{Phys} = \mathsf{C^*\text{-}Alg}$, BFV were able to show that the net $O \mapsto \mathscr{A}^{\mathrm{kin}}(\boldsymbol{M}; O)$ (for nonempty open, relatively compact, globally hyperbolic O) satisfies the Haag–Kastler axioms of algebraic QFT.[6] In general spacetimes, it therefore provides a generalisation of the Haag–Kastler framework. It is worth focussing on one particular issue. Dimock [11] also articulated a version of Haag–Kastler axioms for curved spacetime QFT and also expressed this partly in functorial language. However, Dimock's covariance axiom was global in nature: it required that when spacetimes $\boldsymbol{M}$ and $\boldsymbol{N}$ are isometric then there should be an isomorphism between the nets of algebras on $\boldsymbol{M}$ and $\boldsymbol{N}$. In the BFV framework this idea is localised and extended to situations in which $\boldsymbol{M}$ is hyperbolically embedded in $\boldsymbol{N}$ but not necessarily globally isometric to it. To be specific, suppose that there is a morphism $\psi : \boldsymbol{M} \to \boldsymbol{N}$ and that O is a nonempty, open globally hyperbolic subset of $\boldsymbol{M}$ with finitely many connected components. Then $\psi(O)$ also obeys these conditions in $\boldsymbol{N}$ and, moreover, restricting the domain of ψ to O and the codomain to $\psi(O)$, we obtain an isomorphism $\hat{\psi} : \boldsymbol{M}|_O \to \boldsymbol{N}|_{\psi(O)}$ making the diagram

$$\begin{array}{ccc} \boldsymbol{M}|_O & \xrightarrow[\cong]{\hat{\psi}} & \boldsymbol{N}|_{\psi(O)} \\ {\scriptstyle \iota_{\boldsymbol{M};O}} \downarrow & & \downarrow {\scriptstyle \iota_{\boldsymbol{N};\psi(O)}} \\ \boldsymbol{M} & \xrightarrow{\psi} & \boldsymbol{N} \end{array}$$

[6] More precisely, BFV study the images $\mathscr{A}(\iota_{\boldsymbol{M};O})(\mathscr{A}^{\mathrm{kin}}(\boldsymbol{M}; O))$ as sub-C^*-algebras of $\mathscr{A}(\boldsymbol{M})$.

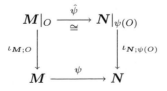

commute. As functors always map commuting diagrams to commuting diagrams, and isomorphisms to isomorphisms, we obtain the commuting diagram

$$\begin{array}{ccc} \mathscr{A}^{\rm kin}(\boldsymbol{M};O) & \xrightarrow[\cong]{\mathscr{A}(\hat{\psi})} & \mathscr{A}^{\rm kin}(\boldsymbol{N};\psi(O)) \\ \alpha^{\rm kin}_{\boldsymbol{M};O}\Big\downarrow & & \Big\downarrow\alpha^{\rm kin}_{\boldsymbol{N};\psi(O)} \\ \mathscr{A}(\boldsymbol{M}) & \xrightarrow{\mathscr{A}(\psi)} & \mathscr{A}(\boldsymbol{N}) \end{array} \tag{2.1}$$

in which, again anticipating future developments, we have written $\alpha^{\rm kin}_{\boldsymbol{M};O}$ for the morphism $\mathscr{A}(\iota_{\boldsymbol{M};O})$ embedding $\mathscr{A}^{\rm kin}(\boldsymbol{M};O)$ in $\mathscr{A}(\boldsymbol{M})$. The significance of this diagram is that it shows that the kinematic description of local physics is truly local: it assigns equivalent descriptions of the physics to isometric subregions of the ambient spacetimes $\boldsymbol{M}$ and $\boldsymbol{N}$. This holds even when there is no hyperbolic embedding of one of these ambient spacetimes in the other, as can be seen if we consider a further morphism $\varphi : \boldsymbol{M} \to \boldsymbol{L}$, thus obtaining an isomorphism $\mathscr{A}(\hat{\psi}) \circ \mathscr{A}(\hat{\varphi}^{-1}) : \mathscr{A}^{\rm kin}(\boldsymbol{L};\varphi(O)) \to \mathscr{A}^{\rm kin}(\boldsymbol{N};\psi(O))$ even though there need be no morphism between $\boldsymbol{L}$ and $\boldsymbol{N}$.

Example: the real scalar field. Take $\mathsf{Phys} = \mathsf{Alg}$. The quantization of the Klein–Gordon equation $(\Box_{\boldsymbol{M}} + m^2)\phi = 0$ is well understood in arbitrary globally hyperbolic spacetimes. To each spacetime $\boldsymbol{M} \in \mathsf{Loc}$ we assign $\mathscr{A}(\boldsymbol{M}) \in \mathsf{Alg}$ with generators $\Phi_{\boldsymbol{M}}(f)$, labelled by test functions $f \in C_0^\infty(\boldsymbol{M})$ and interpreted as smeared fields, subject to the following relations (which hold for arbitrary $f, f' \in C_0^\infty(\boldsymbol{M})$):

- $f \mapsto \Phi_{\boldsymbol{M}}(f)$ is complex linear
- $\Phi_{\boldsymbol{M}}(f)^* = \Phi_{\boldsymbol{M}}(\overline{f})$
- $\Phi_{\boldsymbol{M}}((\Box_{\boldsymbol{M}} + m^2)f) = 0$
- $[\Phi_{\boldsymbol{M}}(f), \Phi_{\boldsymbol{M}}(f')] = iE_{\boldsymbol{M}}(f,f')\mathbf{1}_{\mathscr{A}(\boldsymbol{M})}$.

Here, $E_{\boldsymbol{M}}$ is the advanced-minus-retarded Green function for $\Box_{\boldsymbol{M}} + m^2$, whose existence is guaranteed by global hyperbolicity of $\boldsymbol{M}$. Now if $\psi : \boldsymbol{M} \to \boldsymbol{N}$ is a hyperbolic embedding we have

$$\psi_* \Box_{\boldsymbol{M}} f = \Box_{\boldsymbol{N}} \psi_* f \quad \text{and} \quad E_{\boldsymbol{N}}(\psi_* f, \psi_* f') = E_{\boldsymbol{M}}(f,f') \tag{2.2}$$

for all $f, f' \in C_0^\infty(\boldsymbol{M})$, where

$$(\psi_* f)(p) = \begin{cases} f(\psi^{-1}(p)) & p \in \psi(\boldsymbol{M}) \\ 0 & \text{otherwise.} \end{cases}$$

The second assertion in (2.2) follows from the first, together with the uniqueness of advanced/retarded solutions to the inhomogeneous Klein–Gordon equation. In consequence, the map

$$\mathscr{A}(\psi)(\Phi_{\boldsymbol{M}}(f)) = \Phi_{\boldsymbol{N}}(\psi_* f), \quad \mathscr{A}(\psi)\mathbf{1}_{\mathscr{A}(\boldsymbol{M})} = \mathbf{1}_{\mathscr{A}(\boldsymbol{N})}$$

extends to a $*$-algebra homomorphism $\mathscr{A}(\psi) : \mathscr{A}(\boldsymbol{M}) \to \mathscr{A}(\boldsymbol{N})$. Furthermore, this is a monomorphism because $\mathscr{A}(\boldsymbol{M})$ is simple, so $\mathscr{A}(\psi)$ is indeed a morphism in Alg. It is clear that $\mathscr{A}(\psi \circ \varphi) = \mathscr{A}(\psi) \circ \mathscr{A}(\varphi)$ and $\mathscr{A}(\mathrm{id}_{\boldsymbol{M}}) = \mathrm{id}_{\mathscr{A}(\boldsymbol{M})}$ because these equations hold on the generators, so $\mathscr{A}$

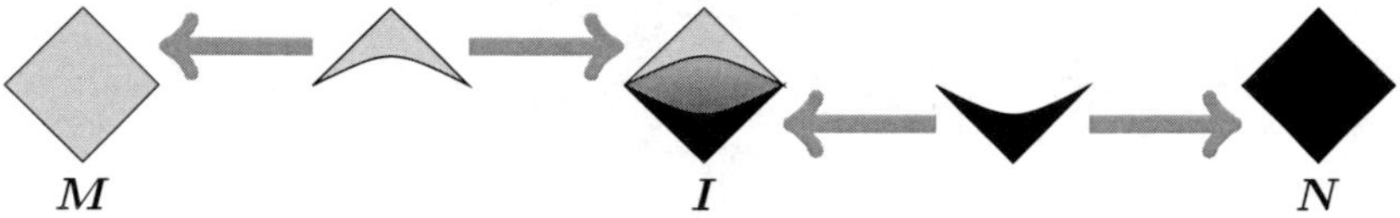

FIGURE 1. Schematic representation of the deformation construction in [18]: globally hyperbolic spacetimes $\boldsymbol{M}$ and $\boldsymbol{N}$, whose Cauchy surfaces are related by an orientation preserving diffeomorphism, can be linked by a chain of Cauchy morphisms and an interpolating spacetime $\boldsymbol{I}$.

indeed defines a functor from Loc to Alg. Finally, if O is a nonempty open globally hyperbolic subset of $\boldsymbol{M}$, $\mathscr{A}^{\rm kin}(\boldsymbol{M}; O)$ is defined as $\mathscr{A}(\boldsymbol{M}|_O)$; its image under $\mathscr{A}(\iota_{\boldsymbol{M};O})$ may be characterized as the unital subalgebra of $\mathscr{A}(\boldsymbol{M})$ generated by those $\Phi_{\boldsymbol{M}}(f)$ with $f \in C_0^\infty(O)$.

3. The time-slice axiom and relative Cauchy evolution

The structures described so far may be regarded as kinematic. To introduce dynamics we need a replacement for the idea that evolution is determined by data on a Cauchy surface. With this in mind, we say that a hyperbolic embedding $\psi : \boldsymbol{M} \to \boldsymbol{N}$ is *Cauchy* if $\psi(\boldsymbol{M})$ contains a Cauchy surface of $\boldsymbol{N}$ and say that a locally covariant theory $\mathscr{A}$ satisfies the *time-slice axiom* if $\mathscr{A}(\psi)$ is an isomorphism whenever ψ is Cauchy. The power of the time-slice axiom arises as follows. Any Cauchy surface naturally inherits an orientation from the ambient spacetime; if two spacetimes $\boldsymbol{M}$ and $\boldsymbol{N}$ have Cauchy surfaces that are equivalent modulo an orientation-preserving diffeomorphism (not necessarily an isometry) then the two spacetimes may be linked by a chain of Cauchy morphisms [18] (see Fig. 1). If the time-slice axiom is satisfied, then the theory $\mathscr{A}$ assigns isomorphisms to each Cauchy morphism; this can often be used to infer that the theory on $\boldsymbol{N}$ obeys some property, given that it holds in $\boldsymbol{M}$.

One of the innovative features of the BFV framework is its ability to quantify the response of the theory to a perturbation in the metric. We write $H(\boldsymbol{M})$ for the set of compactly supported smooth metric perturbations $\boldsymbol{h}$ on $\boldsymbol{M}$ that are sufficiently mild so as to modify $\boldsymbol{M}$ to another globally hyperbolic spacetime $\boldsymbol{M}[\boldsymbol{h}]$. In the schematic diagram of Fig. 2, the metric perturbation lies between two Cauchy surfaces. By taking a globally hyperbolic neighbourhood of each Cauchy surface we are able to find Cauchy morphisms $\iota^\pm$ and $\iota^\pm[\boldsymbol{h}]$ as shown. Now if $\mathscr{A}$ obeys the time-slice axiom, any metric perturbation $\boldsymbol{h} \in H(\boldsymbol{M})$ defines an automorphism of $\mathscr{A}(\boldsymbol{M})$,

$$\mathrm{rce}_{\boldsymbol{M}}[\boldsymbol{h}] = \mathscr{A}(\iota^-) \circ \mathscr{A}(\iota^-[\boldsymbol{h}])^{-1} \circ \mathscr{A}(\iota^+[\boldsymbol{h}]) \circ \mathscr{A}(\iota^+)^{-1}.$$

BFV showed that the functional derivative of the relative Cauchy evolution defines a derivation on the algebra of observables which can be interpreted

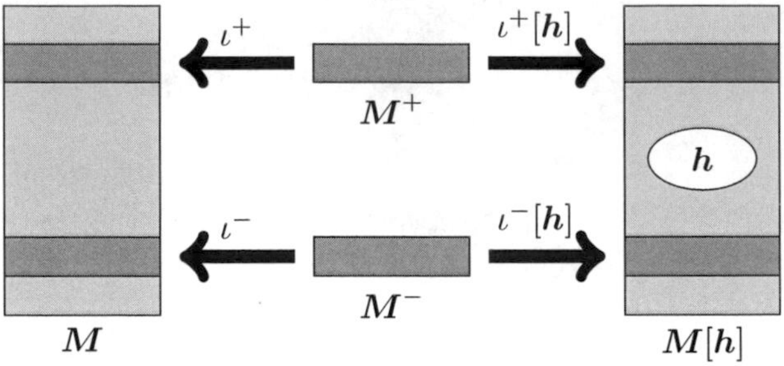

FIGURE 2. Morphisms involved in relative Cauchy evolution

as a commutator with a stress-energy tensor so that

$$[\mathsf{T}_{\boldsymbol{M}}(\boldsymbol{f}), A] = 2i \left.\frac{d}{ds}\mathrm{rce}_{\boldsymbol{M}}[\boldsymbol{h}(s)]A\right|_{s=0},$$

where $\boldsymbol{h}(s)$ is a smooth one-parameter family of metric perturbations in $H(\boldsymbol{M})$ and $\boldsymbol{f} = \dot{\boldsymbol{h}}(0)$. In support of the interpretation of T as a stress-energy tensor we may cite the fact that T is symmetric and conserved. Moreover, the stress-energy tensor in concrete models can be shown to satisfy the above relation (see BFV and [28] for the scalar and Dirac fields respectively) and we will see further evidence for this link below in Sect. 7.

4. The same physics in all spacetimes

4.1. The meaning of SPASs

We can now begin to analyse the question posed at the start of this paper. However, is not yet clear what should be understood by saying that a theory represents the same physics in all spacetimes (SPASs). Rather than give a direct answer we instead posit a property that should be valid for any reasonable notion of SPASs: If *two* theories individually represent the same physics in all spacetimes, and there is *some* spacetime in which the theories coincide then they should coincide in *all* spacetimes. In particular, we obtain the following necessary condition:

Definition 4.1. A class of theories $\mathfrak{T}$ has the SPASs property if no proper subtheory $\mathscr{T}'$ of $\mathscr{T}$ in $\mathfrak{T}$ can fully account for the physics of $\mathscr{T}$ in any single spacetime.

This can be made precise by introducing a new category: the category of locally covariant theories, LCT, whose objects are functors from Loc to Phys and in which the morphisms are natural transformations between such functors. Recall that a natural transformation $\zeta : \mathscr{A} \stackrel{.}{\to} \mathscr{B}$ between theories $\mathscr{A}$ and $\mathscr{B}$ assigns to each $\boldsymbol{M}$ a morphism $\mathscr{A}(\boldsymbol{M}) \xrightarrow{\zeta_{\boldsymbol{M}}} \mathscr{B}(\boldsymbol{M})$ so that for each hyperbolic embedding ψ, the following diagram

$$\begin{array}{ccccc} \boldsymbol{M} & & \mathscr{A}(\boldsymbol{M}) & \xrightarrow{\zeta_{\boldsymbol{M}}} & \mathscr{B}(\boldsymbol{M}) \\ \downarrow{\scriptstyle\psi} & & \downarrow{\scriptstyle\mathscr{A}(\psi)} & & \downarrow{\scriptstyle\mathscr{B}(\psi)} \\ \boldsymbol{N} & & \mathscr{A}(\boldsymbol{N}) & \xrightarrow{\zeta_{\boldsymbol{N}}} & \mathscr{B}(\boldsymbol{N}) \end{array}$$

commutes; that is, $\zeta_{\boldsymbol{N}} \circ \mathscr{A}(\psi) = \mathscr{B}(\psi) \circ \zeta_{\boldsymbol{M}}$. The interpretation is that ζ embeds $\mathscr{A}$ as a sub-theory of $\mathscr{B}$. If every $\zeta_{\boldsymbol{M}}$ is an isomorphism, ζ is an *equivalence* of $\mathscr{A}$ and $\mathscr{B}$; if some $\zeta_{\boldsymbol{M}}$ is an isomorphism we will say that ζ is a *partial equivalence.*

Example 4.2. With $\mathsf{Phys} = \mathsf{Alg}$: given any $\mathscr{A} \in \mathsf{LCT}$, define $\mathscr{A}^{\otimes k}$ $(k \in \mathbb{N})$ by

$$\mathscr{A}^{\otimes k}(\boldsymbol{M}) = \mathscr{A}(\boldsymbol{M})^{\otimes k}, \qquad \mathscr{A}^{\otimes k}(\psi) = \mathscr{A}(\psi)^{\otimes k}$$

i.e., k independent copies of $\mathscr{A}$, where the algebraic tensor product is used. Then

$$\begin{aligned} \eta^{k,l}_{\boldsymbol{M}} : \mathscr{A}^{\otimes k}(\boldsymbol{M}) &\to \mathscr{A}^{\otimes l}(\boldsymbol{M}) \\ A &\mapsto A \otimes \mathbf{1}^{\otimes(l-k)}_{\mathscr{A}(\boldsymbol{M})} \end{aligned}$$

defines a natural $\eta^{k,l} : \mathscr{A}^{\otimes k} \dot{\to} \mathscr{A}^{\otimes l}$ for $k \le l$ and we have the obvious relations that $\eta^{k,m} = \eta^{l,m} \circ \eta^{k,l}$ if $k \le l \le m$.

We can now formulate the issue of SPASs precisely:

Definition 4.3. A class $\mathfrak{T}$ of theories in LCT has the SPASs property if all partial equivalences between theories in $\mathfrak{T}$ are equivalences.

That is, if $\mathscr{A}$ and $\mathscr{B}$ are theories in $\mathfrak{T}$, such that $\mathscr{A}$ is a subtheory of $\mathscr{B}$ with $\zeta : \mathscr{A} \dot{\to} \mathscr{B}$ in LCT, and $\zeta_{\boldsymbol{M}} : \mathscr{A}(\boldsymbol{M}) \to \mathscr{B}(\boldsymbol{M})$ is an isomorphism for some $\boldsymbol{M}$, then ζ is an equivalence.

4.2. Failure of SPASs in LCT

Perhaps surprisingly, theories in LCT need not represent the same physics in all spacetimes. Indeed, a large class of pathological theories may be constructed as follows. Let us take any functor $\varphi : \mathsf{Loc} \to \mathsf{LCT}$, i.e., a locally covariant choice of locally covariant theory [we will give an example below]. Thus for each $\boldsymbol{M}$, $\varphi(\boldsymbol{M}) \in \mathsf{LCT}$ is a choice of theory defined in all spacetimes, and each hyperbolic embedding $\psi : \boldsymbol{M} \to \boldsymbol{N}$ corresponds to an embedding $\varphi(\psi)$ of $\varphi(\boldsymbol{M})$ as a sub-theory of $\varphi(\boldsymbol{N})$.

Given φ as input, we may define a *diagonal theory* $\varphi_\Delta \in \mathsf{LCT}$ by setting

$$\varphi_\Delta(\boldsymbol{M}) = \varphi(\boldsymbol{M})(\boldsymbol{M}) \qquad \varphi_\Delta(\psi) = \varphi(\psi)_{\boldsymbol{N}} \circ \varphi(\boldsymbol{M})(\psi)$$

for each spacetime $\boldsymbol{M}$ and hyperbolic embedding $\psi : \boldsymbol{M} \to \boldsymbol{N}$. The definition of $\varphi_\Delta(\psi)$ is made more comprehensible by realising that it is the diagonal of a natural square

$$
\begin{array}{ccccc}
\boldsymbol{M} & & \varphi(\boldsymbol{M})(\boldsymbol{M}) & \xrightarrow{\varphi(\psi)_{\boldsymbol{M}}} & \varphi(\boldsymbol{N})(\boldsymbol{M}) \\
\downarrow \psi & & \downarrow \varphi(\boldsymbol{M})(\psi) & \searrow^{\varphi_\Delta(\psi)} & \downarrow \varphi(\boldsymbol{N})(\psi) \\
\boldsymbol{N} & & \varphi(\boldsymbol{M})(\boldsymbol{N}) & \xrightarrow[\varphi(\psi)_{\boldsymbol{N}}]{} & \varphi(\boldsymbol{N})(\boldsymbol{N})
\end{array}
$$

arising from the subtheory embedding $\varphi(\psi) : \varphi(\boldsymbol{M}) \dot{\to} \varphi(\boldsymbol{N})$. The key point is that, while φ_Δ is easily shown to be a functor and therefore defines a locally covariant theory, there seems no reason to believe that φ_Δ should represent the same physics in all spacetimes. Nonetheless, φ_Δ may have many reasonable properties. For example, it satisfies the time-slice axiom if each $\varphi(\boldsymbol{M})$ does and in addition φ does (i.e., $\varphi(\psi)$ is an equivalence for every Cauchy morphism ψ).

To demonstrate the existence of nontrivial functors $\varphi : \mathsf{Loc} \to \mathsf{LCT}$ [in the case $\mathsf{Phys} = \mathsf{Alg}$] let us write $\Sigma(\boldsymbol{M})$ to denote the equivalence class of the Cauchy surface of $\boldsymbol{M}$ modulo orientation-preserving diffeomorphisms and suppose that $\lambda : \mathsf{Loc} \to \mathbb{N} = \{1, 2, \ldots\}$ is constant on such equivalence classes and obeys $\lambda(\boldsymbol{M}) = 1$ if $\Sigma(\boldsymbol{M})$ is noncompact. Defining $\mu(\boldsymbol{M}) = \max\{\lambda(\boldsymbol{C}) : \boldsymbol{C} \text{ a component of } \boldsymbol{M}\}$, for example, one can show:

Proposition 4.4. *Fix any theory $\mathscr{A} \in \mathsf{LCT}$ ($\mathsf{Phys} = \mathsf{Alg}$). Then the assignments*

$$
\varphi(\boldsymbol{M}) = \mathscr{A}^{\otimes \mu(\boldsymbol{M})} \qquad \varphi(\boldsymbol{M} \xrightarrow{\psi} \boldsymbol{N}) = \eta^{\mu(\boldsymbol{M}),\mu(\boldsymbol{N})}
$$

for all $\boldsymbol{M} \in \mathsf{Loc}$ and hyperbolic embeddings $\psi : \boldsymbol{M} \to \boldsymbol{N}$ define a functor $\varphi \in \mathsf{Fun}(\mathsf{Loc}, \mathsf{LCT})$, where $\eta^{k,l}$ are the natural transformations in Example 4.2. If $\mathscr{A}$ obeys the time-slice axiom, then so does φ_Δ.

Proof. The key point is to show that μ is monotone with respect to hyperbolic embeddings: if $\psi : \boldsymbol{M} \to \boldsymbol{N}$ then $\mu(\boldsymbol{M}) \le \mu(\boldsymbol{N})$. This in turn follows from a result in Lorentzian geometry which [for connected spacetimes] asserts: if $\psi : \boldsymbol{M} \to \boldsymbol{N}$ is a hyperbolic embedding and $\boldsymbol{M}$ has compact Cauchy surface then ψ is Cauchy and the Cauchy surfaces of $\boldsymbol{N}$ are equivalent to those of $\boldsymbol{M}$ under an orientation preserving diffeomorphism. The functorial properties of φ follow immediately from properties of $\eta^{k,l}$. If $\mathscr{A}$ obeys the time-slice axiom, then so do its powers $\mathscr{A}^{\otimes k}$. Moreover, if ψ is Cauchy, then $\boldsymbol{M}$ and $\boldsymbol{N}$ have oriented-diffeomorphic Cauchy surfaces and therefore $\varphi(\psi)$ is an identity, and $\varphi_\Delta(\psi) = \mathscr{A}^{\otimes \mu(\boldsymbol{M})}(\psi)$ is an isomorphism. □

As a concrete example, which we call the *one-field-two-field model*, if we put

$$
\lambda(\boldsymbol{M}) = \begin{cases} 1 & \Sigma(\boldsymbol{M}) \text{ noncompact} \\ 2 & \text{otherwise} \end{cases}
$$

then the resulting diagonal theory φ_Δ represents one copy of the underlying theory $\mathscr{A}$ in spacetimes whose Cauchy surfaces are purely noncompact

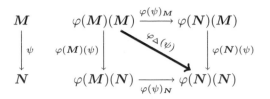

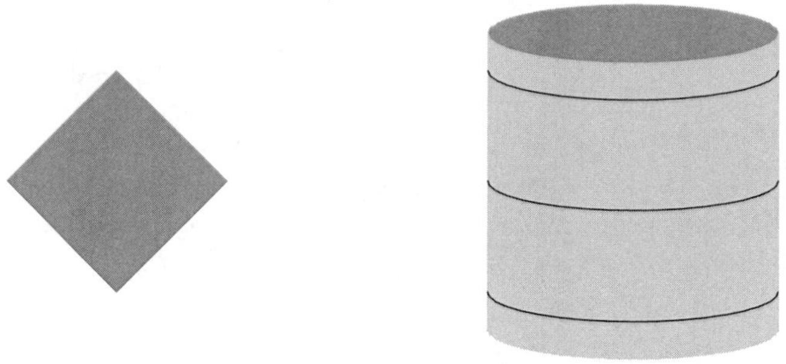

$$\varphi_\Delta(\boldsymbol{D}) = \mathscr{A}(\boldsymbol{D}) \qquad\qquad \varphi_\Delta(\boldsymbol{C}) = \mathscr{A}^{\otimes 2}(\boldsymbol{C})$$
$$\varphi_\Delta(\boldsymbol{C} \sqcup \boldsymbol{D}) = \mathscr{A}^{\otimes 2}(\boldsymbol{C} \sqcup \boldsymbol{D})$$

FIGURE 3. The one-field-two-field model assigns one copy of the theory $\mathscr{A}$ to a diamond spacetime $\boldsymbol{D}$ with noncompact Cauchy surface, but two copies to a spacetime $\boldsymbol{C}$ with compact Cauchy surface, or spacetimes such as the disjoint union $\boldsymbol{C} \sqcup \boldsymbol{D}$ which have at least one component with compact Cauchy surface.

(i.e., have no compact connected components), but two copies in all other spacetimes (see Fig. 3). Moreover, there are obvious subtheory embeddings

$$\mathscr{A} \stackrel{.}{\longrightarrow} \varphi_\Delta \stackrel{.}{\longrightarrow} \mathscr{A}^{\otimes 2} \tag{4.1}$$

which are isomorphisms in some spacetimes but not in others; the left-hand in spacetimes with purely noncompact Cauchy surfaces but not otherwise, and *vice versa* for the right-hand embedding. As we have exhibited partial equivalences that are not equivalences, we conclude that SPASs fails in LCT.

Meditating on this example, we can begin to see the root of the problem. If O is any nonempty open globally hyperbolic subset of any $\boldsymbol{M}$ such that O has noncompact Cauchy surfaces then the kinematic local algebras obey

$$\varphi_\Delta^{\mathrm{kin}}(\boldsymbol{M}; O) = \varphi_\Delta^{\mathrm{kin}}(\boldsymbol{M}|_O) = \mathscr{A}^{\mathrm{kin}}(\boldsymbol{M}; O), \tag{4.2}$$

so the kinematic local algebras are insensitive to the ambient algebra. In one-field-two-field model, for example, the kinematical algebras of relatively compact regions with nontrivial causal complement are always embeddings of the corresponding algebra for a single field, even when the ambient algebra corresponds to two independent fields. This leads to a surprising diagnosis: the framework described so far is insufficiently local, and it is necessary to find another way of describing the local physics.

The problem, therefore, is to detect the existence of the ambient local degrees of freedom that are ignored in the kinematical local algebras. The solution is simple: where there are physical degrees of freedom, there ought to be energy. In other words, we should turn our attention to dynamics.

Some immediate support for the idea that dynamics has a role to play can be obtained by computing the relative Cauchy evolution of a diagonal theory. The result is

$$\mathrm{rce}_{\boldsymbol{M}}^{(\varphi_\Delta)}[\boldsymbol{h}] = (\mathrm{rce}_{\boldsymbol{M}}^{(\varphi)}[\boldsymbol{h}])_{\boldsymbol{M}} \circ \mathrm{rce}_{\boldsymbol{M}}^{(\varphi(\boldsymbol{M}))}[\boldsymbol{h}], \tag{4.3}$$

where $\mathrm{rce}_{\boldsymbol{M}}^{(\varphi)}[\boldsymbol{h}] \in \mathrm{Aut}(\varphi(\boldsymbol{M}))$ is the relative Cauchy evolution of φ in LCT. In known examples this is trivial, but it would be intriguing to find examples in which the choice of theory as a function of the spacetime contributes nontrivially to the total stress-energy tensor of the theory. Provided that $\mathrm{rce}^{(\varphi)}$ is indeed trivial and stress-energy tensors exist, it follows that

$$[\mathsf{T}_{\boldsymbol{M}}^{(\varphi_\Delta)}(\boldsymbol{f}), A] = [\mathsf{T}_{\boldsymbol{M}}^{(\varphi(\boldsymbol{M}))}(\boldsymbol{f}), A]$$

for all $A \in \varphi_\Delta(\boldsymbol{M})$, which means that the stress-energy tensor correctly detects the ambient degrees of freedom that are missed in the kinematic description.

5. Local physics from dynamics

Let $\mathscr{A}$ be a locally covariant theory obeying the time-slice axiom and O be a region in spacetime $\boldsymbol{M}$. The *kinematical* description of the physical content of O was given in terms of the physics assigned to O when considered as a spacetime in its own right, i.e., $\mathscr{A}^{\mathrm{kin}}(\boldsymbol{M}; O) = \mathscr{A}(\boldsymbol{M}|_O)$. For the *dynamical* description, we focus on that portion of the physical content on $\boldsymbol{M}$ that is unaffected by the geometry in the causal complement of O. This can be isolated using the relative Cauchy evolution, which precisely encapsulates the response of the system under a change in the geometry.

Accordingly, for any compact set $K \subset \boldsymbol{M}$, we define $\mathscr{A}^\bullet(\boldsymbol{M}; K)$ to be the maximal subobject of $\mathscr{A}(\boldsymbol{M})$ invariant under $\mathrm{rce}_{\boldsymbol{M}}[\boldsymbol{h}]$ for all $\boldsymbol{h} \in H(\boldsymbol{M}; K^\perp)$, where $H(\boldsymbol{M}; K^\perp)$ is the set of metric perturbations in $H(\boldsymbol{M})$ supported in the causal complement $K^\perp = \boldsymbol{M} \setminus J_{\boldsymbol{M}}(K)$ of K. Subobjects of this type will exist provided the category Phys has equalizers (at least for pairs of automorphisms) and arbitrary set-indexed intersections. If $\mathsf{Phys} = \mathsf{Alg}$, of course, $\mathscr{A}^\bullet(\boldsymbol{M}; K)$ is simply the invariant subalgebra.

From the categorical perspective, it is usually not objects that are of interest but rather the morphisms between them. Strictly speaking, a subobject of an object $\mathcal{A}$ in a general category is an equivalence class of monomorphisms with codomain $\mathcal{A}$, where α and α' are regarded as equivalent if there is a (necessarily unique) isomorphism β such that $\alpha' = \alpha \circ \beta$. In our case, we are really interested in a morphism $\alpha^\bullet_{\boldsymbol{M};K}$ with codomain $\mathscr{A}(\boldsymbol{M})$, characterized (up to isomorphism) by the requirements that (a)

$$\mathrm{rce}_{\boldsymbol{M}}[\boldsymbol{h}] \circ \alpha^\bullet_{\boldsymbol{M};K} = \alpha^\bullet_{\boldsymbol{M};K} \qquad \forall \boldsymbol{h} \in H(\boldsymbol{M}; K^\perp)$$

and (b) if β is any other morphism with this property then there exists a unique morphism $\hat{\beta}$ such that $\beta = \alpha^\bullet_{\boldsymbol{M};K} \circ \hat{\beta}$. The notation $\mathscr{A}^\bullet(\boldsymbol{M}; K)$ denotes the domain of $\alpha^\bullet_{\boldsymbol{M};K}$. For simplicity and familiarity, however, we will ignore this issue and write everything in terms of the objects, rather than the morphisms. This is completely legitimate in categories such as Alg or C^*-Alg, in which the objects are sets with additional structure and $\mathscr{A}^\bullet(\boldsymbol{M}; K)$ can be concretely constructed as the subset of $\mathscr{A}(\boldsymbol{M})$ left fixed by the appropriate relative Cauchy evolution automorphisms. (At the level of detail, however, it is not simply a matter of precision to work with the morphisms: universal definitions, such as that given above for $\alpha^{\bullet}_{\boldsymbol{M};K}$, permit efficient proofs that are 'portable' between different choices of the category Phys.)

The subobjects $\mathscr{A}^{\bullet}(\boldsymbol{M};K)$ have many features reminiscent of a net of local algebras in local quantum physics: for example, isotony ($K_1 \subset K_2$ implies $\mathscr{A}^{\bullet}(\boldsymbol{M};K_1) \subset \mathscr{A}^{\bullet}(\boldsymbol{M};K_2)$) together with

$$\mathscr{A}^{\bullet}(\boldsymbol{M};K) = \mathscr{A}^{\bullet}(\boldsymbol{M};K^{\perp\perp})$$

if $K^{\perp\perp}$ is also compact, and

$$\begin{aligned}\mathscr{A}^{\bullet}(\boldsymbol{M};K_1) \vee \mathscr{A}^{\bullet}(\boldsymbol{M};K_2) &\subset \mathscr{A}^{\bullet}(\boldsymbol{M};K_1 \cup K_2)\\ \mathscr{A}^{\bullet}(\boldsymbol{M};K_1 \cap K_2) &\subset \mathscr{A}^{\bullet}(\boldsymbol{M};K_1) \cap \mathscr{A}^{\bullet}(\boldsymbol{M};K_2)\\ \mathscr{A}^{\bullet}(\boldsymbol{M};\emptyset) &\subset \mathscr{A}^{\bullet}(\boldsymbol{M};K)\end{aligned}$$

for any compact K, K_1, K_2. A rather striking absence is that there is no local covariance property: it is not true in general that a morphism $\psi : \boldsymbol{M} \to \boldsymbol{N}$ induces isomorphisms between $\mathscr{A}^{\bullet}(\boldsymbol{M};K)$ and $\mathscr{A}^{\bullet}(\boldsymbol{N};\psi(K))$ so as to make the diagram

$$\begin{array}{ccc}\mathscr{A}^{\bullet}(\boldsymbol{M};K) & \overset{\cong}{\underset{?}{\dashrightarrow}} & \mathscr{A}^{\bullet}(\boldsymbol{N};\psi(K)) \\ {\scriptstyle \alpha^{\bullet}_{\boldsymbol{M};K}}\Big\downarrow & & \Big\downarrow{\scriptstyle \alpha^{\bullet}_{\boldsymbol{N};\psi(K)}} \\ \mathscr{A}(\boldsymbol{M}) & \xrightarrow{\mathscr{A}(\psi)} & \mathscr{A}(\boldsymbol{N})\end{array} \tag{5.1}$$

commute (unless ψ is itself an isomorphism). Again, this can be seen from the example of diagonal theories, for which we have

$$\varphi^{\bullet}_{\Delta}(\boldsymbol{M};K) = \varphi(\boldsymbol{M})^{\bullet}(\boldsymbol{M};K)$$

as a consequence of (4.3). In the one-field-two-field model, for example, a hyperbolic embedding of $\psi : \boldsymbol{D} \to \boldsymbol{C}$ (where $\boldsymbol{D}$ and $\boldsymbol{C}$ are as in Fig. 3) we have $\varphi^{\bullet}_{\Delta}(\boldsymbol{D};K) = \mathscr{A}^{\bullet}(\boldsymbol{D};K)$ but $\varphi^{\bullet}_{\Delta}(\boldsymbol{C};\psi(K)) = \mathscr{A}^{(\otimes 2)\bullet}(\boldsymbol{C};\psi(K))$, for any compact subset K of $\boldsymbol{D}$. In general, these subobjects will not be isomorphic in the required sense. This situation should be compared with the local covariance property enjoyed by the kinematic description, expressed by the commuting diagram (2.1).

The discussion so far has shown how dynamics allows us to associate a subobject $\mathscr{A}^{\bullet}(\boldsymbol{M};K)$ of $\mathscr{A}(\boldsymbol{M})$ with every compact subset K of $\boldsymbol{M}$. As mentioned above, for simplicity we are taking a concrete viewpoint in which $\mathscr{A}^{\bullet}(\boldsymbol{M};K)$ is a subset of $\mathscr{A}(\boldsymbol{M})$. For purposes of comparison with the kinematic viewpoint, we must also associate subobjects of $\mathscr{A}(\boldsymbol{M})$ with open subsets of $\boldsymbol{M}$. Given an open subset O of $\boldsymbol{M}$, the basic idea will be to take the subobject generated by all $\mathscr{A}^{\bullet}(\boldsymbol{M};K)$ indexed over a suitable class of compact subsets K of $\boldsymbol{M}$. To define the class of K involved, we must introduce some terminology.

Following [7], a *diamond* is defined to be an open relatively compact subset of $\boldsymbol{M}$, taking the form $D_{\boldsymbol{M}}(B)$ where B, the *base* of the diamond, is a subset of a Cauchy surface Σ so that in a suitable chart (U,ϕ) of Σ, $\phi(B)$ is a nonempty open ball in $\mathbb{R}^{n-1}$ whose closure is contained in $\phi(U)$. (A given diamond has many possible bases.) By a *multi-diamond*, we understand any finite union of mutually causally disjoint diamonds; the base of a multi-diamond is formed by any union of bases for each component diamond.

Given these definitions, for any open set O in $\boldsymbol{M}$, let $\mathscr{K}(\boldsymbol{M};O)$ be the set of compact subsets of O that have a multi-diamond neighbourhood with a base contained in O. We then define

$$\mathscr{A}^{\mathrm{dyn}}(\boldsymbol{M};O) = \bigvee_{K\in\mathscr{K}(\boldsymbol{M};O)} \mathscr{A}^{\bullet}(\boldsymbol{M};K) \ ;$$

that is, the smallest Phys-subobject of $\mathscr{A}(\boldsymbol{M})$ that contains all the $\mathscr{A}^{\bullet}(\boldsymbol{M};K)$ indexed by $K \in \mathscr{K}(\boldsymbol{M};O)$. Of course, the nature of the category Phys enters here: if it is Alg (resp., C^*-Alg) then $\mathscr{A}^{\mathrm{dyn}}(\boldsymbol{M};O)$ is the $(C)^*$-subalgebra of $\mathscr{A}(\boldsymbol{M})$ generated by the $\mathscr{A}^{\bullet}(\boldsymbol{M};K)$; if Phys is Sympl, then the linear span is formed. More abstractly, we are employing the categorical join of subobjects, which can be given a universal definition in terms of the morphisms $\alpha^{\bullet}_{\boldsymbol{M};K}$, and gives a morphism (characterized up to isomorphism) $\alpha^{\mathrm{dyn}}_{\boldsymbol{M};O}$ with codomain $\mathscr{A}(\boldsymbol{M})$; $\mathscr{A}^{\mathrm{dyn}}(\boldsymbol{M};O)$ is the domain of some such morphism. In Alg (and more generally, if Phys has pullbacks, and class-indexed intersections) the join has the stronger property of being a categorical union [10], and this is used in our general setup.

It is instructive to consider these algebras for diagonal theories φ_Δ, under the assumption that $\mathrm{rce}^{(\varphi)}$ is trivial. Using (4.3), it follows almost immediately that

$$\varphi^{\bullet}_{\Delta}(\boldsymbol{M};K) = \varphi(\boldsymbol{M})^{\bullet}(\boldsymbol{M};K), \qquad \varphi^{\mathrm{dyn}}_{\Delta}(\boldsymbol{M};O) = \varphi(\boldsymbol{M})^{\mathrm{dyn}}(\boldsymbol{M};O)$$

for all compact $K \subset \boldsymbol{M}$ and open $O \subset \boldsymbol{M}$. This should be contrasted with the corresponding calculations for kinematic algebras in (4.2); we see that the dynamical algebras sense the ambient spacetime in a way that the kinematic algebras do not. In the light of this example, the following definition is very natural.

Definition 5.1. $\mathscr{A}$ obeys *dynamical locality* if

$$\mathscr{A}^{\mathrm{dyn}}(\boldsymbol{M};O) \cong \mathscr{A}^{\mathrm{kin}}(\boldsymbol{M};O)$$

for all nonempty open globally hyperbolic subsets O of $\boldsymbol{M}$ with finitely many components, where the isomorphism should be understood as asserting the existence of isomorphisms $\beta_{\boldsymbol{M};O}$ such that $\alpha^{\mathrm{dyn}}_{\boldsymbol{M};O} = \alpha^{\mathrm{kin}}_{\boldsymbol{M};O} \circ \beta_{\boldsymbol{M};O}$; i.e., equivalence as subobjects of $\mathscr{A}(\boldsymbol{M})$.

In the next section, we will see that the class of dynamically local theories in LCT has the SPASs property, our main focus in this contribution. However, we note that dynamical locality has a number of other appealing consequences, which suggest that it may prove to be a fruitful assumption in other contexts.

First, any dynamically local theory $\mathscr{A}$ is necessarily additive, in the sense that

$$\mathscr{A}(\boldsymbol{M}) = \bigvee_{O} \mathscr{A}^{\rm dyn}(\boldsymbol{M}; O),$$

where the categorical union is taken over a sufficiently large class of open globally hyperbolic sets O; for example, the truncated multi-diamonds (intersections of a multi-diamond with a globally hyperbolic neighbourhood of a Cauchy surface containing its base).

Second, the 'net' $K \mapsto \mathscr{A}^\bullet(\boldsymbol{M}; K)$ is locally covariant: if $\psi : \boldsymbol{M} \to \boldsymbol{N}$ and $K \subset \boldsymbol{M}$ is *outer regular* then there is a unique isomorphism making the diagram (5.1) commute. Here, a compact set K is said to be outer regular if there exist relatively compact nonempty open globally hyperbolic subsets O_n ($n \in \mathbb{N}$) with finitely many components such that (a) $\mathrm{cl}(O_{n+1}) \subset O_n$ and $K \in \mathscr{K}(\boldsymbol{M}; O_n)$ for each n and (b) $K = \bigcap_{n\in\mathbb{N}} O_n$.

Third, in the case $\mathsf{Phys} = \mathsf{Alg}$ or $\mathsf{C^*\text{-}Alg}$,[7] if $\mathscr{A}^\bullet(\boldsymbol{M}; \emptyset) = \mathbb{C}\mathbf{1}_{\mathscr{A}(\boldsymbol{M})}$ and $\mathscr{A}$ is dynamically local then the theory obeys *extended locality* [25], i.e., if O_1, O_2 are causally disjoint nonempty globally hyperbolic subsets then $\mathscr{A}^{\rm kin}(\boldsymbol{M}; O_1) \cap \mathscr{A}^{\rm kin}(\boldsymbol{M}; O_2) = \mathbb{C}\mathbf{1}_{\mathscr{A}(\boldsymbol{M})}$. Conversely, we can infer triviality of $\mathscr{A}^\bullet(\boldsymbol{M}; \emptyset)$ from extended locality.

Fourth, there is an interesting application to the old question of whether there can be any preferred state in general spacetimes. Suppose that $\mathscr{A}$ is a theory in LCT with Phys taken to be Alg or $\mathsf{C^*\text{-}Alg}$. A *natural state* ω of the theory is an assignment $\boldsymbol{M} \mapsto \omega_{\boldsymbol{M}}$ of states $\omega_{\boldsymbol{M}}$ on $\mathscr{A}(\boldsymbol{M})$, subject to the contravariance requirement that $\omega_{\boldsymbol{M}} = \omega_{\boldsymbol{N}} \circ \mathscr{A}(\psi)$ for all $\psi : \boldsymbol{M} \to \boldsymbol{N}$. Hollands and Wald [22] give an argument to show that this is not possible for the free scalar field; likewise, BFV also sketch an argument to this effect (also phrased concretely in terms of the free field). The following brings these arguments to a sharper and model-independent form:

Theorem 5.2. *Suppose $\mathscr{A}$ is a dynamically local theory in LCT equipped with a natural state ω. If there is a spacetime $\boldsymbol{M}$ with noncompact Cauchy surfaces such that $\omega_{\boldsymbol{M}}$ induces a faithful representation of $\mathscr{A}(\boldsymbol{M})$ with the Reeh–Schlieder property, then the relative Cauchy evolution in $\boldsymbol{M}$ is trivial. Moreover, if extended locality holds in $\boldsymbol{M}$ then $\mathscr{A}$ is equivalent to the trivial theory.*

By the Reeh–Schlieder property, we mean that the GNS vector $\Omega_{\boldsymbol{M}}$ corresponding to $\omega_{\boldsymbol{M}}$ is cyclic for the induced representation of each $\mathscr{A}^{\rm kin}(\boldsymbol{M}; O)$ for any open, globally hyperbolic, relatively compact $O \subset \boldsymbol{M}$. This requirement will be satisfied if, for example, the theory in Minkowski space obeys standard assumptions of AQFT with the natural state reducing to the Minkowski vacuum state. The trivial theory $\mathscr{I}$ is the theory assigning the trivial unital algebra to each $\boldsymbol{M}$ and its identity morphism to any morphism in LCT.

---

[7]One may also formulate an analogous statement for more general categories Phys.

Sketch of proof. We first argue that $\omega_{\boldsymbol{M}} \circ \mathrm{rce}_{\boldsymbol{M}}[\boldsymbol{h}] = \omega_{\boldsymbol{M}}$ for all $\boldsymbol{h} \in H(\boldsymbol{M})$ and all $\boldsymbol{M} \in \mathsf{Loc}$ by virtue of the natural state assumption and the definition of the relative Cauchy evolution. In the GNS representation $\pi_{\boldsymbol{M}}$ of $\omega_{\boldsymbol{M}}$, the relative Cauchy evolution can be unitarily represented by unitaries leaving the vacuum vector $\Omega_{\boldsymbol{M}}$ invariant. Choose any $\boldsymbol{h} \in H(\boldsymbol{M})$ and an open, relatively compact, globally hyperbolic O spacelike separated from $\mathrm{supp}\, \boldsymbol{h}$. By general properties of the relative Cauchy evolution, we have $\mathrm{rce}_{\boldsymbol{M}}[\boldsymbol{h}] \circ \alpha^{\mathrm{kin}}_{\boldsymbol{M};O} = \alpha^{\mathrm{kin}}_{\boldsymbol{M};O}$ and hence that the unitary $U_{\boldsymbol{M}}[\boldsymbol{h}]$ implementing $\mathrm{rce}_{\boldsymbol{M}}[\boldsymbol{h}]$ is equal to the identity on the subspace $\pi_{\boldsymbol{M}}(\mathscr{A}(\iota_{\boldsymbol{M};O})A)\Omega_{\boldsymbol{M}}$ $(A \in \mathscr{A}(\boldsymbol{M}|_O))$ which is dense by the Reeh–Schlieder property. As $\pi_{\boldsymbol{M}}$ is faithful, the relative Cauchy evolution is trivial as claimed.

It follows that $\mathscr{A}^\bullet(\boldsymbol{M}; K) = \mathscr{A}(\boldsymbol{M})$ for all compact K; consequently, by dynamical locality, $\mathscr{A}^{\mathrm{kin}}(\boldsymbol{M}; O) = \mathscr{A}^{\mathrm{dyn}}(\boldsymbol{M}; O) = \mathscr{A}(\boldsymbol{M})$ for all nonempty open, globally hyperbolic O. Taking two such O at spacelike separation and applying extended locality, we conclude that $\mathscr{A}(\boldsymbol{M}) = \mathbb{C}\mathbf{1}$. The remainder of the proof will be given in the next section. □

6. SPASs at last

Our aim is to show that the category of dynamically local theories has the SPASs property.

Theorem 6.1. *Suppose $\mathscr{A}$ and $\mathscr{B}$ are theories in* LCT *obeying dynamical locality. Then any partial equivalence $\zeta : \mathscr{A} \dot{\to} \mathscr{B}$ is an equivalence.*

Sketch of proof. The main ideas are as follows. First, we show that if $\zeta_{\boldsymbol{N}}$ is an isomorphism then $\zeta_{\boldsymbol{L}}$ is an isomorphism for every spacetime $\boldsymbol{L}$ for which there is a morphism $\psi : \boldsymbol{L} \to \boldsymbol{N}$. Second, if $\psi : \boldsymbol{L} \to \boldsymbol{N}$ is Cauchy, then $\zeta_{\boldsymbol{L}}$ is an isomorphism if and only if $\zeta_{\boldsymbol{N}}$ is.

As ζ is a partial equivalence, there exists a spacetime $\boldsymbol{M}$ for which $\zeta_{\boldsymbol{M}}$ is an isomorphism. It follows that $\zeta_{\boldsymbol{D}}$ is an isomorphism for every multi-diamond spacetime $\boldsymbol{D}$ that can be embedded in $\boldsymbol{M}$. Using deformation arguments [18] there is a chain of Cauchy morphisms linking each such multi-diamond spacetime to any other multi-diamond spacetime with the same number of connected components. Hence $\zeta_{\boldsymbol{D}}$ is an isomorphism for all multi-diamond spacetimes $\boldsymbol{D}$. Now consider any other spacetime $\boldsymbol{N}$. Owing to dynamical locality, both $\mathscr{A}(\boldsymbol{N})$ and $\mathscr{B}(\boldsymbol{N})$ are generated over subobjects that correspond to diamond spacetimes, which (it transpires) are isomorphic under the restriction of $\zeta_{\boldsymbol{N}}$. By the properties of the categorical union, it follows that $\zeta_{\boldsymbol{N}}$ is an isomorphism. □

We remark that the chain of partial equivalences in (4.1) shows that we cannot relax the condition that *both* $\mathscr{A}$ and $\mathscr{B}$ be dynamically local.

The foregoing result allows us to conclude the discussion of natural states:

End of the proof of Theorem 5.2. Note that there is a natural transformation $\zeta : \mathscr{I} \dot{\to} \mathscr{A}$, such that each $\zeta_{\boldsymbol{N}}$ simply embeds the trivial unital algebra in $\mathscr{A}(\boldsymbol{N})$. For the spacetime $\boldsymbol{M}$ given in the hypotheses, we already know that $\mathscr{A}(\boldsymbol{M}) = \mathbb{C}\mathbf{1}$, so $\zeta_{\boldsymbol{M}}$ is an isomorphism. Hence by dynamical locality and Theorem 6.1, ζ is an equivalence, so $\mathscr{A} \cong \mathscr{I}$. □

Dynamical locality also significantly reduces our freedom to construct pathological theories.

Theorem 6.2. *If φ_Δ is a diagonal theory (with* rce^φ *trivial) such that φ_Δ and every $\varphi(\boldsymbol{M})$ are dynamically local, then all $\varphi(\boldsymbol{M})$ are equivalent. If the $\varphi(\boldsymbol{M})$ have trivial automorphism group then φ_Δ is equivalent to each of them.*

Sketch of proof. Consider any morphism $\psi : \boldsymbol{M} \to \boldsymbol{N}$ in Loc. Using dynamical locality of φ_Δ and $\varphi(\boldsymbol{N})$, we may deduce that $\varphi(\psi)_{\boldsymbol{M}}$ is an isomorphism and hence $\varphi(\psi)$ is an equivalence. As any two spacetimes in Loc can be connected by a chain of (not necessarily composable) morphisms, it follows that all the $\varphi(\boldsymbol{M})$ are equivalent.

In particular, writing $\boldsymbol{M}_0$ for Minkowski space, the previous argument allows us to choose an equivalence $\zeta_{(\boldsymbol{M})} : \varphi(\boldsymbol{M}_0) \dot{\to} \varphi(\boldsymbol{M})$ for each spacetime $\boldsymbol{M}$ (no uniqueness is assumed). Every morphism $\psi : \boldsymbol{M} \to \boldsymbol{N}$ then induces an automorphism $\eta(\psi) = \zeta_{(\boldsymbol{M})}^{-1} \circ \varphi(\psi) \circ \zeta_{(\boldsymbol{N})}$ of $\varphi(\boldsymbol{M}_0)$. As $\mathrm{Aut}(\varphi(\boldsymbol{M}_0))$ is assumed trivial, it follows that $\varphi(\psi) = \zeta_{(\boldsymbol{M})} \circ \zeta_{(\boldsymbol{N})}^{-1}$ and it is easy to deduce that there is an equivalence $\zeta : \varphi(\boldsymbol{M}_0) \dot{\to} \varphi_\Delta$ with components $(\zeta_{(\boldsymbol{M})})_{\boldsymbol{M}}$. □

We remark that the automorphism group of a theory in LCT can be interpreted as its group of global gauge invariances [13] and that a theory in which the $\mathscr{A}(\boldsymbol{M})$ are algebras of observables will have trivial gauge group. The argument just given has a cohomological flavour, which can also be made precise and indicates that a cohomological study of Loc is worthy of further study.

7. Example: Klein–Gordon theory

The abstract considerations of previous sections would be of questionable relevance if they were not satisfied in concrete models. We consider the standard example of the free scalar field, discussed above.

First let us consider the classical theory. Let Sympl be the category of real symplectic spaces with symplectomorphisms as the morphisms. To each $\boldsymbol{M}$, we assign the space $\mathscr{L}(\boldsymbol{M})$ of smooth, real-valued solutions to $(\Box + m^2)\phi = 0$, with compact support on Cauchy surfaces; we endow $\mathscr{L}(\boldsymbol{M})$ with the standard (weakly nondegenerate) symplectic product

$$\sigma_{\boldsymbol{M}}(\phi, \phi') = \int_\Sigma (\phi n^a \nabla_a \phi' - \phi' n^a \nabla_a \phi) d\Sigma$$

for any Cauchy surface Σ. To each hyperbolic embedding $\psi : \boldsymbol{M} \to \boldsymbol{N}$ there is a symplectic map $\mathscr{L}(\psi) : \mathscr{L}(\boldsymbol{M}) \to \mathscr{L}(\boldsymbol{N})$ so that $E_{\boldsymbol{N}} \psi_* f = \mathscr{L}(\psi) E_{\boldsymbol{M}} f$ for all $f \in C_0^\infty(\boldsymbol{M})$ (see BFV or [2]).

The relative Cauchy evolution for $\mathscr{L}$ was computed in BFV (their Eq. (15)) as was its functional derivative with respect to metric perturbations (see pp. 61-62 of BFV). An important observation is that this can be put into a nicer form: writing

$$F_{\boldsymbol{M}}[f]\phi = \frac{d}{ds}\mathrm{rce}^{(\mathscr{L})}_{\boldsymbol{M}}[f]\phi\Big|_{s=0}$$

it turns out that

$$\sigma_{\boldsymbol{M}}(F_{\boldsymbol{M}}[f]\phi, \phi) = \int_{\boldsymbol{M}} f_{ab}T^{ab}[\phi]\, d\mathrm{vol}_{\boldsymbol{M}}, \tag{7.1}$$

where

$$T_{ab}[\phi] = \nabla_a\phi\nabla_b\phi - \frac{1}{2}g_{ab}\nabla^c\phi\nabla_c\phi + \frac{1}{2}m^2 g_{ab}\phi^2$$

is the classical stress-energy tensor of the solution ϕ. (In passing, we note that (7.1) provides an explanation, for the free scalar field and similar theories, as to why the relative Cauchy evolution can be regarded as equivalent to the specification of the classical action.)

We may use these results to compute $\mathscr{L}^\bullet(\boldsymbol{M};K)$ and $\mathscr{L}^{\mathrm{dyn}}(\boldsymbol{M};O)$: they consist of solutions whose stress-energy tensors vanish in $K^\perp$ (resp., $O' = \boldsymbol{M}\backslash\mathrm{cl}\, J_{\boldsymbol{M}}(O)$). For nonzero mass, the solution must vanish wherever the stress-energy does and we may conclude that $\mathscr{L}^{\mathrm{dyn}}(\boldsymbol{M};O) = \mathscr{L}^{\mathrm{kin}}(\boldsymbol{M};O)$, i.e., we have dynamical locality of the classical theory $\mathscr{L}$.

At zero mass, however, we can deduce only that the solution is constant in regions where its stress-energy tensor vanishes, so $\mathscr{L}^\bullet(\boldsymbol{M};K)$ consists of solutions in $\mathscr{L}(\boldsymbol{M})$ that are constant on each connected component of $K^\perp$. In particular, if $K^\perp$ is connected and $\boldsymbol{M}$ has noncompact Cauchy surfaces this forces the solutions to vanish in $K^\perp$ as in the massive case. However, if $\boldsymbol{M}$ has compact Cauchy surfaces this argument does not apply and indeed the constant solution $\phi \equiv 1$ belongs to every $\mathscr{L}^\bullet(\boldsymbol{M};K)$ and $\mathscr{L}^{\mathrm{dyn}}(\boldsymbol{M};O)$, but it does not belong to $\mathscr{L}^{\mathrm{kin}}(\boldsymbol{M};O)$ unless O contains a Cauchy surface of $\boldsymbol{M}$. Hence the massless theory fails to be dynamically local. The source of this problem is easily identified: it arises from the global gauge symmetry $\phi \mapsto \phi + \mathrm{const}$ in the classical action.

At the level of the quantized theory, one may show that similar results hold: the $m > 0$ theory is dynamically local, while the $m = 0$ theory is not. We argue that this should be taken seriously as indicating a (fairly mild) pathology of the massless minimally coupled scalar field, rather than a limitation of dynamical locality. In support of this position we note:

- Taking the gauge symmetry seriously, we can alternatively quantize the theory of currents $j = d\phi$; this turns out to be a well-defined locally covariant and dynamically local theory in dimensions $n > 2$. While dynamical locality fails for this model in $n = 2$ dimensions in the present setting, it may be restored by restricting the scope of the theory to connected spacetimes.
- The constant solution is also the source of another well-known problem: there is no ground state for the theory in ultrastatic spacetimes

with compact spatial section (see, for example [19]). The same problem afflicts the massless scalar field in two-dimensional Minkowski space, where it is commonplace to reject the algebra of fields in favour of the algebra of currents.

- The nonminimally coupled scalar field (which does not have the gauge symmetry) is dynamically local even at zero mass (a result due to Ferguson [12]).

Actually, this symmetry has other interesting aspects: it is spontaneously broken in Minkowski space [29], for example, and the automorphism group of the functor $\mathscr{A}$ is noncompact: $\mathrm{Aut}(\mathscr{A}) = \mathbb{Z}_2 \ltimes \mathbb{R}$ [13].

8. Summary and outlook

We have shown that the notion of the 'same physics in all spacetimes' can be given a formal meaning (at least in part) and can be analysed in the context of the BFV framework of locally covariant theories. While local covariance in itself does not guarantee the SPASs property, the dynamically local theories do form a class of theories with SPASs; moreover, dynamical locality seems to be a natural and useful property in other contexts. Relative Cauchy evolution enters the discussion in an essential way, and seems to be the replacement of the classical action in the axiomatic setting. A key question, given our starting point, is the extent to which the SPASs condition is sufficient as well as necessary for a class of theories to represent the same physics in all spacetimes. As a class of theories that had no subtheory embeddings other than equivalences would satisfy SPASs, there is clearly scope for further work on this issue. In particular, is it possible to formulate a notion of 'the same physics on all spacetimes' in terms of individual theories rather than classes of theories?

In closing, we remark that the categorical framework opens a completely new way of analysing quantum field theories, namely at the functorial level. It is conceivable that all structural properties of QFT should have a formulation at this level, with the instantiations of the theory in particular spacetimes taking a secondary place. Lest this be seen as a flight to abstraction, we emphasize again that this framework is currently leading to new viewpoints and concrete calculations in cosmology and elsewhere. This provides all the more reason to understand why theories based on our experience with terrestrial particle physics can be used in very different spacetime environments while preserving the same physical content.

References

[1] Adámek, J., Herrlich, H., Strecker, G.E.: Abstract and concrete categories: the joy of cats. Repr. Theory Appl. Categ. (17), 1–507 (2006), reprint of the 1990 original [Wiley, New York]

[2] Bär, C., Ginoux, N., Pfäffle, F.: Wave equations on Lorentzian manifolds and quantization. European Mathematical Society (EMS), Zürich (2007)

[3] Bernal, A.N., Sánchez, M.: Globally hyperbolic spacetimes can be defined as causal instead of strongly causal. Class. Quant. Grav. **24**, 745–750 (2007), gr-qc/0611138

[4] Brunetti, R., Fredenhagen, K.: Microlocal analysis and interacting quantum field theories: Renormalization on physical backgrounds. Comm. Math. Phys. **208**(3), 623–661 (2000)

[5] Brunetti, R., Fredenhagen, K., Verch, R.: The generally covariant locality principle: A new paradigm for local quantum physics. Commun. Math. Phys. **237**, 31–68 (2003)

[6] Brunetti, R., Ruzzi, G.: Superselection sectors and general covariance. I. Commun. Math. Phys. **270**, 69–108 (2007)

[7] Brunetti, R., Ruzzi, G.: Quantum charges and spacetime topology: The emergence of new superselection sectors. Comm. Math. Phys. **287**(2), 523–563 (2009)

[8] Dappiaggi, C., Fredenhagen, K., Pinamonti, N.: Stable cosmological models driven by a free quantum scalar field. Phys. Rev. **D77**, 104015 (2008)

[9] Degner, A., Verch, R.: Cosmological particle creation in states of low energy. J. Math. Phys. **51**(2), 022302 (2010)

[10] Dikranjan, D., Tholen, W.: Categorical structure of closure operators, *Mathematics and its Applications*, vol. 346. Kluwer Academic Publishers Group, Dordrecht (1995)

[11] Dimock, J.: Algebras of local observables on a manifold. Commun. Math. Phys. **77**, 219–228 (1980)

[12] Ferguson, M.: In preparation

[13] Fewster, C.J.: In preparation

[14] Fewster, C.J.: Quantum energy inequalities and local covariance. II. Categorical formulation. Gen. Relativity Gravitation **39**(11), 1855–1890 (2007)

[15] Fewster, C.J., Pfenning, M.J.: Quantum energy inequalities and local covariance. I: Globally hyperbolic spacetimes. J. Math. Phys. **47**, 082303 (2006)

[16] Fewster, C.J., Verch, R.: Dynamical locality; Dynamical locality for the free scalar field. In preparation

[17] Fulling, S.A.: Nonuniqueness of canonical field quantization in Riemannian space-time. Phys. Rev. **D7**, 2850–2862 (1973)

[18] Fulling, S.A., Narcowich, F.J., Wald, R.M.: Singularity structure of the two-point function in quantum field theory in curved spacetime. II. Ann. Physics **136**(2), 243–272 (1981)

[19] Fulling, S.A., Ruijsenaars, S.N.M.: Temperature, periodicity and horizons. Phys. Rept. **152**, 135–176 (1987)

[20] Haag, R.: Local Quantum Physics: Fields, Particles, Algebras. Springer-Verlag, Berlin (1992)

[21] Hawking, S.W., Ellis, G.F.R.: The Large Scale Structure of Space-Time. Cambridge University Press, London (1973)

[22] Hollands, S., Wald, R.M.: Local Wick polynomials and time ordered products of quantum fields in curved spacetime. Commun. Math. Phys. **223**, 289–326 (2001)

[23] Hollands, S., Wald, R.M.: Existence of local covariant time ordered products of quantum fields in curved spacetime. Commun. Math. Phys. **231**, 309–345 (2002)

[24] Kay, B.S.: Casimir effect in quantum field theory. Phys. Rev. **D20**, 3052–3062 (1979)

[25] Landau, L.J.: A note on extended locality. Comm. Math. Phys. **13**, 246–253 (1969)

[26] Mac Lane, S.: Categories for the Working Mathematician, 2nd edn. Springer-Verlag, New York (1998)

[27] Sanders, K.: On the Reeh-Schlieder property in curved spacetime. Comm. Math. Phys. **288**(1), 271–285 (2009)

[28] Sanders, K.: The locally covariant Dirac field. Rev. Math. Phys. **22**(4), 381–430 (2010)

[29] Streater, R.F.: Spontaneous breakdown of symmetry in axiomatic theory. Proc. Roy. Soc. Ser. A **287**, 510–518 (1965)

[30] Streater, R.F., Wightman, A.S.: PCT, spin and statistics, and all that. Princeton Landmarks in Physics. Princeton University Press, Princeton, NJ (2000). Corrected third printing of the 1978 edition

[31] Verch, R.: A spin-statistics theorem for quantum fields on curved spacetime manifolds in a generally covariant framework. Commun. Math. Phys. **223**, 261–288 (2001)

[32] Verch, R.: Local covariance, renormalization ambiguity, and local thermal equilibrium in cosmology (2011). These proceeedings and arXiv:1105.6249.

Christopher J. Fewster
Department of Mathematics
University of York
Heslington, York YO10 5DD
United Kingdom
e-mail: chris.fewster@york.ac.uk

Local Covariance, Renormalization Ambiguity, and Local Thermal Equilibrium in Cosmology

Rainer Verch

Abstract. This article reviews some aspects of local covariance and of the ambiguities and anomalies involved in the definition of the stress-energy tensor of quantum field theory in curved spacetime. Then, a summary is given of the approach proposed by Buchholz et al. to define local thermal equilibrium states in quantum field theory, i.e., non-equilibrium states to which, locally, one can assign thermal parameters, such as temperature or thermal stress-energy. The extension of that concept to curved spacetime is discussed and some related results are presented. Finally, the recent approach to cosmology by Dappiaggi, Fredenhagen and Pinamonti, based on a distinguished fixing of the stress-energy renormalization ambiguity in the setting of the semiclassical Einstein equations, is briefly described. The concept of local thermal equilibrium states is then applied, to yield the result that the temperature behaviour of a quantized, massless, conformally coupled linear scalar field at early cosmological times is more singular than that of classical radiation.

Mathematics Subject Classification (2010). 83F05, 81T05, 81T16.

Keywords. Cosmology, stress-energy-tensor, renormalization ambiguity, local thermal equilibrium.

1. Local covariant quantum field theory

The main theme of this contribution is the concept of local thermal equilibrium states and their properties in quantum field theory on cosmological spacetimes. The starting point for our considerations is the notion of local covariant quantum field theory which has been developed in [36, 5, 22] and which is also reviewed in [1] and is, furthermore, discussed in Chris Fewster's contribution to these conference proceedings [12]. I will therefore attempt to be as far as possible consistent with Fewster's notation and shall at several points refer to his contribution for precise definitions and further discussion of matters related to local covariant quantum field theory.

The basic idea of local covariant quantum field theory is to consider not just a quantum field on some fixed spacetime, but simultaneously the "same" type of quantum field theory on all — sufficiently nice — spacetimes at once. In making this more precise, one collects all spacetimes which one wants to consider in a category Loc. By definition, an object in Loc is a four-dimensional, globally hyperbolic spacetime with chosen orientation and time-orientation. An arrow, or morphism, $\boldsymbol{M} \xrightarrow{\psi} \boldsymbol{N}$ of Loc is an isometric, hyperbolic embedding which preserves orientation and time-orientation. (See Fewster's contribution for more details.)

Additionally, we also assume that we have, for each object $\boldsymbol{M}$ of Loc, a quantum field $\Phi_{\boldsymbol{M}}$, thought of as describing some physics happening in $\boldsymbol{M}$. More precisely, we assume that there is a topological $*$-algebra $\mathscr{A}(\boldsymbol{M})$ (with unit) and that $f \mapsto \Phi_{\boldsymbol{M}}(f)$, $f \in C_0^\infty(\boldsymbol{M})$, is an operator-valued distribution taking values in $\mathscr{A}(\boldsymbol{M})$. (This would correspond to a "quantized scalar field" which we treat here for simplicity, but everything can be generalized to more general tensor- and spinor-type fields, as e.g. in [36].) One would usually require that the $\Phi_{\boldsymbol{M}}(f)$ generate $\mathscr{A}(\boldsymbol{M})$ in a suitable sense, if necessary allowing suitable completions. The unital topological $*$-algebras also form a category which shall be denoted by Alg, where the morphisms $\mathscr{A} \xrightarrow{\alpha} \mathscr{B}$ are injective, unital, topological $*$-algebraic morphisms between the objects. Then one says that the family $(\Phi_{\boldsymbol{M}})$, as $\boldsymbol{M}$ ranges over the objects of Loc, is a *local covariant quantum field theory* if the assignments $\boldsymbol{M} \to \mathscr{A}(\boldsymbol{M})$ and $\mathscr{A}(\psi)(\Phi_{\boldsymbol{M}}(f)) = \Phi_{\boldsymbol{N}}(\psi_* f)$, for any morphism $\boldsymbol{M} \xrightarrow{\psi} \boldsymbol{N}$ of Loc, induce a functor $\mathscr{A} : \mathsf{Loc} \to \mathsf{Alg}$. (See again Fewster's contribution [12] for a fuller dish.) Actually, a generally covariant quantum field theory should more appropriately be viewed as a natural transformation and we refer to [5] for a discussion of that point. The degree to which the present notion of local covariant quantum field theory describes the "same" quantum field on all spacetimes is analyzed in Fewster's contribution to these proceedings.

We write $\boldsymbol{\Phi} = (\Phi_{\boldsymbol{M}})$ to denote a local covariant quantum field theory, and we remark that examples known so far include the scalar field, the Dirac field, the Proca field, and — with some restrictions — the electromagnetic field, as well as perturbatively constructed $P(\phi)$ and Yang-Mills models [22, 20, 36]. A *state* of a local covariant quantum field theory $\boldsymbol{\Phi}$ is defined as a family $\boldsymbol{\omega} = (\omega_{\boldsymbol{M}})$, $\boldsymbol{M}$ ranging over the objects of Loc, where each $\omega_{\boldsymbol{M}}$ is a state on $\mathscr{A}(\boldsymbol{M})$ — i.e. an expectation value functional. It might appear natural to assume that there is an invariant state $\boldsymbol{\omega}$, defined by

$$\omega_{\boldsymbol{N}} \circ \mathscr{A}(\psi) = \omega_{\boldsymbol{M}} \tag{1.1}$$

for all morphisms $\boldsymbol{M} \xrightarrow{\psi} \boldsymbol{N}$ of Loc, but it has been shown that this property is in conflict with the dynamics and stability properties one would demand of the $\omega_{\boldsymbol{M}}$ for each $\boldsymbol{M}$ [5, 22]; the outline of an argument against an invariant state with the property (1.1) is given in Fewster's contribution [12].

2. The quantized stress-energy tensor

A general property of quantum field theories is the existence of a causal dynamical law, or time-slice axiom. In a local covariant quantum field theory $\mathbf{\Phi}$, this leads to a covariant dynamics [5] termed "relative Cauchy evolution" in Fewster's contribution [12], to which we refer again for further details. The relative Cauchy evolution consists, for each $\boldsymbol{M}$ in Loc, of an isomorphism

$$\mathrm{rce}_{\boldsymbol{M}}(h) : \mathscr{A}(\boldsymbol{M}) \to \mathscr{A}(\boldsymbol{M})$$

which describes the effect of an additive perturbation of the metric of $\boldsymbol{M}$ by a symmetric 2-tensor field h on the propagation of the quantum field, akin to a scattering transformation brought about by perturbation of the background metric. One may consider the derivation – assuming it exists – which is obtained as $d/ds|_{s=0}\,\mathrm{rce}_{\boldsymbol{M}}(h(s))$ where $h(s)$ is any smooth family of metric perturbations with $h(s{=}0) = 0$. Following the spirit of Bogoliubov's formula [3], this gives rise (under fairly general assumptions) to a local covariant quantum field (T_M) which generates the derivation upon taking commutators, and can be identified (up to some multiplicative constant) with the quantized stress-energy tensor (and has been shown in examples to agree indeed with the derivation induced by forming the commutator with the quantized stress-energy tensor [5]). Assuming for the moment that $\mathsf{T}_{\boldsymbol{M}}(f)$ takes values in $\mathscr{A}(\boldsymbol{M})$, we see that in a local covariant quantum field theory which fulfills the time-slice axiom the stress-energy tensor is characterized by the following properties:

LCSE-1 *Generating property for the relative Cauchy evolution*:

$$2i[\mathsf{T}_{\boldsymbol{M}}(f), A] = \left.\frac{d}{ds}\right|_{s=0} \mathrm{rce}_{\boldsymbol{M}}(h(s))(A)\,, \quad A \in \mathscr{A}(\boldsymbol{M})\,, \quad f = \left.\frac{d}{ds}\right|_{s=0} h(s) \tag{2.1}$$

Note that $f = f^{ab}$ is a smooth, symmetric C_0^∞ two-tensor field on $\boldsymbol{M}$. Using the more ornamental abstract index notation, one would therefore write

$$\mathsf{T}_{\boldsymbol{M}ab}(f^{ab}) = \mathsf{T}_{\boldsymbol{M}}(f)\,.$$

LCSE-2 *Local covariance*:

$$\mathscr{A}(\psi)\,(\mathsf{T}_{\boldsymbol{M}}(f)) = \mathsf{T}_{\boldsymbol{N}}(\psi_* f) \tag{2.2}$$

whenever $\boldsymbol{M} \xrightarrow{\psi} \boldsymbol{N}$ is an arrow in Loc.

LCSE-3 *Symmetry*:

$$\mathsf{T}_{\boldsymbol{M}ab} = \mathsf{T}_{\boldsymbol{M}ba}\,. \tag{2.3}$$

LCSE-4 *Vanishing divergence*:

$$\nabla^a \mathsf{T}_{\boldsymbol{M}ab} = 0\,, \tag{2.4}$$

where ∇ is the covariant derivative of the metric of $\boldsymbol{M}$.

Conditions **LCSE-3,4** are (not strictly, but morally) consequences of the previous two conditions, see [5] for discussion.

The conditions **LCSE-1...4** ought to be taken as characterizing for any stress-energy tensor of a given local covariant quantum field theory $\mathbf{\Phi}$. However, they don't fix $\mathsf{T}_{\boldsymbol{M}}$ completely. Suppose that we have made some choice, $(\mathsf{T}^{[1]}_{\boldsymbol{M}})$, of a local covariant stress-energy tensor for our local covariant quantum field theory $\mathbf{\Phi}$. Now choose a local covariant family $(C_{\boldsymbol{M}})$ of (number-valued) smooth tensor fields $C_{\boldsymbol{M}} = C_{\boldsymbol{M}ab}$ on $\boldsymbol{M}$, and write $C_{\boldsymbol{M}}(f) = \int_{\boldsymbol{M}} C_{\boldsymbol{M}ab} f^{ab}\, d\mathrm{vol}_{\boldsymbol{M}}$ where $d\mathrm{vol}_{\boldsymbol{M}}$ denotes the metric-induced volume form on $\boldsymbol{M}$. If $(C_{\boldsymbol{M}})$ is not only local covariant, but also symmetric and has vanishing divergence, then we may set

$$\mathsf{T}^{[2]}_{\boldsymbol{M}}(f) = \mathsf{T}^{[1]}_{\boldsymbol{M}}(f) + C_{\boldsymbol{M}}(f)\mathbf{1}$$

(where $\mathbf{1}$ is the unit of $\mathscr{A}(\boldsymbol{M})$) to obtain in this way another choice, $(\mathsf{T}^{[2]}_{\boldsymbol{M}})$, of a stress-energy tensor for $\mathbf{\Phi}$ which is as good as the previous one since it satisfies the conditions **LCSE-1...4** equally well. Again under quite general conditions, we may expect that this is actually the complete freedom for the stress-energy tensor that is left by the above conditions, in particular the freedom should in fact be given by a multiple of unity (be state-independent). Since the $C_{\boldsymbol{M}}$ must be local covariant and divergence-free, they should be local functionals of the spacetime metric of $\boldsymbol{M}$, and one can prove this upon adding some technical conditions.

The freedom which is left by the conditions **LCSE-1...4** for the quantized stress-energy tensor of a local covariant quantum field theory can be traced back to the circumstance that even in such a simple theory as the linear scalar field, the stress-energy tensor is a renormalized quantity. Let us explain this briefly, using the example of the minimally coupled linear scalar field, following the path largely developed by Wald [37, 40]. On any spacetime, the classical minimally coupled linear scalar field ϕ obeys the field equation

$$\nabla^a \nabla_a \phi + m^2 \phi = 0\,. \tag{2.5}$$

The stress-energy tensor for a classical solution ϕ of the field equation can be presented in the form

$$T_{ab}(x) = \lim_{y \to x} P_{ab}(x, y; \nabla_{(x)}, \nabla_{(y)})\phi(x)\phi(y)$$

(where x, y are points in spacetime) with some partial differential operator $P_{ab}(x, y; \nabla_{(x)}, \nabla_{(y)})$. This serves as starting point for defining $\mathsf{T}_{\boldsymbol{M}ab}$ through replacing ϕ by the quantized linear scalar field $\Phi_{\boldsymbol{M}}$. However, one finds already in Minkowski spacetime that upon performing this replacement, the behaviour of the resulting expression is singular in the coincidence limit $y \to x$. As usual with non-linear expressions in a quantum field at coinciding points, one must prescribe a renormalization procedure in order to obtain a well-defined quantity. In Minkowski spacetime, this is usually achieved by normal ordering with respect to the vacuum. This makes explicit reference to the vacuum state in Minkowski spacetime, the counterpart of which is not available in case of generic spacetimes. Therefore, one proceeds in a different manner on generic curved spacetimes (but in Minkowski spacetime, the result coincides with what one obtains from normal ordering, up to a renormalization freedom).

Since one is mainly interested in expectation values of the quantized stress-energy tensor — as these are the quantities entering on the right-hand side of the semiclassical Einstein equations, discussed in the next section — one may concentrate on a class of states $\mathcal{S}(\boldsymbol{M})$ on $\mathscr{A}(\boldsymbol{M})$ for which the expectation value of the stress-energy tensor can be defined as unambiguously as possible, and in a manner consistent with the conditions **LCSE-1...4** above. Defining only the expectation value of the stress-energy tensor has the additional benefit that one need not consider the issue if, or in which sense, $\mathsf{T}_{\boldsymbol{M}}(f)$ is contained in $\mathscr{A}(\boldsymbol{M})$ (or its suitable extensions). Of course, it has the drawback that non-linear expressions in $\mathsf{T}_{\boldsymbol{M}}(f)$, as they would appear in the variance of the stress-energy, need extra definition, but we may regard this as an additional issue. Following the idea above for defining the quantized stress-energy tensor, one can see that the starting point for the expectation value of the stress-energy tensor in a state $\omega_{\boldsymbol{M}}$ on $\mathscr{A}(\boldsymbol{M})$ is the corresponding two-point function

$$w_{\boldsymbol{M}}(x,y) = \omega_{\boldsymbol{M}}(\Phi_{\boldsymbol{M}}(x)\Phi_{\boldsymbol{M}}(y))\,,$$

written here symbolically as a function, although properly it is a distribution on $C_0^\infty(\boldsymbol{M}\times\boldsymbol{M})$. One now considers a particular set $\mathcal{S}_{\mu sc}(\boldsymbol{M})$, the states on $\mathscr{A}(\boldsymbol{M})$ whose two-point functions fulfill the *microlocal spectrum condition* [28, 4, 35]. The microlocal spectrum condition specifies the wavefront set of the distribution $w_{\boldsymbol{M}}$ in a particular, asymmetric way, which is reminiscent of the spectrum condition for the vacuum state in Minkowski spacetime. Using the transformation properties of the wavefront set under diffeomorphisms, one can show (1) the sets $\mathcal{S}_{\mu sc}(\boldsymbol{M})$ transform contravariantly under arrows $\boldsymbol{M} \xrightarrow{\psi} \boldsymbol{N}$ in Loc, meaning that every state in $\mathcal{S}_{\mu sc}(\boldsymbol{N})$ restricts to a state of $\mathcal{S}_{\mu sc}(\psi(\boldsymbol{M}))$ and $\mathcal{S}_{\mu sc}(\psi(\boldsymbol{M}))\circ\mathscr{A}(\psi) = \mathcal{S}_{\mu sc}(\boldsymbol{M})$, and moreover (2) that the microlocal spectrum condition is equivalent to the *Hadamard condition* on $w_{\boldsymbol{M}}$ [28, 29]. This condition says [23] that the two-point function $w_{\boldsymbol{M}}$ splits in the form

$$w_{\boldsymbol{M}}(x,y) = \mathrm{H}_{\boldsymbol{M}}(x,y) + u(x,y)\,,$$

where $\mathrm{H}_{\boldsymbol{M}}$ is a Hadamard parametrix for the wave equation (2.5), and u is a C^∞ integral kernel. Consequently, the singular behaviour of $\mathrm{H}_{\boldsymbol{M}}$ and hence of $w_{\boldsymbol{M}}$ is determined by the spacetime geometry of $\boldsymbol{M}$ and is state-independent within the set $\mathcal{S}_{\mu sc}(\boldsymbol{M})$ whereas different states are distinguished by different smooth terms $u(x,y)$. With respect to such a splitting of a two-point function, one can then define the expectation value of the stress-energy tensor in a state $\omega_{\boldsymbol{M}}$ in $\mathcal{S}(\boldsymbol{M})$ as

$$\langle\tilde{\mathsf{T}}_{\boldsymbol{M}ab}(x)\rangle_{\omega_{\boldsymbol{M}}} = \lim_{y\to x} P_{ab}(x,y;\nabla_{(x)},\nabla_{(y)})\,(w_{\boldsymbol{M}} - \mathrm{H}_{\boldsymbol{M}})\,. \tag{2.6}$$

Note that $\langle\tilde{\mathsf{T}}_{\boldsymbol{M}ab}(x)\rangle_{\omega_{\boldsymbol{M}}}$ is, in fact, smooth in x and therefore is a C^∞ tensor field on $\boldsymbol{M}$. The reason for the appearance of the twiddle on top of the symbol for the just defined stress-energy tensor is due to the circumstance that with this definition, the stress-energy tensor is (apart from exceptional cases) not divergence-free. However, Wald [37, 40] has shown that this can be repaired by subtracting the divergence-causing term. More precisely, there is a smooth function $Q_{\boldsymbol{M}}$ on $\boldsymbol{M}$, constructed locally from the metric of $\boldsymbol{M}$ (so that the family $(Q_{\boldsymbol{M}})$ is local covariant), such that

$$\langle \mathsf{T}_{\boldsymbol{M}ab}(x)\rangle_{\omega_{\boldsymbol{M}}} = \langle \tilde{\mathsf{T}}_{\boldsymbol{M}ab}(x)\rangle_{\omega_{\boldsymbol{M}}} - Q_{\boldsymbol{M}}(x) g_{ab}(x)\,, \tag{2.7}$$

where g_{ab} denotes the spacetime metric of $\boldsymbol{M}$, defines an expectation value of the stress-energy tensor which is divergence-free. Moreover, with this definition, the conditions **LCSE-1...4** hold when interpreted as valid for expectation values of states fulfilling the Hadamard condition [40, 5] (possibly after suitable symmetrization in order to obtain **LCSE-4**). This may actually be seen as one of the main motivations for introducing the Hadamard condition in [37], and in the same paper, Wald proposed conditions on the expectation values of the stress-energy tensor which are variants of our **LCSE-1...4** above. Wald [37, 40] also proved that, if there are two differing definitions for the expectation values of the stress-energy tensor complying with the conditions, then the difference is given by a state-independent, local covariant family $(C_{\boldsymbol{M}ab})$ of smooth, symmetric, divergence-free tensor fields, in complete analogy to our discussion above. Actually, the formulation of the local covariance condition for the expectation values of the stress-energy tensor in [40] was an important starting point for the later development of local covariant quantum field theory, so the agreement between Wald's result on the freedom in defining the stress-energy tensor and ours, discussed above, are in no way coincidental.

How does this freedom come about? To see this, note that only the singularities of the Hadamard parametrix $\mathrm{H}_{\boldsymbol{M}}$ are completely fixed. One has the freedom of altering the smooth contributions to that Hadamard parametrix, and as long as this leads to an expected stress-energy tensor which is still consistent with conditions **LCSE-1...4**, this yields an equally good definition of $\langle \mathsf{T}_{\boldsymbol{M}ab}(x)\rangle_{\omega_{\boldsymbol{M}}}$. Even requiring as in [37, 40] that the expectation value of the stress-energy tensor agrees in Minkowski spacetime with the expression obtained by normal ordering — implying that in Minkowski spacetime, the stress-energy expectation value of the vacuum vanishes — doesn't fully solve the problem. For if another definition is chosen so that the resulting difference term $C_{\boldsymbol{M}ab}$ is made of curvature terms, that difference vanishes on Minkowski spacetime. The problem could possibly be solved if there was a distinguished state $\boldsymbol{\omega} = (\omega_{\boldsymbol{M}})$ for which the expectation value of the stress-energy tensor could be specified in some way, but as was mentioned before, the most likely candidate for such a state, the invariant state, doesn't exist (at least not as a state fulfilling the microlocal spectrum condition). This means that apparently the setting of local covariant quantum field theory does not, at least without using further ingredients, specify intrinsically the absolute value of the local stress-energy content of a quantum field state, since the ambiguity of being able to add a difference term $C_{\boldsymbol{M}ab}$ always remains. In fact, this difference term ambiguity has the character of a renormalization ambiguity since it occurs in the process of renormalization (by means of discarding the singularities of a Hadamard parametrix), as is typical in quantum field theory. The fact that local covariant quantum field theory does not fix the local stress-energy content may come as a surprise — and disappointment — since it differs from what one would be inclined to expect from quantum field theory on Minkowski spacetime with a distinguished vacuum state. This just goes to show that one cannot too naively transfer concepts from quantum field theory on Minkowski spacetime to local covariant quantum field theory (a fact which, incidentally, has already been demonstrated by the Hawking and Unruh effects).

Another complication needs to be addressed: Anomalies of the quantized stress-energy tensor. It is not only that the value of (the expectation value of) the quantized stress-energy tensor isn't fixed on curved spacetime due to the appearance of metric- or curvature-dependent renormalization ambiguities, but there are also curvature-induced anomalies. Recall that in quantum field theory one generally says that a certain quantity is subject to an anomaly if the quantized/renormalized quantity fails to feature a property — mostly, a symmetry property — which is fulfilled for the counterpart of that quantity in classical field theory. In the case of the stress-energy tensor, this is the trace anomaly or conformal anomaly. Suppose that ϕ is a C^∞ solution of the conformally coupled, massless linear scalar wave equation, i.e.

$$\left(\nabla^a\nabla_a + \tfrac{1}{6}R\right)\phi = 0$$

on a globally hyperbolic spacetime $\boldsymbol{M}$ with scalar curvature R. Then the classical stress-energy tensor of ϕ has vanishing trace, $T^a{}_a = 0$. This expresses the conformal invariance of the equation of motion. However, as was shown by Wald [38], it is not possible to have the vanishing trace property for the expectation value of the stress-energy tensor in the presence of curvature if one insists on the stress-energy tensor having vanishing divergence: Under these circumstances, one necessarily finds

$$\langle \mathsf{T}^a_{\boldsymbol{M}a}\rangle_{\omega_{\boldsymbol{M}}} = -4Q_{\boldsymbol{M}}\,.$$

Note that this is independent of the choice of Hadamard state $\omega_{\boldsymbol{M}}$. The origin of this trace-anomaly lies in having subtracted the divergence-causing term Qg_{ab} in the definition of the renormalized stress-energy tensor. In a sense, non-vanishing divergence of the renormalized stress-energy tensor could also be viewed as an anomaly, so in the case of the conformally coupled, massless quantized linear scalar, one can trade the non-vanishing "divergence anomaly" of the renormalized stress-energy tensor for the trace anomaly. The condition **LCSE-4** assigns higher priority to vanishing divergence, whence one has to put up with the trace anomaly.

One may wonder why the features of the quantized/renormalized stress-energy tensor have been discussed here at such an extent while our main topic are local thermal equilibrium states. The reason is that the ambiguities and anomalies by which the definition of the stress-energy tensor for quantum fields in curved spacetime is plagued will also show up when trying to generalize the concept of local thermal equilibrium in quantum field theory in curved spacetime. Furthermore, the ambiguities and anomalies of the stress-energy tensor do play a role in semiclassical gravity (and semiclassical cosmology), and we will come back to this point a bit later.

3. Local thermal equilibrium

3.1. LTE states on Minkowski spacetime

Following the idea of Buchholz, Ojima and Roos [7], local thermal equilibrium (LTE, for short) states are states of a quantum field for which local, intensive thermal quantities such as — most prominently — temperature and pressure can be defined and take the values they would assume for a thermal equlibribum state. Here "local" means, in fact, at a collection of spacetime points. We will soon be more specific about this.

Although the approach of [7] covers also interacting fields, for the purposes of this contribution we will restrict attention to the quantized linear scalar field; furthermore, we start by introducing the concept of LTE states on Minkowski spacetime. Consider the quantized massive linear field Φ_0 on Minkowski spacetime, in its usual vacuum representation, subject to the field equation $(\Box + m^2)\Phi_0 = 0$ (to be understood in the sense of operator-valued distributions). Now, at each point x in Minkowski spacetime, one would like to define a set of "thermal observables" $\boldsymbol{\Theta}_x$ formed by observables which are sensitive to intensive thermal quantities at x. If that quantity is, e.g., temperature, then the corresponding quantity in $\boldsymbol{\Theta}_x$ would be a thermometer "located" at spacetime point x. One may wonder if it makes physical sense to idealize a thermometer as being of pointlike "extension" in space and time, but as discussed in [7], this isn't really a problem.

What are typical elements of $\boldsymbol{\Theta}_x$? Let us look at a very particular example. Assume that $m = 0$, so we have the massless field. Moreover, fix some Lorentzian frame consisting of a tetrad $(e_0^a, e_1^a, e_2^a, e_3^a)$ of Minkowski vectors such that $\eta_{ab}e_\mu^a e_\nu^b = \eta_{\mu\nu}$ and with e_0^a future-directed. Relative to these data, let ω_{β,e_0} denote the global thermal equilibrium state — i.e. KMS state — with respect to the time direction $e_0 \equiv e_0^a$ at inverse temperature $\beta > 0$ [19]. Evaluating $:\Phi_0^2:(x)$, the Wick square of Φ_0 at the spacetime point x, in the KMS state gives

$$\langle :\Phi_0^2:(x)\rangle_{\omega_{\beta,e_0}} = \frac{1}{12\beta^2}\,.$$

Recall that the Wick square is defined as

$$:\Phi_0^2:(x) = \lim_{\zeta\to 0} \Phi_0(q_x(\zeta))\Phi_0(q_x(-\zeta)) - \langle \Phi_0(q_x(\zeta))\Phi_0(q_x(-\zeta))\rangle_{\text{vac}}\,, \quad (3.1)$$

where $\langle\,.\,\rangle_{\text{vac}}$ is the vacuum state and we have set

$$q_x(\zeta) = x + \zeta \quad (3.2)$$

for Minkowski space coordinate vectors x and ζ, tacitly assuming that ζ isn't zero or lightlike. Thus, $:\Phi_0^2:(x)$ is an observable localized at x; strictly speaking, without smearing with test functions with respect to x, it isn't an operator but a quadratic form. Evaluating $:\Phi_0^2:(x)$ in the global thermal equilibrium state yields a monotonous function of the temperature. This is also the case for mass $m > 0$, only the monotonous function is a bit more complicated. So $:\Phi_0^2:(x)$ can be taken as a "thermometer observable" at x. As in our simple model the KMS state for Φ_0 is homogeneous and isotropic, $\langle :\Phi_0^2:(x)\rangle_{\omega_{\beta,e_0}}$ is, in fact, independent of x. Let us abbreviate the Wick-square "thermometer observable" by

$$\boldsymbol{\vartheta}(x) = :\Phi_0^2:(x)\,. \tag{3.3}$$

Starting from the Wick square one can, following [7], form many more elements of $\boldsymbol{\Theta}_x$. The guideline was that these elements should be sensitive to intensive thermal quantitites at x. This can be achieved by forming the *balanced derivatives of the Wick square of order* n, defined as

$$\begin{aligned}&\eth_{\mu_1\ldots\mu_n}:\Phi_0^2:(x)\\ &= \lim_{\zeta\to 0}\partial_{\zeta^{\mu_1}}\cdots\partial_{\zeta^{\mu_n}}\Big(\Phi_0(q_x(\zeta))\Phi_0(q_x(-\zeta)) - \langle\Phi_0(q_x(\zeta))\Phi_0(q_x(-\zeta))\rangle_{\text{vac}}\Big).\end{aligned} \tag{3.4}$$

Prominent among these is the second balanced derivative because

$$\boldsymbol{\varepsilon}_{\mu\nu}(x) = -\frac{1}{4}\eth_{\mu\nu}:\Phi_0^2:(x) \tag{3.5}$$

is the *thermal stress-energy tensor* which in the KMS state ω_{β,e_0} takes on the values[1]

$$\varepsilon_{\mu\nu}(x) = \langle\boldsymbol{\varepsilon}_{\mu\nu}(x)\rangle_{\omega_{\beta e_0}} = \frac{1}{(2\pi)^3}\int_{\mathbb{R}^3}\frac{p_\mu p_\nu}{(\mathrm{e}^{\beta p_0}-1)p_0}\,d^3p\,, \tag{3.6}$$

where $p_\mu = p_a e^a_\mu$ are the covariant coordinates of p with respect to the chosen Lorentz frame, and $p_0 = |\underline{p}|$ with $(p_\mu) = (p_0, \underline{p})$. Again, this quantity is independent of x in the unique KMS state of our quantum field model.

Notice that the values of $\varepsilon_{\mu\nu}$ depend not only on the inverse temperature, but also on the time direction e_0 of the Lorentz frame with respect to which ω_{β,e_0} is a KMS state. Since the dependence of the thermal quantities on β and e_0 in expectation values of ω_{β,e_0} is always a function of the timelike, future-directed vector $\boldsymbol{\beta} = \beta\cdot e_0$, it is hence useful to label the KMS states correspondingly as $\omega_{\boldsymbol{\beta}}$, and to define LTE states with reference to $\boldsymbol{\beta}$.

Definition 3.1. Let ω be a state of the linear scalar field Φ_0 on Minkowski spacetime, and let $N \in \mathbb{N}$.

[1] We caution the reader that in previous publications on LTE states, ϑ and $\varepsilon_{\mu\nu}$ are always used to denote the expectation values of our $\boldsymbol{\vartheta}$ and $\boldsymbol{\varepsilon}$ in LTE states. We hope that our use of bold print for the thermal observables is sufficient to distinguish the thermal observables from their expectation values in LTE states.

(i) Let D be a subset of Minkowski spacetime and let $\boldsymbol{\beta} : x \mapsto \boldsymbol{\beta}(x)$ be a (smooth, if D is open) map assigning to each $x \in D$ a future-directed timelike vector $\boldsymbol{\beta}(x)$. Then we say that ω *is a local thermal equilibrium state of order N at sharp temperature for the temperature vector field $\boldsymbol{\beta}$* (for short, ω is a $[D, \boldsymbol{\beta}, N]$-LTE state) if

$$\langle \eth_{\mu_1 \cdots \mu_n} :\Phi_0^2: (x) \rangle_\omega = \langle \eth_{\mu_1 \cdots \mu_n} :\Phi_0^2: (0) \rangle_{\boldsymbol{\beta}(x)} \tag{3.7}$$

holds for all $x \in D$ and $0 \leq n \leq N$. Here, we have written $\langle \,.\, \rangle_{\boldsymbol{\beta}(x)} = \omega_{\boldsymbol{\beta}(x)}(\,.\,)$ for the KMS state of Φ_0 defined with respect to the timelike, future-directed vector $\boldsymbol{\beta}(x)$. The balanced derivatives of the Wick square on the right-hand side are evaluated at the spacetime point 0; since the KMS state on the right-hand side is homogeneous and isotropic, one has the freedom to make this choice.

(ii) Let D be a subset of Minkowski spacetime and let $m : x \mapsto \varrho_x$ be a map which assigns to each $x \in D$ a probability measure compactly supported on V_+, the open set of all future-directed timelike Minkowski vectors. It will be assumed that the map is smooth if D is open. Then we say that ω *is a local thermal equilibrium state of order N with mixed temperature distribution ϱ* (for short, ω is a $[D, \varrho, N]$-LTE state) if

$$\langle \eth_{\mu_1 \cdots \mu_n} :\Phi_0^2: (x) \rangle_\omega = \langle \eth_{\mu_1 \cdots \mu_n} :\Phi_0^2: (0) \rangle_{\varrho_x} \tag{3.8}$$

holds for all $x \in D$ and $0 \leq n \leq N$, where

$$\langle \eth_{\mu_1 \cdots \mu_n} :\Phi_0^2: (0) \rangle_{\varrho_x} = \int_{V_+} \langle \eth_{\mu_1 \cdots \mu_n} :\Phi_0^2: (0) \rangle_{\boldsymbol{\beta}'} \, d\varrho_x(\boldsymbol{\beta}') \,. \tag{3.9}$$

In this definition, Φ_0 can more generally also be taken as the linear scalar field with a finite mass different from zero. Let us discuss a couple of features of this definition and some results related to it.

(A) The set of local thermal observables used for testing thermal properties of ω in (3.7) and (3.8) is $\boldsymbol{\Theta}_x^{(N)}$, formed by the Wick square and all of its balanced derivatives up to order N. One can generalize the condition to unlimited order of balanced derivatives, using as thermal observables the set $\boldsymbol{\Theta}_x^{(\infty)} = \bigcup_{N=1}^\infty \boldsymbol{\Theta}_x^{(N)}$.

(B) The condition (3.7) demands that at each x in D, the expectation values of thermal observables in $\boldsymbol{\Theta}_x^{(N)}$ evaluated in ω and in $\langle \,.\, \rangle_{\boldsymbol{\beta}(x)}$ coincide; in other words, with respect to these thermal observables at x, ω looks just like the thermal equilibrium state $\langle \,.\, \rangle_{\boldsymbol{\beta}(x)}$. Note that $\boldsymbol{\beta}(x)$ can vary with x, so an LTE state can have a different temperature and a different "equilibrium rest frame", given by the direction of $\boldsymbol{\beta}(x)$, at each x.

(C) With increasing order N, the sets $\boldsymbol{\Theta}_x^{(N)}$ become larger; so the higher the order N for which the LTE condition is fulfilled, the more can the state ω be regarded as coinciding with a thermal state at x. In this way, the maximum order N for which (3.7) is fulfilled provides a measure of the deviation from local thermal equilibrium (and similarly, for the condition (3.8)).

(D) Condition (3.7) demands that, on $\boldsymbol{\Theta}_x^{(N)}$, ω coincides with a thermal equilibrium state at sharp temperature, and sharp thermal rest-frame. Condition (3.8) is less restrictive, demanding only that ω coincides with a mixture of thermal equilibrium states, described by the probability measure ϱ_x, on the thermal observables $\boldsymbol{\Theta}_x$. Of course, (3.7) is a special case of (3.8).

(E) Clearly, each global thermal equilibrium state, or KMS state, $\omega_{\boldsymbol{\beta}}$ is an LTE state, with constant inverse temperature vector $\boldsymbol{\beta}$. The interesting feature of the definition of LTE states is that $\boldsymbol{\beta}(x)$ can vary with x in spacetime, and the question of existence of LTE states which are not global KMS states arises. In particular, fixing the order N and the spacetime region D, which functions $\boldsymbol{\beta}(x)$ or ϱ_x can possibly occur? They surely cannot be completely arbitrary, particularly for open D, since the Wick square of Φ_0 and its balanced derivatives are subject to dynamical constraints which are a consequence of the equation of motion for Φ_0. We put on record some of the results which have been established so far.

The hot bang state [7, 6]

A state of Φ_0 (zero mass case) which is an LTE state with a variable, sharp inverse temperature vector field on the open forward lightcone V_+ was constructed in [7] and further investigated in [6]. This state is called the hot bang state ω_{HB}, and for x, y in V_+, its two-point function w_{HB} has the form

$$w_{\mathrm{HB}}(x,y) = \frac{1}{(2\pi)^3}\int_{\mathbb{R}^4} \mathrm{e}^{-i(x-y)^\mu p_\mu}\epsilon(p_0)\delta(p^\mu p_\mu)\frac{1}{1-\mathrm{e}^{-a((x-y)^\mu p_\mu)}}d^4p \quad (3.10)$$

where $a > 0$ is some parameter and ϵ denotes the sign function. To compare, the two-point function of a KMS state with constant inverse temperature vector $\boldsymbol{\beta}'$ is given by

$$w_{\boldsymbol{\beta}'}(x,y) = \frac{1}{(2\pi)^3}\int_{\mathbb{R}^4} \mathrm{e}^{-i(x-y)^\mu p_\mu}\epsilon(p_0)\delta(p^\mu p_\mu)\frac{1}{1-\mathrm{e}^{-\boldsymbol{\beta}'^\mu p_\mu}}d^4p\,. \quad (3.11)$$

Upon comparison, one can see that ω_{HB} has an inverse temperature vector field

$$\boldsymbol{\beta}(x) = 2ax\,, \quad x \in V_+\,.$$

Thus, the temperature diverges at the boundary of V_+, with the thermal rest frame tilting lightlike; moreover, the temperature decreases away from the boundary with increasing coordinate time x^0. This behaviour is sketched in Figure 1.

The hot bang state provides an example of a state where the thermal stress-energy tensor $\varepsilon_{\mu\nu}(x) = -(1/4)\langle \eth_{\mu\nu} :\Phi_0^2:(x)\rangle_{\mathrm{HB}}$ deviates from the expectation values of the full stress-energy tensor $\langle \mathsf{T}_{\mu\nu}(x)\rangle_{\mathrm{HB}}$. In fact, as discussed in [7], for the stress-energy tensor of the massless linear scalar field in Minkowski spacetime one finds generally

$$\mathsf{T}_{\mu\nu}(x) = \boldsymbol{\varepsilon}_{\mu\nu} + \frac{1}{12}(\partial_\mu\partial_\nu - \eta_{\mu\nu}\Box) :\Phi_0^2:(x)\,,$$

FIGURE 1. Sketch of temperature distribution of the hot bang state.

and thus one obtains, for the hot bang state,

$$\varepsilon_{\mu\nu}^{(\mathrm{HB})}(x) = \langle \boldsymbol{\varepsilon}_{\mu\nu}(x) \rangle_{\mathrm{HB}} = \frac{\pi^2}{1440a^4} \frac{4x_\mu x_\nu - \eta_{\mu\nu} x^\lambda x_\lambda}{(x^\kappa x_\kappa)^3} ,$$

whereas

$$\langle \mathsf{T}_{\mu\nu}(x) \rangle_{\mathrm{HB}} = \varepsilon_{\mu\nu}^{(\mathrm{HB})}(x) + \frac{1440a^2}{288\pi^2} \varepsilon_{\mu\nu}^{(\mathrm{HB})}(x) .$$

The first term is due to the thermal stress-energy, while the second term is a convection term which will dominate over the thermal stress-energy when the parameter a exceeds 1 by order of magnitude. Thus, for non-stationary LTE states, the expectation value of the full stress-energy tensor can in general not be expected to coincide with the thermal stress-energy contribution due to transport terms which are not seen in $\varepsilon_{\mu\nu}$.

Maximal spacetime domains for nontrivial LTE states [6]

As mentioned before, the inverse temperature vector field for an LTE state cannot be arbitrary since the dynamical behaviour of the quantum field imposes constraints. A related but in some sense stronger constraint which appears not so immediate comes in form of the following result which was established by Buchholz in [6]: Suppose that ω is a $[D, \boldsymbol{\beta}, \infty]$-LTE state, i.e. an LTE state of infinite order with sharp inverse temperature vector field $\boldsymbol{\beta}$. If D, the region on which ω has the said LTE property, contains a translate of V_+, then for $\boldsymbol{\beta}$ to be non-constant it is necessary that D is contained in some timelike simplicial cone, i.e. an intersection of characteristic half-spaces (which means, in particular, that D cannot contain any complete timelike line).

Existence of non-trivial mixed temperature LTE states [33]

The hot bang state mentioned above has been constructed for the case of the massless linear scalar field. It turned out to be more difficult to construct non-trivial LTE states for the massive linear scalar field, and the examples known to date for the massive case are mixed temperature LTE states. Recently, Solveen [33] proved a general result on the existence of non-trivial mixed temperature LTE states: Given any compact subset D of Minkowski spacetime, there are non-constant probability measure-valued functions $x \mapsto \varrho_x$, $x \in D$, together with states ω which are $[D, \varrho, \infty]$-LTE states.

LTE condition as generalization of the KMS condition [27]

The choice of balanced derivatives of the Wick square as thermal observables fixing the LTE property may appear, despite the motivation given in [7], as being somewhat arbitrary, so that one would invite other arguments for their prominent role in setting up the LTE condition. One attempt in this direction has been made by Schlemmer who pointed at a relation between an Unruh-like detector model and balanced derivatives of the Wick square [31]. On the other hand, recent work by Pinamonti and Verch shows that the LTE condition can be viewed as a generalization of the KMS condition. The underlying idea will be briefly sketched here, for full details see the forthcoming publication [27]. For a KMS state $\langle\,.\,\rangle_{\beta,e_0}$, let

$$\varphi(\tau) = \varphi_x(\tau) = \Big\langle \Phi_0\Big(q(-\tfrac{1}{2}\tau e_0)\Big)\,\Phi_0\Big(q(+\tfrac{1}{2}\tau e_0)\Big)\Big\rangle_{\beta,e_0}.$$

Then the KMS condition implies that there is a function

$$f = f_x : S_\beta = \{\tau + i\sigma : \tau \in \mathbb{R},\ 0 < \sigma < \beta\} \to \mathbb{C}$$

which is analytic on the open strip, defined and continuous on the closed strip except at the boundary points with $\tau = 0$, such that

$$\lim_{\sigma\to 0}(\varphi(\tau) - f(\tau + i\sigma)) = 0 \quad\text{and}\quad \lim_{\sigma'\to\beta}(\varphi(-\tau) - f(\tau + i\sigma')) = 0\,.$$

Now $\langle\,.\,\rangle_\omega$ be a (sufficiently regular) state for the quantum field Φ_0, and let

$$\psi(\tau) = \psi_x(\tau) = \Big\langle \Phi_0\Big(q_x(-\tfrac{1}{2}\tau e_0)\Big)\,\Phi_0\Big(q_x(\tfrac{1}{2}\tau e_0)\Big)\Big\rangle_\omega.$$

Setting $\boldsymbol{\beta} = \beta e_0$ as before, and taking $D = \{x\}$, i.e. the set containing just the point x, it is not difficult to see that $\langle\,.\rangle_\omega$ is an $[\{x\}, \boldsymbol{\beta}, N]$-LTE state iff there is a function

$$f = f_x : S_\beta = \{\tau + i\sigma : \tau \in \mathbb{R},\ 0 < \sigma < \beta\} \to \mathbb{C}$$

which is analytic on the open strip, defined and continuous on the closed strip except at the boundary points with $\tau = 0$, such that

$$\begin{aligned}&\lim_{\tau\to 0}\partial_\tau^n \lim_{\sigma\to 0}(\psi(\tau) - f(\tau + i\sigma)) = 0 \quad\text{and}\\ &\lim_{\tau\to 0}\partial_\tau^n \lim_{\sigma'\to\beta}(\psi(-\tau) - f(\tau + \sigma')) = 0 \quad (n \le N)\,.\end{aligned}$$

In this sense, the LTE condition appears as a generalization of the KMS condition.

Structure of the set of thermal observables

The linearity of the conditions (3.7) and (3.8) implies that they are also fulfilled for linear combinations of elements in $\mathbf{\Theta}_x^{(N)}$. This means that the LTE property of a state extends to elements of the vector space spanned by $\mathbf{\Theta}_x^{(N)}$. That is of some importance since one can show that $\mathrm{span}(\mathbf{\Theta}_x^{(N)})$ is dense in the set of all thermal observables, its closure containing, e.g., the entropy flux density [7, 6].

Furthermore, the equation of motion of Φ_0 not only provides constraints on the possible functions $\boldsymbol{\beta}(x)$ or ϱ_x which can occur for LTE states, but even determines evolution equations for these — and other — thermal observables in LTE states [7, 6]. Let us indicate this briefly by way of an example. For an LTE state, it holds that the trace of the thermal stress-energy tensor vanishes, $\varepsilon^{\nu}{}_{\nu}(x) = 0$, due to the analogous property for KMS states. On the other hand, from relations between Wick products and their balanced derivatives one obtains the equation $\boldsymbol{\varepsilon}^{\nu}_{\nu}(x) = \partial^{\nu}\partial_{\nu} :\Phi_0^2: (x)$, and therefore, for any LTE state which fulfills the LTE condition on an open domain, the differential equation $\partial^{\nu}\partial_{\nu}\vartheta(x) = 0$ must hold, where $\vartheta(x)$ denotes the expectation values of $\boldsymbol{\vartheta}(x)$, i.e. the Wick square, in the LTE state. This is a differential equation for (a function of) $\boldsymbol{\beta}(x)$ or ϱ_x.

3.2. LTE states in curved spacetime

The concept of LTE states, describing situations which are locally approximately in thermal equilibrium, appears promising for quantum field theory in curved spacetime where global thermal equilibrium in general is not at hand. This applies in particular to early stages in cosmological scenarios where thermodynamical considerations play a central role for estimating processes which determine the evolution and structures of the Universe in later epochs. However, usually in cosmology the thermal stress-energy tensor is identified with the full stress-energy tensor, and it is also customary to use an "instantaneous equilibrium" description of the (thermal) stress-energy tensor. In view of our previous discussion, this can at best be correct in some approximation. Additional difficulties arise from the ambiguities and anomalies affecting the quantized stress-energy tensor, and we will see that such difficulties are also present upon defining LTE states in curved spacetime. Let us see how one may proceed in trying to generalize the LTE concept to curved spacetime, and what problems are met in that attempt.

We assume that we are given a local covariant quantum field $\mathbf{\Phi} = (\Phi_{\boldsymbol{M}})$ with a local covariant stress-energy tensor; we also assume that $\mathbf{\Phi}$ admits global thermal equilibrium states for time symmetries on flat spacetime. For concreteness, we will limit our discussion to the case that our local covariant quantum field is a scalar field with curvature coupling ξ. The parameter ξ can be any real number, but the most important cases are $\xi = 0$ (minimal coupling) or $\xi = 1/6$ (conformal coupling). So $\Phi_{\boldsymbol{M}}$ obeys the field equation

$$(\nabla^a\nabla_a + \xi R + m^2)\Phi_{\boldsymbol{M}} = 0\,, \qquad (3.12)$$

where ∇ is the covariant derivative of $\boldsymbol{M}$ and R the associated scalar curvature, with mass $m \geq 0$.

The central objects in the definition of LTE states in flat spacetime were (i) at each spacetime point x a set (or family of sets) $\boldsymbol{\Theta}_x^{(N)}$ containing "thermal observables" localized at x, and (ii) a set of global thermal equilibrium states (KMS states) serving as "reference states" determining the thermal equilibrium values of elements in $\boldsymbol{\Theta}_x$. In a generic curved spacetime one encounters the problem that there aren't any global thermal equilibrium states (unless the spacetime possesses suitable time symmetries) and thus there are in general no candidates for the requisite reference states of (ii). To circumvent this problem, one can take advantage of having assumed that $\boldsymbol{\Phi}$ is a local covariant quantum field theory. Thus, let $\boldsymbol{M}$ be a globally hyperbolic spacetime, and let $\Phi_{\boldsymbol{M}}$ be the quantum field on $\boldsymbol{M}$ given by $\boldsymbol{\Phi}$. There is also a quantum field Φ_0 on Minkowski spacetime given by $\boldsymbol{\Phi}$. One can use the exponential map $\exp_x$ in $\boldsymbol{M}$ at x to identify $\Phi_{\boldsymbol{M}} \circ \exp_x$ and Φ_0, and thereby one can push forward thermal equilibrium states $\omega_{\boldsymbol{\beta}}$ of Φ_0 to thermal reference states on which to evaluate correlations of $\Phi_{\boldsymbol{M}} \circ \exp_x$ infinitesimally close to x. This way of defining thermal reference states at each spacetime point x in $\boldsymbol{M}$ has been proposed by Buchholz and Schlemmer [8].

As a next step, one needs a generalization of balanced derivatives of the Wick square of $\Phi_{\boldsymbol{M}}$ as elements of the $\boldsymbol{\Theta}_{\boldsymbol{M},x}^{(N)}$, where we used the label $\boldsymbol{M}$ to indicate that the thermal observables are defined with respect to the spacetime $\boldsymbol{M}$. So far, mostly the case $N = 2$ has been considered, which is enough to study the thermal stress-energy tensor. In [32], the following definition was adopted: Let $\omega_{\boldsymbol{M}}$ be a state of $\Phi_{\boldsymbol{M}}$ fulfilling the microlocal spectrum condition (in other words, $\omega_{\boldsymbol{M}}$ is a Hadamard state), and let $w_{\boldsymbol{M}}$ denote the corresponding two-point function. We consider the smooth part $u(x,y) = w_{\boldsymbol{M}}(x,y) - \mathrm{H}_{\boldsymbol{M}}(x,y)$ obtained from the two-point function after subtracting the singular Hadamard parametrix $\mathrm{H}_{\boldsymbol{M}}(x,y)$ as in (2.6). Then we take its symmetric part $u_+(x,y) = \frac{1}{2}(u(x,y) + u(y,x))$ and define the expectation value of the Wick square as the coincidence limit

$$\langle :\Phi_{\boldsymbol{M}}^2: (x)\rangle_{\omega_{\boldsymbol{M}}} = \lim_{y \to x} u_+(x,y)\,. \tag{3.13}$$

Furthermore, we define the second balanced derivative of the Wick square of $\Phi_{\boldsymbol{M}}$ in terms of expectation values as

$$\langle \eth_{ab} :\Phi_{\boldsymbol{M}}^2: (x)\rangle_{\omega_{\boldsymbol{M}}} = \lim_{x' \to x} \Big(\nabla_a \nabla_b - \nabla_a \nabla_{b'} - \nabla_{a'} \nabla_b + \nabla_{a'} \nabla_{b'}\Big) u_+(x,x')\,, \tag{3.14}$$

where on the right-hand side unprimed indices indicate covariant derivatives with respect to x, whereas primed indices indicate covariant derivatives with respect to x'. In [32] it is shown that (3.14) amounts to taking the second derivatives of $u_+(\exp_x(\zeta), \exp_x(-\zeta))$ with respect to ζ, and evaluating at $\zeta = 0$. Note that upon using the symmetric part u_+ of u in defining the Wick product, its first balanced derivative vanishes, as it would on Minkowski spacetime for the normal ordering definition. The definitions (3.13) and (3.14) are referred to as *symmetric Hadamard parametrix subtraction (SHP)* prescription, as in [32]. Note, however, that for the massive case, $m > 0$, the SHP prescription differs from the normal ordering prescription by m-dependent universal constants, a fact which must be taken into account when defining the LTE condition; see [32] for details.

By definition, $\mathbf{\Theta}^{(2)}_{\boldsymbol{M},x}$ is then taken to consist of multiples of the unit operator, $:\Phi^2_{\boldsymbol{M}}:(x)$ and $\eth_{ab}:\Phi^2_{\boldsymbol{M}}:(x)$, defined according to the SHP prescription. With these ingredients in place, one can attempt the definition of LTE states on a curved, globally hyperbolic spacetime $\boldsymbol{M}$.

Definition 3.2. Let $\omega_{\boldsymbol{M}}$ be a state of $\Phi_{\boldsymbol{M}}$ fulfilling the microlocal spectrum condition, and let D be a subset of the spacetime $\boldsymbol{M}$.

(i) We say that $\omega_{\boldsymbol{M}}$ is an LTE state with sharp temperature vector field (timelike, future-directed, and smooth if D is open) $\boldsymbol{\beta} : x \mapsto \boldsymbol{\beta}(x)$ ($x \in D$) of order 2 if

$$\langle :\Phi^2_{\boldsymbol{M}}:(x)\rangle_{\omega_{\boldsymbol{M}}} = \langle :\Phi^2_0:(0)\rangle_{\boldsymbol{\beta}(x)} \quad \text{and} \tag{3.15}$$

$$\langle \eth_{\mu\nu}:\Phi^2_{\boldsymbol{M}}:(x)\rangle_{\omega_{\boldsymbol{M}}} = \langle \eth_{\mu\nu}:\Phi^2_0:(0)\rangle_{\boldsymbol{\beta}(x)} \tag{3.16}$$

hold for all $x \in D$. On the right-hand side there appear the expectation values of the thermal equilibrium state $\langle\,.\,\rangle_{\boldsymbol{\beta}(x)}$ of Φ_0 on Minkowski spacetime, where the vector $\boldsymbol{\beta}(x) \in T_x\boldsymbol{M}$ on the left-hand side is identified with a Minkowski space vector $\boldsymbol{\beta}(x)$ on the right-hand side via the exponential map $\exp_x$, using that $\exp_x(0) = x$, and the coordinates refer to a choice of Lorentz frame at x. On the right-hand side, we have now used the definition of Wick product and its second balanced derivative according to the SHP prescription.

(ii) Let $\varrho : x \mapsto \varrho_x$ ($x \in D$) be a map from D to compactly supported probability measures in V_+ (assumed to be smooth if D is open). We say that $\omega_{\boldsymbol{M}}$ is an LTE state with mixed temperature distribution ϱ of order 2 if

$$\langle :\Phi^2_{\boldsymbol{M}}:(x)\rangle_{\omega_{\boldsymbol{M}}} = \langle :\Phi^2_0:(0)\rangle_{\varrho(x)} \quad \text{and} \tag{3.17}$$

$$\langle \eth_{\mu\nu}:\Phi^2_{\boldsymbol{M}}:(x)\rangle_{\omega_{\boldsymbol{M}}} = \langle \eth_{\mu\nu}:\Phi^2_0:(0)\rangle_{\varrho(x)} \tag{3.18}$$

hold for all $x \in D$. The same conventions as in (i) regarding identification of curved spacetime objects (left-hand sides) and Minkowski space objects (right-hand sides) by $\exp_x$ applies here as well. The definition of the ϱ_x-averaged objects is as in (3.9).

Let us discuss some features of this definition and some first results related to it on generic spacetimes. We will present results pertaining to LTE states on cosmological spacetimes in the next section.

($\boldsymbol{\alpha}$) The definition of thermal observables given here is local covariant since the Wick square and its covariant derivatives are local covariant quantum fields [22]. In particular, if $\mathscr{A}$ is the functor describing our local covariant quantum field $\mathbf{\Phi}$, then for any arrow $\boldsymbol{M} \xrightarrow{\psi} \boldsymbol{N}$ one has $\mathscr{A}(\psi)(\mathbf{\Theta}^{(2)}_{\boldsymbol{M},x}) =$

$\mathbf{\Theta}^{(2)}_{\boldsymbol{N},\psi(x)}$ (by a suitable extension of $\mathscr{A}$, see [22]). Obviously, it is desirable to define thermal observables in curved spacetimes in such a way as to be local covariant. Otherwise, they would depend on some global properties of the particular spacetimes, in contrast to their interpretation as local intensive quantities.

(β) As is the case for the quantized stress-energy tensor, also the thermal observables in $\mathbf{\Theta}^{(2)}_{\boldsymbol{M},x}$, i.e. the Wick square of $\Phi_{\boldsymbol{M}}$ and its second balanced derivative, are subject to renormalization ambiguities and anomalies. Supposing that choices of $:\Phi^{2\,[1]}_{\boldsymbol{M}}:(x)$ and $\eth_{\mu\nu}:\Phi^{2\,[1]}_{\boldsymbol{M}}:(x)$ have been made, where [1] serves as label for the particular choice, one has, in principle, the freedom of redefining these observables by adding suitable quantities depending only on the local curvature of spacetime (so as to preserve local covariance), like

$$
\begin{aligned}
:\Phi^{2\,[2]}_{\boldsymbol{M}}:(x) &= \,:\Phi^{2\,[1]}_{\boldsymbol{M}}:(x) + \mathsf{y}^{[2][1]}_{\boldsymbol{M}}(x)\,,\\
\eth_{\mu\nu}:\Phi^{2\,[2]}_{\boldsymbol{M}}:(x) &= \eth_{\mu\nu}:\Phi^{2\,[1]}_{\boldsymbol{M}}:(x) + Y^{[2][1]}_{\boldsymbol{M}\mu\nu}(x)\,.
\end{aligned}
$$

Therefore, on curved spacetime the thermal interpretation of Wick square and its balanced derivatives depends on the choice one makes here, and it is worth contemplating if there are preferred choices which may restrict the apparent arbitrariness affecting the LTE criterion.

(γ) Similarly as observed towards the end of the previous section, there are differential equations to be fulfilled in order that the LTE condition can be consistent. For the case of the linear (minimally coupled, massless) scalar field, Solveen [33] has noted that the condition of the thermal stress-energy tensor having vanishing trace in LTE states leads to a differential equation of the form

$$
\frac{1}{4}\nabla^a\nabla_a\vartheta(x) + \xi R(x)\vartheta(x) + U(x) = 0\,,
$$

where ϑ is the expectation value of the Wick square in an LTE state, R is the scalar curvature of the underlying spacetime $\boldsymbol{M}$, and U is another function determined by the curvature of $\boldsymbol{M}$. In view of the previous item **(β)**, a redefinition of the Wick square will alter the function $U(x)$ of the differential equation that must be obeyed by $\vartheta(x)$. The consequences of that possibility are yet to be determined. It is worth mentioning that for the Dirac field, which we don't treat in these proceedings, an analogous consistency condition leads to an equation which can only be fulfilled provided that — in this case — the first balanced derivative is defined appropriately, i.e. with addition of a distinct curvature term relative to the SHP definition of the Wick square [24]. We admit that so far we do not fully understand the interplay of the LTE condition and the renormalization ambiguity which is present in the definition of Wick products and their balanced derivatives in curved spacetime, but hope to address some aspects of that interplay in greater detail elsewhere [18].

(δ) Finally we mention that one can prove so-called "quantum energy inequalities", i.e. lower bounds on weighted integrals of the energy density in LTE states along timelike curves (see [14] for a review on quantum energy inequalities); the lower bounds depend on the maximal temperature an LTE state attains along the curve. This holds for a wide range of curvature couplings ξ and all mass parameters m [32]. That is of interest since, while state-independent lower bounds on weighted integrals of the energy density have been established for all Hadamard states of the minimally coupled linear scalar field in generic spacetimes [13], such a result fails in this generality for the non-minimally coupled scalar field [15]. We recommend that the reader takes a look at the references for further information on this circle of questions and their possible relevance regarding the occurrence of singularities in solutions to the semiclassical Einstein equations.

4. LTE states on cosmological spacetimes

As mentioned previously, one of the domains where one can apply the concept of LTE states and also examine its utility is early cosmology. Thus, we review the steps which have been taken, or are currently being taken, in investigating LTE states in cosmological scenarios.

The central premise in standard cosmology is that one considers phenomena at sufficiently large scales such that it is a good approximation to assume that, at each instant of time, the geometry of space (and, in order to be consistent with Einstein's equations, the distribution of matter and energy) is isotropic and homogeneous [42]. Making for simplicity the additional assumption (for which there seems to be good observational motivation) that the geometry of space is flat at each time, one obtains

$$I \times \mathbb{R}^3 , \quad ds^2 = dt^2 - a(t)^2 \Big((dx^1)^2 + (dx^2)^2 + (dx^3)^2 \Big) \tag{4.1}$$

for the general form of spacetime manifold and metric, respectively, in standard cosmology. Here, I is an open interval hosting the time coordinate t, and $a(t)$ is a smooth, strictly positive function called the *scale factor*. With the spacetime geometry of the general form (4.1), the only freedom is the time function $a(t)$, to be determined by Einstein's equation together with a matter model and initial conditions.

The time coordinate in (4.1) is called *cosmological time*. Under suitable (quite general) conditions on $a(t)$ one can pass to a new time coordinate, called *conformal time*,

$$\eta = \eta(t) = \int_{t_0}^{t} \frac{dt'}{a(t')} \, dt'$$

for some choice of t_0. Setting

$$\Omega(\eta) = a(t(\eta)) ,$$

the metric (4.1) takes the form

$$ds^2 = \Omega(\eta)^2\Big(d\eta^2 - (dx^1)^2 - (dx^2)^2 - (dx^3)^2\Big) \tag{4.2}$$

with respect to the conformal time coordinate, so it is conformally equivalent to flat Minkowski spacetime.[2]

4.1. Existence of LTE states at fixed cosmological time

Now let Φ_M be the linear, quantized scalar field on the cosmological spacetime (4.1) for some $a(t)$, or equivalently on $\mathbb{R}^4$ with metric (4.2); we assume arbitrary curvature coupling, as in (3.12). The first question one would like to answer is if there are LTE states of order 2 for Φ_M. As one might imagine, this is a very difficult problem, and it seems that the method employed by Solveen [33] to establish existence of non-trivial LTE states on bounded open regions of Minkowski spacetime cannot be used to obtain an analogous result in curved spacetime. In view of the constraints on the temperature evolution it seems a good starting point to see if there are any second-order LTE states at some fixed cosmological time, i.e. on a Cauchy surface — this would postpone the problem of having to establish solutions to the evolution equations of the temperature distribution. Moreover, to fit into the formalism, such states have to fulfill the microlocal spectrum condition, i.e. they have to be Hadamard states. Allowing general $a(t)$, this problem is much harder than it seems, since if one took an "instantaneous KMS state" at some value of cosmological time — such a state, defined in terms of the Cauchy data formulation of the quantized linear scalar field, appears as a natural candidate for an LTE state at fixed time — then that state is in general not Hadamard if $a(t)$ is time-dependent. The highly non-trivial problem was solved by Schlemmer in his PhD thesis [30]. The result he established is as follows.

Theorem 4.1. *Let t_1 be a value of cosmological time in the interval I, and let $e_0^a = (dt)^a$ be the canonical time vector field of the spacetime (4.1). Then there is some $\beta_1 > 0$ (depending on m, ξ and the behaviour of $a(t)$ near t_1) such that, for each $\beta < \beta_1$, there is a quasifree Hadamard state of Φ_M which is a $[\{t_1\}\times\mathbb{R}^3, \beta e_0^a, 2]$-LTE state (at sharp temperature).*

In other words, there is a second-order LTE state at sharp temperature at fixed cosmological time provided the LTE temperature is high enough. This state will, in general, not preserve the sharp temperature second-order LTE property when evolving it by the field equation (3.12) in time away from the t_1 Cauchy surface. In Schlemmer's thesis, this is illustrated by means of a numerical example. Let us take some $[\{t_1\}\times\mathbb{R}^3, \beta e_0^a, 2]$-LTE state ω, and define

$$\theta(x) = \langle \boldsymbol{\vartheta}(x)\rangle_\omega \quad \text{and} \quad \epsilon_{ab}(x) = \langle \boldsymbol{\varepsilon}_{ab}(x)\rangle_\omega$$

as the expectation values of Wick square and thermal stress-energy tensor in that state. One can, at each x, look for a second-order LTE state $\omega_{\boldsymbol{\beta}(x)}$ at x

[2]For simplicity, we assume here that the range of the conformal time coordinate η is all of $\mathbb{R}$; the variations in the following arguments for the — important — case where this is not so should be fairly obvious.

such that $\omega_{\beta(x)}(\vartheta(x)) = \theta(x)$. Likewise, one can look for second-order LTE states $\omega_{\beta_{\mu\nu}(x)}$ at x with $\boldsymbol{\beta}_{\mu\nu}(x) = \beta_{\mu\nu}(x)e_0^a$ and $\omega_{\boldsymbol{\beta}_{\mu\nu}(x)}(\boldsymbol{\varepsilon}_{\mu\nu}(x)) = \epsilon_{\mu\nu}(x)$ (note that there is no sum over the indices) with respect to a tetrad basis at x containing e_0^a. If ω is itself a second-order LTE state at sharp temperature, then all the numbers $\beta(x)$ and $\beta_{\mu\nu}(x)$ must coincide, or rather the associated absolute temperatures $\mathrm{T}(x) = 1/k\beta(x)$ and $\mathrm{T}_{\mu\nu}(x) = 1/k\beta_{\mu\nu}(x)$. Otherwise, the mutual deviation of these numbers can be taken as a measure for the failure of ω to be a sharp temperature, second-order LTE state at x. Schlemmer has investigated such a case, choosing $\xi = 0.1$ and $m = 1.5$ in natural units, and $a(t) = \mathrm{e}^{Ht}$ with $H = 1.3$; he constructed a spatially isotropic and homogeneous state ω which is second-order LTE at conformal time $\eta = \eta_1 = -1.0$, and calculated, as just described, the "would-be LTE" comparison temperatures $\mathrm{T}(x) \equiv \mathrm{T}(\eta)$ and $\mathrm{T}_{\mu\nu}(x) \equiv \mathrm{T}_{\mu\nu}(\eta)$ numerically for earlier and later conformal times. The result is depicted in Figure 2.

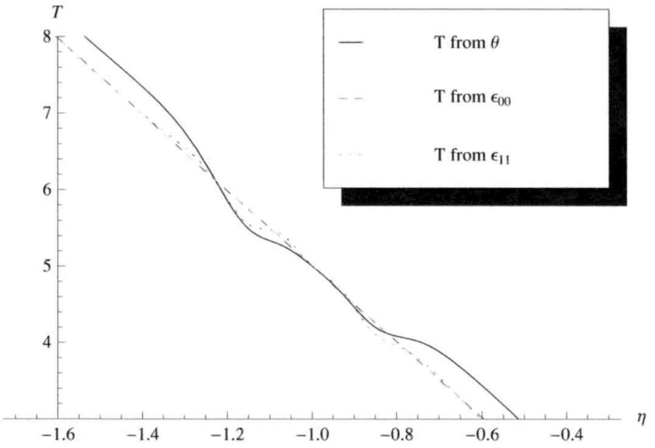

FIGURE 2. LTE comparison temperatures calculated from $\theta(\eta)$ and $\epsilon_{\mu\nu}(\eta)$ for a state constructed as second-order LTE state at $\eta = -1.0$ [30]

Figure 2 shows that the comparison temperatures deviate from each other away from $\eta = \eta_1 = 1.0$. However, they all drop with increasing η as they should for an increasing scale factor. Both the absolute and relative temperature deviations are larger at low temperatures than at high temperatures. This can be taken as an indication of a general effect, namely that the LTE property is more stable against perturbations (which drive the system away from LTE) at high temperature than at low temperature.

4.2. LTE and metric scattering

Let us turn to a situation demonstrating that effect more clearly. We consider the quantized, conformally coupled ($\xi = 1/6$) linear scalar field Φ_M on $\mathbb{R}^4$ with conformally flat metric of the form (4.2), where $\Omega(\eta)$ is chosen with

$$\Omega(\eta)^2 = \lambda + \frac{\Delta\lambda}{2}(1 + \tanh(\rho\eta)), \tag{4.3}$$

where λ, $\Delta\lambda$ and ρ are real positive constants. One can show that the quantum field Φ_M has asymptotic limits as $\eta \to \pm\infty$; in a slightly sloppy notation,

$$\lim_{\eta' \to \mp\infty} \Phi_M(\eta + \eta', \mathbf{x}) = \begin{cases} \Phi_{\text{in}}(\eta, \mathbf{x}) \\ \Phi_{\text{out}}(\eta, \mathbf{x}) \end{cases} \qquad (\eta \in \mathbb{R},\ \mathbf{x} = (x^1, x^2, x^3) \in \mathbb{R}^3)\,,$$

where Φ_{in} and Φ_{out} are copies of the quantized linear scalar field with mass m on Minkowski spacetime, however with different (constant) scale factors. This means that Φ_{in} and Φ_{out} obey the field equations

$$(\lambda\Box + m^2)\Phi_{\text{in}} = 0\,, \qquad ((\lambda + \Delta\lambda)\Box + m^2)\Phi_{\text{out}} = 0\,,$$

where $\Box = \partial_\eta^2 - \partial_{x^1}^2 - \partial_{x^2}^2 - \partial_{x^2}^2$. If ω_{in} is a state of Φ_{in}, it induces a state ω_{out} of Φ_{out} by setting

$$\langle \Phi_{\text{out}}(\eta_1, \mathbf{x}_1) \cdots \Phi_{\text{out}}(\eta_n, \mathbf{x}_n) \rangle_{\omega_{\text{out}}} = \langle \Phi_{\text{in}}(\eta_1, \mathbf{x}_1) \cdots \Phi_{\text{in}}(\eta_n, \mathbf{x}_n) \rangle_{\omega_{\text{in}}}\,.$$

Thus, ω_{out} is the state obtained from ω_{in} through the scattering process the quantum field Φ_M undergoes by propagating on a spacetime with the expanding conformal scale factor as in (4.3). This particular form of the scale factor has the advantage that the scattering transformation taking Φ_{in} into Φ_{out} can be calculated explicitly, and so can the relation between ω_{in} and ω_{out} [2]. It is known that if ω_{in} is a Hadamard state, then so is ω_{out}.

In a forthcoming paper [25], we pose the following question: If we take ω_{in} as a global thermal equilibrium state (with respect to the time coordinate η) of Φ_{in}, is ω_{out} an LTE state of second order for Φ_{out}? The answers, which we will present in considerably more detail in [25], can be roughly summarized as follows.

(1) For a wide range of parameters $m, \lambda, \Delta\lambda$ and ρ, and inverse temperature β_{in} of ω_{in}, ω_{out} is a mixed temperature LTE state (with respect to the time direction of η) of second order.

(2) Numerical calculations of the comparison temperatures T and $\mathrm{T}_{\mu\nu}$ from θ and $\epsilon_{\mu\nu}$ for ω_{out} show that they mutually deviate (so ω_{out} is no sharp temperature LTE state). Numerically one can show that the deviations decrease with increasing temperature T_{in} of ω_{in} (depending, of course, on the other parameters $m, \lambda, \Delta\lambda$ and ρ).

This shows that scattering of the quantum field by the expanding spacetime metric tends to drag the initially global thermal state away from equilibrium. However, this effect is the smaller the higher the initial temperature, as one would expect intuitively. As the initial temperature goes to zero, ω_{in} approaches the vacuum state. In this case, ω_{out} is a non-vacuum state due to quantum particle creation induced by the time-varying spacetime metric which is non-thermal [41]. That may appear surprising in view of the close analogy of the particle-creation-by-metric-scattering effect to the Hawking effect. However, it should be noted that the notion of temperature in the context of the Hawking effect is different form the temperature definition entering the LTE condition. In the framework of quantum fields on de Sitter spacetime, Buchholz and Schlemmer [8] noted that the KMS temperature of a state with respect to a Killing flow differs, in general, from the LTE temperature. To give a basic example, we note that the vacuum state of a quantum field theory in Minkowski spacetime is a KMS state at non-zero temperature with respect to the Killing flow of the Lorentz boosts along a fixed spatial direction (that's the main assertion of the Bisognano-Wichmann theorem, see [19] and references cited there). On the other hand, the vacuum state is also an LTE state of infinite order, but at zero temperature. It appears that the question how to modify the LTE condition such that it might become sensitive to the Hawking temperature has not been discussed so far.

4.3. LTE temperature in the Dappiaggi-Fredenhagen-Pinamonti cosmological approach

In this last part of our article, we turn to an application of the LTE concept in an approach to cosmology which takes as its starting point the semiclassical Einstein equation

$$G_{\boldsymbol{M}ab}(x) = 8\pi G \langle \mathsf{T}_{\boldsymbol{M}ab}(x) \rangle_{\omega_{\boldsymbol{M}}} \,. \tag{4.4}$$

On the left-hand side, we have the Einstein tensor of the spacetime geometry $\boldsymbol{M}$, on the right-hand side the expectation value of the stress-energy tensor of $\Phi_{\boldsymbol{M}}$ which is part of a local covariant quantum field $\boldsymbol{M}$ in a state $\omega_{\boldsymbol{M}}$. Adopting the standing assumptions of standard cosmology, it seems a fair ansatz to assume that a solution can be obtained for an $\boldsymbol{M}$ of the spatially flat Friedmann-Robertson-Walker form (4.1) while taking for $\Phi_{\boldsymbol{M}}$ the quantized linear conformally coupled scalar field whose field equation is

$$\left(\nabla^a \nabla_a + \tfrac{1}{6} R + m^2\right) \Phi_{\boldsymbol{M}} = 0 \,. \tag{4.5}$$

And, in fact, in a formidable work, Pinamonti [26] has recently shown that this assumption is justified: Adopting the metric ansatz (4.1) for $\boldsymbol{M}$, he could establish that, for any given $m \geq 0$, there are an open interval of cosmological time, a scale factor $a(t)$ and a Hadamard state $\omega_{\boldsymbol{M}}$ such that (4.4) holds for the t-values in the said interval (identifying $x = (t, x^1, x^2, x^3)$).

However, we will simplify things here considerably by setting $m = 0$ from now on, so that we have a conformally covariant quantized linear scalar field with field equation

$$\left(\nabla^a \nabla_a + \tfrac{1}{6} R\right) \Phi_{\boldsymbol{M}} = 0 \,. \tag{4.6}$$

As pointed out above in Sec. 2, in the case of conformal covariant $\Phi_{\boldsymbol{M}}$ one has to face the trace anomaly

$$\mathsf{T}_{\boldsymbol{M}}{}^a{}_a(x) = -4 Q_{\boldsymbol{M}}(x) \mathbf{1} \,, \tag{4.7}$$

where $Q_{\boldsymbol{M}}$ is the divergence-compensating term of (2.7), a quantity which is determined by the geometry of $\boldsymbol{M}$ and which hence is state-independent, so (4.7) is to be read as an equation at the level of operators. However, equation (4.7) is not entirely complete as it stands, since it doesn't make explicit that $\mathsf{T}_{\boldsymbol{M}ab}(x)$ is subject to a renormalization ambiguity which lies in the freedom of adding divergence-free, local covariant tensor fields $C_{\boldsymbol{M}ab}(x)$. Therefore, also the trace of the renormalized stress-energy tensor is subject to such an ambiguity. Let us now be a bit more specific about this point. Suppose we define, in a first step, a renormalized $\tilde{\mathsf{T}}_{\boldsymbol{M}ab}(x)$ by the SHP renormalization prescription. Then define

$$\mathsf{T}_{\boldsymbol{M}ab}(x) = \mathsf{T}^{[0]}_{\boldsymbol{M}ab}(x) = \tilde{\mathsf{T}}_{\boldsymbol{M}ab}(x) - Q_{\boldsymbol{M}}(x)g_{ab}(x)\,,$$

where $Q_{\boldsymbol{M}}(x)$ is the divergence-compensating term defined with respect to $\tilde{\mathsf{T}}_{\boldsymbol{M}ab}(x)$. Now, if $\mathsf{T}_{\boldsymbol{M}ab}(x)$ is re-defined as

$$\mathsf{T}^{[C]}_{\boldsymbol{M}ab}(x) = \mathsf{T}^{[0]}_{\boldsymbol{M}ab}(x) + C_{\boldsymbol{M}ab}(x)$$

with $C_{\boldsymbol{M}ab}(x)$ local covariant and divergence-free, one obtains for the trace

$$\mathsf{T}^{[C]a}_{\boldsymbol{M}\ \ a}(x) = \big(-4Q_{\boldsymbol{M}}(x) + C^{a}_{\boldsymbol{M}a}(x)\big)\mathbf{1}\,. \tag{4.8}$$

For generic choice of $C_{\boldsymbol{M}ab}$, the right-hand side of (4.8) contains derivatives of the spacetime metric of higher than second order. However, Dappiaggi, Fredenhagen and Pinamonti [9] have pointed out that one can make a specific choice of $C_{\boldsymbol{M}ab}$ such that the right-hand side of (4.8) contains only derivatives of the metric up to second order. Considering again the semiclassical Einstein equation (4.4) and taking traces on both sides, one gets

$$R_{\boldsymbol{M}}(x) = 8\pi G\big(-4Q_{\boldsymbol{M}}(x) + C^{a}_{\boldsymbol{M}a}(x)\big)\,, \tag{4.9}$$

which (in the case of a conformally covariant $\Phi_{\boldsymbol{M}}$ considered here) determines the spacetime geometry independent of the choice of a state $\omega_{\boldsymbol{M}}$ (after supplying also initial conditions). With our metric ansatz (4.1), equation (4.9) is equivalent to a non-linear differential equation for $a(t)$, of the general form

$$F\big(a^{(n)}(t), a^{(n-1)}(t), \ldots, a^{(1)}(t), a(t)\big) = 0\,, \tag{4.10}$$

with $a^{(j)}(t) = (d^j/dt^j)a(t)$ and with n denoting the highest derivative order in the differential equation for $a(t)$. For generic choices of $C_{\boldsymbol{M}ab}$, n turns out to be greater than 2. On the other hand, if $C_{\boldsymbol{M}ab}$ is specifically chosen so that the trace $\mathsf{T}^{[C]a}_{\boldsymbol{M}\ \ a}$ contains only derivatives up to second order of the spacetime metric, then $n = 2$. Making or not making this choice has drastic consequences for the behaviour of solutions $a(t)$ to (4.10). The case of generic $C_{\boldsymbol{M}ab}$, leading to $n \geq 3$ in (4.10), was investigated in a famous article by Starobinski [34]. He showed that, in this case, the differential equation (4.10) for $a(t)$ has solutions with an inflationary, or accelerating ($a^{(2)}(t) > 0$), phase which is unstable at small time scales after the initial conditions. This provided a natural argument why the inflationary phase of early cosmology would end after a very short timespan. However, recent astronomical observations have shown accelerating phases of the Universe over a large timescale at late cosmic times. The popular explanation for this phenomenon postulates an exotic form of energy to be present in the Universe, termed "Dark Energy" [42]. In contrast, Dappiaggi, Fredenhagen and Pinamonti have shown that the specific choice of $C_{\boldsymbol{M}ab}$ leads to a differential equation (4.10) for $a(t)$ with $n = 2$ which admits stable solutions with a long-term accelerating phase at late cosmological times [9]. In a recent work [10], it was investigated if such solutions could account for the observations of the recently observed accelerated cosmic expansion. Although that issue remains so far undecided, also in view of the fact that linear quantized fields are certainly too simple to describe the full physics of quantum effects in cosmology, it bears the interesting possibility that "cosmic acceleration" and "Dark Energy" could actually be traced back to the renormalization ambiguities and anomalies arising in the definition of the stress-energy tensor of quantum fields in the presence of spacetime curvature. In that light, one can take the point of view that the renormalization ambiguity of $\mathsf{T}_{\boldsymbol{M}ab}$ is ultimately constrained (and maybe fixed) by the behaviour of solutions to the semiclassical Einstein equation: One would be inclined to prefer such $C_{\boldsymbol{M}ab}$ which lead to stable solutions, in spite of the argument of [34]. At any rate, only those $C_{\boldsymbol{M}ab}$ can be considered which lead to solutions to the semiclassical Einstein equations compatible with observational data. After all, the renormalization freedom in quantum field theoretic models of elementary particle physics is fixed in a very similar way.

The differential equation with stable solutions $a(t)$ derived in [9] can be expressed in terms of the following differential equation for the Hubble function $H(t) = a^{(1)}(t)/a(t)$,

$$\dot{H}(H^2 - H_0) = -H^4 + 2H_0H^2\,, \tag{4.11}$$

where H_0 is some universal positive constant and the dot means differentiation with respect to t. There are two constant solutions to (4.11) (obviously $H(t) = 0$ is a solution) as well as non-constant solutions depending on initial conditions. For the non-constant solutions, one finds an asymptotic behaviour as follows [9, 17]: For early cosmological times,

$$H(t) \approx \frac{1}{t - t_0}\,, \qquad a(t) \approx \Gamma(t - t_0)\,, \tag{4.12}$$

with some constants $\Gamma > 0$ and real t_0, for $t > t_0$. On the other hand, for late cosmological times:

$$H(t) \approx \sqrt{2}H_0 \coth\left(2\sqrt{2}H_0t - 1\right). \tag{4.13}$$

Therefore, $H(t)$ and $a(t)$ have, for early cosmological times, a singularity as $t \to t_0$, but different from the behaviour of a Universe filled with classical radiation, which is known to yield [39, 42]

$$H_{\rm rad}(t) \approx \frac{1}{2(t - t_0)}\,, \qquad a_{\rm rad}(t) \approx \Gamma'\sqrt{t - t_0}$$

with some positive constant Γ'. For the temperature behaviour of radiation close to the singularity at $t \to t_0$ one then obtains

$$\mathrm{T}_{\rm rad}(t) \approx \frac{\kappa'}{\sqrt{t - t_0}}$$

with another constant κ'.

Now we wish to compare this to the temperature behaviour, as $t \to t_0$, of any second-order, sharp temperature LTE state $\omega_{\boldsymbol{M}}$ of $\Phi_{\boldsymbol{M}}$ fulfilling the semiclassical Einstein equation in the Dappiaggi-Fredenhagen-Pinamonti approach for $t \to t_0$. Of course, the existence of such LTE states is an assumption, and in view of the results of Subsection 4.1, cf. Figure 2, this assumption is certainly an over-idealization. On the other hand, Figure 2 can also be interpreted as saying that, at early cosmic times, while sharp temperature LTE states might possibly not exist, it is still meaningful to attribute an approximate temperature behaviour to states as $t \to t_0$.

In order to determine the temperature behaviour of the assumed second-order LTE state ω_M, one observes first that [32]

$$\langle \mathsf{T}^{[C]}_{ab}(x)\rangle_{\omega_M} = \varepsilon_{ab}(x) + \frac{1}{12}\nabla^a\nabla_b\vartheta(x) + -\frac{1}{3}g_{ab}(x)\varepsilon^c{}_c(x) + q_M(x)\vartheta(x) + K^{[C]}_{Mab}(x) \tag{4.14}$$

with some state-independent tensor $K^{[C]}_{Mab}$ which depends on C_{Mab} and which has local covariant dependence of the spacetime geometry M; likewise so has q_M. Both $K^{[C]}_{Mab}$ and q_M can be explicitly calculated once C_{Mab} and, consequently, $a(t)$ (respectively, $H(t)$) are specified. Then, $\varepsilon_{ab}(x)$ and $\vartheta(x)$ are functions of $\beta(t)$ (making the usual assumption that ω_M is homogeneous and isotropic). To determine $\beta(t)$, relation (4.14) is plugged into the vanishing-of-divergence equation

$$\nabla^a\langle \mathsf{T}^{[C]}_{ab}(x)\rangle_{\omega_M} = 0\,,$$

thus yielding a non-linear differential equation involving $\beta(t)$ and $a(t)$. Inserting any $a(t)$ coming from the non-constant Dappiaggi-Fredenhagen-Pinamonti solutions, one can derive the behaviour, as $t \to t_0$, of $\beta(t)$, and it turns out [17] that $\beta(t) \approx \gamma a(t)$ with another constant $\gamma > 0$. This resembles the behaviour of classical radiation. But in view of the different behaviour of $a(t)$ as compared to early cosmology of a radiation-dominated Universe, one now obtains a temperature behaviour

$$\mathrm{T}(t) \approx \frac{\kappa}{t - t_0}\,,$$

with yet another constant $\kappa > 0$; this is more singular than $\mathrm{T}_{\mathrm{rad}}(t)$ in the limit $t \to t_0$. We will present a considerably more detailed analysis of the temperature behaviour of LTE states in the context of the Dappiaggi-Fredenhagen-Pinamonti cosmological model elsewhere [18].

5. Summary and outlook

The LTE concept allows it to describe situations in quantum field theory where states are no longer in global thermal equilibrium, but still possess, locally, thermodynamic parameters. In curved spacetime, this concept is intriguingly interlaced with local covariance and the renormalization ambiguities which enter into its very definition via Wick products and their balanced derivatives. Particularly when considering the semiclassical Einstein equations in a cosmological context this plays a role, and the thermodynamic properties of quantum fields in the very early stages of cosmology can turn out to be different from what is usually assumed in considerations based on, e.g., modelling matter as classical radiation. The implications of these possibilities for theories of cosmology remain yet to be explored — so far we have only scratched the tip of an iceberg, or so it seems.

There are several related developments concerning the thermodynamic behaviour of quantum fields in curved spacetime, and in cosmological spacetimes in particular, which we haven't touched upon in the main body of the text. Worth mentioning in this context is the work by Hollands and Leiler on a derivation of the Boltzmann equation in quantum field theory [21]. It should be very interesting to try and explore relations between their approach and the LTE concept. There are also other concepts of approximate thermal equilibrium states [11], and again, the relation to the LTE concept should render interesting new insights. In all, the new light that these recent developments shed on quantum field theory in early cosmology is clearly conceptually fruitful and challenging.

References

[1] C. Bär, N. Ginoux and F. Pfäffle, *Wave equations on Lorentzian manifolds and quantization*, European Mathematical Society, Zürich, 2007

[2] N.D. Birrell and P.C.W. Davies, *Quantum fields in curved space*, Cambridge University Press, Cambridge, 1982

[3] N.N. Bogoliubov and D.V. Shirkov, *Introduction to the theory of quantized fields*, Wiley-Interscience, New York, 1959

[4] R. Brunetti, K. Fredenhagen and M. Köhler, *The microlocal spectrum condition and Wick polynomials of free fields on curved space-times*, Commun. Math. Phys. **180** (1996) 633

[5] R. Brunetti, K. Fredenhagen and R. Verch, *The generally covariant locality principle — a new paradigm for local quantum field theory*, Commun. Math. Phys. **237** (2003) 31

[6] D. Buchholz, *On hot bangs and the arrow of time in relativistic quantum field theory*, Commun. Math. Phys. **237** (2003) 271

[7] D. Buchholz, I. Ojima and H.-J. Roos, *Thermodynamic properties of nonequilibrium states in quantum field theory*, Annals Phys. **297** (2002) 219

[8] D. Buchholz and J. Schlemmer, *Local temperature in curved spacetime*, Class. Quant. Grav. **24** (2007) F25

[9] C. Dappiaggi, K. Fredenhagen and N. Pinamonti, *Stable cosmological models driven by a free quantum scalar field*, Phys. Rev. **D77** (2008) 104015

[10] C. Dappiaggi, T.-P. Hack, J. Möller and N. Pinamonti, *Dark energy from quantum matter*, e-Print: arXiv:1007.5009 [astro-ph.CO]

[11] C. Dappiaggi, T.-P. Hack and N. Pinamonti, *Approximate KMS states for scalar and spinor fields in Friedmann-Robertson-Walker spacetimes*, e-Print: arXiv:1009.5179 [gr-qc]

[12] C.J. Fewster, *On the notion of "the same physics in all spacetimes"*, in these proceedings, e-Print: arXiv:1105.6202 [math-ph]

[13] C.J. Fewster, *A general worldline quantum inequality*, Class. Quantum Grav. **17** (2000) 1897

[14] C.J. Fewster, *Energy inequalities in quantum field theory*, in: J. Zambrini (ed.), *Proceedings of the XIV International Congress on Mathematical Physics, Lisbon, 2003*, World Scientific, Singapore, 2005. e-Print: arXiv:math-ph/0501073

[15] C.J. Fewster and L. Osterbrink, *Quantum energy inequalities for the non-minimally coupled scalar field*, J. Phys. **A 41** (2008) 025402

[16] C.J. Fewster and C.J. Smith, *Absolute quantum energy inequalities in curved spacetime*, Ann. Henri Poincaré **9** (2008) 425

[17] M. Gransee, Diploma Thesis, Institute of Theoretical Physics, University of Leipzig, 2010

[18] A. Knospe, M. Gransee and R. Verch, in preparation

[19] R. Haag, *Local quantum physics*, 2nd ed., Springer-Verlag, Berlin-Heidelberg-New York, 1996

[20] S. Hollands, *Renormalized quantum Yang-Mills fields in curved spacetime*, Rev. Math. Phys. **20** (2008) 1033

[21] S. Hollands and G. Leiler, *On the derivation of the Boltzmann equation in quantum field theory: flat spacetime*, e-Print: arXiv:1003.1621 [cond-mat.stat-mech]

[22] S. Hollands and R.M. Wald, *Local Wick polynomials and time ordered products of quantum fields in curved space-time*, Commun. Math. Phys. **223** (2001) 289

[23] Kay, B.S. and Wald, R.M., *Theorems on the uniqueness and thermal properties of stationary, nonsingular, quasifree states on space-times with a bifurcate Killing horizon*, Phys. Rep. **207** (1991) 49

[24] A. Knospe, Diploma Thesis, Institute of Theoretical Physics, University of Leipzig, 2010

[25] F. Lindner and R. Verch, to be published

[26] N. Pinamonti, *On the initial conditions and solutions of the semiclassical Einstein equations in a cosmological scenario*, e-Print: arXiv:1001.0864 [gr-qc]

[27] N. Pinamonti and R. Verch, in preparation

[28] M.J. Radzikowski, *Micro-local approach to the Hadamard condition in quantum field theory on curved space-time*, Commun. Math. Phys. **179** (1996) 529

[29] K. Sanders, *Equivalence of the (generalised) Hadamard and microlocal spectrum condition for (generalised) free fields in curved spacetime*, Commun. Math. Phys. **295** (2010) 485

[30] J. Schlemmer, PhD Thesis, Faculty of Physics, University of Leipzig, 2010

[31] J. Schlemmer, *Local thermal equilibrium states and Unruh detectors in quantum field theory*, e-Print: hep-th/0702096 [hep-th]

[32] J. Schlemmer and R. Verch, *Local thermal equilibrium states and quantum energy inequalities*, Ann. Henri Poincaré **9** (2008) 945

[33] C. Solveen, *Local thermal equilibrium in quantum field theory on flat and curved spacetimes*, Class. Quant. Grav. **27** (2010) 235002

[34] A. Starobinski, *A new type of isotropic cosmological models without singularity*, Phys. Lett. **B91** (1980) 99

[35] A. Strohmaier, R. Verch and M. Wollenberg, *Microlocal analysis of quantum fields on curved spacetimes: analytic wavefront sets and Reeh-Schlieder theorems*, J. Math. Phys. **43** (2002) 5514

[36] R. Verch, *A spin-statistics theorem for quantum fields on curved spacetime manifolds in a generally covariant framework*, Commun. Math. Phys. **223** (2001) 261

[37] R.M. Wald, *The back reaction effect in particle creation in curved spacetime*, Commun. Math. Phys. **54** (1977) 1

[38] R.M. Wald, *Trace anomaly of a conformally invariant quantum field in curved space-time*, Phys. Rev. **D17** (1978) 1477

[39] R.M. Wald, *General relativity*, University of Chicago Press, Chicago, 1984

[40] R.M. Wald, *Quantum field theory in curved spacetime and black hole thermodynamics*, University of Chicago Press, Chicago, 1994

[41] R.M. Wald, *Existence of the S matrix in quantum field theory in curved spacetime*, Ann. Physics (N.Y.) **118** (1979) 490

[42] S. Weinberg, *Cosmology*, Oxford University Press, Oxford, 2008

Rainer Verch
Institut für Theoretische Physik
Universität Leipzig
Vor dem Hospitaltore 1
D-04103 Leipzig
Germany
e-mail: `verch@itp.uni-leipzig.de`

Shape Dynamics. An Introduction

Julian Barbour

Abstract. Shape dynamics is a completely background-independent universal framework of dynamical theories from which all absolute elements have been eliminated. For particles, only the variables that describe the shapes of the instantaneous particle configurations are dynamical. In the case of Riemannian three-geometries, the only dynamical variables are the parts of the metric that determine angles. The local scale factor plays no role. This leads to a shape-dynamic theory of gravity in which the four-dimensional diffeomorphism invariance of general relativity is replaced by three-dimensional diffeomorphism invariance and three-dimensional conformal invariance. Despite this difference of symmetry groups, it is remarkable that the predictions of the two theories – shape dynamics and general relativity – agree on spacetime foliations by hypersurfaces of constant mean extrinsic curvature. However, the two theories are distinct, with shape dynamics having a much more restrictive set of solutions. There are indications that the symmetry group of shape dynamics makes it more amenable to quantization and thus to the creation of quantum gravity. This introduction presents in simple terms the arguments for shape dynamics, its implementation techniques, and a survey of existing results.

Mathematics Subject Classification (2010). 70G75, 70H45, 83C05, 83C45.

Keywords. Gravity theory, conformal invariance, Mach's principle.

1. Introduction

One of Einstein's main aims in creating general relativity was to implement Mach's idea [1, 2] that dynamics should use only relative quantities and that inertial motion as expressed in Newton's first law should arise, not as an effect of a background absolute space, but from the dynamical effect of the universe as a whole. Einstein called this *Mach's principle* [3]. However, as he explained later [4, 5] (p. 186), he found it impractical to realize Mach's principle directly and was forced to use coordinate systems. This has obscured the extent to which and how general relativity is a background-independent theory. My aim in this paper is to present a universal framework for the *direct* and *explicit* creation of completely background-independent theories.

I shall show that this leads to a theory of gravity, *shape dynamics*, that is distinct from general relativity because it is based on a different symmetry group, according to which only the local shapes of Riemannian 3-geometries are dynamical. Nevertheless, it is remarkable that the two theories have a nontrivial 'intersection', agreeing exactly in spatially closed universes whenever and wherever Einsteinian spacetimes admit foliation by hypersurfaces of constant mean extrinsic curvature. However, many solutions of general relativity that appear manifestly unphysical, such as those with closed timelike curves, are not allowed in shape dynamics. In addition, it appears that the structure of shape dynamics makes it significantly more amenable to quantization than general relativity.

This is not the only reason why I hope the reader will take an interest in shape dynamics. The question of whether motion is absolute or relative has a venerable history [6, 7], going back to long before Newton made it famous when he formulated dynamics in terms of absolute space and time [8]. What is ultimately at stake is the definition of position and, above all, velocity. This has abiding relevance in our restless universe. I shall show that it is possible to eliminate every vestige of Newtonian absolutes except for just one. But this solitary remnant is hugely important: it allows the universe to expand. Shape dynamics highlights this remarkable fact.

This introduction will be to a large degree heuristic and based on Lagrangian formalism. A more rigorous Hamiltonian formulation of shape dynamics better suited to calculations and quantum-gravity applications was recently discovered in [9] (a simplified treatment is in [10]). Several more papers developing the Hamiltonian formulation in directions that appear promising from the quantum-gravity perspective are in preparation. A dedicated website (shapedynamics.org) is under construction; further background information can be found at platonia.com.

The contents list (see the start of this volume) obviates any further introduction, but a word on terminology will help. Two distinct meanings of *relative* are often confused. Mach regarded inter-particle separations as relative quantities; in Einstein's theories, the division of spacetime into space and time is made relative to an observer's coordinate system. To avoid confusion, I use *relational* in lieu of Mach's notion of relative.

2. The relational critique of Newton's dynamics

2.1. Elimination of redundant structure

Newton's First Law states: "Every body continues in its state of rest or uniform motion in a right line unless it is compelled to change that state by forces impressed on it." Since the (absolute) space in which the body's motion is said to be straight and the (absolute) time that measures its uniformity are both invisible, this law as stated is clearly problematic. Newton knew this and argued in his Scholium in the *Principia* [8] that his invisible absolute motions could be deduced from visible relative motions. This can be done but requires more relative data than one would expect if only directly observable initial data governed the dynamics. As we shall see, this fact, which is not widely known, indicates how mechanics can be reformulated with less kinematic structure than Newton assumed and simultaneously be made more predictive. It is possible to create a framework that fully resolves the debate about the nature of motion. In this framework, *the fewest possible observable initial data determine the observable evolution.*[1]

I show first that all candidate relational configurations[2] of the universe have structures determined by a Lie group, which may be termed their *structure group*. The existence of such a group is decisive. It leads directly to a natural way to achieve the aim just formulated and to a characteristic universal structure of dynamics applicable to a large class of systems. It is present in modern gauge theories and, in its most perfect form, in general relativity. However, the relational core of these theories is largely hidden because their formulation retains redundant kinematic structure.

To identify the mismatch that shape dynamics aims to eliminate, the first step is to establish the essential structure that Newtonian dynamics employs. It will be sufficient to consider N, $N \geq 3$, point particles interacting through Newtonian gravity. In an assumed inertial frame of reference, each particle a, $a = 1, ..., N$, has coordinates $x_a^i(t)$, $i = x, y, z$, that depend on t, the Newtonian time. The x_a^i's and t are all assumed to be observable. The particles, assumed individually identifiable, also have constant masses m_a. For the purposes of our discussion, they can be assumed known.

Let us now eliminate potentially redundant structure. Newton granted that only the inter-particle separations r_{ab}, assumed to be 'seen' all at once, are observable. In fact, this presupposes an external (absolute) ruler. Closer to empirical reality are the dimensionless ratios

$$\tilde{r}_{ab} := \frac{r_{ab}}{R_{rmh}}, \qquad R_{rmh} := \sqrt{\sum_{a<b} r_{ab}^2}, \tag{1}$$

where R_{rmh} is the root-mean-harmonic separation. It is closely related to the centre-of-mass moment of inertia I_{cms}:

$$I_{cms} := \sum_a m_a \mathbf{x}^a \cdot \mathbf{x}^a \equiv \frac{1}{M} \sum_{a<b} m_a m_b r_{ab}^2, \quad M := \sum_a m_a. \tag{2}$$

The system has the 'size' $\sqrt{I_{cms}}$ if we grant a scale, but we do not and take the instantaneous sets $\{\tilde{r}_{ab}\}$ of scale-free ratios $\tilde{r}_{ab}$ to be our raw data. They are 'snapshots' of the *instantaneous shapes* of the system. The time t too is unobservable. There is no clock hung up in space, just the particles moving

[1] The notion of what is observable is not unproblematic. For now it will suffice that inter-particle separations are more readily observed than positions in invisible space.

[2] We shall see (Sec. 4) that the foundation of dynamics on instantaneous extended configurations, rather than point events, is perfectly compatible with Einsteinian relativity.

relative to each other. All that we have are the sets $\{\tilde{r}_{ab}\}$. The totality of such sets is *shape space* Q_{ss}^N, which only exists for $N \geq 3$.[3] The number of dimensions of Q_{ss}^N is $3N - 7$: from the $3N$ Cartesian coordinates, six are subtracted because Euclidean translations and rotations do not change the r_{ab}'s and the seventh because the $\{\tilde{r}_{ab}\}$'s are scale invariant.

Shape space is our key concept. Mathematically, we reach it through a succession of spaces, the first being the $3N$-dimensional *Cartesian configuration space* Q^N. In it, all configurations that are carried into each other by translations t in T, the group of Euclidean translations, belong to a common *orbit* of T. Thus, T decomposes Q^N into its group orbits, which are defined to be the points of the $(3N - 3)$-dimensional *quotient space* $\mathrm{T}^N := \mathrm{Q}^N/\mathsf{T}$. This first quotienting to T^N is relatively trivial. More significant is the further quotienting by the rotation group R to the $(3N-6)$-dimensional *relative configuration space* $\mathrm{Q}_{rcs}^N := \mathrm{Q}^N/\mathsf{TR}$ [11]. The final quotienting by the dilatation (scaling) group S leads to shape space $\mathrm{Q}_{ss}^N := \mathrm{Q}^N/\mathsf{TRS}$ [12]. The groups T and R together form the *Euclidean group*, while the inclusion of S yields the *similarity group*. The orbit of a group is a space with as many dimensions as the number of elements that specify a group element. The orbits of S thus have seven dimensions (Fig. 1).

The groups $\mathsf{T}, \mathsf{R}, \mathsf{S}$ are *groups of motion*, or *Lie groups* (groups that are simultaneously manifolds, i.e., their elements are parametrized by continuous parameters). If we have a configuration q of N particles in Euclidean space, $q \in Q^N$, we can 'move it around' with T or R or 'change its size' with S. This intuition was the basis of Lie's work. It formalizes the fundamental geometrical notions of *congruence* and *similarity*. Two figures are *congruent* if they can be brought to exact overlap by a combination of translations and rotations and *similar* if dilatations are allowed as well.

Relational particle dynamics can be formulated in any of the quotient spaces just considered. Intuition suggests that the dynamics of an 'island universe' in Euclidean space should deal solely with its possible shapes. The similarity group is then the fundamental structure group.[4] This leads to particle shape dynamics and by analogy to the conformal geometrodynamics that will be considered in the second part of the paper.

Lie groups and their infinite-dimensional generalizations are fundamental in modern mathematics and theoretical physics. They play a dual role in shape dynamics, first in indicating how potentially redundant structure can be pared away and, second, in providing the tool to create theories that are relationally perfect, i.e., free of the mismatch noted above. Moreover, because

---

[3] A single point is not a shape, and the distance between two particles can be scaled to any value, so nothing dimensionless remains to define a shape. Also the configuration in which all particles coincide is not a shape and does not belong to shape space.

[4] One might want to go further and consider the general linear group, under which angles are no longer invariant. I will consider this possibility later.

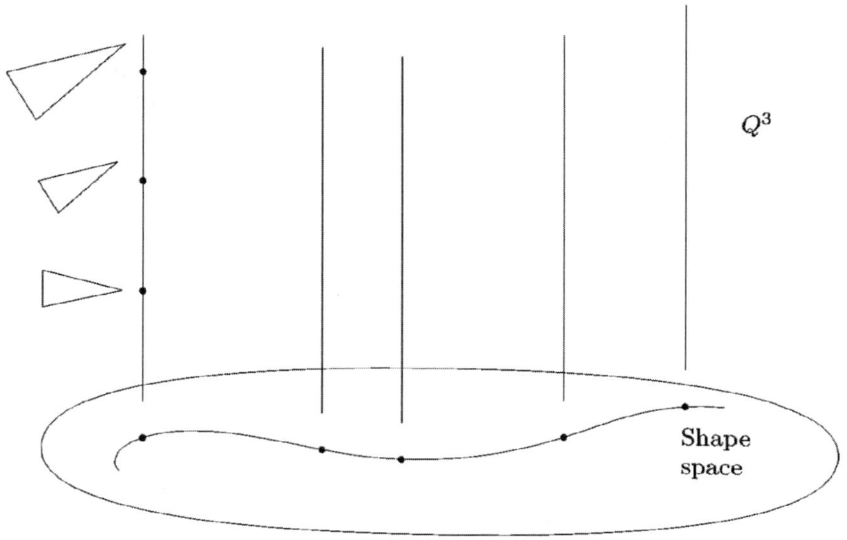

FIGURE 1. Shape space for the 3-body problem is obtained by decomposing the Newtonian configuration space Q^3 into orbits of the similarity group S. The points on any given vertical line (any group orbit) correspond to all possible representations in Euclidean space of one of the possible shapes of the triangle formed by the three particles. Each such shape is represented below its orbit as a point in shape space. Each orbit is actually a seven-dimensional space. The effects of rotation and scaling are shown.

Lie groups, as groups that are simultaneously manifolds, have a common underlying structure and are ubiquitous, they permit essentially identical methods to be applied in many different situations. This is why shape dynamics is a universal framework.

2.2. Newtonian dynamics in shape space

We now identify the role that absolute space and time play in Newtonian dynamics by *projection to shape space.* We have removed structure from the q's in Q^N, reducing them to points $s \in \mathrm{Q}^N_{ss}$. This is projection of q's. We can also project complete Newtonian histories $q(t)$. To include time at the start, we adjoin to Q^N the space T of absolute times t, obtaining the space $\mathrm{Q}^N\mathrm{T}$. Newtonian histories are then (monotonically rising) continuous curves in $\mathrm{Q}^N\mathrm{T}$. However, clocks are parts of the universe; there is no external clock available to provide the reading for the T axis. All the objective information is carried by the successive configurations of the universe. We must therefore remove the T axis and, in the first projection, label the points representing the configurations in Q^N by an arbitrary increasing parameter λ and then make the further projection to the shape space Q^N_{ss}. The history becomes $s(\lambda)$

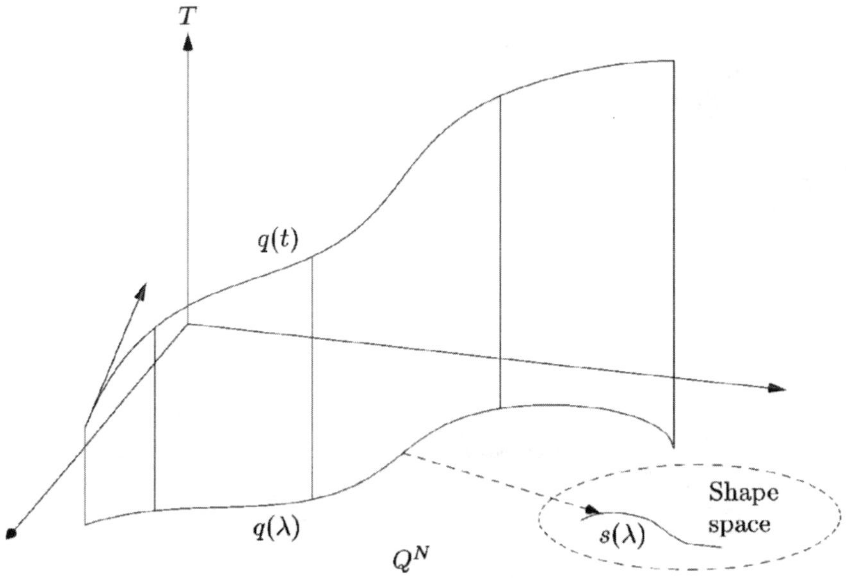

FIGURE 2. In Newtonian dynamics, the history of a system is a monotonically rising curve $q(t)$ in $Q^N T$ or a curve $q(\lambda)$ in Q^N labelled by a monotonic λ. The objective observable history is the projected curve $s(\lambda)$ in shape space Q_{ss}^N.

(Fig. 2). A history is the next most fundamental concept in shape dynamics. There is no 'moving now' in this concept. History is not a spot moving along $s(\lambda)$, lighting up 'nows' as it goes. It is the curve; λ merely labels its points. Newtonian dynamics being time-reversal invariant, there is no past-to-future direction on curves in Q_{ss}^N.

Given a history of shapes s, we can define a shape *velocity*. Suppose first that in fact by some means we can define a distinguished parameter p, or *independent variable*, along a suitably continuous curve in Q_{ss}^N. Then at each point along the curve we have a shape s and its (multi-component) velocity $\mathrm{d}s/\mathrm{d}p$. This is a *tangent vector* to the curve. If we have no p but only an arbitrary λ, we can still define shape velocities $\mathrm{d}s/\mathrm{d}\lambda = s'$, but all we really have is the direction d (in Q_{ss}^N) in which s is changing. The difference between tangent vectors and directions associated with curves in shape space will be important later.

We can now identify the mismatch that, when eliminated, leads to the shape-dynamic ideal. To this end, we recall Laplacian determinism in Newtonian dynamics: given q and $\dot{q}$ at some instant, the evolution of the system is uniquely determined (the particle masses and the force law assumed known). The question is this: given the corresponding shape projections s and d, is the evolution *in shape space* Q_{ss}^N uniquely determined? The answer is no for a purely geometrical reason. The fact is that certain initial velocities which are objectively significant in Newtonian dynamics can be generated by purely

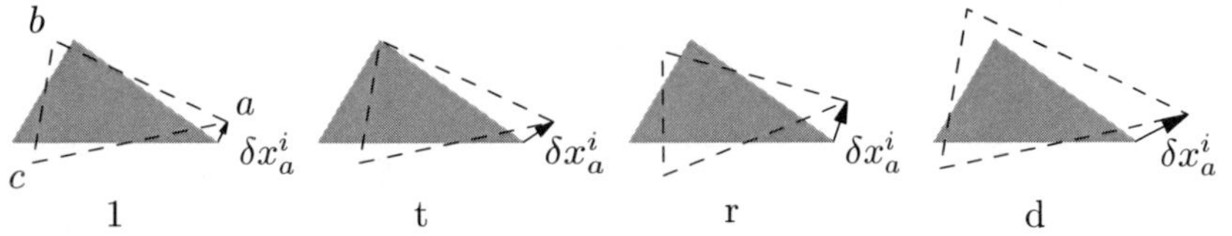

FIGURE 3. The two triangles of slightly different shapes formed by three particles a, b, c define a point s and direction d *uniquely* in shape space, but changes to an original placing (1) in Euclidean space of the dashed triangle relative to the grey one generated by translations (t), rotations (r), and dilatations (d) give rise to different Newtonian initial velocities $\mathrm{d}x_a^i/\mathrm{d}t$.

group actions. To be precise, different Newtonian velocities can be generated from identical data in Q_{ss}^N. This is illustrated for the 3-body problem (in two dimensions) in Fig. 3.

I will not go into the details of the proof (see [13]), but in a Newtonian N-body system the velocities at any given instant can be uniquely decomposed into parts due to an intrinsic change of shape and three further parts due to the three different group actions – translations, rotations, and dilatations – applied as in Fig. 3. These actions are obviously 'invisible' in the shape-space s and d, which define only the shape and the way it is changing.

By Galilean relativity, translations of the system have no effect in Q_{ss}^N. We can ignore them but not rotations and dilatations. Four dimensionless dynamically effective quantities are associated with them. First, two angles determine the direction in space of a rotation axis. Second, from the kinetic energies associated with rotation, T_r, dilatation T_d, and change of shape, T_s, we can form two dimensionless ratios, which it is natural to take to be T_r/T_s and T_d/T_s (since change of shape, represented by T_s, is our 'gold standard'). Thus, the kinematic action of the Lie groups generates *four* parameters that affect the histories in shape space without changing the initial s and d. This is already so for pure inertial motion. If forces are present, there is a fifth parameter, the ratio T/V of the system's kinetic energy T to its potential energy V, that is dynamically significant but is also invisible in the s and d in shape space.

We now see that although Newtonian dynamics seems wonderfully rational and transparent when expressed in an inertial frame of reference, it does not possess perfect Laplacian determinism in shape space. This failure appears especially odd if N is large. Choose some coordinates $s_i, i = 1, 2, ..., 3N-7$, in shape space and take one of them, call it τ, as a surrogate for Newton's t. If only shapes had dynamical effect, then by analogy with inertial-frame Newtonian dynamics, the initial values of $\tau, s_i, \mathrm{d}s_i/\mathrm{d}\tau, i = 1..., 3N-8$ would fix the evolution. They do not.

Five more data are needed and must be taken from among the second derivatives $\mathrm{d}^2 s_i/\mathrm{d}\tau^2$. Moreover, no matter how large N, say a million as in a globular cluster, we always need just five.[5] They make no sense from the shape-dynamic perspective. Poincaré, writing as a philosopher deeply committed to relationalism, found the need for them repugnant [14, 15]. But, in the face of the manifest presence of angular momentum in the solar system, he resigned himself to the fact that there is more to dynamics than, literally, meets the eye.[6] In fact, the extra $\mathrm{d}^2 s_i/\mathrm{d}\tau^2$'s *are* explained by Newton's assumption of an all-controlling but invisible frame for dynamics. They are the evidence, and the sole evidence at that, for absolute space.

For some reason, Poincaré did not consider Mach's suggestion [1, 2] that the universe in its totality might somehow determine the structure of the dynamics observed locally. Indeed, the universe exhibits evidence for angular momentum in innumerable localized systems but none overall. This suggests that, regarded as a closed dynamical system, it has no angular momentum and meets the *Poincaré principle*: either a point and direction (strong form) or a point and a tangent vector (weak form) in the universe's shape space determine its evolution. The stronger form of the principle will hold if the universe satisfies a *geodesic principle* in Q^N_{ss}, since a point and a direction are the initial conditions for a geodesic. The need for the two options, either of which may serve as the definition of Mach's principle [16], will be clarified in the next section.

To summarize: on the basis of Poincaré's analysis and intuition, we would like the universe to satisfy Laplacian determinism in its shape space and not merely in a special frame of reference that employs kinematic structure not present in shape space.

3. The universal structure of shape dynamics

3.1. The elimination of time

In standard dynamical theory, the time t is an independent variable supplied by an external clock. But any clock is a mechanical system. If we wish to treat the universe as a single system, the issue of what clock, if any, to use becomes critical. In fact, it is not necessary to use any clock.

This can be demonstrated already in Q^N. We simply proceed without a clock. Histories of the system are then just curves in Q^N, and we seek a law that determines them. An obvious possibility is to define a metric on Q^N and require histories to be geodesics with respect to it.

---

[5]If $N = 3$ or 4, there are insufficient $\mathrm{d}^2 s_i/\mathrm{d}\tau^2$'s and we need higher derivatives too.

[6]Poincaré's penetrating analysis, on which this subsection is based, only takes into account the role of angular momentum in the 'failure' of Newtonian dynamics when expressed in relational quantities. Despite its precision and clarity, it has been almost totally ignored in the discussion of the absolute vs relative debate in dynamics.

A metric is readily found because the Euclidean geometry of space that defines Q^N in the first place also defines a natural metric on Q^N:

$$\mathrm{d}s_{kin} = \sqrt{\sum_a \frac{m_a}{2} \mathrm{d}\mathbf{x}_a \cdot \mathrm{d}\mathbf{x}_a}. \tag{3}$$

This is called the *kinetic metric* [17]; division of $\mathrm{d}\mathbf{x}_a$ by an external $\mathrm{d}t$ transforms the radicand into the Newtonian kinetic energy. We may call (3) a *supermetric*. We shall see how it enables us to exploit structure defined at the level of Q^N at the shape-space level.

We can generate further such supermetrics from (3) by multiplying its radicand by a function on Q^N, for example $\sum_{a<b} m_a m_b / r_{ab}$. We obtain a whole family of geodesic principles defined by the variational requirement

$$\delta I = 0, \quad I = 2 \int \mathrm{d}\lambda \sqrt{(E - V(q)) T_{kin}}, \quad T_{kin} := \frac{1}{2} \sum_a m_a \frac{\mathrm{d}\mathbf{x}_a}{\mathrm{d}\lambda} \cdot \frac{\mathrm{d}\mathbf{x}_a}{\mathrm{d}\lambda}, \tag{4}$$

where λ is a curve parameter, the 2 is for convenience and, since a constant is a function on Q^N, the constant E reflects its possible presence.

The Euler–Lagrange equations that follow from (4) are

$$\frac{\mathrm{d}}{\mathrm{d}\lambda} \left(\sqrt{\frac{E - V}{T_{kin}}} m_a \frac{\mathrm{d}\mathbf{x}}{\mathrm{d}\lambda} \right) = -\sqrt{\frac{T_{kin}}{E - V}} \frac{\partial V}{\partial \mathbf{x}_a}. \tag{5}$$

This equation simplifies if we choose the freely specifiable λ such that

$$E - V = T_{kin}. \tag{6}$$

If we denote this λ by t, then (5) becomes Newton's second law and (6) becomes the energy theorem. However, in our initially timeless context it becomes the definition of an emergent time, or better *duration*, created by a geodesic principle. In fact, the entire objective content of Newtonian dynamics for a closed system is recovered. It is illuminating to give the explicit expression for the increment of this emergent duration:

$$\delta t = \sqrt{\frac{\sum_a m_a \delta \mathbf{x}_a \cdot \delta \mathbf{x}_a}{2(E - V)}}. \tag{7}$$

This is the first example of the holism of relational dynamics: the time that we take to flow locally everywhere is a distillation of all the changes everywhere in the universe. Since everything in the universe interacts with everything else, every difference must be taken into account to obtain the *exact* measure of time. The universe is its own clock.

The definition of duration through (7) is unique (up to origin and unit) if clocks are to have any utility. Since we use them to keep appointments, they are useless unless they march in step. This leads unambiguously to (7) as the only sensible definition. For suppose an island universe contains within it subsystems that are isolated in the Newtonian sense. We want to use the motions within each to generate a time signal. The resulting signals must all march in step with each other. Now this will happen if, for each system, the signal is generated using (7). The reason is important. Suppose we used only the numerators in (7) to measure time; then subsystems without interactions would generate time signals that march in step, but with interactions one system may be sinking into its potential well as another is rising out of its. Then the 'time' generated by the former will pass faster than the latter's. However the denominators in (7) correct this automatically since $E - V$ increases or decreases with T. Time must be measured by some motion, but for generic systems only the time label that ensures conservation of the energy can meet the marching-in-step criterion. Duration is defined as uniquely as entropy is through the logarithm of probability.

In textbooks, (4) is derived as Jacobi's principle [17] and used to determine the dynamical orbit of systems in Q^N (as, for example, a planet's orbit, which is not to be confused with a group orbit). The speed in orbit is then determined from (6) regarded as the energy theorem. The derivation above provides the deeper interpretation of (6) in a closed system. It is the definition of time. Note that time is eliminated from the initial kinematics by *a square root* in the Lagrangian. This pattern will be repeated in more refined relational settings below, in which we can address the question of what potentials V are allowed in relational dynamics.

A final comment. Time has always appeared elusive. It is represented in dynamics as the real line R^1. Instants are mere points on the line, each identical to the other. This violates the principle that things can be distinguished only by differences. There must be variety. In relational dynamics R^1 is redundant and there are only configurations, but they double as instants of time. The need for variety is met.

3.2. Best matching

The next step is to determine curves in shape space Q^N_{ss} that satisfy the strong or weak form of the Poincaré principle. As already noted, the strong form, with which we begin, will be satisfied by geodesics with respect to a metric defined on Q^N_{ss}. For this, given two nearly identical shapes, s_1, s_2, i.e., neighbouring points in Q^N_{ss}, we need to define a 'distance' between them based on their difference and nothing else. Once again we use the Euclidean geometry that underlies both Q^N and Q^N_{ss}.

Shape s_1 in Q^N_{ss} has infinitely many representations in Q^N: all of the points on its group orbit in Q^N. Pick one with coordinates $\mathbf{x}^1_a$. Pick a nearby point on the orbit of s_2 with coordinates $\mathbf{x}^2_a$. In Newtonian dynamics, the coordinate differences $\mathrm{d}\mathbf{x}_a = \mathbf{x}^2_a - \mathbf{x}^1_a$ are physical displacements, but in shape dynamics they mix physical difference of shape with spurious difference due to the arbitrary positioning of s_1 and s_2 on their orbits. To obtain a measure of the shape difference, hold s_1 fixed in Q^N and move s_2 around in its orbit,[7] for the moment using only Euclidean translations and rotations. This changes

[7] Recall that a group orbit is generically a multi-dimensional space.

$\mathrm{d}\mathbf{x}_a = \mathbf{x}_a^2 - \mathbf{x}_a^1$ and simultaneously

$$\mathrm{d}s_{trial} := \sqrt{(E-V)\sum_a \frac{m_a}{2}\mathrm{d}\mathbf{x}_a \cdot \mathrm{d}\mathbf{x}_a}. \tag{8}$$

Since (8) is positive definite and defines a nonsingular metric on Q^N, it will be possible to move shape s_2 into the unique position in its orbit at which (8) is minimized (for given position of s_1). This unique position can be characterized in two equivalent ways: 1) Shape s_2 has been moved to the position in which it most closely 'covers' s_1, i.e., the two shapes, which are incongruent, have been brought as close as possible to congruence, as measured by (8). This is the *best-matched* position. 2) The $3N$-dimensional vector joining s_1 and s_2 in their orbits in Q^N is *orthogonal* to the orbits. This is true in the first place for the kinetic metric, for which $E - V = 1$, but also for all choices of $E - V$. For each, the best-matched position is the same but there is a different best-matched 'distance' between s_1 and s_2:

$$\mathrm{d}s_{bm} := \text{min of } \sqrt{(E-V)\sum_a \frac{m_a}{2}\mathrm{d}\mathbf{x}_a \cdot \mathrm{d}\mathbf{x}_a} \text{ between orbits} \tag{9}$$

Because orthogonality of two vectors can only be established if all components of both vectors are known, best matching introduces a further degree of holism into relational physics. The two ways of conceptualizing best matching are shown in Fig. 4 for the 3-body problem in two dimensions.

It is important that the orthogonal separation (9) is the same at all points within the orbits of either s_1 or s_2. This is because the metric (8) on Q^N is *equivariant*: if the same group transformations are applied to the configurations in Q^N that represent s_1 and s_2, the value of (8) is unchanged. In differential-geometric terms, equivariance is present because the translation and rotation group orbits are Killing vectors of the kinetic metric in Q^N. The equivariance property only holds if $E - V$ satisfies definite conditions, which I have tacitly assumed so far but shall spell out soon.

In fact, it is already lost if we attempt to include dilational best matching with respect to the kinetic metric in order to determine a 'distance' between shapes rather than only relative configurations as hitherto. For suppose we represent two shapes by configurations of given sizes in Q^N and find their best-matched separation d_{bm} using Euclidean translations and rotations. We obtain some value for d_{bm}. If we now change the scale of one of the shapes, d_{bm} must change because the kinetic metric has dimensions $m^{1/2}l$ and scales too. To correct for this in a natural way, we can divide the kinetic metric by the square root of I_{cms}, the centre-of-mass moment of inertia (2), and then best match to get the inter-shape distance

$$\mathrm{d}s_{sbm} := \text{min of } \sqrt{I_{cms}^{-1}\sum_a m_a \mathrm{d}\mathbf{x}_a \cdot \mathrm{d}\mathbf{x}_a} \text{ between orbits} \tag{10}$$

As it must be, $\mathrm{d}s_{sbm}$ is dimensionless and defines a metric on Q_{ss}^N. It is precisely such a metric that we need in order to implement Poincaré's principle.

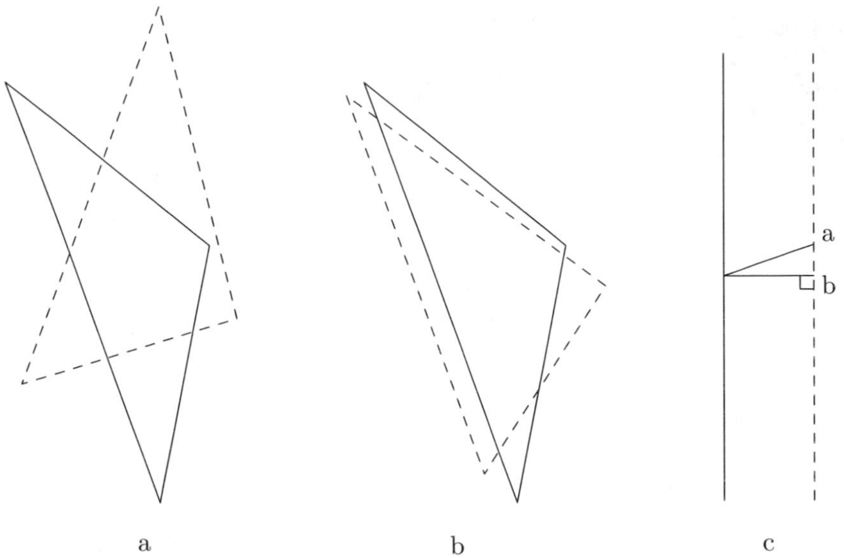

FIGURE 4. *a*) An arbitrary placing of the dashed triangle relative to the undashed triangle; b) the best-matched placing reached by translational and rotational minimization of (8); *c*) the two positions of the triangle configurations on their group orbits in Q^N. The connecting 'strut' is orthogonal with respect to the supermetric on Q^N in the best-matched position. Best matching brings the centres of mass to coincidence and reduces the net rotation to zero.

Terminologically, it will be convenient to call directions in Q^N that lie entirely in group orbits *vertical* and the best-matched orthogonal directions *horizontal.* Readers familiar with fibre bundles will recognize this terminology. A paper presenting best-matching theory *ab initio* in terms of fibre bundles is in preparation.

3.3. The best-matched action principle

We can now implement the strong Poincaré principle. We calculate in Q^N, but the reality unfolds in Q^N_{ss}. The task is this: given two shapes s_a and s_b in Q^N_{ss}, find the geodesic that joins them. The distance along the trial curves between s_a and s_b is to be calculated using the best-matched metric (10) found in Q^N and then 'projected' down to Q^N_{ss}. The projected metric is unique because the best-matching metric in Q^N is equivariant.

The action principle in Q^N has the form

$$\delta I_{bm} = 0, \quad I_{bm} = 2\int \mathrm{d}\lambda\sqrt{WT_{bm}}, \quad T_{bm} = \frac{1}{2}\sum_a \frac{\mathrm{d}\mathbf{x}_a^{bm}}{\mathrm{d}\lambda}\cdot\frac{\mathrm{d}\mathbf{x}_a^{bm}}{\mathrm{d}\lambda}, \tag{11}$$

where $\mathrm{d}\mathbf{x}_a^{bm}/\mathrm{d}\lambda$ is the limit of $\delta\mathbf{x}_a^{bm}/\delta\lambda$ as $\mathrm{d}\lambda \to 0$, and the potential-type term W must be such that equivariance holds. In writing the action in this way, I have taken a short cut. Expressed properly [11], I_{bm} contains the generators of the various group transformations, and the variation with respect to them leads to the best-matched velocities $\mathrm{d}\mathbf{x}_a^{bm}/\mathrm{d}\lambda$. It is assumed in (11) that this variation has already been done.

The action (11) is interpreted as follows. One first fixes a trial curve in Q_{ss}^N between s_a and s_b and represents it by a trial curve in Q^N through the orbits of the shapes in the Q_{ss}^N trial curve. The Q^N trial curve must never 'run vertically'. It may run orthogonally to the orbits, and this is just what we want. For if it does, the $\delta\mathbf{x}_a$ that connect the orbits are best matched. It is these $\delta\mathbf{x}_a^{bm}$, dependent only on the shape differences, that are to determine the action.

To make the trial curve in Q^N orthogonal, we divide it into infinitesimal segments between adjacent orbits $1, 2, 3, ..., m$ (orbits 1 and m are s_a and s_b, respectively). We hold the initial point of segment 1–2 fixed and move the other end into the horizontal best-matched position on orbit 2. We then move the original 2–3 segment into the horizontal with its end 2 coincident with the end of the adjusted 1–2 segment. We do this all the way to the s_b orbit. Making the segment lengths tend to zero, we obtain a smooth horizontal curve. Because the Q^N metric is equivariant, this curve is not unique – its initial point can be moved 'vertically' to any other point on the initial orbit; all the other points on the curve are then moved vertically by the same amount. If M is the dimension of the best-matching group ($M = 7$ for the similarity group), we obtain an M-parameter family of horizontal best-matched curves that all yield a common unique value for the action along the trial curve in Q_{ss}^N. This is illustrated in Fig. 5.

In this way we obtain the action for all trial curves in shape space between s_a and s_b. The best-matching construction ensures that the action depends only on the shapes that are explored by the trial curves and nothing else. It remains to find which trial curve yields the shortest distance between s_a and s_b. This requires us to vary the trial curve in Q_{ss}^N, which of course changes the associated trial curves in Q^N, which, when best matched, give different values for the best-matched action. When we find the (in general unique) curve for which the shape-space action is stationary, we have found the solution that satisfies the strong Poincaré principle. Theories satisfying only the weak principle arise when the equivariance condition imposed on W in (11) is somewhat relaxed, as we shall now see.

3.4. Best-matching constraints and consistency

To obtain a definite representation in the above picture of best matching, we must refer the initial shape s_1 to a particular Cartesian coordinate system with a definite choice of scale. This 'places' shape s_1 at some position on its group orbit in Q^N. If we now place the next, nearly identical shape s_2 on its orbit close to the position chosen for s_1 on its orbit but not in the best-matched position, we obtain certain coordinate differences $\delta\mathbf{x}_a = \mathbf{x}_a^2 - \mathbf{x}_a^1$.

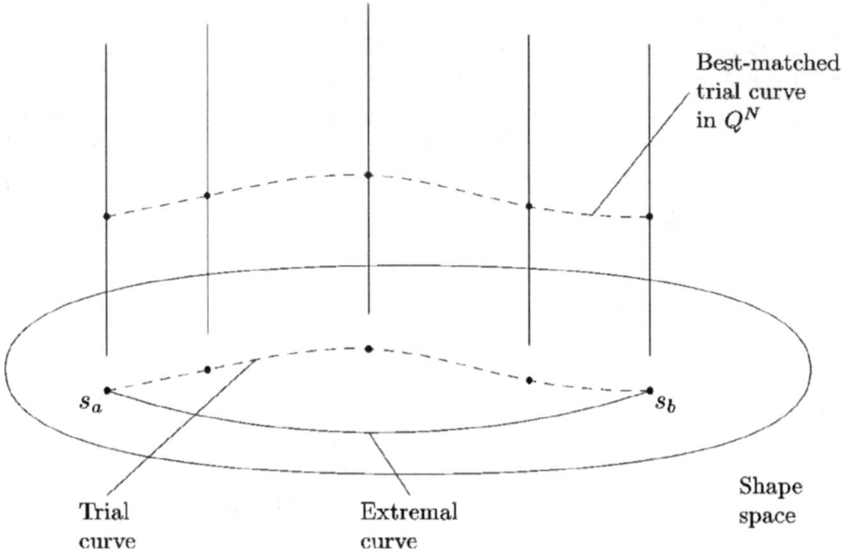

FIGURE 5. The action associated with each trial curve between shapes s_a and s_b in shape space is calculated by finding a best-matched curve in Q^N that runs through the group orbits 'above' the trial curve. The best-matched curve is determined uniquely apart from 'vertical lifting' by the same amount in each orbit, which does not change the best-matched action I_{bm}. The trial curve in shape space for which I_{bm} is extremal is the desired curve in shape space.

Dividing these by a nominal δt, we obtain velocities from which, in Newtonian terms, we can calculate a total momentum $\mathbf{P} = \sum_a m_a \dot{\mathbf{x}}_a$, angular momentum $\mathbf{L} = \sum_a m_a \mathbf{x}_a \times \dot{\mathbf{x}}_a$ and rate of change of I: $\dot{I} = D = 2\sum_a m_a \mathbf{x}_a \cdot \dot{\mathbf{x}}_a$. We can change their values by acting on s_2 with translations, rotations, and dilatations respectively. Indeed, it is intuitively obvious that by choosing these group transformations appropriately we can ensure that

$$\mathbf{P} = 0, \tag{12}$$

$$\mathbf{L} = 0, \tag{13}$$

$$D = 0. \tag{14}$$

It is also intuitively obvious that the fulfilment of these conditions is precisely the indication that the best-matching position has been reached.

Let us now stand back and take an overall view. *The* reality in shape dynamics is simply a curve in Q^N_{ss}, which we can imagine traversed in either direction. There is no rate of change of shapes, just their succession. The only convenient way to represent this succession is in Q^N. However, any one curve in Q^N_{ss}, denote it C_{ss}, is represented by an infinite set $\{C^{Q^N}_{ss}\}$ of curves in Q^N. They all pass through the orbits of the shapes in C_{ss}, within which the $\{C_{ss}^{Q^N}\}$ curves can run anywhere. Prior to the introduction of the best-matching dynamics, all the curves $\{C_{ss}^{Q^N}\}$ are equivalent representations of C_{ss} and no curve parametrization is privileged.

Best matching changes this by singling out curves in the set $\{C_{ss}^{Q^N}\}$ that 'run horizontally'. They are *distinguished representations*, uniquely determined by the best matching up to a seven-parameter freedom of position in one nominally chosen initial shape in its orbit. There is also a distinguished curve parametrization (Sec. 3.1), uniquely fixed up to its origin and unit. When speaking of the distinguished representation, I shall henceforth mean that the curves in Q^N and their parametrization have both been chosen in the distinguished form (modulo the residual freedoms).

Let us now consider how the dynamics that actually unfolds in Q_{ss}^N is seen to unfold in the distinguished representation. From the form of the action (11), knowing that Newton's second law can be recovered from Jacobi's principle by choosing the distinguished curve parameter using (7), we see that we shall recover Newton's second law exactly. We derive not only Newton's dynamics but also the frame and time in which it holds (Fig. 6). There is a further bonus, for the best-matching dynamics is more predictive: the conditions (12), (13) and (14) must hold at any initial point that we choose and be maintained subsequently. Such conditions that depend only on the initial data (but not accelerations) and must be maintained (propagated) are called *constraints*. This is the important topic treated by Dirac [18].

Since the dynamics in the distinguished representation is governed by Newton's second law, we need to establish the conditions under which it will propagate the constraints (12), (13) and (14). In fact, we have to impose conditions on the potential term W in (11). If (12) is to propagate, W must be a function of the coordinate differences $\mathbf{x}_a - \mathbf{x}_b$; if (13) is to propagate, W can depend on only the inter-particle separations r_{ab}. These are both standard conditions in Newtonian dynamics, in which they are usually attributed to the homogeneity and isotropy of space. Here they ensure consistency of best matching wrt the Euclidean group. Propagation of (14) introduces a novel element. It requires W to be homogeneous with length dimension l^{-2}. This requirement is immediately obvious in (11) from the length dimension l^2 of the kinetic term, which the potential must balance out. Note that in this case a constant E, corresponding to a nonzero energy of the system, cannot appear in W. The system must, in Newtonian terms, have total energy zero. However, potentials with dimension l^{-2} are virtually never considered in Newtonian dynamics because they do not appear to be realized in nature.[8] I shall discuss this issue in the next subsection after some general remarks.

[8]It is in fact possible to recover Newtonian gravitational and electrostatic forces exactly from l^{-2} potentials by dividing the l^{-1} Newtonian potentials by the square root of the moment of inertia I_{cms}. This is because I_{cms} is dynamically conserved and is effectively absorbed into the gravitational constant G and charge values. However, the presence of I_{cms} in the action leads to an additional force that has the form of a time-dependent 'cosmological constant' and ensures that I_{cms} remains constant. See [12] for details.

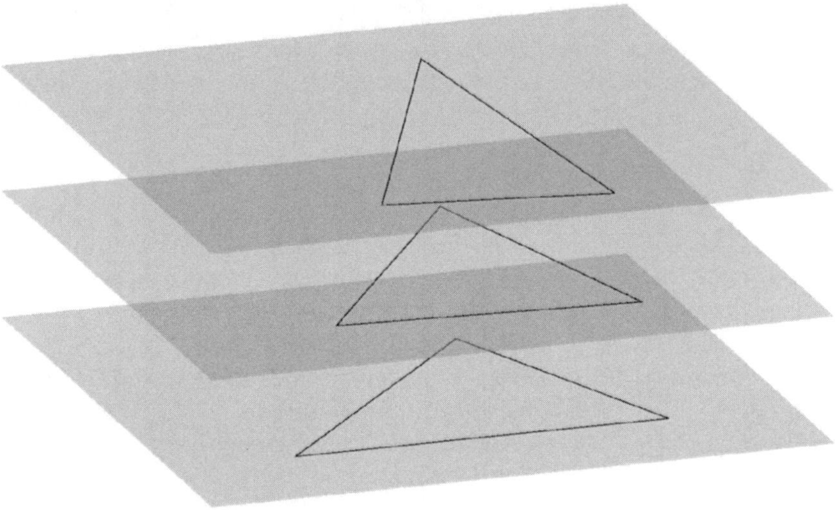

FIGURE 6. The distinguished representation of best-matched shape dynamics for the 3-body problem. For the initial shape, one chooses an arbitrary position in Euclidean space. Each successive shape is placed on its predecessor in the best-matched position ('horizontal stacking'). The 'vertical' separation is chosen in accordance with the distinguished curve parameter t determined by the condition (7). In the framework thus created, the particles behave exactly as Newtonian particles in an inertial frame of reference with total momentum and angular momentum zero.

Best matching is a process that determines a metric on Q_{ss}^N. For this, three things are needed: a supermetric on Q^N, best matching to find the orthogonal inter-orbit separations determined by it, and the equivariance property that ensures identity of them at all positions on the orbits. Nature gives us the metric of Euclidean space, and hence the supermetric on Q^N; the second and third requirements arise from the desire to implement Poincaré type dynamics in Q_{ss}^N. The orbit orthogonality, leading to the constraints (12), (13), and (14), distinguishes best-matched dynamics from Newtonian theory, which imposes no such requirements. Moreover, the constraint propagation needed for consistency of best matching enforces symmetries of the potential that in Newtonian theory have to be taken as facts additional to the basic structure of the theory.

It is important that the constraints (12), (13), and (14) apply only to the 'island universe' of the complete N-body system. Subsystems within it that are isolated from each other, i.e., exert negligible forces on each other, can perfectly well have nonvanishing values of $\mathbf{P}, \mathbf{L}, D$. It is merely necessary that their values for all of the subsystems add up to zero. However, the consistency conditions imposed on the form of the potential must be maintained at the level of the subsystems.

We see that shape dynamics has several advantages over Newtonian dynamics. The two forms of dynamics have Euclidean space in common, but shape dynamics derives all of Newton's additional kinematic structure: absolute space (inertial frame of reference), the metric of time (duration), and the symmetries of the potential. Besides these qualitative advantages, shape dynamics is more powerful: fewer initial data predict the evolution.

This subsection has primarily been concerned with the problem of defining change of position. Newton clearly understood that this requires one to know when one can say a body is *at the same place* at different instants of time. Formally at least he solved this problem by the notion of absolute space. Best matching is *the relational alternative to absolute space.* For when one configuration has been placed relative to another in the best-matched position, every position in one configuration is uniquely paired with a position in the other. If a body is at these paired positions at the two instants, one can say it is at the same place. The two positions are *equilocal.* The image of 'placing' one configuration on another in the best-matched position is clearly more intuitive than the notion of inter-orbit orthogonality. It makes the achievement of relational equilocality manifest. It is also worth noting that the very thing that creates the problem of defining change of position – the action of the similarity group – is used to resolve it in best matching.

3.5. Two forms of scale invariance

We now return to the reasons for the failure of Laplacian determinism of Newtonian dynamics when considered in shape space. This will explain why it is desirable to keep open the option of the weaker form of the Poincaré principle. It will be helpful to consider Newtonian dynamics once more in the form of Jacobi's principle:

$$\delta I = 0, \quad I = 2\int \mathrm{d}\lambda\sqrt{(E - V(q))T_{kin}}, \quad T_{kin} := \frac{1}{2}\sum_a m_a \frac{\mathrm{d}\mathbf{x}_a}{\mathrm{d}\lambda}\cdot\frac{\mathrm{d}\mathbf{x}_a}{\mathrm{d}\lambda}. \tag{15}$$

Typically $V(q)$ is a sum of terms with different, usually integer homogeneity degrees: gravitational and electrostatic potentials have l^{-1}, harmonic-oscillator potentials are l^2. Moreover, since (15) is timeless and only the dimensionless mass ratios have objective meaning, length is the sole significant dimension. Because all terms in V must have the same dimension, dimensionful coupling constants must appear. One can be set to unity because an overall factor multiplying the action has no effect on its extremals. If we take G=1, a fairly general action will have

$$W = E - V = E + \sum_{i<j}\frac{m_i m_j}{r_{ij}} - g_i V_i, \tag{16}$$

where the V_i's have different homogeneity degrees, some perhaps the same (as for gravity and electrostatics). Now the crux: different E and g_i values lead to different curves in shape space, but not to any differences that can be expressed through an initial point and direction in Q^N_{ss}, which cannot encode dimensionful information. Thus, each such E and g_i present in (16) adds a one-parameter degree of uncertainty into the evolution from an initial point and direction in Q^N_{ss}. If the strong Poincaré principle holds, this unpredictability is eliminated. There may still be several different terms in V but they must all have the same homogeneity degree -2 and dimensionless coupling constants; in particular, the constant E cannot be present. Note also that *any* best matching enhances predictability and eliminates potentially redundant structure. But other factors may count. Nature may have reasons not to best match with respect to all conceivable symmetries.

Indeed, the foundation of particle shape dynamics on the similarity group precluded consideration of the larger general linear group. I suspect that this group would leave too little structure to construct dynamics at all easily and that angles are the irreducible minimum needed. Another factor, possibly more relevant, is the difference between velocities (and momenta), which are vectors, and directions, which are not (since multiplication of them by a number is meaningless). Vectors and vector spaces have mathematically desirable properties. In quantum mechanics, the vector nature of momenta ensures that the momentum and configuration spaces have the same number of dimensions, which is important for the equivalence of the position and momentum representations (transformation theory). If we insist on the strong Poincaré principle, the equivalence will be lost for a closed system regarded as an island universe. There are then two possibilities: either equivalence is lost, and transformation theory only arises for subsystems (just as inertial frames of reference arise from shape dynamics), or the strong Poincaré principle is relaxed just enough to maintain equivalence.

There is an interesting way to do this. In the generic N-body problem the energy E and angular momentum L, as dimensionful quantities, are not scale invariant. But they are if $E = L = 0$. Then the behaviour is scale invariant. Further [12], there is a famous qualitative result in the N-body problem, first proved by Lagrange, which is that $\ddot{I} > 0$ if $E \geq 0$. Then the curve of I as a function of time is concave upwards and its time derivative, which is $2D$ (defined just before (14)), is strictly monotonic, increasing from $-\infty$ to ∞ (if the evolution is taken nominally to begin at $D = -\infty$).

Now suppose that, as I conjecture, in its classical limit the quantum mechanics of the universe does require there to be velocities (and with them momenta) in shape space and its geometrodynamical generalization, to which we come soon. Then there must at the least be a one-parameter family of solutions that emanate from a point and a direction in Q^N_{ss}. We will certainly want rotational best matching to enforce $L = 0$. We will then have to relax dilatational best matching in such a way that a one-parameter freedom is introduced. In the N-body problem we can do this, without having a best-matching symmetry argument that enforces it, by requiring $E = 0$. The corresponding one-parameter freedom in effect converts a direction in Q^N_{ss} into a vector. The interesting thing is now that, by Lagrange's result, D is monotonic when $E = 0$. This means that the shape-space dynamics can be monotonically parametrized by the dimensionless ratio D_c/D_0, where D_0 is an initial value of D and D_c is the current value. Thus, D_c/D_0 provides an objective 'time' difference between shapes s_1 and s_2. The scare quotes are used because it does not march in step with the time defined by (6).

An alternative dimensionless parametrization of the shape-space curves in this case is by means of the (not necessarily monotonic) ratio I_c/I_0. Because the moment of inertia measures the 'size' of the universe, this ratio measures 'the expansion of the universe' from an initial size to its current size. One might question whether in this case one should say that the dynamics unfolds on shape space. Size still has some meaning, though not at any one instant but only as a ratio at two instants. Moreover, on shape space this ratio plays the role of 'time' or 'independent variable'. It does not appear as a dependent dynamical variable. This is related to the cosmological puzzle that I highlighted at the end of the introduction: from the shape-dynamic perspective, the expansion of the universe seems to be made possible by a last vestige of Newton's absolute space. I shall return to this after presenting the dynamics of geometry in terms of best matching.

To conclude the particle dynamics, the strong form of the Poincaré principle does almost everything that one could ask. It cannot entirely fix the potential term W but does require all of its terms to be homogeneous of degree l^{-2} with dimensionless coefficients, one of which can always be set to unity. If the strong Poincaré principle fails, the most interesting way the weak form can hold in the N-body problem is if $E = 0$. In this case a one-parameter freedom in the shape-space initial data for given s and d is associated with the ratio T_s/T in Q^N.

4. Conformal geometrodynamics

Although limited to particle dynamics, the previous section has identified the two universal elements of shape dynamics: derivation of time from difference and best matching to obviate the introduction of absolute (nondynamical) structure. However, nothing can come of nothing. The bedrock on which dynamics has been derived is the geometrical structure of individual configurations of the universe. We began with configurations in Euclidean space and removed from them more and more structure by group quotienting. We left open the question of how far such quotienting should be taken, noting that nature must decide that. In this section, we shall see that, with two significant additions, the two basic principles of shape dynamics can be directly applied to the dynamics of geometry, or *geometrodynamics*. This will lead to a novel derivation of, first, general relativity, *then* special relativity (and gauge theory) and after that to the remarkable possibility that gravitational theory introduces a dynamical standard of rest in a closed universe.

In this connection, let me address a likely worry of the reader, anticipated in footnote 2, about the fundamental role given to instantaneous configurations of the universe. Does this not flagrantly contradict the relativity of simultaneity, which is confirmed by countless experiments? In response, let me mention some possibly relevant facts.

When Einstein and Minkowski created special relativity, they did not ask how it is that inertial frames of reference come into existence. They took them as given. Even when creating general relativity, Einstein did not directly address the origin of local inertial frames of reference. Moreover, although he gave a definition of simultaneity at spatially separated points, he never asked how temporally separated durations are to be compared. What does it mean to say that a second today is the same as a second yesterday? Shape dynamics directly addresses both of these omissions of Einstein, to which may be added his adoption of length as fundamental, which Weyl questioned in 1918 [19, 20]. Finally, it is a pure historical accident that Einstein, as he himself said, created general relativity so early, a decade before quantum mechanics was discovered. Now it is an architectonic feature of quantum mechanics that the Schrödinger wave function is defined on configuration space, not (much to Einstein's dismay) on spacetime.

This all suggests that instantaneous spatial configurations of the universe could at the least be considered as the building blocks of gravitational theory. Indeed, they are in the Hamiltonian dynamical form of general relativity introduced by Dirac [21] and Arnowitt, Deser and Misner [22]. However, many relativists regard that formulation as less fundamental than Einstein's original one. In contrast, I shall argue that the shape-dynamical approach might be more fundamental and that the geometrical theory of gravity could have been found rather naturally using it. I ask the reader to keep an open mind.

4.1. Superspace and conformal superspace

Differential geometry begins with the idea of continuity, encapsulated in the notion of a manifold, the rigorous definition of which takes much care. I assume that the reader is familiar with the essentials and also with diffeomorphisms; if not, [23] is an excellent introduction. To model a closed universe, we need to consider closed manifolds. The simplest possibility that matches our direct experience of space is S^3, which can be pictured as the three-dimensional surface of a four-dimensional sphere.

Now suppose that on S^3 we define a Riemannian 3-metric $g_{ij}(x)$. As a 3×3 symmetric matrix at each space point, it can always be transformed at a given point to diagonal form with $1, 1, 1$ on the diagonal. Such a metric does three things. First, it defines the *length* $\mathrm{d}s$ of the line element $\mathrm{d}x^i$ connecting neighbouring points of the manifold: $\mathrm{d}s = \sqrt{g_{ij}\mathrm{d}x^i\mathrm{d}x^j}$. This is well known. However, for shape dynamics it is more important that $g_{ij}(x)$ determines *angles*. Let two curves at x be tangent to the line elements $\mathrm{d}x^i$ and $\mathrm{d}y^i$ and

θ be the angle between them. Then

$$\cos\theta = \frac{g_{ij}\mathrm{d}x^i\mathrm{d}y^j}{\sqrt{g_{kl}\mathrm{d}x^k\mathrm{d}x^l g_{mn}\mathrm{d}y^m\mathrm{d}y^n}}. \tag{17}$$

The third thing that the metric does (implicitly) is give information about the coordinates employed to express the metric relations.

We see here an immediate analogy between a 3-metric and an N-body configuration of particles in Euclidean space. Coordinate information is mixed up with geometrical information, which itself comes in two different forms: distances and angles. Let us take this analogy further and introduce corresponding spaces and structure groups.

Riem(S^3) is the (infinite-dimensional) space of all suitably continuous Riemannian 3-metrics g_{ij} on S^3 (henceforth omitted). Thus, each point in Riem is a 3-metric. However, many of these 3-metrics express identical distance relationships on the manifold that are simply expressed by means of different coordinates, or labels. They can therefore be carried into each other by three-dimensional diffeomorphisms without these distance relations being changed. They form a diffeomorphism equivalence class $\{g_{ij}\}_{\mathit{diff}}$, and the 3-diffeomorphisms form a structure group that will play a role analogous to the Euclidean group in particle dynamics. Each such equivalence class is an orbit of the 3-diffeomorphism group in Riem and is defined as a *three-geometry.* All such 3-geometries form *superspace.* This is a familiar concept in geometrodynamics [24]. Less known is *conformal superspace*, which is obtained from superspace by the further quotienting by conformal transformations:

$$g_{ij}(x) \to \phi(x)^4 g_{ij}(x), \quad \phi(x) > 0. \tag{18}$$

Here, the fourth power of the position-dependent function ϕ is chosen for convenience, since it makes the transformation of the scalar curvature R simple (in four dimensions, the corresponding power is 2); the condition $\phi > 0$ is imposed to stop the metric being transformed to the zero matrix.

The transformations (18) change the distance relations on the manifold but not the angles between curves. Moreover, distances are not directly observable. To measure an interval, we must lay a ruler adjacent to it. If the interval and the ruler subtend the same angle at our eye, we say that they have the same length. This is one reason for thinking that angles are more fundamental than distances; another is that they are dimensionless. We also have the intuition that shape is more basic than size; we generally speak of *the*, not an, equilateral triangle. It is therefore natural to make the combination of the group of 3-diffeomorphisms and the conformal transformations (18) the structure group of conformal geometrodynamics.

Before continuing, I want to mention the subgroup of the transformations (18) that simply multiply the 3-metric by a constant C:

$$g_{ij}(x) \to C g_{ij}(x), \quad C > 0. \tag{19}$$

One can say that the transformations (19) either 'change the size of the universe' or change the unit of distance. Like similarity transformations, they leave all length ratios unchanged, and are conceptually distinct from the general transformations (18), which change the ratios of the geodesic lengths $d(a,b)$ and $d(c,d)$ between point pairs a,b and c,d. As a result, general conformal transformations open up a vastly richer field for study than similarity transformations. Another subgroup consists of the *volume-preserving conformal transformations* (18). They leave the total volume $V = \int \sqrt{g}\mathrm{d}^3x$ of the universe unchanged. We shall see that these transformations play an important role in cosmology. The seemingly minor restriction of the transformations (18) to be volume preserving is the mysterious last vestige of absolute space that I mentioned in the introduction.

The idea of geometrodynamics is nearly 150 years old. Clifford, the translator of Riemann's 1854 paper on the foundations of geometry, conjectured in 1870 that material bodies in motion might be nothing more than regions of empty but differently curved three-dimensional space moving relative to each other [24], p. 1202. This idea is realized in Einstein's general relativity in the vacuum (matter-free) case in the geometrodynamic interpretation advocated by Wheeler [24]. I shall briefly describe his superspace-based picture, before taking it further to conformal superspace.

Consider a matter-free spacetime that is globally hyperbolic. This means that one can slice it by nowhere intersecting spacelike hypersurfaces identified by a monotonic time label t (Fig. 7). Each hypersurface carries a 3-geometry, which can be represented by many different 3-metrics g_{ij}. At any point x on one hypersurface labelled by t one can move in spacetime orthogonally to the $t + \delta t$ hypersurface, reaching it after the proper time $\delta\tau = N\delta t$, where N is called the *lapse*. If the time labelling is changed, N is rescaled in such a way that $N\delta t$ is invariant. In general, the coordinates on successive 3-geometries will be chosen arbitrarily, so that the point with coordinate x on hypersurface $t + \delta t$ will not lie at the point at which the normal erected at point x on hypersurface t pierces hypersurface $t + \delta t$. There will be a lateral displacement of magnitude $\delta x^i = N^i\delta t$. The vector N^i is called the *shift*. The lapse and shift encode the g_{00} and g_{0i} components respectively of the 4-metric: $g_{00} = N_iN^i - N^2$, $g_{0i} = N_i$.

Each 3-metric g_{ij} on the successive hypersurfaces is a point in Riem, and the one-parameter family of successive g_{ij}'s is represented as a curve in Riem parametrized by t. This is just one representation of the spacetime. First, one can change the time label freely on the curve (respecting monotonicity). This leaves the curve in Riem unchanged and merely changes its parametrization. Second, by changing the spatial coordinates on each hypersurface one can change the successive 3-metrics and move the curve around to a considerable degree in Riem. However, each of these curves corresponds to one and the same curve in superspace. But, third, one and the same spacetime can be sliced in many different ways because the definition of simultaneity in general relativity is to a high degree arbitrary (Fig. 8). Thus, an infinity of curves in superspace, and an even greater infinity of curves in Riem, represent the same spacetime. In addition, they can all carry infinitely many different

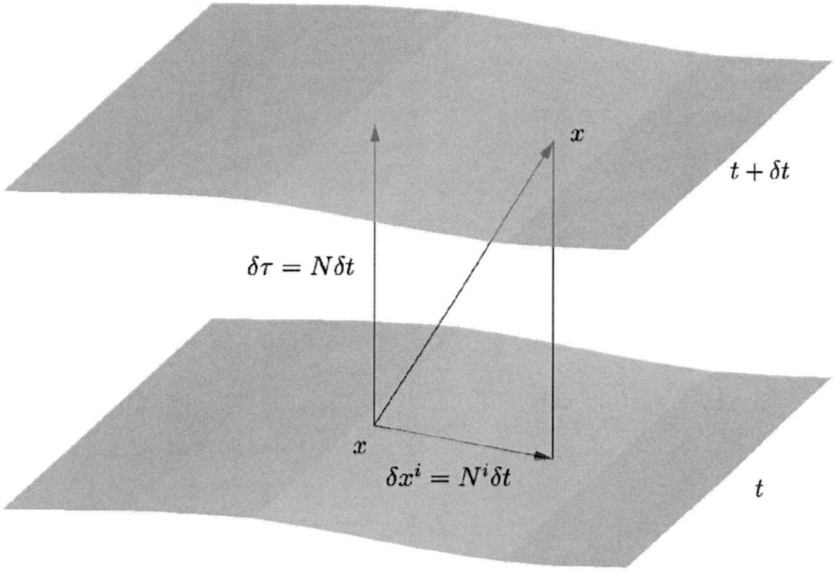

FIGURE 7. The 3 + 1 decomposition of spacetime as explained in the text.

parametrizations by time labels. This huge freedom corresponds to the possibility of making arbitrary four-dimensional coordinate transformations, or equivalently 4-diffeomorphisms, on spacetime.

As long as one insists on the equal status of all different slicings by spacelike hypersurfaces – on *slicing* or *foliation* invariance – it is not possible to represent the evolution of 3-geometry by a unique curve in a geometrical configuration space. This is the widely accepted view of virtually all relativists. Shape dynamics questions this. I shall now sketch the argument.

Purely geometrically, distinguished foliations in spacetime *do exist.* The flat *intrinsic* (two-dimensional) geometry of a sheet of paper is unchanged when it is rolled into a tube and acquires *extrinsic curvature.* By analogy, just as a 3-metric g_{ij} describes intrinsic geometry, a second fundamental form, also a 3×3 symmetric tensor K^{ij}, describes extrinsic curvature. Its trace $K = g_{ij}K^{ij}$ is the *mean extrinsic curvature.* A *constant-mean-curvature* (CMC) hypersurface is one embedded in spacetime in such a way that K is everywhere constant. In three-dimensional Euclidean space two-dimensional soap bubbles have CMC surfaces. Such surfaces are extremal and are therefore associated with 'good' mathematics. At least geometrically, they are clearly distinguished.

A complete understanding of the possibilities for slicing a spatially closed vacuum Einsteinian spacetime, i.e., one that satisfies Einstein's field equations $G_{\mu\nu} = 0$, by CMC hypersurfaces does not yet exist.[9] However, as

[9]There certainly exist spacetimes that satisfy Einstein's field equations and do not admit CMC foliation. However, shape dynamics does not have to yield all solutions allowed by

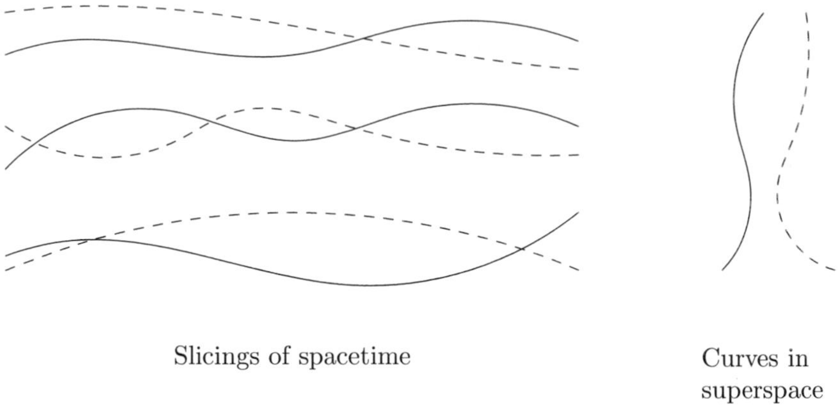

FIGURE 8. Because there is no distinguished definition of simultaneity in general relativity, a spacetime can be sliced in many different ways. This slicing, or foliation, freedom leads to many different representations of the spacetime by curves in superspace. Two slicings and corresponding curves in superspace are shown.

we shall see, there exists a very effective and reliable way to generate 'patches' of CMC-foliated Einsteinian spacetimes. In such a patch CMC-foliated spacetime exists in an open neighbourhood either side of some CMC hypersurface labelled by $t = 0$. A noteworthy property of CMC foliations is that K, which is necessarily a spatial constant on each hypersurface, must change monotonically in the spatially closed case. Moreover, K measures the rate of change of the spatial volume $V = \int \sqrt{g}\, \mathrm{d}^3x$ in unit proper time.[10] In both these respects, K is closely analogous to the quantity D (14) in particle mechanics. We recall that it is the rate of increase of the moment of inertia I, which, like V, characterizes the size of the universe.

Let us now suppose that we do have a vacuum Einsteinian spacetime that is CMC foliated either in its entirety or in some patch. On each leaf (slice) of the foliation there will be some 3-geometry and a uniquely determined conformal 3-geometry, i.e., that part of the 3-geometry that relates only to angle measurements. We can take the successive conformal 3-geometries and plot them as a curve in conformal superspace (CS). Having done this, we could change the slicing in the spacetime, obtaining a different curve of 3-geometries in superspace. They too would have associated conformal 3-geometries, and each different curve of 3-geometries in superspace would generate a different curve of conformal 3-geometries in CS. According to the standard interpretation of general relativity, all these different curves in superspace and in CS

general relativity but only those relevant for the description of the universe. The ability to reproduce nature, not general relativity, is what counts.

[10]The value of K is only defined up to its sign.

are to be regarded as physically equivalent. I believe there are grounds to at least question this.

If we go back to Clifford's original inspiration, note that only angles are observable, and insist on either the strong or weak form of the Poincaré principle, we are led naturally to the desire to create a dynamical theory of conformal geometry in which either a point and a direction in CS or a point and a tangent vector in CS suffice to determine a unique evolution in CS. This will be exactly analogous to the aim of particle shape dynamics and will be implemented in the following subsections. What makes this shape-dynamic approach interesting is that the successions of conformal 3-geometries generated in the weak case correspond exactly to the successions of conformal 3-geometries obtained on CMC foliated Einsteinian spacetimes. Moreover, the best matching by which the dynamic curves in CS are obtained simultaneously generates the complete spacetime as the distinguished representation of the conformal dynamics. Once this spacetime has been generated in the CMC foliation, one can go over to an arbitrary foliation within it and recover all of the familiar results of general relativity. Three distinct ingredients create conformal dynamics. I shall present them one by one.

4.2. The elimination of time

It is easy (Sec. 3.1) to remove time from the kinematics of particle dynamics and recover it as a distinguished parameter from geodesic dynamics. It will help now to look at the structure of the canonical momenta in relational particle dynamics. Given a Lagrangian $L(q_a, q'_a)$ that depends on dynamical variables q_a and their velocities $q'_a = \mathrm{d}q_a/\mathrm{d}\lambda$, the canonical momentum of q_a is $p^a := \partial L/\partial q'_a$. For the best-matched action (11),

$$\mathbf{p}^a := \frac{\partial L^{bm}}{\partial \mathbf{x}'_a} = \sqrt{\frac{W}{T_{bm}}} m_a \frac{\mathrm{d}\mathbf{x}_a^{bm}}{\mathrm{d}\lambda}, \qquad L^{bm} = \sqrt{W T^{bm}}. \tag{20}$$

The distinguished time label t is obtained by choosing λ such that always $W = T_{bm}$ so that the cofactor of $m_a \mathrm{d}\mathbf{x}_a^{bm}/\mathrm{d}\lambda$ is unity. The definition is holistic for two reasons. First, the $\mathrm{d}\mathbf{x}_a^{bm}/\mathrm{d}\lambda$ are obtained by global best matching and are therefore determined by all the changes of the relative separations of the particles. Second, the denominator of the factor $\sqrt{W/T_{bm}}$ is a sum over the displacements of all the particles in the universe. This is seen explicitly in the expression (6). The $\mathbf{p}^a$ have a further important property: they are reparametrization invariant. If one rescales λ, $\lambda \to \bar{\lambda}(\lambda)$, the velocities q'_a scale, but because velocities occur linearly in the numerator and denominator of (20) there is no change in $\mathbf{p}^a$, which is in essence a *direction.* (It is in fact a direction cosine wrt to the conformal metric obtained by multiplying the kinetic metric by W.)

However, one could take the view that, at any instant, one should obtain a *local* measure of time derived from purely local differences. This would still yield an holistic notion of time if the local differences were obtained by best matching. However, in the case of particle dynamics, a local derived time of this kind cannot be obtained for the simple reason that particles are, by definition, structureless. The situation is quite different in field dynamics because fields have several components at each space point. This opens up the possibility of a local measure of time, as I shall now show (deferring the conceptually distinct issue of best matching until later).

Let the action on Riem, the configuration space in which calculations are of necessity made, have the form

$$I = \int \mathrm{d}\lambda L, \quad L = \int \mathrm{d}^3x\sqrt{gWT}, \tag{21}$$

where $g = \sqrt{\det g_{ij}}$ is introduced explicitly to make the integrand a tensor density, the scalar W is a local functional of g_{ij} (that is, it depends on g_{ij} and its spatial derivatives up to some finite order, which will in fact be the second), and T depends quadratically on the metric velocities $g'_{ij} := \mathrm{d}g_{ij}/\mathrm{d}\lambda$ and also quadratically on g^{ij}. It will actually have the form

$$T = G_A^{ijkl} g'_{ij} g'_{kl}, \quad G_A^{ijkl} = g^{ik}g^{jl} + Ag^{ij}g^{kl}. \tag{22}$$

Here, G_A^{ijkl}, in which A is an as yet arbitrary constant, is a supermetric (cf (3)) and appears because one needs to construct from the velocities g'_{ij} a quantity that is a scalar under 3-diffeomorphisms. Because g'_{ij} is, like the 3-metric, a symmetric tensor, there are only two independent scalars that one can form from it by contraction using the inverse metric.[11]

The key thing about (21) is that one first forms a quantity quadratic in the velocities at each space point, takes the square root at each space point, and only then integrates over space. This is a *local square root* and can be justified as follows. First, a square root must be introduced in *some* way to create a theory without any external time variable. Next, there are two ways in which this can be done. The first is by direct analogy with Jacobi's principle (4) or (15), and would lead to an action with 'global' square roots of the form

$$I = \int \mathrm{d}\lambda \sqrt{\int \mathrm{d}^3x\sqrt{g}W} \sqrt{\int \mathrm{d}^3x\sqrt{g}T}. \tag{23}$$

Besides being a direct generalization, the action (23) is on the face of it mathematically more correct than (21) since it defines a proper metric on Riem, which is not the case if the square root is taken, as in (21), before the integration over space. Nevertheless, it turns out that an action of the form (21) does lead to a consistent theory. This will be shown below, but we can already see that in such a case we obtain a theory with a local emergent time. For this, we merely need to calculate the form of the canonical momenta of

[11]The most general supermetric formed from g^{ij} and acting on a general tensor has three terms. For $A = -1$, we obtain the DeWitt supermetric, which will appear later. In principle, one could also consider supermetrics formed with spatial derivatives of g^{ij}, but these would lead to very complicated theories. As Einstein always recommended, it is advisable to look first for the simplest nontrivial realizations of an idea.

the 3-metric g_{ij} that follow from (21):

$$p^{ij} := \frac{\partial L}{\partial g'_{ij}} = \sqrt{\frac{W}{T}} G^{ijkl} g'_{kl}. \tag{24}$$

The similarity of the p^{ij} to the particle canonical momenta $\mathbf{p}^a$ (20) is obvious. First, under $\lambda \to \bar{\lambda}(\lambda)$, the momenta p^{ij} are, like $\mathbf{x}^a$, unchanged. Second, the complex of bare velocities $G^{ijkl}g'_{kl}$ is multiplied by the Jacobi-type factor $\sqrt{W/T}$. However, the key difference is that this factor is no longer a global but a position-dependent local quantity. I will not go into further details yet except to say that when the theory is fully worked out it leads to the appearance of a local increment of proper time given by $\delta\tau = N\delta\lambda$, where $N = \sqrt{T/4R}$ can be identified with the lapse in general relativity.

Whereas the elimination of time in Jacobi's principle and for an action like (23) with global square roots is, at the classical level at least,[12] a trivial matter with no impact on the best matching (and vice versa), the elimination of external time by the local square root has a huge effect and its consequences become intimately interconnected with those of the best matching. Perhaps the most important effect is that it drastically reduces the number of *consistent* actions that one can construct. This was first recognized by my collaborator Niall Ó Murchadha, and its consequences were explored in [28], about which I shall say something after the description of geometrodynamic best matching.

4.3. Geometrodynamic best matching

The basic idea of geometrodynamic best matching is exactly as for particles but leads to a vastly richer theory because a 3-geometry, either Riemannian or conformal, is infinitely more structured than a configuration of particles in Euclidean space. However, the core idea is the same: to 'minimize the incongruence' of two intrinsically distinct configurations. This is done by using the spatial structure groups of the configurations to bring one configuration into the position in which it most closely overlaps the other.

Let us first consider 3-diffeomorphisms. If we make an infinitesimal coordinate transformation on a given 3-metric $g_{ij}(x)$, obtaining new functions of new coordinates, $g_{ij}(x) \to \bar{g}_{ij}(\bar{x})$, and then consider $\bar{g}_{ij}(\bar{x})$ at the old x values, the resulting 3-metric $\bar{g}_{ij}(x)$ is what one obtains by a 3-diffeomorphism generated by some 3-vector field $\xi^i(x)$: $\bar{g}_{ij}(x) = g_{ij}(x) + \xi_{(i;j)}$ (the semicolon denotes the covariant derivative wrt to g_{ij} and the round parentheses symmetrization). The two 3-metrics $g_{ij}(x)$ and $\bar{g}_{ij}(x)$ are diffeomorphically related representations of one and the same 3-geometry. This is analogous to changing the Cartesian coordinates of a particle configuration.

Now suppose that $g_{ij}(x) + \delta g_{ij}(x)$ represents a 3-geometry genuinely distinct from $g_{ij}(x)$, i.e., δg_{ij} cannot be represented in the form $\xi_{(i;j)}$, which

[12]In quantum mechanics, the effect is dramatic, since the quantization of Jacobi's principle leads to a time-independent, and not time-dependent Schrödinger equation. This is one aspect of the famous 'problem of time' in canonical quantum gravity [25, 26, 27].

would indicate a spurious diffeomorphically-induced change. The difficulty that we now face is that, because we are considering intrinsically different 3-geometries, mere identity of the coordinate values x_i does not mean that they specify 'the same point' in the two different 3-geometries. In fact, the problem has nothing to do with coordinates. Given an apple and a pear, there does not appear to be any way to establish a 1-to-1 pairing of all the points on the apple's surface with all those on the pear's. However, best matching does just that if the compared objects differ only infinitesimally. To apply the technique rigourously, one must use rates of change rather than finite differences.

Mathematically, we can always specify a 3-metric $g_{ij}(x)$ and its velocity $g'_{ij} = \mathrm{d}g_{ij}/\mathrm{d}\lambda$. The problem is that $g'_{ij} = \mathrm{d}g_{ij}/\mathrm{d}\lambda$ mixes information about the intrinsic change of the described 3-geometry with arbitrary information about the way in which the coordinates are laid down as the 3-geometry changes. There is an equivalence class of velocities $\{g'_{ij} - \xi'_{(i;j)}\}$ that all represent the same intrinsic change. The task of best matching is to select a unique one among them that can be said to measure the true change.

We note first that there is no objection to fixing coordinates on the original 3-geometry, giving $g_{ij}(x)$, just as we chose an initial Cartesian representation for the particle configurations. To fix the way the coordinates are then laid down, we consider the effect of λ-dependent diffeomorphisms on (21). It becomes

$$I = \int \mathrm{d}\lambda L, \quad L = \int \mathrm{d}^3x\sqrt{gWT}, \quad T = G_A^{ijkl}(g'_{ij} - \xi'_{(i;j)})(g'_{kl} - \xi'_{(k;l)}). \quad (25)$$

The possibility of constructing consistent geometrodynamical theories is considered in [28],[13] to which I refer the reader for details, since I only wish to indicate what the results are.

The basic theoretical structure obtained in geometrodynamics is broadly the same as in particle dynamics. One obtains constraints and conditions under which they propagate consistently. These conditions strongly restrict the set of consistent theories. I shall first identify the constraints and then indicate how they act as 'theory selectors'.

First, there are constraints because of the local square root in (25). Before giving them, I need to draw attention to a similar constraint, or rather identity, in the particle model. It follows from the form (20) of the canonical momenta $\mathbf{p}^a$ that

$$\sum_a \frac{\mathbf{p}^a \cdot \mathbf{p}^a}{2m_a} \equiv W. \quad (26)$$

This is a *square-root identity*, since it follows directly from the square root in the Lagrangian and means that the $\mathbf{p}^a$ are in essence direction cosines. In the Hamiltonian formalism, (26) becomes a constraint and is a single global relation. In contrast, the geometrodynamic action contains an infinity of square

[13]Some of the conclusions reached in [28] are too strong, being based on tacit simplicity assumptions that Anderson identified [29, 30]. I shall report here the most interesting results that are obtained when the suitable caveats are made.

roots, one at each space point. Correspondingly, the canonical momenta satisfy *infinitely* many identities (or Hamiltonian constraints):

$$p_{ij}p^{ij} - \frac{2A}{3A-1}p^2 \equiv gW, \quad p = g_{ij}p^{ij}. \tag{27}$$

Second, constraints arise through the best matching wrt diffeomorphisms. This is implemented by variation of (25) with respect to ξ_i', treated as a Lagrange multiplier. This also leads to a constraint at each space point:

$$p^{ij}_{\ \ ;j} = 0. \tag{28}$$

These linear constraints are closely analogous to the linear constraints (12) and (13) in particle dynamics. For the form (25) of the action, they propagate automatically and do not lead to restrictions. This is because the action (21) was chosen in advance in a form invariant under λ-independent 3-diffeomorphisms, which in turn ensured that (25) is invariant under λ-dependent 3-diffeomorphisms. Had we chosen a more general functional of g_{ij} and its spatial and λ derivatives, propagation of the constraints (28) would have forced us to specialize the general form to (25). This is another manifestation of the power of combining the structure group of the 3-metrics (the 3-diffeomorphism group) with the Poincaré requirement.

There is no analogous control over the quadratic constaints (27) that arise from the local square root in (21) and (25). As is shown in [28], the only action that consistently propagates both the quadratic and linear constraints has the form

$$I_{BSW} = \int \mathrm{d}\lambda \int \mathrm{d}^3x \sqrt{(\Lambda + dR)T_{A=-1}}, \tag{29}$$

where the subscript $A = -1$ of T indicates that the undetermined coefficient in the supermetric is forced to take the DeWitt value. More impressive is the drastic restriction on the possible form of the potential term W, which is restricted to be $\Lambda + dR$, $d = 0$ or ± 1. The action (29) is in fact the Baierlein–Sharp–Wheeler action [31], which is dynamically equivalent to the Einstein–Hilbert action for globally hyperbolic spacetimes. The only freedom is in the choice of the constant Λ, which corresponds to the cosmological constant, and the three options for d. The case $d = 0$ yields so-called strong gravity and is analogous to pure inertial motion for particles. The case d corresponds to a Lorentzian spacetime and hence to the standard form of general relativity, while $d = -1$ gives Euclidean general relativity.

When translated into spacetime terms,[14] the constraints (27) and (28) are respectively the 00 and $0i, i = 1,2,3$, Einstein field equations $G_{\mu\nu} = 0, \mu, \nu = 0,1,2,3$. Whereas the particle dynamics associated with the global Euclidean group leads to global relations, implementation of the Poincaré principle in geometrodynamics by the local elimination of time and best matching wrt local 3-diffeomorphisms leads to local constraints, the propagation of which directly determines the simplest nontrivial realization of the

[14]The 'construction of spacetime' will be described later.

whole idea: general relativity. Of course, the immense power of local symmetry requirements was one of the great discoveries of 20th-century physics. It first became apparent with Einstein's creation of general relativity. If shape dynamics has value, it is not so much in the locality of the symmetries as in their choice and in the treatment of time. I shall compare the shape-dynamic approach with Einstein's at the end of the paper. Here I want to continue with the results of [28].

So far, we have considered pure geometrodynamics. The assumption that the structure of spacetime always reduces locally to the Minkowski-space form of special relativity (a key element in Einstein's approach) played no role in the derivation. The manner in which special relativity arises in [28] is striking. In field theory, the essence of special relativity is a *universal light cone*: all fields must have the same limiting signal propagation velocity. Now vacuum general relativity has a 'light cone'. What happens if we attempt, as the simplest possibility, to couple a scalar field φ to vacuum geometrodynamics described by the action (25)?

The propagation speed of such a field, with action containing the field velocities $\mathrm{d}\varphi/\mathrm{d}\lambda$ and first spatial derivatives $\partial_i\varphi$ quadratically, is determined by a single coefficient C, which fixes the ratio of the contributions of $\mathrm{d}\varphi/\mathrm{d}\lambda$ and $\partial_i\varphi$ to the action. When the scalar field is added to the action (for details see [28]), the constraints (27) and (28) acquire additional terms, and one must verify that the modified constraints are propagated by the equations of motion. It is shown in [28] that propagation of the modified quadratic constraint fixes the coefficient C to be exactly such that it shares the geometrodynamic light cone. Otherwise, the scalar field can have a term in its action corresponding to a mass and other self-interactions.

The effect of attempting to couple a single 3-vector field $\mathbf{A}$ to the geometry is even more remarkable. In this case, there are three possible terms that can be formed from the first spatial (necessarily covariant) derivatives of $\mathbf{A}$. Each may enter in principle with an arbitrary coefficient. The requirement that the modified quadratic constraint propagate not only fixes all three coefficients in such a way that the 3-vector field has the same light cone as the geometry but also imposes the requirement that the canonical momenta $\mathbf{P}$ of $\mathbf{A}$ satisfy the constraint $\mathrm{div}\,\mathbf{P} = 0$. In fact, the resulting field is none other than the Maxwell field interacting with gravity. The constraint $\mathrm{div}\,\mathbf{P} = 0$ is the famous Gauss constraint. This can be taken further [32]. If one attempts to construct a theory of several 3-vector fields that interact with gravity and with each other, they have to be Yang–Mills gauge fields. Unlike the scalar field, all the gauge fields must be massless.

To conclude this subsection, let us indulge in some 'what-might-have-been' history. Clifford's 'dream' of explaining all motion and matter in terms of dynamical Riemannian 3-geometry was in essence a proposal for a new ontology of the world. The history of science shows that new, reasonably clearly defined ontologies almost always precede major advances. A good example is Descartes's formulation of the mechanical world view; it led within a few decades to Newton's dynamics ([7], Chaps. 8–10). Clifford died tragically young; he could have lived to interact with both Mach and Poincaré. Between them, they had the ideas and ability needed to create a relational theory of dynamical geometry (and other fields) along the lines described above. In this way, well before 1905, they could have discovered, first, general relativity in the form of the Baierlein–Sharp–Wheeler action (29), next special relativity through a universal light cone, and even, third, gauge theory. All of this could have happened as part of a programme to realize Clifford's original inspiration in the simplest nontrivial way.

I want to emphasize the role that the concept of time would have played in such a scenario. In 1905, Einstein transformed physics by insisting that the description of motion has no meaning "unless we are quite clear as to what we understand by 'time' " [33]. He had in mind the problem of defining simultaneity at spatially separated points. Resolution of this issue in 1905 was perhaps the single most important thing that then led on to general relativity. However, in 1898, Poincaré [34, 35] had noted the existence of *two* fundamental problems related to time: the definition of simultaneity and the older problem of defining duration: What does it mean to say that a second today is the same as a second tomorrow? Even earlier, in 1883, Mach had said: "It is utterly beyond our power to measure the changes of things by time. Quite the contrary, time is an abstraction at which we arrive by means of the changes of things." Both Mach and Poincaré had clearly recognized the need for a theory of duration along the lines of Sec. 3.1.[15]

There is now an intriguing fact. The structure of dynamics so far presented in this paper has been based on two things: best matching and a theory of duration. Both were initially realized globally, after which a local treatment was introduced. Moreover, entirely different schemes were used to achieve the desired aims of a relational treatment of displacement and of duration (best matching and a square root in the action respectively). Remarkably, Einstein's theory of simultaneity appeared as a *consequence* of these relational inputs. In line with my comments at the end of Sec. 3.1, I believe that the concept of duration as a measure of difference is more fundamental than the definition of simultaneity, so it is reassuring that Einstein's well confirmed results can be recovered starting from what may be deeper foundations. In this connection, there is another factor to consider. In the standard representation of general relativity, spacetime is a four-dimensional block. One is not supposed to think that the Riemannian 3-geometry on the leaves of a 3+1 foliation is more fundamental than the lapse and shift, which tell one how the 3-geometries on the leaves 'fit together' (Fig. 7). The lapse is particularly important: it tells you the orthogonal separation (in spacetime) between the 3-geometries that are the leaves of a 3+1 foliation. However, the G^{00} Einstein field equation enables one to solve algebraically for the lapse in terms of the

[15]Despite a careful search through his papers, I have been unable to find any evidence that Einstein ever seriously considered the definition of duration. As we saw in Sec. 3.1, this is intimately related to the theory of clocks, which Einstein did grant had not been properly included in general relativity. He called the omission a 'sin' [36].

other variables. It is precisely this step that led Baierlein, Sharp and Wheeler to the BSW action (29). It contains no lapse, but, as we have seen, is exactly the kind of action that one would write down to implement (locally) Mach's requirement that time (duration) be derived from differences. Thus, there is an exactly right theory of duration at the heart of general relativity, but it is hidden in the standard representation.

However, this is not the end of the story. Quite apart from the implications of the two aspects of time – duration and simultaneity – for the quantum theory of the universe, there is also what Weyl [37] called the "disturbing question of length": Why does nature seem to violate the principle that size should be relative? We shall now see that a possible answer to this question may add yet another twist to the theory of time.

4.4. Conformal best matching

In best matching wrt 3-diffeomorphisms, we are in effect looking at all possible ways in which all points on one 3-geometry can be mapped bijectively to the points of an intrinsically different 3-geometry and selecting the bijection that extremalizes[16] the quantity chosen to measure the incongruence of the two. So far, we have not considered changing the local scale factor of the 3-metrics in accordance with the conformal transformations (18). But given that only angles are directly observable, we have good grounds for supposing that lengths should not occur as genuine dynamical degrees of freedom in the dynamics of geometry. If we best match wrt conformal transformations, only the angle-determining part of 3-metrics can play a dynamical role. Moreover, we have already noted (Sec. 4.1) the possibility that we might wish to best match only wrt volume-preserving conformal transformations.

At this point, it is helpful to recall the geometrical description of best matching in the particle model. It relies on a supermetric on the 'large' configuration space Q^N, which is foliated by the orbits of whatever group one is considering. Each orbit represents the intrinsic physical configuration of the system. Hitherto the supermetric chosen on the 'large' space (Q^N or Riem) has been equivariant, so that the orthogonal separation ds between neighbouring orbits is the same at all points on the orbits. This made it possible to calculate the orthogonal ds anywhere between the orbits and, knowing that the same value would always be obtained, project any such ds down to the physical quotient space. This met the key aim – to define a metric on the physical space.

Now there is in principle a different way in which this aim can be met. It arises if the orthogonal separation between the orbits is not constant *but*

[16] We have to extremalize rather than minimize because the DeWitt supermetric ($G^{ijkl}_{A=-1}$ in (22)) is indefinite. Einsteinian gravity is unique among all known physical fields in that its kinetic energy is not positive definite. The part associated with expansion of space – the second term in (25) – enters with the opposite sign to the part associated with the change of the conformal part of the 3-metric, i.e., its shape.

has a unique extremum at some point between any two considered orbits.[17] This unique extremal value can then be taken to define the required distance on the physical quotient space. I shall now indicate how this possibility can be implemented. Since the equations become rather complicated, I shall not attempt to give them in detail but merely outline what happens.

We start with the BSW action (29), since our choices have already been restricted to it by the local square root and the diffeomorphism best matching (neither of which we wish to sacrifice, though we will set $\Lambda = 0$ for simplicity). As just anticipated, we immediately encounter a significant difference from the best-matching wrt to 3-diffeomorphisms, for which we noted that (21) is invariant under λ-independent diffeomorphisms. In the language of gauge theory, (21) has a global (wrt λ) symmetry that is subsequently gauged by replacing the bare velocity g'_{ij} by the corrected velocity $g'_{ij} - \xi'_{(i;j)}$. It is the global symmetry which ensures that the inter-orbit separation in Riem is everywhere constant (equivariance). In contrast to the invariance of (21) under λ-independent diffeomorphisms, there is no invariance of (21) under λ-independent conformal transformations of the form (18). The kinetic term by itself is invariant, but $\sqrt{g}R$ is not. Indeed,

$$\sqrt{g}R \to \sqrt{g}\phi^4\sqrt{R - 8\frac{\nabla^2\phi}{\phi}}. \tag{30}$$

It should however be stressed that when (21) is 'conformalized' in accordance with (18) the resulting action is invariant under the combined gauge-type transformation

$$g_{ij} \to \omega^4 g_{ij}, \quad \phi \to \frac{\phi}{\omega}, \tag{31}$$

where $\omega = \omega(x, \lambda)$ is an arbitrary function. This exactly matches the invariance of (25) under 3-diffeomorphisms that arises because the transformation of g'_{ij} is offset by a compensating transformation of the best-matching correction $\xi'_{(i;j)}$. The only difference is that under the diffeomorphisms the velocities alone are transformed because of the prior choice of an action that is invariant under λ-independent transformations, whereas (31) generates transformations of both the dynamical variables and their velocities.

We now note that if we best match (25) wrt unrestricted conformal transformations, we run into a problem since we can make the action ever smaller by taking the value of ϕ ever smaller. Thus, we have no chance of finding an extremum of the action. There are two ways in which this difficulty can be resolved. The first mimics what we did in particle dynamics in order to implement the strong Poincaré principle on shape space, namely use a Lagrangian that overall has length dimension zero.

In the particle model we did this by dividing the kinetic metric ds by $\sqrt{I_{cms}}$, where I_{cms} is the cms moment of inertia. The analog of I_{cms} in

[17] To the best of my knowledge, this possibility (which certainly does not occur in gauge theory) was first considered by Ó Murchadha, who suggested it as a way to implement conformal best matching in [38].

geometrodynamics is V, the total volume of the universe, and division of the Lagrangian in (21) by $V^{2/3}$ achieves the desired result. This route is explored in [39]. It leads to a theory on conformal superspace that satisfies the strong Poincaré principle and is very similar to general relativity, except for an epoch-dependent emergent cosmological constant. This has the effect of enforcing $V =$ constant, with the consequence that the theory is incapable of explaining the diverse cosmological phenomena that are all so well explained by the theory of the expanding universe. The theory is not viable.

An alternative is to satisfy the weak Poincaré principle by restricting the conformal transformations (18) to be such that they leave the total volume unchanged. At the end of Sec. 4.1, I briefly described the consequences. Let me now give more details; for the full theory, see [40]. The physical space is initially chosen to be conformal superspace (CS), to which the space V of possible volumes V of the universe is adjoined, giving the space CS+V. One obtains a theory that in principle yields a unique curve between any two points in CS+V. These two points are specified by giving two conformal geometries c_1 and c_2, i.e., two points in CS, and associated volumes V_1 and V_2. However, there are two caveats. First, one cannot guarantee monotonicity of V. This difficulty can be avoided by passing from V to its canonically conjugate variable; in spacetime terms, this turns out to be K, the constant mean curvature of CMC hypersurfaces. Second, both V and K have dimensions and as such have no direct physical significance. Only the curves projected from CS+V to CS correspond to objective reality. In fact, a two-parameter family of curves in CS+V projects to a single-parameter family of curves in CS labelled by the dimensionless values of V_2/V_1 or, better, the monotonic K_2/K_1.[18]

A comparison with the standard variational principle for the N-body problem is here helpful. In it one specifies initial and final configurations in Q^N, i.e., $2 \times 3N$ numbers, together with a time *difference* $t_2 - t_1$. Thus, the variational problem is defined by $6N + 1$ numbers. However, the initial value problem requires only $6N$ numbers: a point in Q^N and the $3N$ numbers required to specify the (unconstrained) velocities at that point. In a geodesic problem, one requires respectively $6N$ and $6N-1$ numbers in the two different but essentially equivalent formulations. In the conformal theory, we thus have something very like a monotonic 'time', but it does not enter as a difference $t_2 - t_1$ but as the ratio K_2/K_1. This result seems to me highly significant because it shows (as just noted in the footnote) that in the shape-dynamic description of gravity one can interpret the local shapes of space as the true degrees of freedom and K_2/K_1 as an independent variable. As K_2/K_1 varies, the shapes interact with each other. This mirrors the interaction of particle positions in Newtonian dynamics as time, or, as we saw earlier, T/V changes.

[18]This fact escaped notice in [40]. Its detection led to [41], which shows that a point and a tangent vector in CS are sufficient to determine the evolution in CS. In turn, this means that the evolution is determined by exactly four local Hamiltonian shape degrees of freedom per space point. The paper [40] was written in the mistaken belief that one extra global degree of freedom, the value of V, also plays a true dynamical role.

However, the closer analogy in Newtonian dynamics is with the system's change of the shape as D_2/D_1 changes.

The only input data in this form of the shape-dynamic conformal theory are the initial point and tangent vector in CS. There is no trace of local inertial frames of reference, local proper time, or local proper distance. In the standard derivation of general relativity these are all presupposed in the requirement that locally spacetime can be approximated in a sufficiently small region by Minkowski space.[19] In contrast, in the conformal approach, this entire structure emerges from specification of a point and direction in conformal superspace.

One or two points may be made in this connection. First, as the reader can see in [40], the manner in which the theory selects a distinguished 3-geometry in a theory in which only conformal 3-geometry is presupposed relies on intimate interplay of the theory's ingredients. These are the local square root and the two different best matchings: wrt to diffeomorphisms and conformal transformations. Second, the construction of spacetime in a CMC foliation is fixed to the minutest detail from input that can in no way be reduced. Expressed in terms of two infinitesimally differing conformal 3-geometries C_1 and C_2, the outcome of the best matchings fixes the local scale factor $\sqrt{g}$ on C_1 and C_2, making them into 3-geometries G_1 and G_2. Thus, it takes one to a definite position in the conformal orbits. This is the big difference from the best matching with respect to diffeomorphisms alone and what happens in the particle model and gauge theory. In these cases the position in the orbit is not fixed. Next, the best matching procedure pairs each point on G_1 with a unique point on G_2 and determines a duration between them. In the spacetime that the theory 'constructs' the paired points are connected by spacetime vectors orthogonal to G_1 and G_2 and with lengths equal to definite (position-dependent) proper times. These are determined on the basis of the expression (24) for the canonical momenta, in which $W = gR$. The lapse N is $N = \sqrt{T/4R}$ and the amount of proper time $\delta\tau$ between the paired points is $\delta\tau = N\delta\lambda$. It is obvious that $\delta\tau$ is the outcome of a huge holistic process: the two best matchings together determine not only which points are to be paired but also the values at the paired points of all the quantities that occur in the expression $N = \sqrt{T/4R}$.

We can now see that there are two very different ways of interpreting general relativity. In the standard picture, spacetime is assumed from the beginning and it must locally have precisely the structure of Minkowski space. From the structural point of view, this is almost identical to an amalgam of Newton's absolute space and time. This near identity is reflected in the essential identity locally of Newton's first law and Einstein's geodesic law for the motion of an idealized point particle. In both cases, it must move in

[19] The 4-metric $g_{\mu\nu}$ has 10 components, of which four correspond to coordinate freedom. If one takes the view, dictated by general covariance, that all the remaining six are equally physical, then the entire theory rests on Minkowski space. One merely allows it to be bent, as is captured in the 'comma goes to semicolon' rule.

a straight line at a uniform speed. As I already mentioned, this very rigid initial structure is barely changed by Einstein's theory in its standard form. In Wheeler's aphorism [24], "Space tells matter how to move, matter tells space how to bend." But what we find at the heart of this picture is Newton's first law barely changed. No explanation for the law of inertia is given: it is a – one is tempted to say *the* – first principle of the theory. The wonderful structure of Einstein's theory as he constructed it rests upon it as a pedestal. I hope that the reader will at least see that there is another way of looking at the law of inertia: it is not the point of departure but the destination reached after a journey that takes into account all possible ways in which the configuration of the universe could change.

This bears on the debate about reductionism vs holism. I believe that the standard spacetime representation of general relativity helps to maintain the plausibility of a reductionist approach. Because Minkowski's spacetime seems to be left essentially intact in local regions, I think many people (including those working in quantum field theory in external spacetimes) unconsciously assume that the effect of the rest of the universe can be ignored. Well, for some things it largely can. However, I feel strongly that the creation of quantum gravity will force us to grasp the nettle. What happens locally is the outcome of everything in the universe. We already have a strong hint of this from the classical theory, which shows that the 'reassuring' local Minkowskian framework is determined – through elliptic equations in fact – by every last structural detail in the remotest part of the universe.

4.5. Shape dynamics or general relativity?

There is no question that general relativity has been a wonderful success and as yet has passed every experimental test. The fact that it predicts singularities is not so much a failure of the theory as an indication that quantum gravity must at some stage come into play and 'take over'. A more serious criticism often made of general relativity is that its field equations $G_{\mu\nu} = T_{\mu\nu}$ allow innumerable solutions that strike one as manifestly unphysical, for example, the ones containing closed timelike curves. There is a good case for seeking a way to limit the number of solutions. Perhaps the least controversial is the route chosen by Dirac [21] and Arnowitt, Deser, and Misner (ADM) [22]. The main justification for their 3+1 dynamical approach is the assumption that gravity can be described in the Hamiltonian framework, which is known to be extremely effective in other branches of physics and especially in quantum mechanics.

If a Hamiltonian framework is adopted, it then becomes especially attractive to assume that the universe is spatially closed. This obviates the need for arbitrary boundary conditions, and, as Einstein put it when discussing Mach's principle ([42], p. 62), "the series of causes of mechanical phenomena [is] closed".

The main difficulty in suggesting that the spacetime picture should be replaced by the more restrictive Hamiltonian framework arises from the relativity principle, i.e., the denial of any distinguished definition of simultaneity.

In the ideal form of Hamiltonian theory, one seeks to have the dynamics represented by a *unique* curve in a phase space of *true Hamiltonian degrees of freedom*. This is equivalent to having a unique curve in a corresponding configuration space of true geometrical degrees of freedom even if mathematical tractability means that the calculations must always be made in Riem. Dirac and ADM showed that dynamics in Riem could be interpreted in superspace, thereby reducing the six degrees of freedom per space point in a 3-metric to the three in a 3-geometry. But the slicing freedom within spacetime means that a single spacetime still corresponds to an infinity of curves in superspace. The Hamiltonian ideal is not achieved. The failure is tantalizing, because much evidence suggests that gravity has only *two* degrees of freedom per space point, hinting at a configuration space smaller than superspace.

As long as relativity of simultaneity is held to be sacrosanct, there is no way forward to the Hamiltonian ideal. York and Wheeler came close to suggesting that it was to be found in conformal superspace, but ultimately balked at jettisoning the relativity principle.[20] In this connection, it is worth pointing out that Einstein's route to general relativity occurred at a particular point in history and things could have been approached differently. I think it entirely possible that Einstein's discovery of his theory of gravitation in spacetime form could be seen as a glorious historical accident. In particular, Einstein could easily have looked differently at certain fundamental issues related to the nature of space, time, and motion. Let me end this introduction to shape dynamics with some related observations on each.

Space. Riemann based his generalization of Euclidean geometry on *length* as fundamental. It was only in 1918, three years after the creation of general relativity, that Weyl [19, 20] challenged this and identified – in a four-dimensional context – *angles* as more fundamental. I will argue elsewhere that Weyl's attempt to generalize general relativity to eliminate the correctly perceived weakness of Riemann's foundations failed because it was not sufficiently radical – instead of eliminating length completely from the foundations, Weyl retained it in a less questionable form.

Time. As I noted earlier, in 1898 Poincaré [34, 35] identified *two* equally fundamental problems related to time: how is one to define *duration* and how is one to define *simultaneity* at spatially separated points? Einstein attacked the second problem brilliantly but made no attempt to put a solution to the second into the foundations of general relativity.

Motion. In the critique of Newtonian mechanics that was such a stimulus to general relativity, Mach argued that only relative velocities should

[20]York's highly important work on the initial-value problem of general relativity [43, 44] is intimately related to the shape-dynamic programme and was one of its inspirations. For a discussion of the connections, see [40]. One of the arguments for the shape-dynamic approach is that it provides a first-principles *derivation* of York's method, which in its original form was found by trial and error. It may also be noted here that York's methods, which were initially developed for the vacuum (matter-free) case, can be extended to include matter [45, 46]. This suggests that the principles of shape dynamics will extend to the case in which matter is present.

occur in dynamics. Einstein accepted this aspiration, but did not attempt to put it directly into the foundations of general relativity, arguing that it was impractical ([3, 5], p. 186). Instead it was necessary to use coordinate systems and achieve Mach's ideal by putting them all on an equal footing (general covariance).

All three alternatives in approach listed above are put directly into shape dynamics. I think that this has been made adequately clear with regard to the treatment of space and motion. I wish to conclude with a comment on the treatment of time, which is rather more subtle.

It is well known that Einstein regarded special relativity as a principles theory like thermodynamics, which was based on human experience: heat energy never flows spontaneously from a cold to a hot body. Similarly, uniform motion was always found to be indistinguishable – within a closed system – from rest. Einstein took this fact as the basis of relativity and never attempted to explain effects like time dilatation at a microscopic level in the way Maxwell and Boltzmann developed the atomic statistical theory of thermodynamics. Since rods and clocks are ultimately quantum objects, I do not think such a programme can be attempted before we have a better idea of the basic structure of quantum gravity. However, I find it interesting and encouraging that *a microscopic theory of duration* is built in at a very basic level in shape dynamics. This is achieved in particle dynamics using Jacobi's principle, which leads to a global definition of duration, and in conformal dynamics using the local-square-root action (21). I also find it striking that, as already noted, the simple device of eliminating the lapse from Einstein's spacetime theory immediately transforms his theory from one created without any thought of a microscopic theory of duration into one (based on the Baierlein–Sharp–Wheeler action (29)) that has such a theory at its heart. A theory of duration was there all along. It merely had to be uncovered by removing some of the structure that Einstein originally employed – truly a case of less is more.

The effect of the local square root is remarkable. At the level of theory creation in superspace, in which length is taken as fundamental, the local square root acts as an extremely powerful selector of consistent theories and, as we have seen, enforces the appearance of the slicing freedom, universality of the light cone, and gauge fields as the simplest bosonic fields that couple to dynamic geometry. As I have just noted, it also leads to a microscopic theory of local duration (local proper time). Thus, the mere inclusion of the local square root goes a long way to establishing a constructive theory of special-relativistic effects. It is not the whole way, because quantum mechanics must ultimately explain why physically realized clocks measure the local proper time created by the local theory of duration.

The effect of the local square root is even more striking when applied in theory creation in conformal superspace. It still enforces universality of the light cone and the appearance of gauge fields but now does two further things. First, it leads to a *microscopic theory of length.* For the conformal best matching, in conjunction with the constraints that follow from the local square root, fixes a distinguished scale factor of the 3-metric. Second, it introduces the distinguished CMC foliation within spacetime without changing any of the classical predictions of general relativity. It leads to a *theory of simultaneity*.

Thus, the conformal approach to geometrodynamics suggests that there are two candidate theories of gravity that can be derived from different first principles. Einstein's general relativity is based on the idea that spacetime is the basic ontology; its symmetry group is four-dimensional diffeomorphism invariance. But there is also an alternative *dual* theory based on three-dimensional diffeomorphism invariance and conformal best matching [9, 10, 40, 41]. The set of allowed solutions of the conformal theory is significantly smaller than the general relativity set. In principle, this is a good feature, since it makes the conformal theory more predictive, but it cannot be ruled out that, being tied to CMC foliations, the conformal theory will be unable to describe physically observable situations that are correctly described by general relativity.

I will end with two comments. First, shape dynamics in conformal superspace is a new and mathematically well-defined framework of dynamics. Second, its physical applications are most likely to be in quantum gravity.

Acknowledgements

The work reported in this paper grew out of collaboration with (in chronological order) Bruno Bertotti, Niall Ó Murchadha, Brendan Foster, Edward Anderson, Bryan Kelleher, Henrique Gomes, Sean Gryb, and Tim Koslowski. Discussions over many years with Karel Kuchař and Lee Smolin were also very helpful, as were extended discussions with Jimmy York in 1992. I am much indebted to them all. Thanks also to Boris Barbour for creating the figures. This work was funded by a grant from the Foundational Questions Institute (FQXi) Fund, a donor advised fund of the Silicon Valley Community Foundation on the basis of proposal FQXi-RFP2-08-05 to the Foundational Questions Institute.

References

[1] E Mach. *Die Mechanik in ihrer Entwickelung historisch-kritisch dargestellt*. 1883.

[2] E Mach. *The Science of Mechanics*. Open Court, 1960.

[3] A Einstein. Prinzipielles zur allgemeinen Relativitätstheorie. *Annalen der Physik*, 55:241–244, 1918.

[4] A Einstein. Dialog über Einwände gegen die Relativitätstheorie. *Die Naturwissenschaften*, 6:697–702, 1918.

[5] J Barbour and H Pfister, editors. *Mach's Principle: From Newton's Bucket to Quantum Gravity*, volume 6 of *Einstein Studies*. Birkhäuser, Boston, 1995.

[6] J Barbour. *Absolute or Relative Motion? Volume 1. The Discovery of Dynamics*. Cambridge University Press, 1989.

[7] J Barbour. *The Discovery of Dynamics.* Oxford University Press, 2001.

[8] I Newton. *Sir Isaac Newton's Mathematical Principles of Natural Philosophy.* University of California Press, 1962.

[9] H Gomes, S Gryb, and T Koslowski. Einstein gravity as a 3D conformally invariant theory (arXiv:1010.2481). *Class. Quant. Grav.*, 28:045005, 2011.

[10] H Gomes and T Koslowski. The link between general relativity and shape dynamics. arXiv:1101.5974.

[11] J Barbour and B Bertotti. Mach's principle and the structure of dynamical theories (downloadable from platonia.com). *Proceedings of the Royal Society London A*, 382:295–306, 1982.

[12] J Barbour. Scale-invariant gravity: particle dynamics. *Class. Quantum Grav.*, 20:1543–1570, 2003, gr-qc/0211021.

[13] D Saari. *Collisions, Rings, and Other Newtonian N-Body Problems.* American Mathematical Society, Providence, Rhode Island, 2005.

[14] H Poincaré. *Science et Hypothèse.* Paris, 1902.

[15] H Poincaré. *Science and Hypothesis.* Walter Scott, London, 1905.

[16] J Barbour. The definition of Mach's principle (arXiv:1007.3368). *Found. Phys.*, 40:1263–1284, 2010.

[17] C Lanczos. *The Variational Principles of Mechanics.* University of Toronto Press, 1949.

[18] P A M Dirac. *Lectures on Quantum Mechanics.* Belfer Graduate School of Science, Yeshiva University, New York, 1964.

[19] H Weyl. Gravitation und Elektrizität. *Sitzungsber. Preuss. Akad. Berlin*, pages 465–480, 1918.

[20] H Weyl. Gravitation and electricity. In Ó Raifeartaigh, editor, *The dawning of gauge theory*, pages 24–37. Princeton University Press, 1997.

[21] P A M Dirac. Generalized Hamiltonian dynamics. *Proc. R. Soc. (London)*, A246:326–343, 1958.

[22] R Arnowitt, S Deser, and C W Misner. The dynamics of general relativity. In L Witten, editor, *Gravitation: An Introduction to Current Research*, pages 227–265. Wiley, New York, 1962.

[23] B F Schutz. *Geometrical Methods of Mathematical Physics.* Cambridge University Press, Cambridge, 1980.

[24] C W Misner, K S Thorne, and J A Wheeler. *Gravitation.* W H Freeman and Company, San Francisco, 1973.

[25] K Kuchař. Time and interpretations of quantum gravity. In G Kunstatter, D Vincent, and J Williams, editors, *Proceedings 4th Canadian Conf. General Relativity and Relativistic Astrophysics*, pages 211–314. World Scientific, Singapore, 1992.

[26] C J Isham. Canonical quantum gravity and the problem of time. In L A Ibort and M A Rodríguez, editors, *Integrable Systems, Quantum Groups, and Quantum Field Theory*, pages 157–287. Kluwer, Dordrecht, 1993.

[27] J Barbour. *The End of Time.* Weidenfeld and Nicolson, London; Oxford University Press, New York, 1999.

[28] J Barbour, B Z Foster, and N Ó Murchadha. Relativity without relativity. *Class. Quantum Grav.*, 19:3217–3248, 2002, gr-qc/0012089.

[29] E Anderson. On the recovery of geometrodynamics from two different sets of first principles. *Stud. Hist. Philos. Mod. Phys.*, 38:15, 2007. arXiv:gr-qc/0511070.

[30] E Anderson. Does relationalism alone control geometrodynamics with sources? 2007. arXiv:0711.0285.

[31] R Baierlein, D Sharp, and J Wheeler. Three-dimensional geometry as a carrier of information about time. *Phys. Rev.*, 126:1864–1865, 1962.

[32] E Anderson and J Barbour. Interacting vector fields in relativity without relativity. *Class. Quantum Grav.*, 19:3249–3262, 2002, gr-qc/0201092.

[33] A Einstein. Zur Elektrodynamik bewegter Körper. *Ann. Phys.*, 17:891–921, 1905.

[34] H Poincaré. La mesure du temps. *Rev. Métaphys. Morale*, 6:1, 1898.

[35] H Poincaré. The measure of time. In *The Value of Science*. 1904.

[36] A Einstein. Autobiographical notes. In P Schilpp, editor, *Albert Einstein: Philosopher–Scientist*. Harper and Row, New York, 1949.

[37] H Weyl. *Symmetry*. Princeton University Press, 1952.

[38] J Barbour and N Ó Murchadha. Classical and quantum gravity on conformal superspace. 1999, gr-qc/9911071.

[39] E Anderson, J Barbour, B Z Foster, and N Ó Murchadha. Scale-invariant gravity: geometrodynamics. *Class. Quantum Grav.*, 20:1571, 2003, gr-qc/0211022.

[40] E Anderson, J Barbour, B Z Foster, B Kelleher, and N Ó Murchadha. The physical gravitational degrees of freedom. *Class. Quantum Grav.*, 22:1795–1802, 2005, gr-qc/0407104.

[41] J Barbour and N Ó Murchadha. Conformal Superspace: the configuration space of general relativity, arXiv:1009.3559.

[42] A Einstein. *The Meaning of Relativity*. Methuen and Co Ltd, London, 1922.

[43] J W York. Gravitational degrees of freedom and the initial-value problem. *Phys. Rev. Letters*, 26:1656–1658, 1971.

[44] J W York. The role of conformal 3-geometry in the dynamics of gravitation. *Phys. Rev. Letters*, 28:1082–1085, 1972.

[45] J Isenberg, N Ó Murchadha, and J W York. Initial-value problem of general relativity. III. *Phys. Rev. D*, 12:1532–1537, 1976.

[46] J Isenberg and J Nester. Extension of the York field decomposition to general gravitationally coupled fields. *Ann. Phys.*, 108:368–386, 1977.

Julian Barbour
College Farm, South Newington
Banbury, Oxon, OX15 4JG
UK
e-mail: `Julian.Barbour@physics.ox.ac.uk`
`julian@platonia.com`

On the Motion of Point Defects in Relativistic Fields

Michael K.-H. Kiessling

Abstract. We inquire into classical and quantum laws of motion for the point charge sources in the nonlinear Maxwell–Born–Infeld field equations of classical electromagnetism in flat and curved Einstein spacetimes.

Mathematics Subject Classification (2010). 78A02, 78A25, 83A05, 83C10, 83C50, 83C75.

Keywords. Electromagnetism, point charges, Maxwell–Born–Infeld field equations, laws of motion.

1. Introduction

Shortly after it was realized that there was a problem, quantum physics was invented. Since then we have been doing quantum physics, and the problem was forgotten. But it is still with us!

Detlef Dürr and Sergio Albeverio, late 1980s

The "forgotten problem" that Dürr and Albeverio were talking about some 20+ years ago is the construction of a consistent classical theory for the joint evolution of electromagnetic fields and their point charge sources. Of course the problem was not completely forgotten, but it certainly has become a backwater of mainstream physics with its fundamental focus on quantum theory: first quantum field theory and quantum gravity, then string, and in recent years now M-theory. Unfortunately, more than a century of research into quantum physics has not yet produced a consistent quantum field theory of the electromagnetic interactions without artificial irremovable mathematical regularizers; the incorporation of the weak and strong interactions in the standard model has not improved on this deficiency. The consistent theory of quantum gravity has proved even more elusive, and nobody (presumably) knows whether M-theory is ever going to see the light of the day. So it may yet turn out that the "forgotten classical problem" will be solved first.

In the following, I will report on some exciting recent developments towards the solution of the "forgotten classical problem" in general-relativistic spacetimes, in terms of the coupling of the Einstein–Maxwell–Born–Infeld theory for an electromagnetic spacetime with point defects (caused by point charge sources) to a Hamilton–Jacobi theory of motion for these point defects. Mostly I will talk about the special-relativistic spacetime limit, though. Since I want to emphasize the evolutionary aspects of the theory, I will work in a space+time splitting of spacetime rather than using the compact formalism of spacetime geometry. Furthermore I will argue that this putative solution to the classical problem also teaches us something new about the elusive consistent quantum theory of electromagnetism with point sources, and its coupling to gravity. Namely, while the spacetime structure and the electromagnetic fields will still be treated at the classical level, replacing our classical Hamilton–Jacobi law of motion for the electromagnetic point defects by a de Broglie–Bohm–Dirac quantum law of motion for the point defects yields a reasonable "first quantization with spin" of the classical theory of motion of point defects in the fields, which has the additional advantage that it doesn't suffer from the infamous measurement problem. In all of these approaches, the structure of spacetime is classical. I will have to leave comments on the pursuit of the photon and the graviton, and quantum spacetimes to a future contribution.

I now begin by recalling the "forgotten classical problem."

2. Lorentz electrodynamics with point charges

I briefly explain why the formal equations of classical Lorentz electrodynamics with point charges fail to yield a well-defined classical theory of electromagnetism.[1]

2.1. Maxwell's field equations

I prepare the stage by recalling Maxwell's field equations of electromagnetism in Minkowski spacetime, written with respect to any convenient flat foliation (a.k.a. Lorentz frame) into space points $\boldsymbol{s} \in \mathbb{R}^3$ at time $t \in \mathbb{R}$. Suppose a relativistic theory of matter has supplied an electric charge density $\rho(t, \boldsymbol{s})$ and an electric current vector-density $\boldsymbol{j}(t, \boldsymbol{s})$, satisfying the local law of charge conservation,

$$\frac{\partial}{\partial t}\rho(t, \boldsymbol{s}) + \boldsymbol{\nabla} \cdot \boldsymbol{j}(t, \boldsymbol{s}) = 0. \tag{2.1}$$

Maxwell's electromagnetic field equations comprise the two evolution equations

$$\frac{1}{c}\frac{\partial}{\partial t}\boldsymbol{\mathcal{B}}_{\mathrm{M}}(t, \boldsymbol{s}) = -\boldsymbol{\nabla} \times \boldsymbol{\mathcal{E}}_{\mathrm{M}}(t, \boldsymbol{s}), \tag{2.2}$$

$$\frac{1}{c}\frac{\partial}{\partial t}\boldsymbol{\mathcal{D}}_{\mathrm{M}}(t, \boldsymbol{s}) = +\boldsymbol{\nabla} \times \boldsymbol{\mathcal{H}}_{\mathrm{M}}(t, \boldsymbol{s}) - 4\pi\frac{1}{c}\boldsymbol{j}(t, \boldsymbol{s}), \tag{2.3}$$

[1] I hope that this also dispels the perennial myth in the plasma physics literature that these ill-defined equations were "the fundamental equations of a classical plasma."

and the two constraint equations

$$\nabla \cdot \boldsymbol{\mathcal{B}}_{\mathrm{M}}(t, \boldsymbol{s}) = 0\,, \tag{2.4}$$

$$\nabla \cdot \boldsymbol{\mathcal{D}}_{\mathrm{M}}(t, \boldsymbol{s}) = 4\pi\rho(t, \boldsymbol{s})\,. \tag{2.5}$$

These field equations need to be supplemented by a relativistic "constitutive law" which expresses the electric and magnetic fields $\boldsymbol{\mathcal{E}}_{\mathrm{M}}$ and $\boldsymbol{\mathcal{H}}_{\mathrm{M}}$ in terms of the magnetic induction field $\boldsymbol{\mathcal{B}}_{\mathrm{M}}$ and the electric displacement field $\boldsymbol{\mathcal{D}}_{\mathrm{M}}$. The constitutive law reflects the "constitution of matter" and would have to be supplied by the theory of matter carrying ρ and $\boldsymbol{j}$. (Later we will adopt a different point of view.)

For matter-free space Maxwell proposed

$$\boldsymbol{\mathcal{H}}_{\mathrm{M}}(t, \boldsymbol{s}) = \boldsymbol{\mathcal{B}}_{\mathrm{M}}(t, \boldsymbol{s})\,, \tag{2.6}$$

$$\boldsymbol{\mathcal{E}}_{\mathrm{M}}(t, \boldsymbol{s}) = \boldsymbol{\mathcal{D}}_{\mathrm{M}}(t, \boldsymbol{s})\,. \tag{2.7}$$

The system of Maxwell field equations (2.2), (2.3), (2.4), (2.5) with $\rho \equiv 0$ and $\boldsymbol{j} \equiv \mathbf{0}$, supplemented by Maxwell's "law of the pure aether" (2.6) and (2.7), will be called the *Maxwell–Maxwell field equations.* They feature a large number of conserved quantities [AnTh2005], including the field energy, the field momentum, and the field angular momentum, given by, respectively (cf. [Abr1905], [Jac1975]),

$$E_{\mathrm{f}} = \frac{1}{8\pi} \int_{\mathbb{R}^3} \left(|\boldsymbol{\mathcal{E}}_{\mathrm{MM}}(t, \boldsymbol{s})|^2 + |\boldsymbol{\mathcal{B}}_{\mathrm{MM}}(t, \boldsymbol{s})|^2 \right) \mathrm{d}^3 s, \tag{2.8}$$

$$\boldsymbol{P}_{\mathrm{f}} = \frac{1}{4\pi c} \int_{\mathbb{R}^3} \boldsymbol{\mathcal{E}}_{\mathrm{MM}}(t, \boldsymbol{s}) \times \boldsymbol{\mathcal{B}}_{\mathrm{MM}}(t, \boldsymbol{s})\, \mathrm{d}^3 s\,, \tag{2.9}$$

$$\boldsymbol{L}_{\mathrm{f}} = \frac{1}{4\pi c} \int_{\mathbb{R}^3} \boldsymbol{s} \times \left(\boldsymbol{\mathcal{E}}_{\mathrm{MM}}(t, \boldsymbol{s}) \times \boldsymbol{\mathcal{B}}_{\mathrm{MM}}(t, \boldsymbol{s}) \right) \mathrm{d}^3 s\,. \tag{2.10}$$

These integrals will retain their meanings also in the presence of point sources.

2.2. The Maxwell–Lorentz field equations

Although Maxwell pondered atomism — think of Maxwell's velocity distribution and the Maxwell–Boltzmann equation in the kinetic theory of gases —, it seems that he did not try to implement atomistic notions of matter into his electromagnetic field equations. This step had to wait until the electron was discovered, by Wiechert [Wie1897] and Thomson [Tho1897] (see [Pip1997] and [Jos2002]). Assuming the electron to be a point particle with charge $-e$, the Maxwell field equations for a single electron embedded in Maxwell's "pure aether" at $\boldsymbol{Q}(t) \in \mathbb{R}^3$ at time t become the *Maxwell–Lorentz field equations* (for a single electron),

$$\tfrac{1}{c}\tfrac{\partial}{\partial t} \boldsymbol{\mathcal{B}}_{\mathrm{ML}}(t, \boldsymbol{s}) = -\nabla \times \boldsymbol{\mathcal{E}}_{\mathrm{ML}}(t, \boldsymbol{s})\,, \tag{2.11}$$

$$\tfrac{1}{c}\tfrac{\partial}{\partial t} \boldsymbol{\mathcal{E}}_{\mathrm{ML}}(t, \boldsymbol{s}) = +\nabla \times \boldsymbol{\mathcal{B}}_{\mathrm{ML}}(t, \boldsymbol{s}) + 4\pi e \delta_{\boldsymbol{Q}(t)}(\boldsymbol{s}) \tfrac{1}{c} \dot{\boldsymbol{Q}}(t)\,, \tag{2.12}$$

$$\nabla \cdot \boldsymbol{\mathcal{B}}_{\mathrm{ML}}(t, \boldsymbol{s}) = 0\,, \tag{2.13}$$

$$\nabla \cdot \boldsymbol{\mathcal{E}}_{\mathrm{ML}}(t, \boldsymbol{s}) = -4\pi e \delta_{\boldsymbol{Q}(t)}(\boldsymbol{s})\,, \tag{2.14}$$

where "$\delta(\,\cdot\,)$" is Dirac's delta function, and $\dot{\boldsymbol{Q}}(t)$ the velocity of the point electron. Note that the point charge "density" $\rho(t,\boldsymbol{s}) \equiv -e\delta_{\boldsymbol{Q}(t)}(\boldsymbol{s})$ and current vector-"density" $\boldsymbol{j}(t,\boldsymbol{s}) \equiv -e\delta_{\boldsymbol{Q}(t)}(\boldsymbol{s})\dot{\boldsymbol{Q}}(t)$ jointly satisfy the continuity equation (2.1) in the sense of distributions; of course, the Maxwell–Lorentz field equations have to be interpreted in the sense of distributions, too. As is well-known, the Maxwell–Lorentz field equations are covariant under the Poincaré group; of course, the position and velocity of all the point charges are transformed accordingly as well.

Given any twice continuously differentiable, subluminal ($|\dot{\boldsymbol{Q}}(t)| < c$) motion $t \mapsto \boldsymbol{Q}(t)$, the Maxwell–Lorentz field equations are solved by

$$\boldsymbol{\mathcal{E}}^{\mathrm{ret}}_{\mathrm{LW}}(t,\boldsymbol{s}) = -e\frac{1}{(1-\boldsymbol{n}\cdot\dot{\boldsymbol{Q}}/c)^3}\Big(\frac{\boldsymbol{n}-\dot{\boldsymbol{Q}}/c}{\gamma^2 r^2} + \frac{\boldsymbol{n}\times[(\boldsymbol{n}-\dot{\boldsymbol{Q}}/c)\times\ddot{\boldsymbol{Q}}/c^2]}{r}\Big)\Big|_{\mathrm{ret}} \tag{2.15}$$

$$\boldsymbol{\mathcal{B}}^{\mathrm{ret}}_{\mathrm{LW}}(t,\boldsymbol{s}) = \boldsymbol{n}\big|_{\mathrm{ret}} \times \boldsymbol{\mathcal{E}}^{\mathrm{ret}}_{\mathrm{LW}}(t,\boldsymbol{s})\,, \tag{2.16}$$

where $\boldsymbol{n} = (\boldsymbol{s}-\boldsymbol{Q}(t))/r$ with $r = |\boldsymbol{s}-\boldsymbol{Q}(t)|$, $\gamma^2 = 1/(1-|\dot{\boldsymbol{Q}}(t)|^2/c^2)$, and where "ret" means that the t-dependent functions $\boldsymbol{Q}(t)$, $\dot{\boldsymbol{Q}}(t)$, and $\ddot{\boldsymbol{Q}}(t)$ are to be evaluated at the retarded time t_{ret} defined implicitly by $c(t-t_{\mathrm{ret}}) = |\boldsymbol{s}-\boldsymbol{Q}(t_{\mathrm{ret}})|$; see [Lié1898], [Wie1900]. By linearity, the general solution of the Maxwell–Lorentz equations with many point sources is now obtained by adding all their pertinent Liénard–Wiechert fields to the general solution to the Maxwell–Maxwell equations.

As the Maxwell–Lorentz field equations are consistent with *any* smooth subluminal motion $t \mapsto \boldsymbol{Q}(t)$, to determine the physical motions a *law of motion* for the point electron has to be supplied. For the physicist of the late 19th and early 20th centuries this meant a Newtonian law of motion.

2.3. The Lorentz force

2.3.1. The test charge approximation: all is well! Lorentz [Lor1904], Poincaré [Poi1905/6], and Einstein [Ein1905a] showed that at the level of the *test particle* approximation, Newton's law for the rate of change of its mechanical momentum, equipped with the Lorentz force [Lor1892]

$$\frac{\mathrm{d}\boldsymbol{P}(t)}{\mathrm{d}t} = -e\Big[\boldsymbol{\mathcal{E}}(t,\boldsymbol{Q}(t)) + \tfrac{1}{c}\dot{\boldsymbol{Q}}(t)\times\boldsymbol{\mathcal{B}}(t,\boldsymbol{Q}(t))\Big], \tag{2.17}$$

combined with the relativistic law between mechanical momentum and velocity,

$$\frac{1}{c}\frac{\mathrm{d}\boldsymbol{Q}(t)}{\mathrm{d}t} = \frac{\boldsymbol{P}(t)}{\sqrt{m_{\mathrm{e}}^2c^2 + |\boldsymbol{P}(t)|^2}}, \tag{2.18}$$

provides an empirically highly accurate law of motion for the point electron; in (2.17), $\boldsymbol{\mathcal{E}} = \boldsymbol{\mathcal{E}}_{\mathrm{MM}}$ and $\boldsymbol{\mathcal{B}} = \boldsymbol{\mathcal{B}}_{\mathrm{MM}}$ are solutions of the Maxwell–Maxwell field equations, and the m_{e} in (2.18) is the empirical inert rest mass of the physical electron, which is *defined* through this test particle law! This law of test particle motion is globally well-posed as a Cauchy problem for the map

$t \mapsto (\boldsymbol{P}(t), \boldsymbol{Q}(t))$ *given* any Lipschitz continuous so-called *external fields* $\boldsymbol{\mathcal{E}}_{\rm MM}$ and $\boldsymbol{\mathcal{B}}_{\rm MM}$.

2.3.2. The Lorentz self-force: infinite in all directions! Clearly, fundamentally there is no such thing as a "test charge" in "external fields;" only the total fields, the solutions to the Maxwell–Lorentz field equations with the point charge as source, can be fundamental in this theory. However, the law of motion (2.17), (2.18) is *a priori undefined* when formally coupled with (2.11), (2.12), (2.13), (2.14), so that $\boldsymbol{\mathcal{E}} = \boldsymbol{\mathcal{E}}_{\rm ML} \equiv \boldsymbol{\mathcal{E}}_{\rm LW}^{\rm ret} + \boldsymbol{\mathcal{E}}_{\rm MM}$ and $\boldsymbol{\mathcal{B}} = \boldsymbol{\mathcal{B}}_{\rm ML} \equiv \boldsymbol{\mathcal{B}}_{\rm LW}^{\rm ret} + \boldsymbol{\mathcal{B}}_{\rm MM}$ in (2.17); indeed, inspection of (2.15), (2.16) makes it plain that "$\boldsymbol{\mathcal{E}}_{\rm LW}^{\rm ret}(t, \boldsymbol{Q}(t))$" and "$\boldsymbol{\mathcal{B}}_{\rm LW}^{\rm ret}(t, \boldsymbol{Q}(t))$" are "infinite in all directions," by which I mean that for any limit $\boldsymbol{s} \to \boldsymbol{Q}(t)$ of $\boldsymbol{\mathcal{E}}_{\rm LW}^{\rm ret}(t, \boldsymbol{s})$ and $\boldsymbol{\mathcal{B}}_{\rm LW}^{\rm ret}(t, \boldsymbol{s})$, the field magnitudes diverge to infinity while their limiting directions, whenever such exist, depend on how the limit is taken.

Lorentz and his peers interpreted the infinities to mean that the electron is not really a point particle. To uncover its structure by computing details of the motion which depend on its structure became the goal of "classical electron theory" [Lor1895, Wie1896, Abr1903, Lor1904, Abr1905, Lor1915]. This story is interesting in its own right, see [Roh1990, Yag1992], but would lead us too far astray from our pursuit of a well-defined theory of electromagnetism with point charges.[2]

2.3.3. Regularization and renormalization? Precisely because "$\boldsymbol{\mathcal{E}}_{\rm LW}^{\rm ret}(t, \boldsymbol{Q}(t))$" and "$\boldsymbol{\mathcal{B}}_{\rm LW}^{\rm ret}(t, \boldsymbol{Q}(t))$" are "infinite *in all directions*," it is tempting to inquire into the possibility of *defining* r.h.s.(2.17) for solutions of the Maxwell–Lorentz field equations through a point limit $\varrho(\boldsymbol{s}|\boldsymbol{Q}(t)) \to \delta_{\boldsymbol{Q}(t)}(\boldsymbol{s})$ of a Lorentz force field $\boldsymbol{\mathcal{E}}_{\rm ML}(t, \boldsymbol{s}) + \frac{1}{c}\dot{\boldsymbol{Q}}(t) \times \boldsymbol{\mathcal{B}}_{\rm ML}(t, \boldsymbol{s})$ averaged over some normalized regularizing density $\varrho(\boldsymbol{s}|\boldsymbol{Q}(t))$. Of course, for the Maxwell–Maxwell part of the Maxwell–Lorentz fields this procedure yields $\boldsymbol{\mathcal{E}}_{\rm MM}(t, \boldsymbol{Q}(t)) + \frac{1}{c}\dot{\boldsymbol{Q}}(t) \times \boldsymbol{\mathcal{B}}_{\rm MM}(t, \boldsymbol{Q}(t))$. For the Liénard–Wiechert fields on the other hand such a definition via regularization cannot be expected to be unique (if it leads to a finite result at all); indeed, one should expect that one can obtain *any* limiting averaged force vector by choosing a suitable family of averages around the location of

---

[2]However, I do take the opportunity to advertise a little-known but very important fact. By postulating energy-momentum (and angular momentum) conservation of the fields alone, Abraham and Lorentz derived an effective Newtonian test particle law of motion from their "purely electromagnetic models" in which both the external Lorentz force *and the particle's inertial mass* emerged through an expansion in a small parameter.

Yet, in [ApKi2001] Appel and myself showed that the Abraham–Lorentz proposal is mathematically *inconsistent*: generically their "fundamental law of motion" does not admit a solution at all! But how could Abraham and Lorentz arrive at the correct approximate law of motion in the leading order of their expansion? The answer is: if you take an inconsistent nonlinear equation and *assume* that it has a solution which provides a small parameter, then formally expand the equation and its hypothetical solution in a power series w.r.t. this parameter, then truncate the expansion and treat the retained expansion coefficients as free parameters to be matched to empirical data, *then* you may very well end up with an accurate equation — all the inconsistencies are hidden in the pruned-off part of the expansion! But you haven't *derived* anything, seriously.

the point charge, so that such a definition of the Lorentz force is quite arbitrary. In the best case one could hope to find some physical principle which selects a unique way of averaging the Lorentz force field and of taking the point limit of it. Meanwhile, in the absence of any such principle, one may argue that anything else but taking the limit $R \to 0$ of a uniform average of $\boldsymbol{\mathcal{E}}^{\rm ret}_{\rm LW}(t,\boldsymbol{s}) + \frac{1}{c}\dot{\boldsymbol{Q}}(t) \times \boldsymbol{\mathcal{B}}^{\rm ret}_{\rm LW}(t,\boldsymbol{s})$ over a sphere of radius R centered at $\boldsymbol{Q}(t)$ *in the instantaneous rest frame* of the point charge, followed by a boost back into the original Lorentz frame, would be *perverse.*

Unfortunately, work by Dirac [Dir1938] has revealed that the point limit of such a "spherical averaging" does *not* produce a finite r.h.s.(2.17) with $\boldsymbol{\mathcal{E}}^{\rm ret}_{\rm LW} + \boldsymbol{\mathcal{E}}_{\rm MM}$ and $\boldsymbol{\mathcal{B}}^{\rm ret}_{\rm LW} + \boldsymbol{\mathcal{B}}_{\rm MM}$ in place of $\boldsymbol{\mathcal{E}}$ and $\boldsymbol{\mathcal{B}}$. Instead, to leading order in powers of R the spherical average of the Lorentz "self-force" field $\boldsymbol{\mathcal{E}}^{\rm ret}_{\rm LW}(t,\boldsymbol{s}) + \frac{1}{c}\dot{\boldsymbol{Q}}(t) \times \boldsymbol{\mathcal{B}}^{\rm ret}_{\rm LW}(t,\boldsymbol{s})$ is given by[3] $-\frac{e^2}{2c^2}R^{-1}\frac{\partial}{\partial t}(\gamma(t)\dot{\boldsymbol{Q}}(t))$. Unless the acceleration vanishes at time t, this term diverges $\uparrow \infty$ in magnitude when $R \downarrow 0$. Incidentally, the fact that this term vanishes when the point particle is unaccelerated shows that the infinities of the Lorentz self-force in stationary motion have been removed by spherical averaging.

In addition to the divergence problems of the Lorentz self-force on a point charge, the field energy integral (2.8) diverges for the Liénard–Wiechert fields because of their local singularity at $\boldsymbol{s} = \boldsymbol{Q}(t)$ (a "classical UV divergence").[4] This confronts us also with the problem that by Einstein's $E = mc^2$ [Ein1905b] the electromagnetic field of a point charge should attach an infinite inert mass to the point charge. In a sense, this is precisely what the in leading order divergent Lorentz self-force term "$\lim_{R\downarrow 0} -\frac{e^2}{2c^2}R^{-1}\frac{\partial}{\partial t}(\gamma(t)\dot{\boldsymbol{Q}}(t))$" expresses. But how does that fit in with the finite $m_{\rm e}$ in (2.18)? There is no easy way out of this dilemma.

In [Dir1938] Dirac proposed to assign an R-dependent bare mass $m_{\rm b}(R)$ to the averaging sphere of radius R, such that $m_{\rm b}(R) + \frac{e^2}{2c^2}R^{-1} \to m_{\rm e}$ as $R \downarrow 0$, where $m_{\rm e}$ is the empirical rest mass of the electron (see above). It seems that this was the first time such a radical step was proposed, a precursor to what eventually became "renormalization theory;" cf. [GKZ1998]. Dirac's procedure in effect removes the divergent "self-force" from r.h.s.(2.17), with $\boldsymbol{\mathcal{E}}^{\rm ret}_{\rm LW} + \boldsymbol{\mathcal{E}}_{\rm MM}$ and $\boldsymbol{\mathcal{B}}^{\rm ret}_{\rm LW} + \boldsymbol{\mathcal{B}}_{\rm MM}$ in place of $\boldsymbol{\mathcal{E}}$ and $\boldsymbol{\mathcal{B}}$, and produces the so-called

[3]The fact that $e^2/2R$ coincides with the electrostatic field energy of a surface-charged sphere of radius R is a consequence of Newton's theorem. The regularization of the Lorentz force field of a point charge by "spherical averaging" should *not* be confused with setting up a dynamical Lorentz model of a "surface-charged sphere," e.g. [ApKi2001], [ApKi2002].

[4]The field energy integral (2.8) for the Liénard–Wiechert fields diverges also because of their slow decay as $\boldsymbol{s} \to \infty$, but this "classical IR divergence" can be avoided by adding a suitable solution of the Maxwell–Maxwell field equations.

Abraham–Lorentz–Dirac equation, viz.[5]

$$\begin{aligned}\frac{\mathrm{d}\boldsymbol{P}(t)}{\mathrm{d}t} &= -e\Big[\boldsymbol{\mathcal{E}}_{\mathrm{MM}}(t,\boldsymbol{Q}(t)) + \tfrac{1}{c}\dot{\boldsymbol{Q}}(t)\times\boldsymbol{\mathcal{B}}_{\mathrm{MM}}(t,\boldsymbol{Q}(t))\Big] \\ &\quad + \tfrac{2e^2}{3c^3}\Big[\mathsf{I}+\gamma^2\tfrac{1}{c^2}\dot{\boldsymbol{Q}}\otimes\dot{\boldsymbol{Q}}\Big]\cdot\Big[3\gamma^4\tfrac{1}{c^2}(\dot{\boldsymbol{Q}}\cdot\ddot{\boldsymbol{Q}})\ddot{\boldsymbol{Q}}+\gamma^2\dddot{\boldsymbol{Q}}\Big](t)\end{aligned} \tag{2.19}$$

coupled with (2.18); here I is the identity operator. The occurrence in (2.19) of the third time derivative of $\boldsymbol{Q}(t)$ signals that our troubles are not over yet. Viewed as a Cauchy problem, $\ddot{\boldsymbol{Q}}(0)$ has now to be prescribed in addition to the familiar initial data $\boldsymbol{Q}(0)$ and $\dot{\boldsymbol{Q}}(0)$, and most choices will lead to self-accelerating run-away solutions. To eliminate these pathological solutions amongst all solutions with the same initial data $\boldsymbol{Q}(0)$ and $\dot{\boldsymbol{Q}}(0)$, Dirac [Dir1938] integrated (2.19) once with the help of an integrating factor, obtaining an integro-differential equation which is of second order in the time derivative of $\boldsymbol{Q}(t)$ but now with the applied force integrated over the whole future of the trajectory. This implies that the remaining solutions are "pre-accelerated,"[6] and they can be non-unique [CDGS1995].

Note that Dirac's ad hoc procedure actually means a step in the direction of taking a renormalized point-particle limit in a family of "extended electron" models, though Dirac made no attempt (at this point) to come up with a consistent dynamical model of an extended electron. If he had, he would have found that for a family of dynamically consistent models of a "relativistically spinning Lorentz sphere" one cannot take the renormalized point charge limit, cf. [ApKi2001].

2.4. The effective equation of motion of Landau–Lifshitz

Despite the conceptual problems with its third-order derivative, the Abraham–Lorentz–Dirac equation has served Landau and Lifshitz [LaLi1962] (and Peierls) as point of departure to arrive at an effective second-order equation of motion which is free of runaway solutions and pre-acceleration. They argued that whenever the familiar second-order equation of test charge motion is empirically successful, the second line at r.h.s.(2.19) should be treated as a *tiny* correction term to its first line. As a rule of thumb this should be true for test particle motions with trajectories whose curvature κ is much smaller than $m_e c^2/e^2$, the reciprocal of the "classical electron radius." But then, in leading order of a formal expansion in powers of the small parameter $\kappa e^2/m_e c^2$, the third time derivative of $\boldsymbol{Q}(t)$ in (2.19) can be expressed in terms of $\boldsymbol{Q}(t)$ and $\dot{\boldsymbol{Q}}(t)$, obtained from the equations of test particle motion

[5] The complicated correction term to the "external Lorentz force" at r.h.s.(2.19) is simply the space part of the Laue [Lau1909] four-vector $(2e^2/3c^3)(\mathrm{g}+\mathbf{u}\otimes\mathbf{u})\cdot\overset{\circ\circ}{\mathbf{u}}$, divided by $\gamma(t)$. Here, $\mathbf{u}$ is the dimensionless four-velocity of the point charge, $\circ$ means derivative w.r.t. "$c\times$proper time," and g is the metric tensor for Minkowski spacetime with signature $(-,+,+,+)$. Note that $\mathrm{g}+\mathbf{u}\otimes\mathbf{u}$ projects onto the subspace four-orthogonal to $\mathbf{u}$.

[6] "It is to be hoped that some day the real solution of the problem of the charge-field interaction will look different, and the equations describing nature will not be so highly unstable that the balancing act can only succeed by having the system correctly prepared ahead of time by a convenient coincidence." Walter Thirring, p. 379 in [Thi1997].

(2.17) and (2.18) as follows: invert (2.18) to obtain $\boldsymbol{P}(t) = m_{\mathrm{e}}\gamma(t)\dot{\boldsymbol{Q}}(t)$, then take the time derivative of this equation and solve for $\ddot{\boldsymbol{Q}}(t)$ to get

$$\ddot{\boldsymbol{Q}}(t) = \tfrac{1}{m_{\mathrm{e}}\gamma(t)}\Big[\mathsf{I} - \tfrac{1}{c^2}\dot{\boldsymbol{Q}}(t)\otimes\dot{\boldsymbol{Q}}(t)\Big]\cdot\dot{\boldsymbol{P}}(t). \tag{2.20}$$

Now substitute the first line at r.h.s.(2.19) for $\dot{\boldsymbol{P}}(t)$ at r.h.s.(2.20), then take another time derivative of the so rewritten equation (2.20) to obtain $\dddot{\boldsymbol{Q}}(t)$ in terms of $\boldsymbol{Q}, \dot{\boldsymbol{Q}}, \ddot{\boldsymbol{Q}}$; for the $\ddot{\boldsymbol{Q}}$ terms in this expression, now resubstitute (2.20), with the first line at r.h.s.(2.19) substituted for $\dot{\boldsymbol{P}}(t)$, to get $\dddot{\boldsymbol{Q}}(t)$ in terms of $\boldsymbol{Q}, \dot{\boldsymbol{Q}}$. Inserting this final expression for $\dddot{\boldsymbol{Q}}(t)$ into the second line at r.h.s.(2.19) yields the so-called Landau–Lifshitz equation of motion of the physical electron,[7] coupled with (2.18). The error made by doing so is of higher order in the small expansion parameter than the retained terms. I refrain from displaying the Landau–Lifshitz equation because its intimidating appearance does not add anything illuminating here.

Of course, the combination of Dirac's ad hoc renormalization procedure with the heuristic Landau–Lifshitz approximation step does not qualify as a satisfactory "derivation" of the Landau–Lifshitz equation of motion from "first principles." A rigorous derivation from an extended particle model has been given by Spohn and collaborators, cf. [Spo2004]. This derivation establishes the status of the Landau–Lifshitz equation as an effective equation of motion for the geometrical center of a particle with structure, not the point particle hoped for by Dirac. Whether it will ever have a higher (say, at least asymptotic) status for a proper theory of charged point-particle motion remains to be seen. In any event, this equation seems to yield satisfactory results for many practical purposes.

3. Wheeler–Feynman electrodynamics

Before I can move on to the next stage in our quest for a proper theory of motion of the point charge sources of the electromagnetic fields, I need to mention the radical solution to this classical problem proposed by Fokker, Schwarzschild, and Tetrode (see [Baru1964], [Spo2004]) and its amplification by Wheeler–Feynman [WhFe1949]. This "action-at-a-distance" theory takes over from formal Lorentz electrodynamics the Liénard–Wiechert fields associated to each point charge, but there are no additional external Maxwell–Maxwell fields. Most importantly, a point charge is not directly affected by its own Liénard–Wiechert field. Therefore the main problem of formal Lorentz

[7]This tedious calculation becomes simplified in the four-vector formulation mentioned in footnote 5. The Landau–Lifshitz reasoning for the leading order $\overset{\circ\circ}{\mathbf{u}}$ yields the proper time derivative of the external Lorentz–Minkowski force, divided by m_{e}, in which in turn the external Lorentz–Minkowski force, divided by m_{e}, is resubstituted for the $\overset{\circ}{\mathbf{u}}$ term. Projection onto the subspace orthogonal to $\mathbf{u}$ and multiplication by $\frac{2e^2}{3c^3}$ yields the Landau–Lifshitz approximation to the Laue four-vector. Its space part divided by $\gamma(t)$ gives the pertinent approximation to the second line at r.h.s.(2.19) in terms of $\boldsymbol{Q}(t)$ and $\dot{\boldsymbol{Q}}(t)$.

electrodynamics, the self-interaction of a point charge with its own Liénard–Wiechert field, simply does not exist!

In the Fokker–Schwarzschild–Tetrode theory, the law of motion of an electron is given by equations (2.17), (2.18), though with $\mathcal{E}$ and $\mathcal{B}$ now standing for the arithmetic means $\frac{1}{2}(\mathcal{E}_{\text{LW}}^{\text{ret}} + \mathcal{E}_{\text{LW}}^{\text{adv}})$ and $\frac{1}{2}(\mathcal{B}_{\text{LW}}^{\text{ret}} + \mathcal{B}_{\text{LW}}^{\text{adv}})$ of the *retarded and advanced* Liénard–Wiechert fields summed over *all the other* point electrons. Nontrivial motions can occur only in the many-particle Fokker–Schwarzschild–Tetrode theory.

While the Fokker–Schwarzschild–Tetrode electrodynamics does not suffer from the infinities of formal Lorentz electrodynamics, it raises another formidable problem: its second-order equations of motion for the system of point charges do not pose a Cauchy problem for the traditional classical state variables $\boldsymbol{Q}(t)$ and $\dot{\boldsymbol{Q}}(t)$ of these point charges. Instead, given the classical state variables $\boldsymbol{Q}(t_0)$ and $\dot{\boldsymbol{Q}}(t_0)$ of each electron at time t_0, the computation of their accelerations at time t_0 requires the knowledge of the states of motion of all point electrons at infinitely many instances in the past *and in the future*. While it would be conceivable in principle, though certainly not possible in practice, to find some historical records of all those past events, how could we anticipate the future without computing it from knowing the present and the past?

Interestingly enough, though, Wheeler and Feynman showed that the Fokker–Schwarzschild–Tetrode equations of motion can be recast as Abraham–Lorentz–Dirac equations of motion for each point charge, though with the external Maxwell–Maxwell fields replaced by the retarded Liénard–Wiechert fields of all the other point charges, *provided* the *Wheeler–Feynman absorber identity* is valid. While this is still not a Cauchy problem for the classical state variables $\boldsymbol{Q}(t)$ and $\dot{\boldsymbol{Q}}(t)$ of each electron, at least one does not need to anticipate the future anymore.

The Fokker–Schwarzschild–Tetrode and Wheeler–Feynman theories are mathematically fascinating in their own right, but pose very difficult problems. Rigorous studies [Bau1998], [BDD2010], [Dec2010] have only recently begun.

4. Nonlinear electromagnetic field equations

Gustav Mie [Mie1912/13] worked out the special-relativistic framework for fundamentally nonlinear electromagnetic field equations without point charges (for a modern treatment, see [Chri2000]). Twenty years later Mie's work became the basis for Max Born's assault on the infinite self-energy problem of a point charge.

4.1. Nonlinear self-regularization

Born [Bor1933] argued that the dilemma of the infinite electromagnetic self-energy of a point charge in formal Lorentz electrodynamics is caused by

Maxwell's "law of the pure aether,"[8] which he suspected to be valid only asymptotically, viz. $\boldsymbol{\mathcal{E}}_{\rm M} \sim \boldsymbol{\mathcal{D}}_{\rm M}$ and $\boldsymbol{\mathcal{H}}_{\rm M} \sim \boldsymbol{\mathcal{B}}_{\rm M}$ in the weak field limit. In the vicinity of a point charge, on the other hand, where the Coulombic $\boldsymbol{\mathcal{D}}_{\rm ML}$ field diverges in magnitude as $1/r^2$, nonlinear deviations of the true law of the "pure aether" from Maxwell's law would become significant and ultimately remove the infinite self-field-energy problems. The sought-after nonlinear "aether law" has to satisfy the requirements that the resulting electromagnetic field equations derive from a Lagrangian which

(P) is covariant under the Poincaré group;
(W) is covariant under the Weyl (gauge) group;
(M) reduces to the Maxwell–Maxwell Lagrangian in the weak field limit;
(E) yields finite field-energy solutions with point charge sources.

The good news is that there are "aether laws" formally satisfying criteria (P), (W), (M), (E). The bad news is that there are too many "aether laws" which satisfy criteria (P), (W), (M), (E) formally, so that additional requirements are needed.

Since nonlinear field equations tend to have solutions which form singularities in finite time (think of shock formation in compressible fluid flows), it is reasonable to look for the putatively least troublesome nonlinearity. In 1970, Boillat [Boi1970], and independently Plebanski [Ple1970], discovered that adding to (P), (W), (M), (E) the requirement that the electromagnetic field equations

(D) are linearly completely degenerate,

a *unique* one-parameter family of field equations emerges, indeed the one proposed by Born and Infeld [BoIn1933b, BoIn1934][9] in "one of those amusing cases of serendipity in theoretical physics" ([BiBi1983], p.37).

4.2. The Maxwell–Born–Infeld field equations

The Maxwell–Born–Infeld field equations, here for simplicity written only for a single (negative) point charge, consist of Maxwell's general field equations with the point charge source terms $\rho(t,\boldsymbol{s}) \equiv -e\delta_{\boldsymbol{Q}(t)}(\boldsymbol{s})$ and $\boldsymbol{j}(t,\boldsymbol{s}) \equiv -e\delta_{\boldsymbol{Q}(t)}(\boldsymbol{s})\dot{\boldsymbol{Q}}(t)$,

$$\tfrac{1}{c}\tfrac{\partial}{\partial t}\boldsymbol{\mathcal{B}}_{\rm MBI}(t,\boldsymbol{s}) = -\boldsymbol{\nabla}\times\boldsymbol{\mathcal{E}}_{\rm MBI}(t,\boldsymbol{s}), \tag{4.1}$$

$$\tfrac{1}{c}\tfrac{\partial}{\partial t}\boldsymbol{\mathcal{D}}_{\rm MBI}(t,\boldsymbol{s}) = +\boldsymbol{\nabla}\times\boldsymbol{\mathcal{H}}_{\rm MBI}(t,\boldsymbol{s}) + 4\pi\tfrac{1}{c}e\delta_{\boldsymbol{Q}(t)}(\boldsymbol{s})\dot{\boldsymbol{Q}}(t), \tag{4.2}$$

$$\boldsymbol{\nabla}\cdot\boldsymbol{\mathcal{B}}_{\rm MBI}(t,\boldsymbol{s}) = 0, \tag{4.3}$$

$$\boldsymbol{\nabla}\cdot\boldsymbol{\mathcal{D}}_{\rm MBI}(t,\boldsymbol{s}) = -4\pi e\delta_{\boldsymbol{Q}(t)}(\boldsymbol{s}), \tag{4.4}$$

[8] After its demolition by Einstein, "aether" will be recycled here as shorthand for what physicists call "electromagnetic vacuum."

[9] While the unique characterization of the Maxwell–Born–Infeld field equations in terms of (P), (W), (M), (E), (D) was apparently not known to Born and his contemporaries, the fact that these field equations satisfy, beside (P), (W), (M), (E), also (D) is mentioned in passing already on p. 102 in [Schr1942a] as the absence of birefringence (double refraction), meaning that the speed of light [sic!] is independent of the polarization of the wave fields. The Maxwell–Lorentz equations for a point charge satisfy (P), (W), (M), (D), but not (E).

together with the electromagnetic "aether law"[10] of Born–Infeld [BoIn1934],

$$\boldsymbol{\mathcal{E}}_{\mathrm{MBI}} = \frac{\boldsymbol{\mathcal{D}}_{\mathrm{MBI}} - \frac{1}{b^2}\boldsymbol{\mathcal{B}}_{\mathrm{MBI}} \times (\boldsymbol{\mathcal{B}}_{\mathrm{MBI}} \times \boldsymbol{\mathcal{D}}_{\mathrm{MBI}})}{\sqrt{1 + \frac{1}{b^2}(|\boldsymbol{\mathcal{B}}_{\mathrm{MBI}}|^2 + |\boldsymbol{\mathcal{D}}_{\mathrm{MBI}}|^2) + \frac{1}{b^4}|\boldsymbol{\mathcal{B}}_{\mathrm{MBI}} \times \boldsymbol{\mathcal{D}}_{\mathrm{MBI}}|^2}}, \tag{4.5}$$

$$\boldsymbol{\mathcal{H}}_{\mathrm{MBI}} = \frac{\boldsymbol{\mathcal{B}}_{\mathrm{MBI}} - \frac{1}{b^2}\boldsymbol{\mathcal{D}}_{\mathrm{MBI}} \times (\boldsymbol{\mathcal{D}}_{\mathrm{MBI}} \times \boldsymbol{\mathcal{B}}_{\mathrm{MBI}})}{\sqrt{1 + \frac{1}{b^2}(|\boldsymbol{\mathcal{B}}_{\mathrm{MBI}}|^2 + |\boldsymbol{\mathcal{D}}_{\mathrm{MBI}}|^2) + \frac{1}{b^4}|\boldsymbol{\mathcal{B}}_{\mathrm{MBI}} \times \boldsymbol{\mathcal{D}}_{\mathrm{MBI}}|^2}}, \tag{4.6}$$

where $b \in (0, \infty)$ is *Born's field strength*, a hypothetical new "constant of nature," which Born determined [Bor1934, BoIn1933a, BoIn1933b, BoIn1934] as follows.

4.2.1. Born's determination of b. In the absence of any charges, these *source-free* Maxwell–Born–Infeld field equations conserve the field energy, the field momentum, and the field angular momentum, given by the following integrals, respectively (cf. [BiBi1983]),

$$E_{\mathrm{f}} = \frac{b^2}{4\pi}\int_{\mathbb{R}^3}\Big(\sqrt{1 + \tfrac{1}{b^2}(|\boldsymbol{\mathcal{B}}^{\mathrm{sf}}_{\mathrm{MBI}}|^2 + |\boldsymbol{\mathcal{D}}^{\mathrm{sf}}_{\mathrm{MBI}}|^2) + \tfrac{1}{b^4}|\boldsymbol{\mathcal{B}}^{\mathrm{sf}}_{\mathrm{MBI}} \times \boldsymbol{\mathcal{D}}^{\mathrm{sf}}_{\mathrm{MBI}}|^2} - 1\Big)(t, \boldsymbol{s})\,\mathrm{d}^3 s, \tag{4.7}$$

$$\boldsymbol{P}_{\mathrm{f}} = \frac{1}{4\pi c}\int_{\mathbb{R}^3}(\boldsymbol{\mathcal{D}}^{\mathrm{sf}}_{\mathrm{MBI}} \times \boldsymbol{\mathcal{B}}^{\mathrm{sf}}_{\mathrm{MBI}})(t, \boldsymbol{s})\,\mathrm{d}^3 s\,, \tag{4.8}$$

$$\boldsymbol{L}_{\mathrm{f}} = \frac{1}{4\pi c}\int_{\mathbb{R}^3}\boldsymbol{s} \times (\boldsymbol{\mathcal{D}}^{\mathrm{sf}}_{\mathrm{MBI}} \times \boldsymbol{\mathcal{B}}^{\mathrm{sf}}_{\mathrm{MBI}})(t, \boldsymbol{s})\,\mathrm{d}^3 s\,. \tag{4.9}$$

Supposing that these integrals retain their meanings also in the presence of sources, Born computed the energy of the field pair $(\boldsymbol{\mathcal{B}}_{\mathrm{Born}}, \boldsymbol{\mathcal{D}}_{\mathrm{Born}})$, with $\boldsymbol{\mathcal{B}}_{\mathrm{Born}} = \mathbf{0}$ and

$$\boldsymbol{\mathcal{D}}_{\mathrm{Born}}(\boldsymbol{s}) = \boldsymbol{\mathcal{D}}_{\mathrm{Coulomb}}(\boldsymbol{s}) \equiv -e\boldsymbol{s}\,/|\boldsymbol{s}|^3, \tag{4.10}$$

which is the unique electrostatic finite-energy solution to the Maxwell–Born–Infeld equations with a single[11] negative point charge source at the origin of (otherwise) empty space; see [Pry1935b], [Eck1986]. The field energy of Born's solution is

$$E_{\mathrm{f}}(\mathbf{0}, \boldsymbol{\mathcal{D}}_{\mathrm{Born}}) = \frac{b^2}{4\pi}\int_{\mathbb{R}^3}\Big(\sqrt{1 + \tfrac{1}{b^2}|\boldsymbol{\mathcal{D}}_{\mathrm{Born}}(\boldsymbol{s})|^2} - 1\Big)\mathrm{d}^3 s = \tfrac{1}{6}\,\mathrm{B}\big(\tfrac{1}{4}, \tfrac{1}{4}\big)\sqrt{be^3}, \tag{4.11}$$

where $\mathrm{B}(p, q)$ is Euler's Beta function; numerical evaluation gives

$$\tfrac{1}{6}\,\mathrm{B}\big(\tfrac{1}{4}, \tfrac{1}{4}\big) \approx 1.2361. \tag{4.12}$$

Finally, inspired by the idea of the later 19th century that the electron's inertia a.k.a. mass has a purely electromagnetic origin, Born [Bor1934] now

[10] Note that Born and Infeld viewed their nonlinear relationship between $\boldsymbol{\mathcal{E}}, \boldsymbol{\mathcal{H}}$ on one side and $\boldsymbol{\mathcal{B}}, \boldsymbol{\mathcal{D}}$ on the other no longer as a constitutive law in the sense of Maxwell, but as the *electromagnetic law of the classical vacuum.*

[11] The only other explicitly known electrostatic solution with point charges was found by Hoppe [Hop1994], describing an infinite crystal with finite energy per charge [Gib1998].

argued that $E_{\mathrm{f}}(\mathbf{0}, \boldsymbol{\mathcal{D}}_{\mathrm{Born}}) = m_{\mathrm{e}}c^2$, thus finding for his field strength constant

$$b_{\mathrm{Born}} = 36\,\mathrm{B}\big(\tfrac{1}{4}, \tfrac{1}{4}\big)^{-2} m^2 c^4 e^{-3}. \tag{4.13}$$

Subsequently Born and Schrödinger [BoSchr1935] argued that this value has to be revised because of the electron's magnetic moment, and they came up with a very rough, purely magnetic estimate. Born then asked Madhava Rao [Rao1936] to improve on their estimate by computing the energy of the electromagnetic field of a charged, stationary circular current density,[12] but Rao's computation is too approximate to be definitive.

We will have to come back to the determination of b at a later time.

4.2.2. The status of (P), (W), (M), (E), (D). The Maxwell–Born–Infeld field equations *formally* satisfy the postulates (P), (W), (M), (E), (D), and there is no other set of field equations which does so. However, in order to qualify as a proper mathematical realization of (P), (W), (M), (E), (D) they need to *generically* have well-behaved solutions. In this subsubsection I briefly review what is rigorously known about generic solutions.

Source-free fields. For the special case of source-free fields the mentioning of point charge sources in (E) is immaterial. The finite-energy requirement remains in effect, of course.

Brenier [Bre2004] has recently given a very ingenious proof that the source-free electromagnetic field equations (4.1)–(4.6) are hyperbolic and pose a Cauchy problem; see also [Ser2004]. The generic existence and uniqueness of global classical solutions realizing (P), (W), (M), (E), (D) for initial data which are sufficiently *small* in a Sobolev norm was only recently shown,[13] by Speck [Spe2010a], who also extended his result to the general-relativistic setting [Spe2010b].

The restriction to small initial data is presumably vital because the works by Serre [Ser1988] and Brenier [Bre2004] have shown that arbitrarily regular plane wave initial data can lead to formation of a singularity in finite time. Now, plane wave initial data trivially have an infinite energy, but in his Ph.D. thesis Speck also showed that the relevant plane-wave initial data can be suitably cut off to yield finite-energy initial data which still lead to a singularity in finite time. The singularity is a divergent field-energy density, not of the shock-type. However, it is still not known whether the initial data that lead to a singularity in finite time form an open neighborhood. If they do, then formation of a singularity in finite time is a generic phenomenon of source-free Maxwell–Born–Infeld field evolutions. In this case it is important to find out how large, in terms of b, the field strengths of the initial data are allowed to be in order to launch a unique global evolution. Paired with empirical data this information should yield valuable bounds on b.

[12] This corresponds to a ring singularity in both the electric and magnetic fields.

[13] This result had been claimed in [ChHu2003], but their proof contained a fatal error.

Fields with point charge sources. Alas, hardly anything is rigorously known about the Maxwell–Born–Infeld field equations with one (or more) point charge source term(s) for *generic* smooth subluminal motions $t \mapsto \boldsymbol{Q}(t)$. Hopefully in the not too distant future it will be shown that this Cauchy problem is locally well-posed, thereby realizing (P), (W), (M), (E), (D) at least for short times. A global well-posedness result would seem unlikely, given the cited works of Serre and Brenier.

Only for the special case where all point charges remain at rest there are generic existence and uniqueness results for *electrostatic* solutions. By applying a Lorentz boost these electrostatic solutions map into unique *traveling* electromagnetic solutions;[14] of course, this says nothing about solutions for generic subluminal motions.

The first generic electrostatic results were obtained for charges with sufficiently small magnitudes in unbounded domains with boundary [KlMi1993, Kly1995]. Recently it has been proved [Kie2011a] that a unique electrostatic solution realizing (P), (W), (M), (E), (D) exists for arbitrary placements, signs and magnitudes of arbitrarily (though finitely) many point charges in $\mathbb{R}^3$; the solutions have C^∞ regularity away from the point charges.[15]

Incidentally, the existence of such electrostatic solutions can be perplexing. Here is Gibbons (p.19 in [Gib1998]): "[W]hy don't the particles accelerate under the influence of the mutual forces? The reason is that they are pinned to their fixed position ... by external forces." A more sober assessment of this situation will be offered in our next section.

5. Classical theory of motion

We now turn to the quest for a well-defined classical law of motion for the point charge sources of the electromagnetic Maxwell–Born–Infeld fields. In the remainder of this presentation I assume that the Cauchy problem for the Maxwell–Born–Infeld field equations with point charge sources which move smoothly at subluminal speeds is *generically* locally well-posed.

5.1. Orthodox approaches: a critique

In the beginning there was an intriguing conjecture, that because of their nonlinearity the Maxwell–Born–Infeld field equations *alone* would yield a (locally) well-posed Cauchy problem for *both*, the fields *and* the point charges. In a sense this idea goes back to the works by Born and Infeld [Bor1934, BoIn1934] who had argued that the law of motion is already contained in the differential law of field energy-momentum conservation obtained as a consequence of the source-free equations (or, equivalently, as a local consequence of

[14] Incidentally, traveling electromagnetic solutions satisfying (P), (W), (M), (E), (D) cannot be source-free if they travel at speeds less than c [Kie2011b].

[15] In [Gib1998] it is suggested that such a result would follow from the results on maximal hypersurfaces described in [Bart1987]. Results by Bartnik and Simon [BaSi1982] and by Bartnik [Bart1989] are indeed important ingredients to arrive at our existence and regularity result, but in themselves not sufficient to do so.

the Maxwell–Born–Infeld field equations away from point sources). A related sentiment can also be found in the work of Einstein, Infeld, and Hoffmann [EIH1938], who seemed to also derive further inspiration from Helmholtz' extraction of the equations of point vortex motions out of Euler's fluid-dynamical equations.

However, if this conjecture were correct, then the existence and uniqueness of well-behaved electrostatic field solutions with fixed point charges, announced in the previous section, would allow us to immediately dismiss the Maxwell–Born–Infeld field equations as unphysical. For example, consider just two identical point charges *initially at rest* and far apart, and consider field *initial data* identical to the pertinent *electrostatic* two-charge solution. If the field equations alone would uniquely determine the (local) future evolution of both, fields and point charges, they inevitably would have to produce the unique electrostatic solution with the point charges remaining at rest, whereas a physically acceptable theory must yield that these two charges begin to (approximately) perform a degenerate Kepler motion, moving away from each other along a straight line.

The upshot is: either the Maxwell–Born–Infeld field equations alone do form a complete classical theory of electromagnetism, and then this theory is *unphysical*, or they are incomplete as a theory of electromagnetism, in which case they may very well be part of a *physically acceptable* classical theory of electromagnetism. As I've stated at the end of the previous section, I expect that it will be shown that the Cauchy problem for the Maxwell–Born–Infeld field equations is locally well-posed for *generic* prescribed smooth subluminal motions of their point charge sources. Assuming this to pan out, the Maxwell–Born–Infeld field equations would have to be complemented by an additional set of dynamical equations which characterize the classical physical motions amongst all possible ones.

While we have been using electrostatic solutions in three dimensions to argue for the incompleteness of the Maxwell–Born–Infeld field equations as a classical theory of electromagnetism, the first person, apparently, to have realized the flaw in the "intriguing conjecture" was Born's son-in-law Maurice Pryce who, after finding analogous electrostatic solutions in two dimensions [Pry1935a], wrote ([Pry1936], p.597): "It was clear from the start of the New Field Theory (although not fully appreciated in I and II [i.e. [Bor1934] and [BoIn1934]]) that the motion of the charges was not governed by the field equations alone, and that some further condition had to be added." But then Pryce continued: "It was also clear from physical considerations of conservation of energy and momentum what this condition had to be; namely, that the total force (...) on each charge must vanish." His proposal is truthful to the revival, by Born and Infeld [BoIn1933a], [BoIn1934], of the old idea that the inertial mass of the electron has a purely electromagnetic origin, and therefore it follows Abraham's and Lorentz' proposal for the equation of motion in a purely electromagnetic model ("the total force vanishes"). However, since, as mentioned earlier, in [ApKi2001] it was shown that the purely electromagnetic Abraham–Lorentz model is overdetermined and generically has no solution, we have the benefit of hindsight and should be apprehensive of Pryce's proposal. Yet, until [ApKi2001] nobody had found anything mathematically suspicious in Abraham's and Lorentz' manipulations up to second order,[16] which got repeated verbatim at least until [Jac1975], so it should come as no surprise that Pryce's adaptation of the Abraham–Lorentz reasoning to the Born–Infeld setting was accepted by most peers. In particular, Schrödinger [Schr1942b] picked up on Pryce's proposal and tried to push the approximate evaluation by including more terms in an expansion using spherical harmonics.

Interestingly, Dirac [Dir1960] used a refinement of Born's original approach [Bor1934] (recanted in [BoIn1933b, BoIn1934]) to arrive at the Newtonian law of motion (2.17), (2.18), with $\boldsymbol{\mathcal{E}}_{\mathrm{MBI}}$ and $\boldsymbol{\mathcal{B}}_{\mathrm{MBI}}$ for, respectively, $\boldsymbol{\mathcal{E}}$ and $\boldsymbol{\mathcal{B}}$ in (2.17), then went on to define the Lorentz force with the total fields through regularization which involved manipulations of the energy-momentum-stress tensor approach used also by Pryce. Dirac also had a "non-electromagnetic mass M" for m_{e} in (2.18), but remarked: "We may have $M = 0$, which is probably the case for an electron."

I will now first explain why the Lorentz self-force still cannot be well-defined, and why postulating field energy-momentum conservation is not going to help, neither in the basic version as proposed first by Born and Infeld [BoIn1934] nor in the amended manner discussed by Pryce [Pry1936]. Our analysis will also bear on some more recent discussions, e.g. [Gib1998].

I will then describe a well-defined Hamilton–Jacobi law of motion by following [Kie2004a], in fact improving over the one proposed in [Kie2004a].

5.2. The inadequacy of the Lorentz self-force concept

5.2.1. Ambiguity of the regularized Lorentz self-force. Even if the Maxwell–Born–Infeld field equations with point charge sources have solutions realizing (P), (W), (M), (E), (D) for generic smooth subluminal motions, as conjectured to be the case, the formal expression for the Lorentz force, "$\boldsymbol{\mathcal{E}}(t,\boldsymbol{Q}(t)) + \frac{1}{c}\dot{\boldsymbol{Q}}(t)\times\boldsymbol{\mathcal{B}}(t,\boldsymbol{Q}(t))$," is still undefined *a priori* when $\boldsymbol{\mathcal{E}}(t,\boldsymbol{s}) = \boldsymbol{\mathcal{E}}_{\mathrm{MBI}}(t,\boldsymbol{s})$ and $\boldsymbol{\mathcal{B}}(t,\boldsymbol{s}) = \boldsymbol{\mathcal{B}}_{\mathrm{MBI}}(t,\boldsymbol{s})$ are the total electric and magnetic fields. Also, it cannot be defined at the locations of the point charge(s) by a limit $\boldsymbol{s}\to\boldsymbol{Q}(t)$ of $\boldsymbol{\mathcal{E}}_{\mathrm{MBI}}(t,\boldsymbol{s})$ and $\boldsymbol{\mathcal{B}}_{\mathrm{MBI}}(t,\boldsymbol{s})$ at time t, because these fields cannot be continuously extended to $\boldsymbol{s} = \boldsymbol{Q}(t)$. Again one can hope to define the total Lorentz force by taking a point limit $\varrho(\boldsymbol{s}|\boldsymbol{Q}(t)) \to \delta_{\boldsymbol{Q}(t)}(\boldsymbol{s})$ of a Lorentz force field $\boldsymbol{\mathcal{E}}_{\mathrm{MBI}}(t,\boldsymbol{s}) + \frac{1}{c}\dot{\boldsymbol{Q}}(t)\times\boldsymbol{\mathcal{B}}_{\mathrm{MBI}}(t,\boldsymbol{s})$ averaged over some normalized regularizing density $\varrho(\boldsymbol{s}|\boldsymbol{Q}(t))$. While in contrast to our experience with this procedure when applied to the solutions of the Maxwell–Lorentz field equations one may now obtain a finite result, because the point charge's self-energy is now finite, this is still not a satisfactory definition because even such a finite result will still depend on the specific details of the regularization ϱ and its

[16] Of course, the third-order term always was a source of much discussion and confusion. And as candidates for a fundamental model of the physical electron these classical approaches were abandoned long ago.

removal. For instance, Dirac [Dir1960] proposed that the procedure should be such as to yield a *vanishing* contribution from the discontinuous part of the electromagnetic fields, but this to me seems as arbitrary as proposing that the regularization should produce the minimal or the maximal possible force, or some other "distinguished" vector. Since Dirac invoked the energy-momentum-stress tensor for his regularization argument, I will comment on a few more details in the next subsection. Here the upshot is: unless a *compelling principle* is found which resolves the ambiguity of the self-field contribution to the total Lorentz force on a point charge, the Lorentz force has to be purged from the list of fundamental classical concepts.

5.2.2. The postulate of local energy-momentum conservation. As I have already claimed, there exists a unique *electrostatic* solution of the Maxwell–Born–Infeld field equations with N point charges *remaining at rest* at their initial positions. Since the integrals of field energy and field momentum (and field angular momentum) are all conserved by any static solution of the Maxwell–Born–Infeld field equations, this shows that simply postulating their conservation does not suffice to produce the right motions. Of course, postulating only the conservation of these global quantities is an infinitely much weaker requirement than postulating *detailed local balance* $\partial_\nu T^{\mu\nu}_{\text{MBI}} = 0^\mu$ *everywhere*, where $T^{\mu\nu}_{\text{MBI}}$ are the components of the symmetric energy(-density)-momentum(-density)-stress tensor of the electromagnetic Maxwell–Born–Infeld field; the local law for field angular-momentum conservation follows from this one and needs not to be postulated separately. Note that in the absence of point charges the law $\partial_\nu T^{\mu\nu}_{\text{MBI}} = 0^\mu$ is a consequence of the field equations and not an independent postulate. Only in the presence of point charges, when $\partial_\nu T^{\mu\nu}_{\text{MBI}} = 0^\mu$ is valid a priori only *away from* these point charges, its continuous extension into the locations of the point charges amounts to a new postulate, indeed. Yet also this "local conservation of field energy-momentum" postulate does not deliver what it promises, and here is why.

Consider the example of fields with two identical point charges (charged $-e$, say). The charges move smoothly at subluminal speed, and suppose the fields solve the Maxwell–Born–Infeld field equations. Introducing the electromagnetic stress tensor of the Maxwell–Born–Infeld fields,

$$\Theta_{\text{MBI}} = \tfrac{1}{4\pi}\Big[\boldsymbol{\mathcal{E}}\otimes\boldsymbol{\mathcal{D}} + \boldsymbol{\mathcal{H}}\otimes\boldsymbol{\mathcal{B}} - b^2\Big(\sqrt{1+\tfrac{1}{b^2}(|\boldsymbol{\mathcal{B}}|^2+|\boldsymbol{\mathcal{D}}|^2)+\tfrac{1}{b^4}|\boldsymbol{\mathcal{B}}\times\boldsymbol{\mathcal{D}}|^2}-1\Big)\mathrm{I}\Big], \tag{5.1}$$

with $\boldsymbol{\mathcal{E}} = \boldsymbol{\mathcal{E}}_{\text{MBI}}$ etc., when viewing $T^{\mu\nu}_{\text{MBI}}(t,\boldsymbol{s})$ as a t-family of distributions over $\mathbb{R}^3$, the space components of $\partial_\nu T^{\mu\nu}_{\text{MBI}}$ then yield the *formal* identity

$$\begin{aligned}\tfrac{1}{4\pi c}\tfrac{\partial}{\partial t}&(\boldsymbol{\mathcal{D}}_{\text{MBI}}\times\boldsymbol{\mathcal{B}}_{\text{MBI}})(t,\boldsymbol{s}) - \boldsymbol{\nabla}\cdot\Theta_{\text{MBI}}(t,\boldsymbol{s})\\ &= e[\text{“}\boldsymbol{\mathcal{E}}_{\text{MBI}}(t,\boldsymbol{Q}_1(t))\text{”} + \tfrac{1}{c}\dot{\boldsymbol{Q}}_1(t)\times\text{“}\boldsymbol{\mathcal{B}}_{\text{MBI}}(t,\boldsymbol{Q}_1(t))\text{”}]\delta_{\boldsymbol{Q}_1(t)}(\boldsymbol{s})\\ &\quad + e[\text{“}\boldsymbol{\mathcal{E}}_{\text{MBI}}(t,\boldsymbol{Q}_2(t))\text{”} + \tfrac{1}{c}\dot{\boldsymbol{Q}}_2(t)\times\text{“}\boldsymbol{\mathcal{B}}_{\text{MBI}}(t,\boldsymbol{Q}_2(t))\text{”}]\delta_{\boldsymbol{Q}_2(t)}(\boldsymbol{s}),\end{aligned} \tag{5.2}$$

where the quotes around the field symbols in the second line of (5.2) remind us that the electric and magnetic fields are generally ill-defined at $\boldsymbol{Q}_1$ and

$\boldsymbol{Q}_2$, indicating that our problem has caught up with us again. At this point, to restore complete electromagnetic energy-momentum conservation, Pryce [Pry1936] and Schrödinger [Schr1942b] rationalized that in effect one has to postulate r.h.s.(5.2)$= \mathbf{0}$. This in essence is what Pryce meant by "the total force on each charge must vanish." (The law of energy conservation is dealt with analogously.) They go on and extract from this postulate the familiar Newtonian test particle law of motion in weak applied fields, the rest mass of a point charge given by its electrostatic field energy/c^2.

For Dirac, on the other hand, allowing an extra non-electromagnetic mass M, r.h.s.(5.2)$\neq \mathbf{0}$ because a time-dependent extra kinetic energy associated with M had to be taken into account, and only total energy-momentum should be conserved. Hence he made a different postulate to remove the ambiguity highlighted by the quotes around the electromagnetic field at the locations of the point charges. Eventually also Dirac obtained the familiar Newtonian test particle law of motion, but when $M = 0$ at the end of the day, the rest mass of the point charge became its purely electrostatic field energy/c^2, too.

To exhibit the *subtle mathematical and conceptual issues* in the reasonings of Pryce, Schrödinger, and Dirac, we return (briefly) to the special case where the two charges are initially at rest, with electrostatic field initial data.

In this particular case, by continuous and *consistent* extension into $\boldsymbol{Q}_1$ and $\boldsymbol{Q}_2$, we find that $\frac{\partial}{\partial t}(\boldsymbol{\mathcal{D}}_{\text{MBI}} \times \boldsymbol{\mathcal{B}}_{\text{MBI}})(t, \boldsymbol{s}) = \mathbf{0}$ for all t and $\boldsymbol{s}$. Now pretending r.h.s.(5.2) were well-defined as a vector-valued distribution, i.e. if "$\boldsymbol{\mathcal{E}}_{\text{MBI}}(0, \boldsymbol{Q}_1)$" and "$\boldsymbol{\mathcal{E}}_{\text{MBI}}(0, \boldsymbol{Q}_2)$" were actual vectors (note that $\dot{\boldsymbol{Q}}_k = \mathbf{0}$ now), we can use Gauss' divergence theorem to actually compute these vectors. Thus, integrating (5.2) over *any* smooth, bounded, simple open domain Λ containing, say, $\boldsymbol{Q}_1$ but not $\boldsymbol{Q}_2$ then yields $-e$"$\boldsymbol{\mathcal{E}}_{\text{MBI}}(0, \boldsymbol{Q}_1)$"$= \int_{\partial\Lambda} \Theta_{\text{MBI}} \cdot \boldsymbol{n} d\sigma$. The surface integral at the right-hand side is well-defined for any such Λ, but since the left-hand side of this equation is independent of Λ, also the surface integral must be independent of Λ. Happily it *is* independent of Λ (as long as Λ does not contain $\boldsymbol{Q}_2$) because for the electrostatic field the distribution $\boldsymbol{\nabla} \cdot \Theta_{\text{MBI}}(0, \boldsymbol{s})$ is supported only at $\boldsymbol{Q}_1$ and $\boldsymbol{Q}_2$. In the absence of any explicit formula for the electrostatic two-point solution one so-far relies on an approximate evaluation. Gibbons [Gib1998] suggests that for large separations between the point charges the answer is Coulomb's force formula. (In [Pry1935a] the two-dimensional electrostatic field with two point charges is computed exactly, and the closed line integral of the stresses around a charge shown to yield Coulomb's formula for distant charges.) If proven rigorously correct in three dimensions, and presumably it can be shown rigorously, this would seem to invalidate Pryce's (and Schrödinger's) line of reasoning that "$\boldsymbol{\mathcal{E}}_{\text{MBI}}(0, \boldsymbol{Q}_1)$" and "$\boldsymbol{\mathcal{E}}_{\text{MBI}}(0, \boldsymbol{Q}_2)$" could be postulated to vanish.

However, Pryce and Schrödinger were no fools. They would have pointed out that what we just explained in the previous paragraph would be an unambiguous definition of "$\boldsymbol{\mathcal{E}}_{\text{MBI}}(0, \boldsymbol{Q}_1)$" and "$\boldsymbol{\mathcal{E}}_{\text{MBI}}(0, \boldsymbol{Q}_2)$" for the *electrostatic field solution* to the Maxwell–Born–Infeld field equations, because we used that for all t and $\boldsymbol{s}$ we have $\frac{\partial}{\partial t}(\boldsymbol{\mathcal{D}}_{\text{MBI}} \times \boldsymbol{\mathcal{B}}_{\text{MBI}})(t, \boldsymbol{s}) = \mathbf{0}$ for the electrostatic solution (actually, this is only needed for $t \in (-\epsilon, \epsilon)$). But, they would have continued, the electrostatic solution is unphysical and has to be ruled out, and this is precisely one of the things which postulating "$\boldsymbol{\mathcal{E}}_{\text{MBI}}(0, \boldsymbol{Q}_1)$"$= \mathbf{0}$ and "$\boldsymbol{\mathcal{E}}_{\text{MBI}}(0, \boldsymbol{Q}_2)$"$= \mathbf{0}$ for the *physical solution* accomplishes, thanks to the mathematical results of the previous paragraph which show that for the *physical solution* to the Maxwell–Born–Infeld field equations with electrostatic initial data for fields and particles one cannot have $\frac{\partial}{\partial t}(\boldsymbol{\mathcal{D}}_{\text{MBI}} \times \boldsymbol{\mathcal{B}}_{\text{MBI}})(0, \boldsymbol{s}) = \mathbf{0}$ for all $\boldsymbol{s}$. This in turn implies that the point charges for the *physical solution* cannot remain at rest but are accelerated by the electrostatic forces so defined. Furthermore, they would have insisted, since the physical $\frac{1}{4\pi c}\frac{\partial}{\partial t}(\boldsymbol{\mathcal{D}}_{\text{MBI}} \times \boldsymbol{\mathcal{B}}_{\text{MBI}})(t, \boldsymbol{s})$ has to exactly offset the distribution $\boldsymbol{\nabla} \cdot \Theta_{\text{MBI}}(t, \boldsymbol{s})$, one obtains an equation of motion for the positions of the point charges for all times.

Brilliant! But does it work? There are two issues to be addressed.

First, there is the issue as to the definition of the forces on the point charges in general dynamical situations. Since $\int_{\partial \Lambda_k} \Theta_{\text{MBI}}(t, \boldsymbol{s}) \cdot \boldsymbol{n} d\sigma$, with Λ_k containing only $\boldsymbol{Q}_k(t)$, will now generally depend on Λ_k, one can at best define the force on the kth charge at time t by taking the limit $\Lambda_k \to \{\boldsymbol{Q}_k(t)\}$, provided the limit exists and is independent of the particular shapes of the shrinking Λ_k. Whether this is possible is a mathematical issue, regarding the behavior of the field solutions near the point charges that move at generic, smooth subluminal speeds. Thus, $\boldsymbol{\mathcal{D}}_{\text{MBI}}(t, \boldsymbol{s})$ and $\boldsymbol{\mathcal{H}}_{\text{MBI}}(t, \boldsymbol{s})$ must not diverge stronger than $1/|\boldsymbol{s} - \boldsymbol{Q}_k(t)|^2$ at each $\boldsymbol{Q}_k(t)$; to get nontrivial forces, they must in fact diverge exactly at this rate in leading order. This is an open problem, but it is not unreasonable to assume, as Pryce did, that this will be proven true, at least for sufficiently short times.

The second issue is more problematic. The distribution $\boldsymbol{\nabla} \cdot \Theta_{\text{MBI}}(t, \boldsymbol{s})$ would, for sufficiently short times $t > 0$ after the initial instant, be of the form $\boldsymbol{f}(t, \boldsymbol{s}) + \sum_k \boldsymbol{F}_k(t)\delta_{\boldsymbol{Q}_k(t)}(\boldsymbol{s})$, where $\boldsymbol{f}(t, \boldsymbol{s})$ is a regular force density field, while $\boldsymbol{F}_k(t)$ is the above defined force vector on the kth point charge. The field $\boldsymbol{f}(t, \boldsymbol{s})$ will be precisely offset by the regular part of $\frac{1}{4\pi c}\frac{\partial}{\partial t}(\boldsymbol{\mathcal{D}}_{\text{MBI}} \times \boldsymbol{\mathcal{B}}_{\text{MBI}})(t, \boldsymbol{s})$ thanks to the local conservation law of electromagnetic energy-momentum away from the charges, as implied by the field equations alone. Thus, in order to get an equation of motion in line with Newtonian classical physical notions, the singular (distributional) part of $\frac{\partial}{\partial t}(\boldsymbol{\mathcal{D}}_{\text{MBI}} \times \boldsymbol{\mathcal{B}}_{\text{MBI}})(t, \boldsymbol{s})$ at $\boldsymbol{Q}_k(t)$ now must be of the form $\delta_{\boldsymbol{Q}_k(t)}(\boldsymbol{s})$ times a vector which depends on $\boldsymbol{Q}_k(t)$ and its first two time-derivatives — note that also the initial source terms for the field equations require $\boldsymbol{Q}_k(0)$ and $\dot{\boldsymbol{Q}}_k(0)$ to be prescribed, suggesting a *second-order* equation of motion for the $\boldsymbol{Q}_k(t)$. In particular, for the initial data obtained from the electrostatic field solution, with charges initially at rest and very far apart, the "physical" $\frac{1}{4\pi c}\frac{\partial}{\partial t}(\boldsymbol{\mathcal{D}}_{\text{MBI}} \times \boldsymbol{\mathcal{B}}_{\text{MBI}})(0, \boldsymbol{s})$ has to be asymptotic to $\sum_k m_e \ddot{\boldsymbol{Q}}_k(0)\delta_{\boldsymbol{Q}_k(0)}(\boldsymbol{s})$ as $|\boldsymbol{Q}_1(0) - \boldsymbol{Q}_2(0)| \to \infty$, if it is to reproduce the physically correct equation of slow and gently accelerated Kepler motions of two physical electrons in the classical regime.

I do not see how the second issue could be resolved favorably. In fact, it should be worthwhile to try to come up with a proof that the putative equation of motion is overdetermined, in the spirit of [ApKi2001], but I haven't tried this yet.

But how could Born, Infeld, Pryce, and Schrödinger, all have convinced themselves that this procedure will work? It is illuminating to see the answer to this question, because it will sound an alarm.

Consider once again the electrostatic initial data with two identical point charges initially at rest and far apart. Let $(\mathbb{R}^2 \cong)\Sigma \subset \mathbb{R}^3$ be the symmetry plane for this electrostatic field, and let Λ_k be the open half-space containing $\boldsymbol{Q}_k(0)$, $k = 1, 2$; thus, $\mathbb{R}^3 = \Lambda_1 \cup \Lambda_2 \cup \Sigma$. Then, by integrating their postulated equation of motion over Λ_k, and with $k + k' = 3$, we find

$$\begin{aligned} \frac{1}{4\pi c}\frac{\mathrm{d}}{\mathrm{d}t}\int_{\Lambda_k}(\boldsymbol{\mathcal{D}}_{\mathrm{MBI}} \times \boldsymbol{\mathcal{B}}_{\mathrm{MBI}})(t, \boldsymbol{s})\,\mathrm{d}^3 s\Big|_{t=0} &= \int_{\partial\Lambda_k}\Theta_{\mathrm{MBI}}(0, \boldsymbol{s})\cdot\boldsymbol{n}\,\mathrm{d}\sigma \\ &\sim e^2\,\frac{\boldsymbol{Q}_k(0) - \boldsymbol{Q}_{k'}(0)}{|\boldsymbol{Q}_k(0) - \boldsymbol{Q}_{k'}(0)|^3} \end{aligned} \tag{5.3}$$

with "$\sim$" as the distance between the charges tends $\to \infty$ (the asymptotic result is assumed to be true, and presumably rigorously provable as I've written already). Pryce and his peers next argued that for gently accelerated motions we should be allowed to replace the field momentum integral $\frac{1}{4\pi c}\int_{\Lambda_k}(\boldsymbol{\mathcal{D}}_{\mathrm{MBI}} \times \boldsymbol{\mathcal{B}}_{\mathrm{MBI}})(t, \boldsymbol{s})\mathrm{d}^3 s$ by $m_{\mathrm{f}}(\boldsymbol{Q}_1, \boldsymbol{Q}_2)\gamma_k(t)\dot{\boldsymbol{Q}}_k(t)$, where $\gamma_k^2(t) = 1/(1-|\dot{\boldsymbol{Q}}_k(t)|^2/c^2)$ and $m_{\mathrm{f}}(\boldsymbol{Q}_1, \boldsymbol{Q}_2)c^2 = \frac{b^2}{4\pi}\int_{\Lambda_k}\big(\sqrt{1 + b^{-2}|\boldsymbol{\mathcal{D}}_{\mathrm{MBI}}|^2} - 1\big)(0, \boldsymbol{s})\mathrm{d}^3 s$, with $\boldsymbol{Q}_k$ standing for $\boldsymbol{Q}_k(0)$. Lastly, we should have $\boldsymbol{\mathcal{D}}_{\mathrm{MBI}}(0, \boldsymbol{s} - \boldsymbol{Q}_k) \to \boldsymbol{\mathcal{D}}_{\mathrm{Born}}(\boldsymbol{s})$ as $|\boldsymbol{Q}_k - \boldsymbol{Q}_{k'}| \to \infty$, with (the relevant) $\boldsymbol{Q}_k$ at the origin, and in this sense, and with $b = b_{\mathrm{Born}}$, we would then also have $m_{\mathrm{f}}(\boldsymbol{Q}_1, \boldsymbol{Q}_2) \to m_{\mathrm{e}}$ as $|\boldsymbol{Q}_1 - \boldsymbol{Q}_2| \to \infty$. Thus, in this asymptotic regime, at the initial time, the reasoning of Pryce and his peers yields

$$\tfrac{\mathrm{d}}{\mathrm{d}t}\big(m_{\mathrm{e}}\gamma_k(t)\dot{\boldsymbol{Q}}_k(t)\big)\big|_{t=0} = e^2\,\tfrac{\boldsymbol{Q}_k(0) - \boldsymbol{Q}_{k'}(0)}{|\boldsymbol{Q}_k(0) - \boldsymbol{Q}_{k'}(0)|^3}. \tag{5.4}$$

Surely this looks very compelling, but it is clear that the heuristic replacement of $\frac{1}{4\pi c}\int_{\Lambda_k}(\boldsymbol{\mathcal{D}}_{\mathrm{MBI}} \times \boldsymbol{\mathcal{B}}_{\mathrm{MBI}})(t, \boldsymbol{s})\mathrm{d}^3 s$ by $m_{\mathrm{f}}(\boldsymbol{Q}_1, \boldsymbol{Q}_2)\gamma_k(t)\dot{\boldsymbol{Q}}_k(t)$ as just explained is only a first approximation. By going one step further Schrödinger [Schr1942b] argued that the (in)famous third-order radiation reaction term will appear, and so, if the approximation were consistent, we would now need a third initial condition, namely on $\ddot{\boldsymbol{Q}}_k(0)$. We are "back to square one" with our problems. One might argue that the third-order term is not yet the consistent approximation and invoke the Landau–Lifshitz reasoning to get an effective second-order equation. However, going on to higher orders would successively bring in higher and higher derivatives of $\boldsymbol{Q}_k(t)$. This looks just like the situation in the old purely electromagnetic classical electron theory of Abraham and Lorentz. All alarm bells should be going off by now, because their equation of motion is overdetermined [ApKi2001].

Dirac, on the other hand, obtains as putatively exact equation of motion

$$\frac{\mathrm{d}}{\mathrm{d}t}\big(M\gamma_k(t)\dot{\boldsymbol{Q}}_k(t)\big) = e_k\Big[\boldsymbol{\mathcal{E}}^{\mathrm{reg}}_{\mathrm{MBI}}(t,\boldsymbol{Q}_k(t)) + \tfrac{1}{c}\dot{\boldsymbol{Q}}_k(t)\times\boldsymbol{\mathcal{B}}^{\mathrm{reg}}_{\mathrm{MBI}}(t,\boldsymbol{Q}_k(t))\Big], \tag{5.5}$$

where the superscripts ^{reg} indicate his regularization procedure, which also involves the integration of the stress tensor over a small domain Λ_k containing $\boldsymbol{Q}_k(t)$, plus Gauss' theorem, followed by the limit $\Lambda_k \to \{\boldsymbol{Q}_k(t)\}$. However, Dirac uses this only to split off a singular term from the ill-defined Lorentz force, arguing that the remaining electromagnetic force field is regular; this field enters in (5.5). As far as I can tell, Dirac's remainder field is generally still singular; the critical passage is on the bottom of page 36 in [Dir1960].

I end here with a comment on Dirac's suggestion that $M = 0$ for an electron. In this case, even with a regular force field, his (5.5) would be overdetermined, because setting the coefficient of the highest derivative equal to zero amounts to a singular limit.

5.2.3. On Newton's law for the rate of change of momentum. I have argued that no matter how you cut the cake, the Lorentz force formula r.h.s.(2.17) for the electromagnetic force on a point charge cannot be well-defined when $\boldsymbol{\mathcal{E}}$ and $\boldsymbol{\mathcal{B}}$ are the total fields. Since only the total fields can possibly be fundamental, while "external field" and "self-field" are only auxiliary notions, it follows that the Lorentz force cannot play a fundamental dynamical role. This now inevitably raises the question as to the status of Newton's law for the rate of change of momentum, $\dot{\boldsymbol{P}}(t) = \boldsymbol{F}(t)$, of which (2.17) pretends to be the particular realization in the context of classical electrodynamics.

If one insists that Newton's law $\dot{\boldsymbol{P}}(t) = \boldsymbol{F}(t)$ remains fundamental throughout classical physics, including relativistic point charge motion coupled to the electromagnetic fields, then one is obliged to continue the quest for a well-defined expression for the fundamental electromagnetic force $\boldsymbol{F}$ on a point charge.

The alternative is to relegate Newton's law $\dot{\boldsymbol{P}}(t) = \boldsymbol{F}(t)$ to the status of an effective law, emerging in the regime of relativistic charged test particle motions. In this case one needs to look elsewhere for the fundamental relativistic law of motion of point charges. Of course, the effective concept of the *external Lorentz force* acting on a *test* particle remains a beacon which any fundamental theory must keep in sight.

This is the point of view taken in [Kie2004a], and also here.

5.3. Hamilton–Jacobi theory of motion

Although we have not only abandoned (2.17) but actually Newton's $\dot{\boldsymbol{P}}(t) = \boldsymbol{F}(t)$ altogether, for now we will hold on to the second one of the two equations (2.17), (2.18) of the *formal* relativistic Newtonian law of motion. But then, since (2.18) expresses the rate of change of position, $\dot{\boldsymbol{Q}}(t)$, in terms of the "mechanical momentum" vector $\boldsymbol{P}(t)$, we need to find a new type of law which gives us $\boldsymbol{P}(t)$ in a well-defined manner. Keeping in mind the moral that "formal manipulations are not to be trusted until they can be vindicated rigorously," in this section we will argue that Hamilton–Jacobi theory supplies a classical law for $\boldsymbol{P}(t)$ which in fact is well-defined, provided the solutions realizing (P), (W), (M), (E), (D) of the Maxwell–Born–Infeld field equations with point charge sources are as well-behaved for generic smooth subluminal motions as they are for Born's static solution, at least locally in time.

Indeed, the electric field $\boldsymbol{\mathcal{E}}_{\mathrm{Born}}$ associated to $(\boldsymbol{\mathcal{B}}_{\mathrm{Born}}, \boldsymbol{\mathcal{D}}_{\mathrm{Born}})$, Born's static field pair for a single (negative) point charge at the origin, is undefined at the origin but uniformly bounded elsewhere; it exhibits a *point defect* at $\boldsymbol{s} = \mathbf{0}$. For $\boldsymbol{s} \neq \mathbf{0}$, it is given by $\boldsymbol{\mathcal{E}}_{\mathrm{Born}}(\boldsymbol{s}) = -\boldsymbol{\nabla}\phi_{\mathrm{Born}}(\boldsymbol{s})$, where

$$\phi_{\mathrm{Born}}(\boldsymbol{s}) = -\sqrt{be}\int_{|\boldsymbol{s}|\sqrt{b/e}}^{\infty} \frac{\mathrm{d}x}{\sqrt{1+x^4}}\,. \tag{5.6}$$

Note, $\phi_{\mathrm{Born}}(\boldsymbol{s}) \sim -e|\boldsymbol{s}|^{-1}$ for $|\boldsymbol{s}| \gg \sqrt{e/b}$, and $\lim_{|\boldsymbol{s}|\downarrow 0}\phi_{\mathrm{Born}}(\boldsymbol{s}) =: \phi_{\mathrm{Born}}(\mathbf{0}) < \infty$. So, away from the origin the electrostatic potential $\phi_{\mathrm{Born}}(\boldsymbol{s})$ is (even infinitely) differentiable, and it can be Lipschitz-continuously extended into the origin. We conjecture that this regularity is *typical* for the electromagnetic *potentials* of the Maxwell–Born–Infeld fields for *generic* smooth subluminal point source motions, in the sense that it should be so in the Lorenz–Lorentz gauge (see below), and remain true under any subsequent *smooth* gauge transformation. Gauge transformations with less regularity would have to be ruled out.

5.3.1. The electromagnetic potentials. Given a solution $t \mapsto (\boldsymbol{\mathcal{B}}_{\mathrm{MBI}}, \boldsymbol{\mathcal{D}}_{\mathrm{MBI}})(t, \boldsymbol{s})$ of the Maxwell–Born–Infeld field equations for some smooth subluminal motion $t \mapsto \boldsymbol{Q}(t)$ of a point charge (or several of them), we can algebraically compute the field pair $t \mapsto (\boldsymbol{\mathcal{B}}_{\mathrm{MBI}}, \boldsymbol{\mathcal{E}}_{\mathrm{MBI}})(t, \boldsymbol{s})$ from the Born–Infeld "aether law." For any such map $t \mapsto (\boldsymbol{\mathcal{B}}_{\mathrm{MBI}}, \boldsymbol{\mathcal{E}}_{\mathrm{MBI}})(t, \boldsymbol{s})$, we define the magnetic vector potential $\boldsymbol{\mathcal{A}}_{\mathrm{MBI}}(t, \boldsymbol{s})$ and the electric potential $\phi_{\mathrm{MBI}}(t, \boldsymbol{s})$ in terms of the following PDE. Namely, $\boldsymbol{\mathcal{A}}_{\mathrm{MBI}}(t, \boldsymbol{s})$ satisfies the evolution equation

$$\tfrac{1}{c}\tfrac{\partial}{\partial t}\boldsymbol{\mathcal{A}}_{\mathrm{MBI}}(t, \boldsymbol{s}) = -\boldsymbol{\nabla}\phi_{\mathrm{MBI}}(t, \boldsymbol{s}) - \boldsymbol{\mathcal{E}}_{\mathrm{MBI}}(t, \boldsymbol{s}) \tag{5.7}$$

and the constraint equation

$$\boldsymbol{\nabla} \times \boldsymbol{\mathcal{A}}_{\mathrm{MBI}}(t, \boldsymbol{s}) = \boldsymbol{\mathcal{B}}_{\mathrm{MBI}}(t, \boldsymbol{s})\,, \tag{5.8}$$

while the evolution of $\phi_{\mathrm{MBI}}(t, \boldsymbol{s})$ is governed by

$$\tfrac{\partial}{\partial t}\phi_{\mathrm{MBI}}(t, \boldsymbol{s}) = -\boldsymbol{\nabla}\cdot\boldsymbol{\mathcal{A}}_{\mathrm{MBI}}(t, \boldsymbol{s}), \tag{5.9}$$

unconstrained by any other equation.

Equation (5.9) is known as the V. Lorenz–H. A. Lorentz gauge condition (see [HaEl1973, JaOk2001]) postulated here for no other reason than that it is simple, invariant under the Poincaré group, and *presumably* compatible with our regularity conjecture for the potentials. While it renders the Maxwell–Lorentz field equations with prescribed point sources as a decoupled set of non-homogeneous wave equations for the four-vector field $(\phi_{\mathrm{ML}}, \boldsymbol{\mathcal{A}}_{\mathrm{ML}})$, readily solved by the Liénard–Wiechert potentials [Lié1898, Wie1900], (5.9) achieves no such simplification for the Maxwell–Born–Infeld equations with point sources.

The Lorenz–Lorentz condition fixes the gauge freedom of the relativistic four-vector potential field $\big(\phi, \boldsymbol{\mathcal{A}}\big)(t, \boldsymbol{s})$ to some extent, yet the equations (5.7), (5.8), and (5.9) are still invariant under the gauge transformations

$$\phi(t, \boldsymbol{s}) \to \phi(t, \boldsymbol{s}) - \tfrac{1}{c}\tfrac{\partial}{\partial t}\Upsilon(t, \boldsymbol{s}), \tag{5.10}$$

$$\boldsymbol{\mathcal{A}}(t, \boldsymbol{s}) \to \boldsymbol{\mathcal{A}}(t, \boldsymbol{s}) + \boldsymbol{\nabla}\Upsilon(t, \boldsymbol{s}), \tag{5.11}$$

with any relativistic scalar field $\Upsilon : \mathbb{R}^{1,3} \to \mathbb{R}$ satisfying the wave equation

$$\tfrac{1}{c^2}\tfrac{\partial^2}{\partial t^2}\Upsilon(t, \boldsymbol{s}) = \boldsymbol{\nabla}^2\Upsilon(t, \boldsymbol{s}), \tag{5.12}$$

with $\boldsymbol{\nabla}^2 = \Delta$, the Laplacian on $\mathbb{R}^3$. Since a (sufficiently regular) solution of (5.12) in $\mathbb{R}_+ \times \mathbb{R}^3$ is uniquely determined by the initial data for Υ and its time derivative $\frac{\partial}{\partial t}\Upsilon$, the gauge freedom that is left concerns the initial conditions of ϕ, $\boldsymbol{\mathcal{A}}$.

5.3.2. Canonical momenta of point defects with intrinsic mass m. As per our (plausible but as of yet unproven) hypothesis, for generic smooth subluminal point charge motions the electromagnetic potential fields ϕ_{MBI} and $\boldsymbol{\mathcal{A}}_{\text{MBI}}$ have Lipschitz continuous extensions to all of space, at any time t. In the following we shall always mean these extensions when we speak of the electromagnetic-field potentials. With our electromagnetic-field potentials unambiguously defined at each location of a field point defect, we are now able to define the so-called canonical momentum of a point defect of the electromagnetic fields associated with a point charge moving along a smooth trajectory $t \mapsto \boldsymbol{Q}(t)$ with subluminal speed. Namely, given a smooth trajectory $t \mapsto \boldsymbol{Q}(t)$, consider (2.18) though now with "intrinsic inert mass m" in place of m_{e}. Inverting (2.18) with intrinsic inert mass m we obtain the "intrinsic" momentum of the point defect,

$$\boldsymbol{P}(t) = m\frac{\dot{\boldsymbol{Q}}(t)}{\sqrt{1 - |\dot{\boldsymbol{Q}}(t)|^2/c^2}}. \tag{5.13}$$

Also, per our conjecture, $\boldsymbol{\mathcal{A}}_{\text{MBI}}(t, \boldsymbol{Q}(t))$ is well-defined at each t, and so, for a negative charge, the *canonical momentum*

$$\boldsymbol{\Pi}(t) := \boldsymbol{P}(t) - \tfrac{1}{c}e\boldsymbol{\mathcal{A}}_{\text{MBI}}(t, \boldsymbol{Q}(t)) \tag{5.14}$$

is well-defined for all t.

We now turn (5.14) around and, for a negative point charge, take

$$\boldsymbol{P}(t) = \boldsymbol{\Pi}(t) + \tfrac{1}{c}e\boldsymbol{\mathcal{A}}_{\text{MBI}}(t, \boldsymbol{Q}(t)) \tag{5.15}$$

as the formula for $\boldsymbol{P}(t)$ that has to be coupled with (2.18). Thus, next we need to find an expression for $\boldsymbol{\Pi}(t)$. Precisely this is supplied by Hamilton–Jacobi theory.

5.3.3. Hamilton–Jacobi laws of motion. In Hamilton–Jacobi theory of single-point motion one introduces the *single-point configuration space* of *generic positions* $\boldsymbol{q} \in \mathbb{R}^3$ of the point defect. A *Hamilton–Jacobi law of motion* consists of two parts: (i) an *ordinary differential equation* for the actual position $\boldsymbol{Q}(t)$ of the point defect, equating its actual velocity with the evaluation — at its actual position — of a velocity field on configuration space; (ii) a *partial differential equation* for this velocity field. The correct law should reduce to the test particle theory in certain regimes, so we begin with the latter.

The test charge approximation: all well again, so far! In advanced textbooks on mathematical classical physics [Thi1997] one finds the equations of relativistic Hamilton–Jacobi theory for test charge motion in the potentials $\boldsymbol{\mathcal{A}}(t,\boldsymbol{s}) = \boldsymbol{\mathcal{A}}_{\text{MM}}(t,\boldsymbol{s})$ and $\phi(t,\boldsymbol{s}) = \phi_{\text{MM}}(t,\boldsymbol{s})$ for the *actual field* solutions $\boldsymbol{\mathcal{E}}_{\text{MM}}(t,\boldsymbol{s})$ and $\boldsymbol{\mathcal{B}}_{\text{MM}}(t,\boldsymbol{s})$ of the Maxwell–Maxwell field equations, serving as "external" fields. This reproduces the test particle motions computed from (2.17), (2.18) with $\boldsymbol{\mathcal{E}} = \boldsymbol{\mathcal{E}}_{\text{MM}}$ and $\boldsymbol{\mathcal{B}} = \boldsymbol{\mathcal{B}}_{\text{MM}}$. This setup is locally well-defined, for "externally generated potentials" are independent of where and how the test charge moves, and so they can be assumed to be smooth functions of space and time. Provided the solutions $\boldsymbol{\mathcal{E}}^{\text{sf}}_{\text{MBI}}(t,\boldsymbol{s})$ and $\boldsymbol{\mathcal{B}}^{\text{sf}}_{\text{MBI}}(t,\boldsymbol{s})$ of the *source-free* Maxwell–Born–Infeld field equations are smooth, the Hamilton–Jacobi law of test particle motion remains locally well-defined if we set $\boldsymbol{\mathcal{A}}(t,\boldsymbol{q}) = \boldsymbol{\mathcal{A}}^{\text{sf}}_{\text{MBI}}(t,\boldsymbol{q})$ and $\phi(t,\boldsymbol{q}) = \phi^{\text{sf}}_{\text{MBI}}(t,\boldsymbol{q})$, the generic-$\boldsymbol{q}$-evaluation of these source-free Maxwell–Born–Infeld potentials.

The relativistic Hamilton–Jacobi guiding equation

Suppose the actual canonical momentum of the point test charge is given by

$$\boldsymbol{\Pi}(t) = \nabla_{\boldsymbol{q}} S_{\text{HJ}}(t,\boldsymbol{Q}(t)), \tag{5.16}$$

where $\boldsymbol{q} \mapsto S_{\text{HJ}}(t,\boldsymbol{q})$ is a time-dependent differentiable scalar field on configuration space. By virtue of (5.16), (5.13), (5.15), with $\boldsymbol{\mathcal{A}}_{\text{MBI}}(t,\boldsymbol{Q}(t))$ replaced by $\boldsymbol{\mathcal{A}}^{\text{sf}}_{\text{MBI}}(t,\boldsymbol{Q}(t))$, we can eliminate $\boldsymbol{\Pi}(t)$ in favor of $\nabla_{\boldsymbol{q}} S_{\text{HJ}}(t,\boldsymbol{Q}(t))$ which, for a negative test charge of mass m, yields the *relativistic Hamilton–Jacobi guiding equation*

$$\frac{1}{c}\frac{\mathrm{d}\boldsymbol{Q}(t)}{\mathrm{d}t} = \frac{\nabla_{\boldsymbol{q}} S_{\text{HJ}}(t,\boldsymbol{Q}(t)) + \frac{1}{c}e\boldsymbol{\mathcal{A}}^{\text{sf}}_{\text{MBI}}(t,\boldsymbol{Q}(t))}{\sqrt{m^2c^2 + \left|\nabla_{\boldsymbol{q}} S_{\text{HJ}}(t,\boldsymbol{Q}(t)) + \frac{1}{c}e\boldsymbol{\mathcal{A}}^{\text{sf}}_{\text{MBI}}(t,\boldsymbol{Q}(t))\right|^2}}\,. \tag{5.17}$$

The relativistic Hamilton–Jacobi partial differential equation

The requirement that the test charge velocity is the space component of a (future-directed) four-velocity vector divided by the relativistic γ factor quite naturally leads to the following *relativistic Hamilton–Jacobi partial differential equation*,

$$\tfrac{1}{c}\tfrac{\partial}{\partial t} S_{\text{HJ}}(t,\boldsymbol{q}) = -\sqrt{m^2c^2 + \left|\nabla_{\boldsymbol{q}} S_{\text{HJ}}(t,\boldsymbol{q}) + \tfrac{1}{c}e\boldsymbol{\mathcal{A}}^{\text{sf}}_{\text{MBI}}(t,\boldsymbol{q})\right|^2} + \tfrac{1}{c}e\phi^{\text{sf}}_{\text{MBI}}(t,\boldsymbol{q}). \tag{5.18}$$

Lorentz and Weyl invariance

As to Lorentz invariance, any solution to (5.18) obviously satisfies

$$\left(\tfrac{1}{c}\tfrac{\partial}{\partial t}S_{\mathrm{HJ}}(t,\boldsymbol{q})-\tfrac{1}{c}e\phi^{\mathrm{sf}}_{\mathrm{MBI}}(t,\boldsymbol{q})\right)^2-\left|\nabla_{\boldsymbol{q}}S_{\mathrm{HJ}}(t,\boldsymbol{q})+\tfrac{1}{c}e\boldsymbol{\mathcal{A}}^{\mathrm{sf}}_{\mathrm{MBI}}(t,\boldsymbol{q})\right|^2=m^2c^2, \quad (5.19)$$

a manifestly relativistically Lorentz scalar equation.

Although $S_{\mathrm{HJ}}(t,\boldsymbol{q})$ is a scalar configuration spacetime field, it cannot be gauge-invariant, for the four-vector field $\big(\phi,\boldsymbol{\mathcal{A}}\big)(t,\boldsymbol{s})$ is not; recall that the Lorenz–Lorentz gauge condition alone does not fix the potentials completely. Instead, if the potentials $(\phi,\boldsymbol{\mathcal{A}})$ are transformed under the gauge transformations (5.11) with any relativistic scalar field $\Upsilon:\mathbb{R}^{1,3}\to\mathbb{R}$ satisfying the wave equation (5.12), then, for a negative charge, S_{HJ} needs to be transformed as

$$S_{\mathrm{HJ}}(t,\boldsymbol{q})\to S_{\mathrm{HJ}}(t,\boldsymbol{q})-\tfrac{1}{c}e\Upsilon(t,\boldsymbol{q}). \quad (5.20)$$

This gauge transformation law also holds more generally for Υ not satisfying (5.12), meaning a change of gauge from Lorenz–Lorentz to something else.

Many-body test charge theory

The generalization to many point charges with either sign is obvious. Since test charges do not "talk back" to the "external" potentials, there is a guiding equation (5.17) coupled with a partial differential equation (5.18) for the guiding velocity field *for each test charge.* Of course, they are just identical copies of the single-particle equations, yet it is important to keep in mind that the many-body theory is to be formulated on many-body configuration space.

Upgrading test particle motions: self-force problems déjà vu! Since the electromagnetic potentials for the *actual* electromagnetic Maxwell–Born–Infeld fields with point charge sources are supposedly defined everywhere, it could now seem that in order to get a well-defined theory of motion of their point charge sources all that needs to be done is to replace $\boldsymbol{\mathcal{A}}^{\mathrm{sf}}_{\mathrm{MBI}}(t,\boldsymbol{q})$ and $\phi^{\mathrm{sf}}_{\mathrm{MBI}}(t,\boldsymbol{q})$ by $\boldsymbol{\mathcal{A}}_{\mathrm{MBI}}(t,\boldsymbol{q})$ and $\phi_{\mathrm{MBI}}(t,\boldsymbol{q})$ in (5.18), which yields the partial differential equation

$$\tfrac{1}{c}\tfrac{\partial}{\partial t}S_{\mathrm{HJ}}(t,\boldsymbol{q})=-\sqrt{m^2c^2+\left|\nabla_{\boldsymbol{q}}S_{\mathrm{HJ}}(t,\boldsymbol{q})+\tfrac{1}{c}e\boldsymbol{\mathcal{A}}_{\mathrm{MBI}}(t,\boldsymbol{q})\right|^2}+\tfrac{1}{c}e\phi_{\mathrm{MBI}}(t,\boldsymbol{q}), \quad (5.21)$$

and to replace $\boldsymbol{\mathcal{A}}^{\mathrm{sf}}_{\mathrm{MBI}}(t,\boldsymbol{Q}(t))$ by $\boldsymbol{\mathcal{A}}_{\mathrm{MBI}}(t,\boldsymbol{Q}(t))$ in (5.17) to get the guiding equation

$$\frac{1}{c}\frac{\mathrm{d}\boldsymbol{Q}(t)}{\mathrm{d}t}=\frac{\nabla_{\boldsymbol{q}}S_{\mathrm{HJ}}(t,\boldsymbol{Q}(t))+\tfrac{1}{c}e\boldsymbol{\mathcal{A}}_{\mathrm{MBI}}(t,\boldsymbol{Q}(t))}{\sqrt{m^2c^2+\left|\nabla_{\boldsymbol{q}}S_{\mathrm{HJ}}(t,\boldsymbol{Q}(t))+\tfrac{1}{c}e\boldsymbol{\mathcal{A}}_{\mathrm{MBI}}(t,\boldsymbol{Q}(t))\right|^2}}. \quad (5.22)$$

So the *actual* electromagnetic potentials as functions of space and time are evaluated at the generic position $\boldsymbol{q}$ in (5.21) and at the actual position $\boldsymbol{Q}(t)$ in (5.22).

Note that almost all flow lines of the gradient field $\nabla_{\boldsymbol{q}}S_{\mathrm{HJ}}(t,\boldsymbol{q})$ would still correspond to test particle motions in the actual $\phi_{\mathrm{MBI}}(t,\boldsymbol{s})$ and $\boldsymbol{\mathcal{A}}_{\mathrm{MBI}}(t,\boldsymbol{s})$ potential fields (simply because almost all generic positions $\boldsymbol{q}$ are not identical to the actual position $\boldsymbol{Q}(t)$ of the point charge source of the actual $\phi_{\mathrm{MBI}}(t,\boldsymbol{s})$

and $\mathcal{A}_{\text{MBI}}(t, s)$), so one may hope that by suitably iterating the given actual motion one can make precisely one of these test particle motions coincide with the actual motion — which is meant by "upgrading test-particle motion."

However, this does not lead to a well-defined theory of motion of point charge sources! The reason is that $\phi_{\text{MBI}}(t, s)$ and $\mathcal{A}_{\text{MBI}}(t, s)$ have non-differentiable "kinks" at $s = Q(t)$. The function $S_{\text{HJ}}(t, q)$ picks up this non-differentiability at $q = Q(t)$ through (5.21). More precisely, (5.21) is only well-defined *away* from the actual positions of the point charges. Trying to extend the definition of $\nabla_q S_{\text{HJ}}(t, q)$ to the actual positions now leads pretty much to the same mathematical problems as encountered when trying to define the "Lorentz self-force" on the point charge sources of the Maxwell–Born–Infeld field equations. In particular, we could regularize the actual potentials $\phi_{\text{MBI}}(t, s)$ and $\mathcal{A}_{\text{MBI}}(t, s)$ by averaging, thereby obtaining a regularized "upgraded test-particle Hamilton–Jacobi theory" which does yield the actual "regularized motion" amongst all "regularized test particle motions" as a nonlinear fixed point problem. Unfortunately, subsequent removal of the regularization generally does not yield a unique limit, so that any so-defined limiting theory of point charge motion would, once again, not be well-defined.

Fortunately, Hamilton–Jacobi theory offers another option. Recall that for the non-relativistic problem of motion of N widely separated point charges interacting through their Coulomb pair interactions, Hamilton–Jacobi theory yields a gradient flow on N-particle configuration space of which *each flow line represents a putative actual trajectory* of the N body problem: there are no test particle trajectories! In this vein, we should focus on a formulation of Hamilton–Jacobi theory which "parallel-processes" putative actual point charge motions.

Parallel processing of putative actual motions: success! While nontrivial motions in a strictly non-relativistic Coulomb problem without "external" fields can occur only when $N \geq 2$ (Kepler motions if $N = 2$), a system with a single point charge source for the electromagnetic Maxwell–Born–Infeld fields generally should feature non-trivial motions on single-particle configuration space because of the dynamical degrees of freedom of the electromagnetic fields. So in the following we focus on the $N = 1$ point charge problem, although eventually we have to address the general N-body problem.

Setting up a Hamilton–Jacobi law which "parallel-processes" putative actual single point source motions in the Maxwell–Born–Infeld field equations is only possible if there exists a generic velocity field on configuration space (here: for a negative point charge), denoted by $v(t, q)$, which varies smoothly with q and t, and which is related to the family of putative actual motions by the guiding law

$$\frac{\mathrm{d}Q(t)}{\mathrm{d}t} = v(t, Q(t)) , \tag{5.23}$$

yielding the actual position $Q(t)$ for each actual initial position $Q(0)$.

Assuming such a velocity field exists, one next needs to construct configuration space fields $\phi_1(t, q)$ and $\mathcal{A}_1(t, q)$ which are "generic-q-sourced" potential fields $\phi^\sharp(t,\boldsymbol{s},\boldsymbol{q})$ and $\boldsymbol{\mathcal{A}}^\sharp(t,\boldsymbol{s},\boldsymbol{q})$ evaluated at $\boldsymbol{s}=\boldsymbol{q}$, their generic point source;[17] i.e. [Kie2004a]:

$$\phi_1(t,\boldsymbol{q}) \equiv \phi^\sharp(t,\boldsymbol{q},\boldsymbol{q}) \qquad \text{and} \qquad \boldsymbol{\mathcal{A}}_1(t,\boldsymbol{q}) \equiv \boldsymbol{\mathcal{A}}^\sharp(t,\boldsymbol{q},\boldsymbol{q}). \tag{5.24}$$

The "canonical" set of partial differential equations for $\phi^\sharp(t,\boldsymbol{s},\boldsymbol{q})$, $\boldsymbol{\mathcal{A}}^\sharp(t,\boldsymbol{s},\boldsymbol{q})$, and their derived fields, which are compatible with the Maxwell–Born–Infeld field equations for the actual fields of a single negative point charge, reads

$$\tfrac{1}{c}\tfrac{\partial}{\partial t}\phi^\sharp(t,\boldsymbol{s},\boldsymbol{q}) = -\tfrac{1}{c}\boldsymbol{v}(t,\boldsymbol{q})\cdot\nabla_{\boldsymbol{q}}\phi^\sharp(t,\boldsymbol{s},\boldsymbol{q}) - \boldsymbol{\nabla}\cdot\boldsymbol{\mathcal{A}}^\sharp(t,\boldsymbol{s},\boldsymbol{q}), \tag{5.25}$$

$$\tfrac{1}{c}\tfrac{\partial}{\partial t}\boldsymbol{\mathcal{A}}^\sharp(t,\boldsymbol{s},\boldsymbol{q}) = -\tfrac{1}{c}\boldsymbol{v}(t,\boldsymbol{q})\cdot\nabla_{\boldsymbol{q}}\boldsymbol{\mathcal{A}}^\sharp(t,\boldsymbol{s},\boldsymbol{q}) - \boldsymbol{\nabla}\phi^\sharp(t,\boldsymbol{s},\boldsymbol{q}) - \boldsymbol{\mathcal{E}}^\sharp(t,\boldsymbol{s},\boldsymbol{q}), \tag{5.26}$$

$$\tfrac{1}{c}\tfrac{\partial}{\partial t}\boldsymbol{\mathcal{D}}^\sharp(t,\boldsymbol{s},\boldsymbol{q}) = -\tfrac{1}{c}\boldsymbol{v}(t,\boldsymbol{q})\cdot\nabla_{\boldsymbol{q}}\boldsymbol{\mathcal{D}}^\sharp(t,\boldsymbol{s},\boldsymbol{q}) + \boldsymbol{\nabla}\times\boldsymbol{\mathcal{H}}^\sharp(t,\boldsymbol{s},\boldsymbol{q}) + 4\pi e\tfrac{1}{c}\boldsymbol{v}(t,\boldsymbol{q})\delta_{\boldsymbol{q}}(\boldsymbol{s}); \tag{5.27}$$

furthermore, $\boldsymbol{\mathcal{D}}^\sharp(t,\boldsymbol{s},\boldsymbol{q})$ obeys the constraint equation[18]

$$\boldsymbol{\nabla}\cdot\boldsymbol{\mathcal{D}}^\sharp(t,\boldsymbol{s},\boldsymbol{q}) = -4\pi e\delta_{\boldsymbol{q}}(\boldsymbol{s}). \tag{5.28}$$

The fields $\boldsymbol{\mathcal{E}}^\sharp(t,\boldsymbol{s},\boldsymbol{q})$ and $\boldsymbol{\mathcal{H}}^\sharp(t,\boldsymbol{s},\boldsymbol{q})$ in (5.26), (5.27) are given in terms of $\boldsymbol{\mathcal{D}}^\sharp(t,\boldsymbol{s},\boldsymbol{q})$ and $\boldsymbol{\mathcal{B}}^\sharp(t,\boldsymbol{s},\boldsymbol{q})$ in the same way as the actual fields $\boldsymbol{\mathcal{E}}_{\mathrm{MBI}}(t,\boldsymbol{s})$ and $\boldsymbol{\mathcal{H}}_{\mathrm{MBI}}(t,\boldsymbol{s})$ are defined in terms of $\boldsymbol{\mathcal{D}}_{\mathrm{MBI}}(t,\boldsymbol{s})$ and $\boldsymbol{\mathcal{B}}_{\mathrm{MBI}}(t,\boldsymbol{s})$ through the Born–Infeld aether law (4.5), (4.6), while $\boldsymbol{\mathcal{B}}^\sharp(t,\boldsymbol{s},\boldsymbol{q})$ in turn is given in terms of $\boldsymbol{\mathcal{A}}^\sharp(t,\boldsymbol{s},\boldsymbol{q})$ in the same way as the actual $\boldsymbol{\mathcal{B}}_{\mathrm{MBI}}(t,\boldsymbol{s})$ is given in terms of the actual $\boldsymbol{\mathcal{A}}_{\mathrm{MBI}}(t,\boldsymbol{s})$ in (5.8). It is straightforward to verify that by substituting the actual $\boldsymbol{Q}(t)$ for the generic $\boldsymbol{q}$ in the "generic-$\boldsymbol{q}$-sourced" $\sharp$-fields satisfying the above field equations, we obtain the actual electromagnetic potentials, fields, and charge-current densities satisfying the Maxwell–Born–Infeld field equations (in Lorenz–Lorentz gauge). That is,

$$\phi_{\mathrm{MBI}}(t,\boldsymbol{s}) \equiv \phi^\sharp(t,\boldsymbol{s},\boldsymbol{Q}(t)) \qquad \text{etc.} \tag{5.29}$$

Next we need to stipulate a law for $\boldsymbol{v}(t,\boldsymbol{q})$.

The Hamilton–Jacobi velocity field

The naïvely obvious thing to try is the generic velocity law

$$\boldsymbol{v}(t,\boldsymbol{q}) = c\frac{\nabla_{\boldsymbol{q}}S_{\mathrm{HJ}}(t,\boldsymbol{q}) + \frac{1}{c}e\boldsymbol{\mathcal{A}}_1(t,\boldsymbol{q})}{\sqrt{m^2c^2 + \left|\nabla_{\boldsymbol{q}}S_{\mathrm{HJ}}(t,\boldsymbol{q}) + \frac{1}{c}e\boldsymbol{\mathcal{A}}_1(t,\boldsymbol{q})\right|^2}}, \tag{5.30}$$

corresponding to the Hamilton–Jacobi PDE

$$\tfrac{1}{c}\tfrac{\partial}{\partial t}S_{\mathrm{HJ}}(t,\boldsymbol{q}) = -\sqrt{m^2c^2 + \left|\nabla_{\boldsymbol{q}}S_{\mathrm{HJ}}(t,\boldsymbol{q}) + \tfrac{1}{c}e\boldsymbol{\mathcal{A}}_1(t,\boldsymbol{q})\right|^2} + \tfrac{1}{c}e\phi_1(t,\boldsymbol{q}), \tag{5.31}$$

[17] The notation is inherited from the N-point-charge problem. In that case there are fields $\phi^\sharp(t,\boldsymbol{s},\boldsymbol{q}_1,\dots,\boldsymbol{q}_N)$ etc. which, when evaluated at $\boldsymbol{s}=\boldsymbol{q}_k$, give configuration space fields $\phi_k(t,\boldsymbol{q}_1,\dots,\boldsymbol{q}_N)$ etc. For a system with a single point charge, $k=1$.

[18] Since the generic charge density $-e\delta_{\boldsymbol{q}}(\boldsymbol{s})$ is t-independent and $\nabla_{\boldsymbol{q}}\delta_{\boldsymbol{q}}(\boldsymbol{s}) = -\boldsymbol{\nabla}\delta_{\boldsymbol{q}}(\boldsymbol{s})$, the reformulation of the continuity equation for charge conservation (in spacetime), $\frac{\partial}{\partial t}\rho^\sharp(t,\boldsymbol{s},\boldsymbol{q}) = -\boldsymbol{v}(t,\boldsymbol{q})\cdot\nabla_{\boldsymbol{q}}\rho^\sharp(t,\boldsymbol{s},\boldsymbol{q}) - \boldsymbol{\nabla}\cdot\boldsymbol{j}^\sharp(t,\boldsymbol{s},\boldsymbol{q})$, is an identity, not an independent equation.

which replaces (5.18). Since $\boldsymbol{\mathcal{A}}_1(t,\boldsymbol{Q}(t)) = \boldsymbol{\mathcal{A}}_{\text{MBI}}(t,\boldsymbol{Q}(t))$, the guiding law (5.23) with velocity field $\boldsymbol{v}$ given by (5.30) is superficially identical to (5.22), yet note that S_{HJ} in (5.22) is not the same S_{HJ} as in (5.30), (5.23) because $\phi(t,\boldsymbol{q})$ and $\boldsymbol{\mathcal{A}}(t,\boldsymbol{q})$ are now replaced by $\phi_1(t,\boldsymbol{q})$ and $\boldsymbol{\mathcal{A}}_1(t,\boldsymbol{q})$. Note also that our single-particle law of motion has a straightforward extension to the N-body problem, which I also presented in [Kie2004a].

It is a reasonable conjecture that the maps $\boldsymbol{q} \mapsto \phi_1(t,\boldsymbol{q})$ and $\boldsymbol{q} \mapsto \boldsymbol{\mathcal{A}}_1(t,\boldsymbol{q})$ are *generically* differentiable,[19] in which case one obtains the first well-defined self-consistent law of motion of a classical point charge source in the Maxwell–Born–Infeld field equations [Kie2004a]. It has an immediate generalization to N-particle systems.

It is straightforward to show that this law readily handles the simplest situation: the trivial motion (i.e., rest) of the point charge source in Born's static solution. Note that no averaging or renormalization has to be invoked!

Since the nonlinearities make it extremely difficult to evaluate the model in nontrivial situations, only asymptotic results are available so far. It is shown in [Kie2004a] and [Kie2004b] that a point charge in Maxwell–Born–Infeld fields which are "co-sourced" jointly by this charge and another one that, in a single-particle setup, is assumed to be immovable (a Born–Oppenheimer approximation to a dynamical two-particle setup), when the charges are far apart, carries out the Kepler motion in leading order, as it should. Moreover, at least formally one can also show that in general the slow motion and gentle acceleration regime of a point charge is governed in leading order by a law of test charge motion as introduced at the beginning of this subsection. Whether this will pan out rigorously, and if so, whether the one-body setup yields physically correct motions if we go beyond the slow motion and gentle acceleration regime has yet to be established.

5.3.4. Conservation laws: re-assessing the value of b. In [Kie2004a] I explained that the system of Maxwell–Born–Infeld field equations with a negative point charge source moving according to our parallel-processing Hamilton–Jacobi laws furnishes the following conserved total energy:

$$\begin{aligned}\mathcal{E} = c\sqrt{m^2c^2 + \left|\nabla_{\boldsymbol{q}} S_{\text{HJ}}(t,\boldsymbol{Q}(t)) + \tfrac{1}{c}e\boldsymbol{\mathcal{A}}_{\text{MBI}}(t,\boldsymbol{Q}(t))\right|^2}\\ + \frac{b^2}{4\pi}\int_{\mathbb{R}^3}\Big(\sqrt{1+\tfrac{1}{b^2}\left(|\boldsymbol{\mathcal{B}}_{\text{MBI}}|^2+|\boldsymbol{\mathcal{D}}_{\text{MBI}}|^2\right)+\tfrac{1}{b^4}|\boldsymbol{\mathcal{B}}_{\text{MBI}}\times\boldsymbol{\mathcal{D}}_{\text{MBI}}|^2}-1\Big)(t,\boldsymbol{s})\,\mathrm{d}^3s\,.\end{aligned} \tag{5.32}$$

In [Kie2004a] I had assumed from the outset that $m = m_{\text{e}}$, but that was somewhat hidden because of the dimensionless units I chose. The assumption $m = m_{\text{e}}$ caught up with me when the *total* rest mass of the point defect plus the electrostatic field around it, with $b = b_{\text{Born}}$, became $2m_{\text{e}}$. With hindsight, I should have allowed the "intrinsic mass of the defect" m to be a parameter,

[19] Normally, a Cauchy problem is locally well-posed if there exists a unique solution, locally in time, which depends Lipschitz-continuously on the initial data. We here expect, and need, a little more regularity than what suffices for basic well-posedness.

as I have done here, because then this bitter pill becomes bittersweet: there is a whole range of combinations of m and b for which $\mathcal{E} = m_{\mathrm{e}}c^2$; yet it is also evident that with $m > 0$, Born's proposal $b = b_{\mathrm{Born}}$ is untenable. More precisely, b_{Born} is an *upper bound* on the admissible b values obtained from adapting Born's argument that the empirical rest mass of the physical electron should now be the total energy over c^2 of a single point defect in its static field.

What these considerations do not reveal is the relative distribution of mass between m and b. My colleague Shadi Tahvildar-Zadeh has suggested that m is possibly the only surviving remnant of a general relativistic treatment, and thereby determined. I come to general relativistic issues in the next subsection.

Before I get to there, I should complete the listing of the traditional conservation laws. Namely, with a negative point charge, the total momentum,

$$\mathcal{P} = \big[\nabla_{\boldsymbol{q}} S_{\mathrm{HJ}} + \tfrac{1}{c} e \boldsymbol{\mathcal{A}}_{\mathrm{MBI}}\big](t, \boldsymbol{Q}(t)) + \tfrac{1}{4\pi c} \int_{\mathbb{R}^3} (\boldsymbol{\mathcal{D}}_{\mathrm{MBI}} \times \boldsymbol{\mathcal{B}}_{\mathrm{MBI}})(t, \boldsymbol{s})\, \mathrm{d}^3 s\,, \qquad (5.33)$$

and the total angular momentum,

$$\mathcal{L} = \ \boldsymbol{Q}(t) \times \big[\nabla_{\boldsymbol{q}} S_{\mathrm{HJ}} + \tfrac{1}{c} e \boldsymbol{\mathcal{A}}_{\mathrm{MBI}}\big](t, \boldsymbol{Q}(t)) + \tfrac{1}{4\pi c} \int_{\mathbb{R}^3} \boldsymbol{s} \times (\boldsymbol{\mathcal{D}}_{\mathrm{MBI}} \times \boldsymbol{\mathcal{B}}_{\mathrm{MBI}})(t, \boldsymbol{s})\, \mathrm{d}^3 s\,, \qquad (5.34)$$

are conserved as well. In addition there are a number of less familiar conservation laws, but this would lead us too far from our main objective.

5.4. General-relativistic spacetimes with point defects

Ever since the formal papers by Einstein, Infeld, and Hoffmann [EIH1938], there have been quite many attempts to prove that Einstein's field equations imply the equations of motion for "point singularities." Certainly they imply the evolution equations of continuum matter when the latter is the source of spacetime geometry, but as to true point singularities the jury is still out. For us this means a clear imperative to investigate this question rigorously when Einstein's field equations are coupled with the Maxwell–Born–Infeld field equations of electromagnetism. Namely, if Einstein's field equations imply the equations of motion for the point charges, as Einstein et al. would have it, then all the developments described in the previous subsections have been in vain. If on the other hand it turns out that Einstein's field equations do not imply the equations of motion for the point charges, then we have the need for supplying such — in that case the natural thing to do, for us, is to adapt the Hamilton–Jacobi type law of motion from flat to curved spacetimes.

Fortunately, the question boils down to a static problem: Does the Einstein–Maxwell–Born–Infeld PDE system with two point charge sources have static, axisymmetric classically regular solutions away from the two worldlines of the point charges, no matter where they are placed? If the answer is "Yes," then Einstein's equations fail to deliver the equations of motion for the charges, for empirically we know that two *physical* point charges in the classical regime would not remain motionless. Shadi Tahvildar-Zadeh and myself have begun to rigorously study this question. I hope to report its answer in the not too distant future.

Meanwhile, I list a few facts that by now are known and which make us quite optimistic. Namely, while the Einstein–Maxwell–Maxwell equations with point charges produce solutions with horrible naked singularities (think of the Reissner–Nordström spacetime with charge and mass parameter chosen to match the empirical electron data), the Einstein–Maxwell–Born–Infeld equations with point charge source are much better behaved. Tahvildar-Zadeh [TaZa2011] recently showed that they not only admit a static spacetime corresponding to a single point charge whose ADM mass equals its electrostatic field energy/c^2, he also showed that the spacetime singularity is of the mildest possible form, namely a conical singularity. Conical singularities are so mild that they lend us hope that the nuisance of "struts" between "particles," known from multiple-black-hole solutions of Einstein's equations, can be avoided. Tahvildar-Zadeh's main theorem takes more than a page to state, after many pages of preparation. Here I will have to leave it at that.

6. Quantum theory of motion

Besides extending the classical flat spacetime theory to curved Lorentz manifolds, I have been working on its extension to the quantum regime. In [Kie2004b] I used a method which I called *least invasive quantization* of the one-charge Hamilton–Jacobi law for parallel processing of putative actual motions. Although I didn't see it this way at the time, by now I have realized that this least invasive quantization can be justified elegantly in the spirit of the quest for unification in physics!

6.1. Quest for unification: least invasive quantization

If we accept as a reasonably well-established working hypothesis that dynamical physical theories derive from an action principle, we should look for an action principle for the Hamilton–Jacobi equation. Because of the first order time derivative for S_{HJ} such an action principle for the classical S_{HJ} can be formulated only at the price of introducing a scalar companion field R_{HJ} which complements S_{HJ}. To illustrate this explicitly it suffices to consider a representative, nonrelativistic Hamilton–Jacobi PDE, written as $\frac{\partial}{\partial t}S_{\text{HJ}}(t,\boldsymbol{q}) + H(\boldsymbol{q}, \boldsymbol{\nabla}_{\boldsymbol{q}} S_{\text{HJ}}(t,\boldsymbol{q})) = 0$. Multiplying this equation by some positive function $R^2(t,\boldsymbol{q})$ and integrating over $\boldsymbol{q}$ and t (the latter over a finite interval I) gives the "action" integral

$$A(R, S_{\text{HJ}}) = \int_I \int_{\mathbb{R}^3} R^2(t,\boldsymbol{q}) \left[\frac{\partial}{\partial t} S_{\text{HJ}}(t,\boldsymbol{q}) + H(\boldsymbol{q}, \boldsymbol{\nabla}_{\boldsymbol{q}} S_{\text{HJ}}(t,\boldsymbol{q}))\right] \mathrm{d}^3\boldsymbol{q}\, \mathrm{d}t = 0\,.$$

Now replacing also S_{HJ} by a generic S in A and seeking the stationary points of $A(R,S)$, denoted by R_{HJ} and S_{HJ}, under variations with fixed end points, we obtain the Euler-Lagrange equations $\frac{\partial}{\partial t}S_{\text{HJ}}(t,\boldsymbol{q}) + H(\boldsymbol{q}, \boldsymbol{\nabla} S_{\text{HJ}}(t,\boldsymbol{q})) = 0$ and $\frac{\partial}{\partial t}R^2_{\text{HJ}}(t,\boldsymbol{q}) + \boldsymbol{\nabla}_{\boldsymbol{q}} \cdot [R^2_{\text{HJ}} \frac{1}{m} \boldsymbol{\nabla}_{\boldsymbol{q}} S_{\text{HJ}}](t,\boldsymbol{q}) = 0$. Clearly, the S_{HJ} equation is just the Hamilton–Jacobi equation we started from, while the R_{HJ} equation is a passive evolution equation: a continuity equation.

The passive evolution of R_{HJ} somehow belies the fact that R_{HJ} is needed to formulate the variational principle for S_{HJ} in the first place. This suggests that R_{HJ} is really a field of comparable physical significance to S_{HJ}. So in the spirit of unification, let's try to find a small modification of the dynamics to symmetrize the roles of R and S at the critical points.

Interestingly enough, by adding an R-dependent penalty term (a Fisher entropy, $\propto \hbar^2$) to the action functional $A(R,S)$, one can obtain (even in the N-body case) a Schrödinger equation for its critical points, denoted $R_{\text{QM}}e^{iS_{\text{QM}}/\hbar} = \psi$, where the suffix HJ has been replaced by QM to avoid confusion with "$R_{\text{HJ}}e^{iS_{\text{HJ}}/\hbar}$." The important point here is that the real and imaginary parts of $R_{\text{QM}}e^{iS_{\text{QM}}/\hbar}$ now satisfy a nicely symmetrical dynamics! In this sense the R_{QM} and S_{QM} fields have been really *unified* into a complex field ψ, whereas $R_{\text{HJ}}e^{iS_{\text{HJ}}/\hbar}$, while clearly complex, is not representing a unification of R_{HJ} and S_{HJ}. Equally important: the guiding equation, and the ontology of points that move, is unaffected by this procedure!

6.1.1. A de Broglie–Bohm–Klein–Gordon law of motion. The same type of argument works for the relativistic Hamilton–Jacobi theory and yields a Klein–Gordon equation. The Klein–Gordon PDE for the complex scalar configuration space field $\psi(t,\boldsymbol{q})$ reads

$$\left(i\hbar\tfrac{1}{c}\tfrac{\partial}{\partial t} + e\tfrac{1}{c}\phi_1\right)^2\psi = m^2c^2\psi + \left(-i\hbar\nabla_{\boldsymbol{q}} + e\tfrac{1}{c}\boldsymbol{A}_1\right)^2\psi \tag{6.1}$$

where ϕ_1 and $\boldsymbol{A}_1$ are the potential fields defined as in our parallel-processing single-charge Hamilton–Jacobi law.

To wit, least invasive quantization does not affect the underlying purpose of the theory to provide a law of motion for the point defects. For a Klein–Gordon PDE on configuration space the velocity field $\boldsymbol{v}$ for the guiding equation $\dot{\boldsymbol{Q}}(t) = \boldsymbol{v}(t,\boldsymbol{Q}(t))$ is now given by the ratio of quantum current vector density to density, $\boldsymbol{j}^{\text{qu}}(t,\boldsymbol{q})/\rho^{\text{qu}}(t,\boldsymbol{q})$, with

$$\rho^{\text{qu}} = \Im\left(\overline{\psi}\left(-\tfrac{\hbar}{mc^2}\tfrac{\partial}{\partial t} + i\tfrac{e}{mc^2}\phi_1\right)\psi\right), \qquad \boldsymbol{j}^{\text{qu}} = \Im\left(\overline{\psi}\left(\tfrac{\hbar}{m}\nabla_{\boldsymbol{q}} + i\tfrac{e}{mc}\boldsymbol{A}_1\right)\psi\right), \tag{6.2}$$

where $\Im$ means imaginary part, and $\overline{\psi}$ is the complex conjugate of ψ; thus

$$\boldsymbol{v}(t,\boldsymbol{q}) \equiv c\frac{\Im\left(\overline{\psi}\left(\hbar\nabla_{\boldsymbol{q}} + ie\tfrac{1}{c}\boldsymbol{A}_1\right)\psi\right)}{\Im\left(\overline{\psi}\left(-\hbar\tfrac{1}{c}\tfrac{\partial}{\partial t} + ie\tfrac{1}{c}\phi_1\right)\psi\right)}(t,\boldsymbol{q}). \tag{6.3}$$

This is a familiar de Broglie–Bohm–Klein–Gordon law of motion [DüTe2009, Hol1993], except that $\boldsymbol{A}_1, \phi_1$ are not external fields, of course.

6.1.2. A de Broglie–Bohm–Dirac law of motion. It is only a small step from a Klein–Gordon to a Dirac equation for spinor-valued ψ coupled to the generic q-sourced potential fields for a negative charge,

$$i\hbar\tfrac{1}{c}\tfrac{\partial}{\partial t}\psi = mc\beta\psi + \boldsymbol{\alpha}\cdot\left(-i\hbar\nabla_{\boldsymbol{q}} + e\tfrac{1}{c}\boldsymbol{A}_1\right)\psi - e\tfrac{1}{c}\phi_1\psi; \tag{6.4}$$

here $\boldsymbol{\alpha}$ and β are the familiar Dirac matrices. The guiding equation for the actual point charge motion is still (5.23), once again with $v = j^{\mathrm{qu}}/\rho^{\mathrm{qu}}$, but now with the quantum density and quantum current vector density given by the Dirac expressions, yielding the de Broglie–Bohm–Dirac guiding equation

$$\frac{1}{c}\frac{\mathrm{d}\boldsymbol{Q}(t)}{\mathrm{d}t} = \frac{\psi^\dagger \boldsymbol{\alpha} \psi}{\psi^\dagger \psi}(t, \boldsymbol{Q}(t)), \tag{6.5}$$

where $\mathbb{C}^4$ inner product is understood in the bilinear terms at the r.h.s. This is a familiar de Broglie–Bohm–Dirac law of motion [DüTe2009, Hol1993] except, once again, that $\boldsymbol{\mathcal{A}}_1, \phi_1$ are not external fields. Presumably ψ has to be restricted to an $\boldsymbol{\mathcal{A}}$-dependent "positive energy subspace," which is tricky, and we do not have space here to get into the details.

6.2. Born–Infeld effects on the Hydrogen spectrum

The two-charge model with an electron and a nuclear charge in Born–Oppenheimer approximation is formally a dynamical one-charge model with an additional charge co-sourcing the Maxwell–Born–Infeld fields. It can be used to investigate Born–Infeld effects on the Hydrogen spectrum.

The hard part is to find the electric potential $\phi^\sharp(\boldsymbol{s}, \boldsymbol{q}, \boldsymbol{q}_n)$ of the electrostatic Maxwell–Born–Infeld field of an electron at $\boldsymbol{q}$ and the nucleus at $\boldsymbol{q}_n = \boldsymbol{0}$ in otherwise empty space. The conceptual benefits offered by the nonlinearity of the Maxwell–Born–Infeld field equations come at a high price: in contrast to the ease with which the general solution to the Maxwell–Lorentz field equations can be written down, there is no general formula to explicitly represent the solutions to the Maxwell–Born–Infeld field equations. So far only stationary solutions with regular sources can be written down systematically with the help of convergent perturbative series expansions [CaKi2010, Kie2011c].

In [Kie2004b] I presented an explicit integral formula for an approximation to $\phi^\sharp(\boldsymbol{q}, \boldsymbol{q}, \boldsymbol{0}) = \phi_1(\boldsymbol{q})$. If the point charges are slightly smeared out and b^{-2} is not too big, then this formula gives indeed the electric potential for the leading order term in the perturbative series expansion in powers of b^{-2} for the displacement field $\boldsymbol{\mathcal{D}}$ developed in [CaKi2010, Kie2011c]. Assuming that the formula for the total electrostatic potential at the location of the electron is giving the leading contribution also for point charges, Born–Infeld effects on the Schrödinger spectrum of Hydrogen were computed[20] in [CaKi2006, FrGa2011]. In [FrGa2011] also the Dirac spectrum was studied. The interesting tentative conclusion from these studies is that Born's value of b gives spectral distortions which are too large to be acceptable. More refined two-body studies are still needed to confirm this finding, but the research clearly indicates that atomic spectral data may well be precise enough to test the viability of the Born–Infeld law for electromagnetism.

[20]The ground state energies as functions of Born's b parameter agree nicely in both numerical studies, but some of the excited states don't, hinting at a bug in our program. I thank Joel Franklin for pointing this out.

7. Closing remarks

In the previous sections I have slowly built up a well-defined theory of motion for point defects in the Maxwell–Born–Infeld fields, both in the classical regime, using Hamilton–Jacobi theory, and also in the quantum regime, using wave equations without and with spin. In either case the important notion is the parallel processing of motions, not test particle motions or their upgrade to a fixed point problem.

Unfortunately, while the nonlinearity of the Maxwell–Born–Infeld equations makes the introduction of such laws of motion possible in the first place, it is also an obstacle to any serious progress in computing the motions actually produced by these laws. But I am sure that it is only a matter of time until more powerful techniques are brought in which will clarify many of the burning open questions.

So far basically everything I discussed referred to the one-charge problem. This is perfectly adequate for the purpose of studying the self-interaction problem of a point charge which lies at the heart of the problem of its motion. But any acceptable solution to this self-interaction problem also has to be generalized to the N-charge situation, and this is another active field of inquiry. While the jury is still out on the correct format of the many charge theory, one aspect of it is presumably here to stay. Namely, a many-charge formulation in configuration space clearly requires synchronization of the various charges; by default one would choose to work with a particular Lorentz frame, but any other choice should be allowed as well. Actually, even the single-charge formulation I gave here tacitly uses the synchronization of the time components in the four-vectors $(ct, \boldsymbol{s})$ and $(q^0, \boldsymbol{q})$. In the test charge approximation synchronization is inconsequential, but in this active charge formulation the many-charge law would seem to depend on the synchronization. Whether the motion will depend on the foliation can naturally be investigated. Even if it does, the law of motion would not automatically be in conflict with Lorentz covariance. What is needed is simply a covariant foliation equation, as used in general relativity [ChKl1993]. A distinguished foliation could be interpreted as restoring three-dimensionality to physical reality. This would be against the traditional spirit of relativity theory, i.e. Einstein's interpretation of it as meaning that physical reality is four-dimensional, but that's OK.

Acknowledgement

I am very grateful to the organizers Felix Finster, Olaf Müller, Marc Nardmann, Jürgen Tolksdorf, and Eberhard Zeidler of the excellent "Quantum Field Theory and Gravity" conference in Regensburg (2010), for the invitation to present my research, and for their hospitality and support. The material reported here is based on my conference talk, but I have taken the opportunity to address in more detail some questions raised in response to my presentation, there and elsewhere. I also now mentioned the definitive status of some results that had been in the making at the time of the conference. The research has been developing over many years, funded by NSF grants

DMS-0406951 and DMS-0807705, which is gratefully acknowledged. I benefited greatly from many discussions with my colleagues, postdocs and students, in particular: Walter Appel, Holly Carley, Sagun Chanillo, Demetrios Christodoulou, Detlef Dürr, Shelly Goldstein, Markus Kunze, Tim Maudlin, Jared Speck, Herbert Spohn, Shadi Tahvildar-Zadeh, Rodi Tumulka, Nino Zanghì. Lastly, I thank the anonymous referee for many helpful suggestions.

References

[Abr1903] Abraham, M., *Prinzipien der Dynamik des Elektrons*, Phys. Z. **4**, 57–63 (1902); Ann. Phys. **10**, pp. 105–179 (1903).

[Abr1905] Abraham, M., *Theorie der Elektrizität, II*, Teubner, Leipzig (1905).

[AnTh2005] Anco, S. C., and The, D., *Symmetries, conservation laws, and cohomology of Maxwell's equations using potentials*, Acta Appl. Math. **89**, 1–52 (2005).

[ApKi2001] Appel, W., and Kiessling, M. K.-H., *Mass and spin renormalization in Lorentz electrodynamics*, Annals Phys. (N.Y.) **289**, 24–83 (2001).

[ApKi2002] Appel, W., and Kiessling, M. K.-H., *Scattering and radiation damping in gyroscopic Lorentz electrodynamics*, Lett. Math. Phys. **60**, 31–46 (2002).

[Bart1987] Bartnik, R., *Maximal surfaces and general relativity*, pp. 24–49 in "Miniconference on Geometry/Partial Differential Equations, 2" (Canberra, June 26-27, 1986), J. Hutchinson and L. Simon, Ed., Proceedings of the Center of Mathematical Analysis, Australian National Univ., **12** (1987).

[Bart1989] Bartnik, R., *Isolated points of Lorentzian mean-curvature hypersurfaces*, Indiana Univ. Math. J. **38**, 811–827 (1988).

[BaSi1982] Bartnik, R., and Simon, L., *Spacelike hypersurfaces with prescribed boundary values and mean curvature*, Commun. Math. Phys. **87**, 131–152 (1982).

[Baru1964] Barut, A. O., *Electrodynamics and classical theory of fields and particles*, Dover, New York (1964).

[Bau1998] Bauer, G., *Ein Existenzsatz für die Wheeler–Feynman-Elektrodynamik*, Doctoral Dissertation, Ludwig Maximilians Universität, München, (1998).

[BDD2010] Bauer, G., Deckert, D.-A., and Dürr, D., *Wheeler–Feynman equations for rigid charges — Classical absorber electrodynamics. Part II.* arXiv:1009.3103 (2010)

[BiBi1983] Białynicki-Birula, I., *Nonlinear electrodynamics: variations on a theme by Born and Infeld*, pp. 31–48 in "Quantum theory of particles and fields," special volume in honor of Jan Łopuszański; eds. B. Jancewicz and J. Lukierski, World Scientific, Singapore (1983).

[Boi1970] Boillat, G., *Nonlinear electrodynamics: Lagrangians and equations of motion*, J. Math. Phys. **11**, 941–951 (1970).

[Bor1933] Born, M., *Modified field equations with a finite radius of the electron*, Nature **132**, 282 (1933).

[Bor1934] Born, M., *On the quantum theory of the electromagnetic field*, Proc. Roy. Soc. **A143**, 410–437 (1934).

[Bor1937] Born, M., *Théorie non-linéaire du champ électromagnétique*, Ann. Inst. H. Poincaré **7**, 155–265 (1937).

[BoIn1933a] Born, M., and Infeld, L., *Electromagnetic mass*, Nature **132**, 970 (1933).

[BoIn1933b] Born, M., and Infeld, L., *Foundation of the new field theory*, Nature **132**, 1004 (1933).

[BoIn1934] Born, M., and Infeld, L., *Foundation of the new field theory*, Proc. Roy. Soc. London **A 144**, 425–451 (1934).

[BoSchr1935] Born, M., and Schrödinger, E., *The absolute field constant in the new field theory*, Nature **135**, 342 (1935).

[Bre2004] Brenier, Y., *Hydrodynamic structure of the augmented Born–Infeld equations*, Arch. Rat. Mech. Anal., **172**, 65–91 (2004).

[CDGS1995] Carati, A., Delzanno, P., Galgani, L., and Sassarini, J., *Nonuniqueness properties of the physical solutions of the Lorentz–Dirac equation*, Nonlinearity **8**, pp. 65–79 (1995).

[CaKi2006] Carley, H., and Kiessling, M. K.-H., *Nonperturbative calculation of Born–Infeld effects on the Schrödinger spectrum of the Hydrogen atom*, Phys. Rev. Lett. **96**, 030402 (1–4) (2006).

[CaKi2010] Carley, H., and Kiessling, M. K.-H., *Constructing graphs over $\mathbb{R}^n$ with small prescribed mean-curvature*, arXiv:1009.1435 (math.AP) (2010).

[ChHu2003] Chae, D., and Huh, H., *Global existence for small initial data in the Born-Infeld equations*, J. Math. Phys. **44**, 6132–6139 (2003).

[Chri2000] Christodoulou, D., *The action principle and partial differential equations*, Annals Math. Stud. **146**, Princeton Univ. Press, Princeton (2000).

[ChKl1993] Christodoulou, D., and Klainerman, S., *The global nonlinear stability of the Minkowski space*, Princeton Math. Ser. **41**, Princeton Univ. Press, Princeton (1993).

[Dec2010] Deckert, D.-A., *Electrodynamic absorber theory: a mathematical study*, Ph.D. dissertation LMU Munich, 2010.

[Dir1938] Dirac, P. A. M., *Classical theory of radiating electrons*, Proc. Roy. Soc. A **167**, 148–169 (1938).

[Dir1960] Dirac, P. A. M., *A reformulation of the Born–Infeld electrodynamics*, Proc. Roy. Soc. A **257**, 32–43 (1960).

[DüTe2009] Dürr, D., and Teufel, S. *Bohmian mechanics as the foundation of quantum mechanics*, Springer (2009).

[Eck1986] Ecker, K. *Area maximizing hypersurfaces in Minkowski space having an isolated singularity*, Manuscr. Math. **56**, 375–397 (1986).

[Ein1905a] Einstein, A., *Zur Elektrodynamik bewegter Körper*, Ann. Phys. **17**, 891–921 (1905).

[Ein1905b] Einstein, A., *Ist die Trägheit eines Körpers von seinem Energieinhalt abhängig?*, Ann. Phys. **18**, pp. 639–641 (1905).

[EIH1938] Einstein, A., Infeld, L., and Hoffmann, B., *The gravitational equations and the problem of motion*, Annals Math. **39**, 65–100 (1938).

[FrGa2011] Franklin, J., and Garon, T., *Approximate calculations of Born–Infeld effects on the relativistic hydrogen spectrum*, Phys. Lett. **A 375** 1391–1395 (2011).

[Gib1998] Gibbons, G. W., *Born-Infeld particles and Dirichlet p-branes*, Nucl. Phys. **B 514**, 603–639 (1998).

[GKZ1998] Gittel, H.-P., Kijowski, J., and Zeidler, E., *The relativistic dynamics of the combined particle-field system in renormalized classical electrodynamics*, Commun. Math. Phys. **198**, 711–736 (1998).

[HaEl1973] Hawking, S. W., and Ellis, G. F. R., *The large scale structure of space-time*, Cambridge Univ. Press, Cambridge (1973).

[Hol1993] Holland, P., *The quantum theory of motion*, Cambridge University Press (1993).

[Hop1994] Hoppe, J., *Some classical solutions of relativistic membrane equations in 4 space-time dimensions*, Phys. Lett. **B 329**, 10–14 (1994).

[Jac1975] Jackson, J. D., *Classical electrodynamics*, J. Wiley & Sons, New York 2nd ed. (1975).

[Jac1999] Jackson, J. D., *Classical electrodynamics*, J. Wiley & Sons, New York 3rd ed. (1999).

[JaOk2001] Jackson, J. D., and Okun, L. B., *Historical roots of gauge invariance*, Rev. Mod. Phys. **73**, 663–680 (2001).

[Jos2002] Jost, R., *Das Märchen vom elfenbeinernen Turm. (Reden und Aufsätze)*, Springer Verlag, Wien (2002).

[Kie2004a] Kiessling, M. K.-H., *Electromagnetic field theory without divergence problems. 1. The Born legacy*, J. Stat. Phys. **116**, 1057–1122 (2004).

[Kie2004b] Kiessling, M. K.-H., *Electromagnetic field theory without divergence problems. 2. A least invasively quantized theory*, J. Stat. Phys. **116**, 1123–1159 (2004).

[Kie2011a] Kiessling, M. K.-H., *On the quasi-linear elliptic PDE* $-\nabla\cdot(\nabla u/\sqrt{1-|\nabla u|^2}) = 4\pi\sum_k a_k\delta_{s_k}$ *in physics and geometry*, Rutgers Univ. preprint (2011).

[Kie2011b] Kiessling, M. K.-H., *Some uniqueness results for stationary solutions of the Maxwell–Born–Infeld field equations and their physical consequences*, submitted to Phys. Lett. A (2011).

[Kie2011c] Kiessling, M. K.-H., *Convergent perturbative power series solution of the stationary Maxwell–Born–Infeld field equations with regular sources*, J. Math. Phys. **52**, art. 022902, 16 pp. (2011).

[KlMi1993] Klyachin, A. A., and Miklyukov, V. M., *Existence of solutions with singularities for the maximal surface equation in Minkowski space*, Mat. Sb. **184**, 103-124 (1993); English transl. in Russ. Acad. Sci. Sb. Math. **80**, 87–104 (1995).

[Kly1995] Klyachin, A. A., *Solvability of the Dirichlet problem for the maximal surface equation with singularities in unbounded domains*, Dokl. Russ. Akad. Nauk **342**, 161–164 (1995); English transl. in Dokl. Math. **51**, 340–342 (1995).

[LaLi1962] Landau, L., and Lifshitz, E. M., *The theory of classical fields*, Pergamon Press, Oxford (1962).

[Lau1909] von Laue, M., *Die Wellenstrahlung einer bewegten Punktladung nach dem Relativitätsprinzip*, Ann. Phys. **28**, 436–442 (1909).

[Lié1898] Liénard, A., *Champ électrique et magnétique produit par une charge concentrée en un point et animée d'un mouvement quelconque*, L'Éclairage électrique **16** p. 5; ibid. p. 53; ibid. p. 106 (1898).

[Lor1892] Lorentz, H. A., *La théorie électromagnetique de Maxwell et son application aux corps mouvants*, Arch. Néerl. Sci. Exactes Nat. **25**, 363–552 (1892).

[Lor1895] Lorentz, H. A., *Versuch einer Theorie der elektrischen und optischen Erscheinungen in bewegten Körpern*, Teubner, Leipzig (1909) (orig. Leyden: Brill, 1895.)

[Lor1904] Lorentz, H. A., *Weiterbildung der Maxwell'schen Theorie: Elektronentheorie.*, Encyklopädie d. Mathematischen Wissenschaften **V**2, Art. 14, 145–288 (1904).

[Lor1915] Lorentz, H. A., *The theory of electrons and its applications to the phenomena of light and radiant heat*, 2nd ed., 1915; reprinted by Dover, New York (1952).

[Rao1936] Madhava Rao, B. S., *Ring singularity in Born's unitary theory - 1*, Proc. Indian Acad. Sci. **A4**, 355–376 (1936).

[Mie1912/13] Mie, G., *Grundlagen einer Theorie der Materie*, Ann. Phys. **37**, 511–534 (1912); ibid. **39**, 1–40 (1912); ibid. **40**, 1–66 (1913).

[Pip1997] Pippard, A. B., *J. J. Thomson and the discovery of the electron*, pp. 1–23 in "Electron—a centenary volume," M. Springford, ed., Cambridge Univ. Press (1997).

[Ple1970] Plebański, J., *Lecture notes on nonlinear electrodynamics*, NORDITA (1970) (quoted in [BiBi1983]).

[Poi1905/6] Poincaré, H., *Sur la dynamique de l'électron*, Comptes-Rendus **140**, 1504–1508 (1905); Rendiconti del Circolo Matematico di Palermo **21**, 129–176 (1906).

[Pry1935a] Pryce, M. H. L., *The two-dimensional electrostatic solutions of Born's new field equations*, Proc. Camb. Phil. Soc. **31**, 50–68 (1935).

[Pry1935b] Pryce, M. H. L., *On a uniqueness theorem*, Proc. Camb. Phil. Soc. **31**, 625–628 (1935).

[Pry1936] Pryce, M. H. L., *On the new field theory*, Proc. Roy. Soc. London **A 155**, 597–613 (1936).

[Roh1990] Rohrlich, F., *Classical charged particles*, Addison Wesley, Redwood City, CA (1990).

[Schr1942a] Schrödinger, E., *Non-linear optics*, Proc. Roy. Irish Acad. **A 47**, 77–117 (1942).

[Schr1942b] Schrödinger, E., *Dynamics and scattering-power of Born's electron*, Proc. Roy. Irish Acad. **A 48**, 91–122 (1942).

[Ser1988] Serre, D., *Les ondes planes en électromagnétisme non-linéaire*, Physica **D31**, 227–251 (1988).

[Ser2004] Serre, D., *Hyperbolicity of the nonlinear models of Maxwell's equations*, Arch. Rat. Mech. Anal. **172**, 309–331 (2004).

[Spe2010a] Speck, J., *The nonlinear stability of the trivial solution to the Maxwell–Born–Infeld system*, arXiv:1008.5018 (2010).

[Spe2010b] Speck, J., *The global stability of the Minkowski spacetime solution to the Einstein-nonlinear electromagnetic system in wave coordinates*, arXiv:1009.6038 (2010).

[Spo2004] Spohn, H., *Dynamics of charged particles and their radiation field*, Cambridge University Press, Cambridge (2004).

[TaZa2011] Tahvildar-Zadeh, A. S., *On the static spacetime of a single point charge*, Rev. Math. Phys. **23**, 309–346 (2011).

[Thi1997] Thirring, W. E., *Classical mathematical physics*, (Dynamical systems and field theory), 3rd ed., Springer Verlag, New York (1997).

[Tho1897] Thomson, J. J., *Cathode rays*, Phil. Mag. **44**, 294–316 (1897).

[WhFe1949] Wheeler, J. A., and Feynman, R., *Classical electrodynamics in terms of direct particle interactions*, Rev. Mod. Phys. **21**, 425–433 (1949).

[Wie1896] Wiechert, E., *Die Theorie der Elektrodynamik und die Röntgen'sche Entdeckung*, Schriften d. Physikalisch-Ökonomischen Gesellschaft zu Königsberg in Preussen **37**, 1-48 (1896).

[Wie1897] Wiechert, E., *Experimentelles über die Kathodenstrahlen*, Schriften d. Physikalisch-Ökonomischen Gesellschaft zu Königsberg in Preussen **38**, 3–16 (1897).

[Wie1900] Wiechert, E., *Elektrodynamische Elementargesetze*, Arch. Néerl. Sci. Exactes Nat. **5**, 549–573 (1900).

[Yag1992] Yaghjian, A. D., *Relativistic dynamics of a charged sphere*, Lect. Notes Phys. Monographs **11**, Springer, Berlin (1992).

Michael K.-H. Kiessling
Department of Mathematics
Rutgers University
110 Frelinghuysen Rd.
Piscataway, NJ 08854
USA
e-mail: miki@math.rutgers.edu

How Unique Are Higher-dimensional Black Holes?

Stefan Hollands

Abstract. In this article, we review the classification and uniqueness of stationary black hole solutions having large abelian isometry groups in higher-dimensional general relativity. We also point out some consequences of our analysis concerning the possible topologies that the black hole exteriors may have.

Mathematics Subject Classification (2010). 35Q75, 53C43, 58D19.

Keywords. General relativity, higher dimensions, black hole uniqueness theorems, differential geometry, manifolds with torus action, non-linear sigma models.

The idea that there might exist extra dimensions beyond the evident three space and one time dimension was already brought up in the early days of General Relativity. It has since become a major ingredient in many fundamental theories of nature, especially those with an eye on a "unification of forces". To explain why the extra dimensions—if indeed they exist—have so far gone unnoticed, it is typically supposed that they are either extremely small in size, or are large but somehow unable to communicate with us directly.

Either way, it seems to be very difficult to probe any of these ideas experimentally, and one does not feel that such unreliable concepts as "beauty" or "naturalness" alone can be trusted as the right means to find the correct theory. But one can at least try to understand in more detail the status and consequences of such theories. A concrete and sensible starting point is to consider theories of the same general type as general relativity, such as e.g. various Kaluza–Klein, or supergravity theories, on a higher-dimensional manifold, and to ask about the nature of their solutions. Of particular interest are solutions describing black holes, and among these, stationary black hole solutions are of special interest, because they might be the end point of the (classical) evolution of a dynamical black hole spacetime.

So, what are for example all stationary black hole spacetimes satisfying the vacuum Einstein equations on a D-dimensional manifold, with say, asymptotically flat boundary conditions at infinity? In $D = 4$ dimensions the answer to this question is provided by the famous black hole uniqueness theorems[1], which state that the only such solutions are provided by the Kerr family of metrics, and these are completely specified by their angular momentum and mass. Unfortunately, already in $D = 5$, the analogous statement is demonstrably false, as there exist different asymptotically flat black holes—having even event horizons of different topology—with the same values of the angular momenta and charges[2]. Nevertheless, one might hope that a classification is still possible if a number of further invariants of the solutions are also incorporated. This turns out to be possible [22, 23], at least if one restricts attention to solutions which are not only stationary, but moreover have a comparable amount of symmetry as the Kerr family, namely the symmetry group[3] $\mathbf{R} \times U(1)^{D-3}$. Such a spacetime cannot be asymptotically flat in $D > 5$, but it can be asymptotically Kaluza-Klein, i.e. asymptotically a direct product e.g. of the form $\mathbf{R}^{4,1} \times \mathbf{T}^{D-5}$. The purpose of this contribution is to outline the nature of this classification.

Because the symmetry group has $D - 2$ dimensions, the metric will, in a sense, depend non-trivially only on two remaining coordinates, and the Einstein equations will consequently reduce to a coupled system of PDE's in these variables. However, before one can study these equations, one must understand more precisely the nature of the two remaining coordinates, or, mathematically speaking, the nature of the orbit space $M/[\mathbf{R} \times U(1)^{D-3}]$. The quotient by $\mathbf{R}$ simply gets rid of a global time coordinate, so one is left with the quotient of a spatial slice Σ by $U(1)^{D-3}$. To get an idea about the possible topological properties of this quotient, we consider the following two simple, but characteristic, examples in the case $\dim \Sigma = 4$, i.e. $D = 5$.

The first example is $\Sigma = \mathbf{R}^4$, with one factor of $U(1) \times U(1)$ acting by rotations in the 12-plane and the other in the 34-plane. Introducing polar coordinates (R_1, ϕ_1) and (R_2, ϕ_2) in each of these planes, the group shifts ϕ_1 resp. ϕ_2, and the quotient is thus given simply by the first quadrant $\{(R_1, R_2) \in \mathbf{R}^2 \mid R_1 \geq 0, R_2 \geq 0\}$, which is a 2-manifold whose boundary consists of the two semi-axes and the corner where the two axes meet. The first axis corresponds to places in $\mathbf{R}^4$ where the Killing field $m_1 = \partial/\partial\phi_1$ vanishes, the second axis to places where $m_2 = \partial/\partial\phi_2$ vanishes. On the corner, both Killing fields vanish and the group action has a fixed point. The second example is the cartesian product of a plane with a 2-torus, $\Sigma = \mathbf{R}^2 \times \mathbf{T}^2$. Letting (x_1, x_2) be cartesian coordinates on the plane, and (ϕ_1, ϕ_2) angles on the torus, the group action is generated by the vector fields $m_1 = \partial/\partial\phi_1$ and by $m_2 = \alpha\partial/\partial\phi_2 + \beta(x_1\partial/\partial x_2 - x_2\partial/\partial x_1)$, where α, β are integers. These

[1] For a recent proof dealing properly with all the mathematical technicalities, see [5]. Original references include [3, 36, 2, 28, 16, 17, 24].

[2] For a review of exact black hole solutions in higher dimensions, see e.g. [9].

[3] In $D = 4$, this is not actually a restriction, because the rigidity theorem [16, 4, 11, 35] shows that any stationary black hole solution has the additional $U(1)$-symmetry. In higher dimensions, there is a similar theorem [18, 19, 31], but it guarantees only one additional $U(1)$-factor, and not $D - 3$.

vector fields do not vanish anywhere, but there are discrete group elements leaving certain points invariant. The quotient is now a cone with deficit angle $2\pi/\alpha$.

The general case turns out to be locally the same as in these examples [34, 22]. In fact, one can show that the quotient $\Sigma/[U(1) \times U(1)]$ is a 2-dimensional conifold with boundaries and corners. Each boundary segment is characterized by a different pair (p, q) of integers such that $pm_1 + qm_2 = 0$ at corresponding points of Σ, see fig. 1.

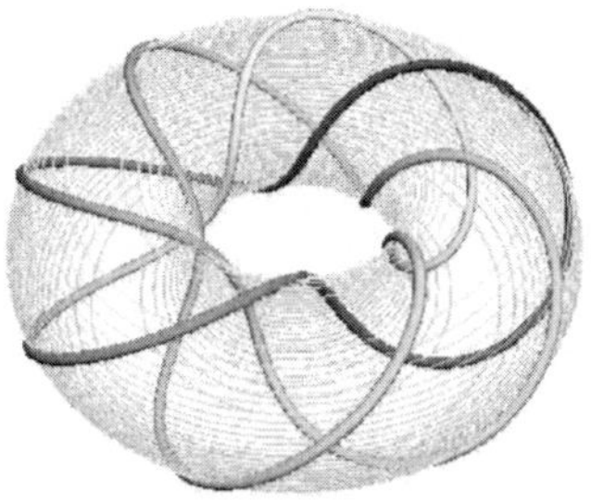

FIGURE 1. The numbers (p, q) may be viewed as winding numbers associated with the generators of the 2-torus generated by the two axial Killing fields. In such a torus, an $U(1)$-orbit winds around the first S^1 generator n times as it goes p times around the other S^1-direction. Here $qn \equiv 1 \bmod p$. The figure shows the situation for $p = 3$, $n = 7$.

FIGURE 2. A more artistic version of the previous figure, from "La Pratica della Perspectiva" (1569), by D. Barbaro.

For subsequent boundary segments adjacent on a corner labeled by (p_i, q_i) and (p_{i+1}, q_{i+1}), we have the condition

$$\begin{vmatrix} p_i & p_{i+1} \\ q_i & q_{i+1} \end{vmatrix} = \pm 1 . \tag{1}$$

Each conical singularity is characterized by a deficit angle, i.e. another integer. In higher dimensions, there is a similar result [23]; now a boundary segment is e.g. characterized by a $(D-3)$-tuple of integers ("winding numbers"), and the compatibility condition at the corners is somewhat more complicated.

In the case where Σ is the spatial section of a black hole spacetime, there are further constraints coming from Einstein's equations and the orientability of the spacetime. The topological censorship theorem [12, 7] is seen to imply [23] that the 2-dimensional orbit space $\hat{\Sigma} = \Sigma/U(1)^{D-3}$ can neither have any conifold points, nor holes, nor handles, and therefore has to be diffeomorphic to an upper half-plane $\hat{\Sigma} \cong \{(r, z) \mid r > 0\}$. The boundary segments correspond to intervals on the boundary ($r = 0$) of this upper half-plane and are places ("axes") in the manifold M where a linear combination of the rotational Killing fields vanishes, or to a horizon. Here the Killing fields do not vanish except where an "axis" meets a horizon. Furthermore, with each boundary segment, one can associate its length $l_i \geq 0$ in an invariant way. It can be shown (see e.g. [6]) that the metric of a black hole spacetime with the indicated symmetries globally takes the Weyl–Papapetrou form

$$g = -\frac{r^2\,\mathrm{d}t^2}{\det f} + \mathrm{e}^{-\nu}(\mathrm{d}r^2 + \mathrm{d}z^2) + f_{ij}(\mathrm{d}\phi^i + w^i\,\mathrm{d}t)(\mathrm{d}\phi^j + w^j\,\mathrm{d}t)\,, \qquad (2)$$

where the metric coefficients only depend on r, z, and where ϕ^i are 2π-periodic coordinates. The vacuum Einstein equations lead to two sets of differential equations. One set can be written [27] as a sigma-model field equation for a matrix field Φ defined by

$$\Phi = \begin{pmatrix} (\det f)^{-1} & -(\det f)^{-1}\chi_i \\ -(\det f)^{-1}\chi_i & f_{ij} + (\det f)^{-1}\chi_i\chi_j \end{pmatrix}\,, \qquad (3)$$

where χ_i are certain potentials that are defined in terms of the metric coefficients. The other set can be viewed as determining the conformal factor ν from Φ. The matrix field Φ obeys certain boundary conditions at $r = 0$ related to the winding numbers and moduli $l_i \in \mathbf{R}_+$ labeling the i-th boundary interval. Hence, it is evident[4] that any uniqueness theorem for the black holes under consideration would have to involve these data in addition to the usual invariants, such as mass and angular momentum.

In fact, what one can prove is the following theorem in $D = 5$ (modulo technical assumptions about analyticity and certain global causal constraints) [22]:

Theorem 1. *Given are two asymptotically flat, vacuum black hole solutions, each with a single non-extremal horizon. If the angular momenta J_1, J_2 coincide, and if all the integers ("winding numbers") $\langle (p_1, q_1), \ldots, (p_N, q_N)\rangle$ and real numbers $\langle l_1, \ldots, l_N\rangle$ ("moduli") coincide, then the solutions are isometric.*

The theorem has a generalization to higher dimensions D (with our symmetry assumption) [23]; here the asymptotically flat condition must be

[4]This observation seems to have been made first in [15]. In this paper, the importance of the quantities l_i, (p_i, q_i), and similarly in higher dimensions $D > 5$, was emphasized, but their properties and relation to the topology and global properties of M were not yet fully understood.

replaced by a suitable asymptotic Kaluza–Klein condition, and the winding numbers become vectors in $\mathbf{Z}^{D-3}$, etc. The idea behind the proof is to construct from the matrix fields Φ for the two solutions a scalar function on the upper half-plane, which measures the pointwise distance between these matrix fields. This function, u, is essentially the geodesic distance in non-compact coset-space in which one can think of Φ as taking its values. It is non-negative, satisfies a suitable Laplace-type equation with positive source, as well as appropriate boundary conditions governed by the moduli and winding numbers. This information suffices to show that $u = 0$ on the upper half-plane, and this in turn implies that the metrics are isometric.

The simplest case when the winding numbers and moduli are trivial (in essence the same as in $D = 4$) was treated in $D = 5$ first by [32]. The uniqueness theorem has subsequently been generalized to extremal black holes in $D = 4, 5$ dimensions [1, 10], using the idea to employ the "near horizon geometries" [25, 26, 20] to resolve the situation at the horizon, which shrinks to a single point in the above upper half-plane picture.

The topological structure of the orbit space is easily seen to imply that the horizon H has to be, in $D = 5$, a 3-sphere S^3, a ring $S^2 \times S^1$, or a lens space $L(p, q)$, but examples of regular black hole spacetimes are known only for the first two cases. In higher dimensions, there is a similar classification. In general, it is not clear whether, for a given set (p_i, q_i) satisfying the constraint (5) at each corner, and given angular momenta J_i and moduli l_i, there actually exists a corresponding black hole solution. Thus, in this sense the above uniqueness theorem is much weaker than that known in 4 dimensions, where there is a known black hole solution for each choice of J and l (the winding numbers are trivial in $D = 4$). One can trade the parameter l for the mass of the black hole if desired, or its horizon area, and a similar remark applies to one of the parameters l_i in higher dimensions. In order to attack the existence questions for black holes in higher dimensions, one can either try to find the solution corresponding to a given set of angular momenta, winding numbers and moduli explicitly. Here it is of great help that the equations for ν and Φ are of a very special nature and amenable to powerful methods from the theory of integrable systems. Another approach would be to try to find an abstract existence proof based on the boundary value problem for ν and Φ. Either way, the solution of the problem does not seem to be simple, and it is certainly not clear—and maybe even unlikely—that a solution exists for the most general set of data J_i, l_i and winding numbers.

The collection of winding numbers can nicely be put in correspondence with the topology of Σ, and hence of $M \cong \Sigma \times \mathbf{R}$. By exploiting the constraints (5) for adjacent winding numbers, one can prove that Σ can be decomposed into simpler pieces. For example, in the case when $D = 5$ and the spacetime is spin[5], one can show [34] that for spherical horizons

$$\Sigma \cong (\mathbf{R}^4 \setminus B^4) \,\#\, (N-2)(S^2 \times S^2) \,. \tag{4}$$

[5]This is of course physically well-motivated. In the general case, the decomposition would have additional factors of $\pm\mathbf{CP}^2$.

The black hole horizon is the boundary $S^3 = \partial B^4$ of a 4-dimensional ball that has been removed. For a ringlike horizon, one has to remove $S^2 \times B^2$ instead; the lens space case is somewhat more complicated. For all known solutions, the "handles" $S^2 \times S^2$ are in fact absent. Furthermore, if we introduce for each triple of adjacent boundary segments of $\hat{\Sigma}$ labeled by $(p_i, q_i), (p_{i+1}, q_{i+1})$ and (p_{i+2}, q_{i+2}) the determinant

$$e_i = \begin{vmatrix} p_i & p_{i+2} \\ q_i & q_{i+2} \end{vmatrix} \in 2\mathbf{Z} , \tag{5}$$

then the signature resp. Euler characteristic of the 4-manifold Σ (with black hole glued back in) are given by $3\tau(\Sigma) = e_1 + \cdots + e_N$ and $\chi(\Sigma) = N$, where $N + 2$ is the number of adjacent boundary segments ("intervals"), see [29]. If it was e.g. known that Σ could carry *some* Einstein metric, then the Hitchin–Thorpe inequality $2\chi(\Sigma) \geq 3|\tau(\Sigma)|$ would give the further constraint $2N \geq |e_1 + \cdots + e_N|$.

In dimension $D \geq 6$, a similar analysis also appears to be possible, but it is more complicated. For example, when $D = 7$ and we consider spherical black holes, we have, using the results of [33]:

$$\Sigma \cong (\mathbf{R}^6 \setminus B^6) \,\#\, (N-4)(S^2 \times S^4) \,\#\, (N-3)(S^3 \times S^3), \tag{6}$$

where it has been assumed that the second Stiefel–Whitney class is $w_2(\Sigma) = 0$.

In summary, in higher dimensions, there are indications [8] that the variety of possible vacuum stationary black hole solutions could be much greater than in $D = 4$. To date it remains largely unexplored by rigorous methods. However, if one makes more stringent symmetry assumptions than seem to be justified on just purely dynamical grounds—i.e., than what is guaranteed by the rigidity theorem [18, 19, 31]—, then a partial classification of stationary black holes is possible, as we have described. Also, if one assumes that the black hole is (asymptotically flat and) even static, then a different kind of argument [14] shows that the solution must be a higher-dimensional analogue of Schwarzschild. These results are not as powerful in the stationary (non-static) case as the corresponding statements in $D = 4$, but they go at least some way in the direction of exploring the landscape of black hole solutions in higher dimensions.

If one is not willing to make any by-hand symmetry assumptions beyond what is guaranteed by the rigidity theorem [18, 19, 31] in higher dimensions [which implies the existence of *one* Killing field ψ generating $U(1)$], then at present not much can be said about uniqueness of higher-dimensional stationary (but non-static), asymptotically flat vacuum black holes. However, certain qualitative statements of general nature can still be made. For example, by a theorem of [13], the horizon manifold must be of "positive Yamabe type", meaning essentially that it can carry a metric of pointwise positive scalar curvature. In $D = 5$, the horizon manifold H is a 3-dimensional compact Riemannian manifold, and the positive Yamabe type condition then imposes some fairly strong conditions on its topology. Actually, in that case, if the positive Yamabe type is combined with the results of the rigidity theorem in $D = 5$, then even stronger conclusions can be drawn, namely [21]:

Theorem 2. *In $D = 5$ vacuum general relativity, the topology of the event horizon H of a compact black hole can be one of the following:*

1. *If ψ has a zero on H, then the topology of H must be*[6]

$$H \cong \#\,(l-1)\cdot(S^2 \times S^1)\,\#\,L(p_1, q_1)\,\#\cdots\#\,L(p_k, q_k)\,. \tag{7}$$

   Here, k is the number of exceptional orbits of the action of $U(1)$ on H that is generated by ψ, and l is the number of connected components of the zero set of ψ.

2. *If ψ does not have a zero on H, then $H \cong S^3/\Gamma$, where Γ can be certain finite subgroups of $SO(4)$, or $H \cong S^2 \times S^1$. This class of manifolds includes again the lens spaces, but also prism manifolds, the Poincaré homology sphere, and various other quotients. All manifolds in this class are certain Seifert fibred spaces over S^2. The precise classification of the possibilities is given in ref. [21].*

References

[1] Amsel, A. J., Horowitz, G. T., Marolf, D. and Roberts, M. M.: arXiv:0906.2367 [gr-qc]

[2] Bunting, G. L.: PhD Thesis, Univ. of New England, Armidale, N.S.W., 1983

[3] Carter, B.: Phys. Rev. Lett. **26**, 331-333 (1971)

[4] Chruściel, P. T.: Commun. Math. Phys. **189**, 1-7 (1997)

[5] Chruściel, P. T. and Lopes Costa, J.: arXiv:0806.0016 [gr-qc]

[6] Chruściel, P. T.: J. Math. Phys. **50**, 052501 (2009)

[7] Chruściel, P. T., Galloway, G. J. and Solis, D.: Annales Henri Poincaré **10**, 893 (2009)

[8] Emparan, R., Harmark, T., Niarchos V. and Obers, N. A.: arXiv:1106.4428 [hep-th] and references therein

[9] Emparan, R. and Reall, H. S.: Living Rev. Rel. **11**, 6 (2008)

[10] Figueras, P. and Lucietti, J.: arXiv:0906.5565 [hep-th]

[11] Friedrich, H., Rácz, I. and Wald, R. M.: Commun. Math. Phys. **204**, 691-707 (1999)

[12] Galloway, G. J., Schleich, K., Witt, D. M. and Woolgar, E.: Phys. Rev. D **60**, 104039 (1999)

[13] Galloway, G. J. and Schoen, R: Commun. Math. Phys. **266**, 571576 (2006)

[14] Gibbons, G. W., Ida, D. and Shiromizu, T.: Phys. Rev. Lett. **89**, 041101 (2002), arXiv:hep-th/0206049

[15] Harmark, T.: Phys. Rev. D **70**, 124002 (2004)

[16] Hawking, S. W. and Ellis, G. F. R.: Cambridge, Cambridge University Press, 1973

[17] Hawking, S. W.: Commun. Math. Phys. **25**, 152-166 (1972)

[6] Here we allow that the lens space be $L(0,1) := S^3$.

[18] Hollands, S., Ishibashi, A. and Wald, R. M.: Commun. Math. Phys. **271**, 699 (2007)

[19] Hollands, S. and Ishibashi, A.: Commun. Math. Phys. **291**, 403 (2009)

[20] Hollands, S. and Ishibashi, A.: arXiv:0909.3462 [gr-qc]

[21] Hollands, S., Holland, J. and Ishibashi, A.: arXiv:1002.0490 [gr-qc]

[22] Hollands, S. and Yazadjiev, S.: Commun. Math. Phys. **283**, 749 (2008)

[23] Hollands, S. and Yazadjiev, S.: arXiv:0812.3036 [gr-qc].

[24] Israel, W.: Phys. Rev. **164**, 1776-1779 (1967)

[25] Kunduri, H. K., Lucietti, J. and Reall, H. S.: Class. Quant. Grav. **24**, 4169 (2007)

[26] Kunduri, H. K. and Lucietti, J.: J. Math. Phys. **50**, 082502 (2009)

[27] Maison, D.: Gen. Rel. Grav. **10**, 717 (1979)

[28] Mazur, P. O.: J. Phys. A **15**, 3173-3180 (1982)

[29] Melvin, P.: Math. Proc. Camb. Phil. Soc. **91**, 305-314 (1982)

[30] Moncrief, V. and Isenberg, J.: Commun. Math. Phys. **89**, 387-413 (1983)

[31] Moncrief, V. and Isenberg, J.: Class. Quant. Grav. **25**, 195015 (2008)

[32] Morisawa, Y. and Ida, D.: Phys. Rev. D **69**, 124005 (2004)

[33] Oh, H. S.: Topology Appl. **13**, 137-154 (1982)

[34] Orlik, P. and Raymond, F.: Topology **13**, 89-112 (1974)

[35] Rácz, I.: Class. Quant. Grav. **17**, 153 (2000)

[36] Robinson, D. C.: Phys. Rev. Lett. **34**, 905-906 (1975)

Stefan Hollands
School of Mathematics
Cardiff University
Cardiff, CF24 4AG
UK
e-mail: HollandsS@Cardiff.ac.uk

Equivalence Principle, Quantum Mechanics, and Atom-Interferometric Tests

Domenico Giulini

Abstract. That gravitation can be understood as a purely metric phenomenon depends crucially on the validity of a number of hypotheses which are summarised by the Einstein Equivalence Principle, the least well tested part of which being the Universality of Gravitational Redshift. A recent and currently widely debated proposal (Nature 463 (2010) 926-929) to re-interpret some 10-year old experiments in atom interferometry would imply, if tenable, substantial reductions on upper bounds for possible violations of the Universality of Gravitational Redshift by four orders of magnitude. This interpretation, however, is problematic and raises various compatibility issues concerning basic principles of General Relativity and Quantum Mechanics. I review some relevant aspects of the equivalence principle and its import into quantum mechanics, and then turn to the problems raised by the mentioned proposal. I conclude that this proposal is too problematic to warrant the claims that were launched with it.

Mathematics Subject Classification (2010). 83C99, 81Q05.

Keywords. General relativity, atom interferometry, equivalence principle.

1. Introduction

That gravitation can be understood as a purely metric phenomenon depends crucially on the validity of a number of hypotheses which are summarised by the Einstein Equivalence Principle, henceforth abbreviated by EEP. These assumptions concern contingent properties of the physical world that may

I thank Felix Finster, Olaf Müller, Marc Nardmann, Jürgen Tolksdorf and Eberhard Zeidler for inviting me to the conference on *Quantum field theory and gravity* (September 27 – October 2, 2010). This paper is the written-up version of my talk: *Down-to-Earth issues in atom interferometry*, given on Friday, October 1st. I thank Claus Lämmerzahl and Ernst Rasel for discussions concerning the issues and views presented here. Last not least I sincerely thank the QUEST cluster for making this work possible.

well either fail to hold in the quantum domain, or simply become meaningless. If we believe that likewise Quantum Gravity *is* Quantum Geometry, we should be able to argue for it by some sort of extension or adaptation of EEP into the quantum domain. As a first attempt in this direction one might ask for the status of EEP if the matter used to probe it is described by ordinary non-relativistic Quantum Mechanics. Can the quantum nature of matter be employed to push the bounds on possible violations of EEP to hitherto unseen lower limits?

In this contribution I shall discuss some aspects related to this question and, in particular, to a recent claim [15], according to which atom interferometric gravimeters have actually already tested the weakest part of EEP, the universality of gravitational redshift, and thereby improved the validity of EEP by about four orders of magnitude! I will come to the conclusion that this claim is unwarranted.[1] But before I do this in some detail, I give a general discussion of the Einstein Equivalence principle, its separation into various sub principles and the logical connection between them, and the import of one of these sub principles, the Universality of Free Fall (UFF), into Quantum Mechanics.

2. Some background

The theory of General Relativity rests on a number of hypotheses, the most fundamental of which ensure, first of all, that gravity can be described by a metric theory [27, 30]. Today these hypotheses are canonised in the Einstein Equivalence Principle (EEP).[2] EEP consists of three parts:

UFF: The Universality of Free Fall. UFF states that free fall of "test particles" (further remarks on that notion will follow) only depend on their initial position and direction in spacetime. Hence test particles define a path structure on spacetime in the sense of [6, 3] which, at this stage, need not necessarily be that of a linear connection. In a Newtonian setting UFF states that the quotient of the inertial and gravitational mass is a universal constant, i.e. independent of the matter the test particle is made of. UFF is also often called the *Weak Equivalence Principle*, abbreviated by WEP, but we shall stick to the label UFF which is more telling.

Possible violations of UFF are parametrised by the Eötvös factor, η, which measures the difference in acceleration of two test masses made of materials A and B:

$$\eta(A,B) = 2\cdot\frac{|a(A)-a(B)|}{a(A)+a(B)} \approx \sum_\alpha \eta_\alpha\left(\frac{E_\alpha(A)}{m_i(A)c^2} - \frac{E_\alpha(B)}{m_i(B)c^2}\right). \tag{2.1}$$

[1]This is the view that I expressed in my original talk for the same reasons as those laid out here. At that time the brief critical note [32] and the reply [16] by the original proponents had appeared in the Nature issue of September 2nd. In the meantime more critique has been voiced [33, 23], though the original claim seems to be maintained by and large [11].

[2]Note that EEP stands for "the Einstein Equivalence Principle" and not "Einstein's Equivalence Principle" because Einstein never expressed it in the modern canonised form.

The second and approximate equality arises if one supposes that violations occur in a specific fashion, for each fundamental interaction (labelled by α) separately. More specifically, one expresses the gravitational mass of the test particle made of material A in terms of its inertial mass and a sum of corrections, one for each interaction α, each being proportional to the fraction that the α's interaction makes to the total rest energy (cf. Sect. 2.4 of [30]):

$$m_g(A) = m_i(A) + \sum_\alpha \eta_\alpha \frac{E_\alpha(A)}{m_i(A)c^2} \,. \tag{2.2}$$

Here the η_α are universal constants depending only on the interaction but not on the test particle. Typical numbers from modern laboratory tests, using rotating torsion balances, are below the 10^{-12} level. Already in 1971 Braginsky and Panov claimed to have reached an accuracy $\eta(\mathrm{Al}, \mathrm{Pt}) < 9 \times 10^{-13}$ for the element pair Aluminium and Platinum [1]. Currently the lowest bound is reached for the elements Beryllium and Titanium [25].[3]

$$\eta(\mathrm{Be}, \mathrm{Ti}) < 2.1 \times 10^{-13} \,. \tag{2.3}$$

Resolutions in terms of η_α's of various tests are discussed in [30]. Future tests like MICROSCOPE ("MICRO-Satellite à traînée Compensée pour l'Observation du Principe d'Equivalence", to be launched in 2014) aim at a lower bound of 10^{-15}. It is expected that freely falling Bose-Einstein condensates will also allow precision tests of UFF, this time with genuine quantum matter [28].

LLI: Local Lorentz Invariance. LLI states that local non-gravitational experiments exhibit no preferred directions in spacetime, neither timelike nor spacelike. Possible violations of LLI concern, e.g., orientation-dependent variations in the speed of light, measured by $\Delta c/c$, or the spatial orientation-dependence of atomic energy levels. In experiments of the Michelson-Morley type, where c is the mean for the round-trip speed, the currently lowest bound from laboratory experiments based on experiments with rotating optical resonators is [10]:

$$\frac{\Delta c}{c} < 3.2 \times 10^{-16} \,. \tag{2.4}$$

Possible spatial orientation-dependencies of atomic energy levels have also been constrained by impressively low upper bounds in so-called Hughes-Drever type experiments.

LLP: Local Position Invariance. LPI is usually expressed by saying that "The outcome of any local non-gravitational experiment is independent of where and when in the universe it is performed" ([31], Sect. 2.1). However, in almost all discussions this is directly translated into the more concrete *Universality of Clock Rates (UCR)* or the *Universality of Gravitational redshift (UGR)*, which state that the rates of standard clocks agree if taken along the same world line (relative comparison) and that they show the standard redshift if

[3] Besides for technical experimental reasons, these two elements were chosen to maximise the difference in baryon number per unit mass.

taken along different worldlines and intercompared by exchange of electromagnetic signals. Suppose a field of light rays intersect the timelike worldlines $\gamma_{1,2}$ of two clocks, the four-velocities of which are $u_{1,2}$. Then the ratio of the instantaneous frequencies measured at the intersection points of one integral curve of k with γ_1 and γ_2 is

$$\frac{\nu_2}{\nu_1} = \frac{g(u_2,k)|_{\gamma_2}}{g(u_1,k)|_{\gamma_1}}\,. \tag{2.5}$$

Note that this does not distinguish between gravitational and Doppler shifts, which would be meaningless unless a local notion of "being at rest" were introduced. The latter requires a distinguished timelike vector field, as e.g. in stationary spacetimes with Killing field K. Then the purely gravitational part of (2.5) is given in case both clocks are at rest, i.e. $u_{1,2} = K/\|K\||_{\gamma_{1,2}}$, where $\gamma_{1,2}$ are now two different integral lines of K and $\|K\| := \sqrt{g(K,K)}$:

$$\frac{\nu_2}{\nu_1} := \frac{g\big(k, K/\|K\|\big)|_{\gamma_2}}{g\big(k, K/\|K\|\big)|_{\gamma_1}} = \sqrt{\frac{g(K,K)|_{\gamma_1}}{g(K,K)|_{\gamma_2}}}\,. \tag{2.6}$$

The last equality holds since $g(k,K)$ is constant along the integral curves of k, so that $g(k,K)|_{\gamma_1} = g(k,K)|_{\gamma_2}$ in (2.6), as they lie on the same integral curve of k. Writing $g(K,K) =: 1 + 2U/c^2$ and assuming $U/c^2 \ll 1$, we get

$$\frac{\Delta\nu}{\nu} := \frac{\nu_2 - \nu_1}{\nu_1} = -\frac{U_2 - U_1}{c^2}\,. \tag{2.7}$$

Possible deviations from this result are usually parametrised by multiplying the right-hand side of (2.7) with $(1+\alpha)$, where $\alpha = 0$ in GR. In case of violations of UCR/UGR, α may depend on the space-time point and/or on the type of clock one is using. The lowest upper bound on α to date for comparing (by electromagnetic signal exchange) clocks on *different* worldlines derives from an experiment made in 1976 (so-called "Gravity Probe A") by comparing a hydrogen-maser clock in a rocket, that during a total experimental time of 1 hour and 55 minutes was boosted to an altitude of about 10 000 km, to a similar clock on the ground. It led to [29]

$$\alpha_{\mathrm{RS}} < 7 \times 10^{-5}\,. \tag{2.8}$$

The best relative test, comparing different clocks (a ^{199}Hg-based optical clock and one based on the standard hyperfine splitting of ^{133}Cs) along the (almost) *same* worldline for six years gives [8]

$$\alpha_{\mathrm{CR}} < 5.8 \times 10^{-6}\,. \tag{2.9}$$

Here and above "RS" and "CR" refer to "redshift" and "clock rates", respectively, a distinction that we prefer to keep from now on in this paper, although it is not usually made. To say it once more: α_{RS} parametrises possible violations of UGR by comparing identically constructed clocks moving along different worldlines, whereas α_{CR} parametrises possible violations of UCR by comparing clocks of different construction and/or composition moving more or less on the same worldline. An improvement in putting upper bounds on

α_{CR}, aiming for at least 2×10^{-6}, is expected from ESA's ACES mission (ACES = Atomic Clock Ensemble in Space), in which a Caesium clock and a H-maser clock will be flown to the Columbus laboratory at the International Space Station (ISS), where they will be compared for about two years [2].

Remark 2.1. The notion of "test particles" essentially used in the formulation of UFF is not without conceptual dangers. Its intended meaning is that of an object free of the "obvious" violations of UFF, like higher multipole moments in its mass distribution and intrinsic spin (both of which would couple to the spacetime curvature) and electric charge (in order to avoid problems with radiation reaction). Moreover, the test mass should not significantly back-react onto the curvature of spacetime and should not have a significant mass defect due to its own gravitational binding. It is clear that the simultaneous fulfilment of these requirements will generally be context-dependent. For example, the Earth will count with reasonable accuracy as a test particle as far as its motion in the Sun's gravitational field is concerned, but certainly not for the Earth-Moon system. Likewise, the notion of "clock" used in UCR/UGR intends to designate a system free of the "obvious" violations. In GR a "standard clock" is any system that allows to measure the length of timelike curves. If the curve is accelerated it is clear that some systems cease to be good clocks (pendulum clocks) whereas others are far more robust. An impressive example for the latter is muon decay, where the decay time is affected by a fraction less than 10^{-25} at an acceleration of $10^{18}g$ [7]. On the other hand, if coupled to an accelerometer, eventual disturbances could in principle always be corrected for. At least as far as classical physics is concerned, there seems to be no serious lack of real systems that classify as test particles and clocks in contexts of interest. But that is a contingent property of nature that is far from self-evident.

Remark 2.2. The lower bounds for UFF, LLI, and UCR/UGR quoted above impressively show how much better UFF and LLI are tested in comparison to UCR/UGR. This makes the latter the weakest member in the chain that constitutes EEP. It would therefore be desirable to significantly lower the upper bounds for violations of the latter. Precisely this has recently (February 2010) been claimed in [15] by remarkable four orders in magnitude - and without doing a single new experiment! This will be analysed in detail below.

It can be carefully argued for (though not on the level of a mathematical theorem) that only metric theories can comply with EEP. In particular, the additional requirements in EEP imply that the path structure implied by UFF alone must be that of a linear connection. Metric theories, on the other hand, are defined by the following properties (we state them with slightly different wordings as compared to [31], Sect. 2.1):

M1. Spacetime is a four-dimensional differentiable manifold, which carries a metric (symmetric non-degenerate bilinear form) of Lorentzian signature, i.e. $(-,+,+,+)$ or $(+,-,-,-)$, depending on convention[4].

[4]Our signature convention will be the "mostly minus" one, i.e. $(+,-,-,-)$.

M2. The trajectories of freely falling test bodies are geodesics of that metric.
M3. With reference to freely falling frames, the non-gravitational laws of physics are those known from Special Relativity.

This canonisation of EEP is deceptive insofar as it suggests an essential logical independence of the individual hypotheses. But that is far from true. In fact, in 1960 the surprising suggestion has been made by Leonard Schiff that UFF should imply EEP, and that hence UFF and EEP should, in fact, be equivalent; or, expressed differently, UFF should already imply LLI and LPI. This he suggested in a "note added in proof" at the end of his classic paper [24], in the body of which he asked the important question whether the three classical "crucial tests" of GR were actually sensitive to the precise form of the field equations (Einstein's equations) or whether they merely tested the more general equivalence principle. He showed that the gravitational red-shift and the deflection of light could be deduced from EEP and that only the correct evaluation of the precession of planetary orbits needed an input from Einstein's equations. If true, it follows that any discrepancy between theory and experiment would have to be reconciled with the experimentally well-established validity of the equivalence principle and special relativity. Hence Schiff concludes:

> "By the same token, it will be extremely difficult to design a terrestrial or satellite experiment that really tests general relativity, and does not merely supply corroborative evidence for the equivalence principle [meaning UFF; D.G.] and special relativity. To accomplish this it will be necessary either to use particles of finite rest mass so that the geodesic equation my be confirmed beyond the Newtonian approximation, or to verify the exceedingly small time or distance changes of order $(GM/c^2r)^2$. For the latter the required accuracy of a clock is somewhat better than one part in 10^{18}."

Note that this essentially says that testing GR means foremost to test UFF, i.e. to perform Eötvös-type experiments.

This immediately provoked a contradiction by Robert Dicke in [5], who read Schiff's assertions as "serious indictment of the very expensive government-sponsored program to put an atomic clock into an artificial satellite". For, he reasoned, "If Schiff's basic assumptions are as firmly established as he believes, then indeed this project is a waste of government funds." Dicke goes on to point out that for several reasons UFF is not as well tested by past Eötvös-type experiments as Schiff seems to assume and hence argues in strong favour of the said planned tests.

As a reaction to Dicke, Schiff added in proof the justifying note already mentioned above. In it he said:

> "The Eötvös experiment show with considerable accuracy that the gravitational and inertial masses of normal matter are equal. This means that the ground-state eigenvalue of the Hamiltonian for this matter appears equally in the inertial mass and in the interaction

> of this mass with a gravitational field. It would be quite remarkable if this could occur without the entire Hamiltonian being involved in the same way, in which case a clock composed of atoms whose motions are determined by this Hamiltonian would have its rate affected in the expected manner by a gravitational field."

This is the origin of what is called *Schiff's conjecture* in the literature. Attempts have been made to "prove" it in special situations [14], but it is well known not to hold in mathematical generality. For example, consider gravity and electromagnetism coupled to point charges just as in GR, but now make the single change that the usual Lagrangian density $-\frac{1}{4}F_{ab}F^{ab}$ for the free electromagnetic field is replaced by $-\frac{1}{4}C^{ab\,cd}F_{ab}F_{cd}$, where the tensor field C (usually called the constitutive tensor; it has the obvious symmetries of the Riemann tensor) can be any function of the metric. It is clear that this change implies that for general C the laws of (vacuum) electrodynamics in a freely falling frame will not reduce to those of Special Relativity and that, accordingly, Schiff's conjecture cannot hold for all C. In fact, Ni proved [18] that Schiff's conjecture holds iff

$$C^{ab\,cd} = \tfrac{1}{2}\big(g^{ac}g^{bd} - g^{ad}g^{bc}\big) + \phi\varepsilon^{abcd}\,, \tag{2.10}$$

where ϕ is some scalar function of the metric.

Another and simpler reasoning, showing that UFF cannot by itself imply that gravity is a metric theory in the semi-Riemannian sense (rather than, say, of Finslerian type) is the following: Imagine the ratio of electric charge and inertial mass were a universal constant for all existing matter and that a fixed electromagnetic field existed throughout spacetime. Test particles would move according to the equation

$$\ddot{x}^a + \Gamma^a_{bc}\dot{x}^b\dot{x}^c = (q/m)F^a_b\dot{x}^b\,, \tag{2.11}$$

where the Γ's are the Christoffel symbols for the metric and (q/m) is the said universal constant. This set of four ordinary differential equations for the four functions x^a clearly defines a path structure on spacetime, but for a general F_{ab} there will be no semi-Riemannian metric with respect to which (2.11) is the equation for a geodesic. Hence Schiff's conjecture should at best be considered as a selection criterion.

2.1. LLI and UGR

We consider a static homogeneous and downward-pointing gravitational field $\vec{g} = -g\vec{e}_z$. We follow Section 2.4 of [30] and assume the validity of UFF and LLI but allow for violations of LPI. Then UFF guarantees the local existence of a freely-falling frame with coordinates $\{x^\mu_f\}$, whose acceleration is the same as that of test particles. For a rigid acceleration we have

$$
\begin{aligned}
ct_f &= (z_s + c^2/g)\sinh(gt_s/c)\,,\\
x_f &= x_s\,,\\
y_f &= y_s\,,\\
z_f &= (z_s + c^2/g)\coth(gt_s/c)\,.
\end{aligned}
$$

LLI guarantees that, *locally*, time measured by, e.g., an atomic clock is proportional to Minkowskian proper length in the freely falling frame. If we consider violations of LPI, the constant of proportionality might depend on the space-time point, e.g. via dependence on the gravitational potential ϕ, as well as the type of clock:

$$
\begin{aligned}
c^2\,d\tau^2 &= F^2(\phi)\big[c^2dt_f^2 - dx_f^2 - dy_f^2 - dz_f^2\big] &\quad (2.12)\\
&= F^2(\phi)\left[\left(1+\frac{gz_s}{c^2}\right)^2 c^2dt_s^2 - dx_s^2 - dy_s^2 - dz_s^2\right]. &\quad (2.13)
\end{aligned}
$$

The *same* time interval $dt_s = dt_s(z_s^{(1)}) = dt_s(z_s^{(2)})$ on the two static clocks at rest wrt. $\{x_s^\mu\}$, placed at different heights $z_s^{(1)}$ and $z_s^{(2)}$, correspond to *different* intervals $d\tau^{(1)}, d\tau^{(2)}$ of the inertial clock, giving rise to the redshift (all coordinates are $\{x_s^\mu\}$ now, so we drop the subscript s):

$$
\zeta := \frac{d\tau^{(2)} - d\tau^{(1)}}{d\tau^{(1)}} = \frac{F(z^{(2)})(1+gz^{(2)}/c^2)}{F(z^{(1)})(1+gz^{(1)}/c^2)} - 1\,. \tag{2.14}
$$

For small $\Delta z = z^{(2)} - z^{(1)}$ this gives to first order in Δz

$$
\Delta\zeta = (1+\alpha)g\Delta z/c^2\,, \tag{2.15}
$$

where

$$
\alpha = \frac{c^2}{g}\,\left(\vec{e}_z\cdot\vec{\nabla}\ln(F)\right) \tag{2.16}
$$

parametrises the deviation from the GR result. α may depend on position, gravitational potential, and the type of clock one is using.

2.2. Energy conservation, UFF, and UGR

In this subsection we wish to present some well-known gedanken-experiment-type arguments [19, 9] according to which there is a link between violations of UFF and UGR, provided energy conservation holds. Here we essentially present Nordtvedt's version; compare Figure 1.

We consider two copies of a system that is capable of 3 energy states A, B, and B' (white, light grey, grey), with $E_A < E_B < E_{B'}$, placed into a vertical downward-pointing homogeneous gravitational field. Initially system 2 is in state B and placed at height h above system 1, which is in state A. At time T_1 system 2 makes a transition $B \to A$ and sends out a photon of energy $h\nu = E_B - E_A$. At time T_2 system 1 absorbs this photon, which is now blue-shifted due to its free fall in the downward-pointing gravitational field, and makes a transition $A \to B'$. At T_3 system 2 has been dropped from height h with an acceleration of modulus g_A that possibly depends on its

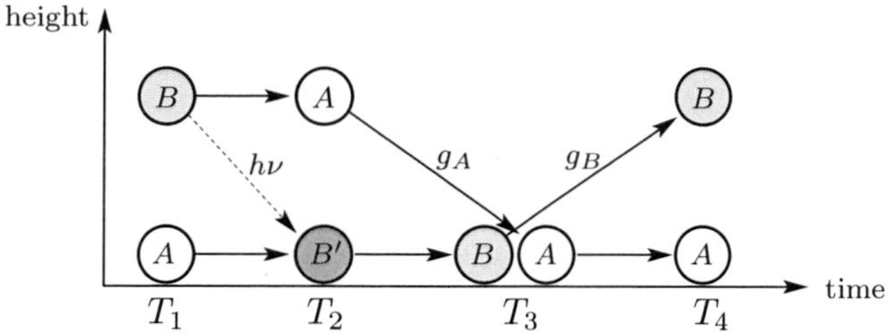

FIGURE 1. Nordtvedt's gedanken experiment. Two systems in three different energy states A, B, and C are considered at four different times T_1, T_2, T_3, and T_4. The initial state at T_1 and the final state at T_4 are identical, which, by means of energy conservation, leads to an interesting quantitative relation between possible violations of UFF and UGR.

inner state A and has hit system 1 inelastically, leaving one system in state A and at rest, and the other system in state B with an upward motion. By energy conservation this upward motion has a kinetic energy of

$$E_{\text{kin}} = M_A g_A h + (E_{B'} - E_B)\,. \tag{2.17}$$

This upward motion is a free fall in a gravitational field and since the system is now in an inner state B, it is decelerated with modulus g_B, which may differ from g_A. At T_4 the system in state B has climbed to the height h, which must be the same as the height at the beginning, again by energy conservation. Hence we have $E_{\text{kin}} = M_B g_B h$, and since moreover $E_{B'} - E_B = (M_{B'} - M_B)c^2$, we get

$$M_A g_A h + M_{B'} c^2 = M_B c^2 + M_B g_B h\,. \tag{2.18}$$

Therefore

$$\begin{aligned} \frac{\delta\nu}{\nu} &:= \frac{(M_{B'} - M_a) - (M_B - M_A)}{M_B - M_A} \\ &= \frac{g_B h}{c^2}\left[1 + \frac{M_A}{M_B - M_A}\,\frac{g_B - g_A}{g_B}\right], \end{aligned} \tag{2.19}$$

so that

$$\alpha = \frac{M_A}{M_B - M_A}\,\frac{g_B - g_A}{g_B} =: \frac{\delta g/g}{\delta M/M}\,. \tag{2.20}$$

This equation gives a quantitative link between violations of UFF, here represented by δg, and violations of UGR, here represented by α. The strength of the link depends on the fractional difference of energies/masses $\delta M/M$, which varies according to the type of interaction that is responsible for the transitions. Hence this equation can answer the question of how accurate a test of UGR must be in order to test the metric nature of gravity to the same level of accuracy than Eötvös-type experiments. Given that for the latter we have $\delta g/g < 10^{-13}$, this depends on the specific situation (interaction) through $\delta M/M$. For atomic clocks the relevant interaction is the magnetic one, since the energies rearranged in hyperfine transitions correspond to magnetic interactions. The variation of the magnetic contribution to the overall self-energy between pairs of chemical elements (A, B) for which the Eötvös factor has been strongly bounded above, like Aluminium and Gold (or Platinum), have been (roughly) estimated to be $|\delta M/M| \approx 5 \times 10^{-5}$ [19]. Hence one needed a precision of $\alpha_{\mathrm{RS}} < 5 \times 10^{-9}$ for a UGR test to match existing UFF tests.

3. UFF in Quantum Mechanics

In classical mechanics the universality of free fall is usually expressed as follows: We consider Newton's Second Law for a point particle of inertial mass m_i,

$$\vec{F} = m_i \ddot{\vec{x}}\,, \tag{3.1}$$

and specialise it to the case in which the external force is gravitational, i.e. $\vec{F} = \vec{F}_{\mathrm{grav}}$, where

$$\vec{F} = m_g\,\vec{g}\,. \tag{3.2}$$

Here m_g denotes the passive gravitational mass and $\vec{g} : \mathbb{R}^4 \to \mathbb{R}^3$ is the (generally space and time dependent) gravitational field. Inserting (3.2) into (3.1) we get

$$\ddot{\vec{x}}(t) = \left(\frac{m_g}{m_i}\right) \vec{g}\big(t, \vec{x}(t)\big)\,. \tag{3.3}$$

Hence the solution of (3.3) only depends on the initial time, spatial position, and spatial velocity iff m_g/m_i is a universal constant, which by appropriate choices of units can be made unity.

This reasoning is valid for point particles only. But it clearly generalises to the centre-of-mass motion of an extended mass distribution in case of spatially homogeneous gravitational fields, where m_i and m_g are then the total inertial and total (passive) gravitational masses. For this generalisation to hold it need not be the case that the spatial distributions of inertial and gravitational masses are proportional. If they are not proportional, the body will deform as it moves under the influence of the gravitational field. If they are proportional and the initial velocities of all parts of the body are the same, the trajectories of the parts will all be translates of one another. If the initial velocities are not the same, the body will disperse without the action of internal cohesive forces in the same way as it would without gravitational field.

There is no pointlike-supported wave packet in quantum mechanics. Hence we ask for the analogy to the situation just described: How does a wave packet fall in a homogeneous gravitational field? The answer is given by the following result, the straightforward proof of which we suppress.

Proposition 3.1. *ψ solves the Schrödinger Equation*

$$i\hbar\partial_t\psi = \left(-\frac{\hbar^2}{2m_i}\Delta - \vec{F}(t)\cdot\vec{x}\right)\psi \tag{3.4}$$

iff

$$\psi = \big(\exp(i\alpha)\,\psi'\big) \circ \Phi^{-1}\,, \tag{3.5}$$

where ψ' solves the free Schrödinger equation (i.e. without potential). Here $\Phi : \mathbb{R}^4 \to \mathbb{R}^4$ is the following spacetime diffeomorphism (preserving time)

$$\Phi(t,\vec{x}) = \big(t, \vec{x} + \vec{\xi}(t)\big)\,, \tag{3.6}$$

where $\vec{\xi}$ is a solution to

$$\ddot{\vec{\xi}}(t) = \vec{F}(t)/m_i \tag{3.7}$$

with $\vec{\xi}(0) = \vec{0}$, and $\alpha : \mathbb{R}^4 \to \mathbb{R}$ given by

$$\alpha(t,\vec{x}) = \frac{m_i}{\hbar}\left\{\dot{\vec{\xi}}(t)\cdot\big(\vec{x} + \vec{\xi}(t)\big) - \frac{1}{2}\int^t dt'\|\dot{\vec{\xi}}(t')\|^2\right\}. \tag{3.8}$$

To clearly state the simple meaning of (3.5) we first remark that changing a trajectory $t \mapsto |\psi(t)\rangle$ of Hilbert-space vectors to $t \mapsto \exp\big(i\alpha(t)\big)\,|\psi(t)\rangle$ results in the *same* trajectory of states, since the state at time t is faithfully represented by the ray in Hilbert space generated by the vector (unobservability of the global phase). As our Hilbert space is that of square-integrable functions on $\mathbb{R}^2$, only the $\vec{x}$-dependent parts of the phase (3.8) change the instantaneous state. Hence, in view of (3.8), the meaning of (3.5) is that the state ϕ at time t is obtained from the freely evolving state at time t with the same initial data by 1) a boost with velocity $\vec{v} = \dot{\vec{\xi}}(t)$ and 2) a spatial displacement by $\vec{\xi}(t)$. In particular, the spatial probability distribution $\rho(t,\vec{x}) := \psi^*(t,\vec{x})\psi(t,\vec{x})$ is of the form

$$\rho = \rho' \circ \Phi^{-1}\,, \tag{3.9}$$

where ρ' is the freely evolving spatial probability distribution. This implies that the spreading of ρ is entirely that due to the free evolution.

Now specialise to a homogeneous and static gravitational field $\vec{g}$, such that $\vec{F} = m_g\vec{g}$; then

$$\vec{\xi}(t) = \vec{v}t + \tfrac{1}{2}\vec{a}t^2 \tag{3.10}$$

with

$$\vec{a} = (m_g/m_i)\,\vec{g}\,. \tag{3.11}$$

In this case the phase (3.8) is

$$\alpha(t,\vec{x}) = \frac{m_i}{\hbar}\Big\{\vec{v}\cdot\vec{x} + \big(\tfrac{1}{2}v^2 + \vec{a}\cdot\vec{x}\big)\,t + \vec{v}\cdot\vec{a}\,t^2 + \tfrac{1}{3}a^2t^3\Big\}\,, \tag{3.12}$$

where $v := \|\vec{v}\|$ and $a := \|\vec{a}\|$.

As the spatial displacement $\vec{\xi}(t)$ just depends on m_g/m_i, so does that part of the spatial evolution of ρ that is due to the interaction with the gravitational field. This is a quantum-mechanical version of UFF. Clearly, the inevitable spreading of the free wave packet, which depends on m_i alone, is just passed on to the solution in the gravitational field. Recall also that the evolution of the full state involves the $\vec{x}$-dependent parts of the phase, which correspond to the gain in momentum during free fall. That gain due to acceleration is just the classical $\delta\vec{p} = m_i\vec{a}t$ which, in view of (3.11), depends on m_g alone.

Other dependencies of physical features on the pair (m_g, m_i) are also easily envisaged. To see this, we consider the stationary case of (3.4), where $i\hbar\partial_t$ is replaced by the energy E, and also take the external force to correspond to a constant gravitational field in negative z-direction: $\vec{F} = -m_g g\vec{e}_z$. The Schrödinger equation then separates, implying free motion perpendicular to the z-direction. Along the z-direction one gets

$$\left(\frac{d^2}{d\zeta^2} - \zeta\right)\psi = 0, \tag{3.13}$$

with

$$\zeta := \kappa z - \varepsilon\,, \tag{3.14}$$

where

$$\kappa := \left[\frac{2m_i m_g g}{\hbar^2}\right]^{\frac{1}{3}}, \quad \varepsilon := E\cdot\left[\frac{2m_i}{m_g^2 g^2\hbar^2}\right]^{\frac{1}{3}}. \tag{3.15}$$

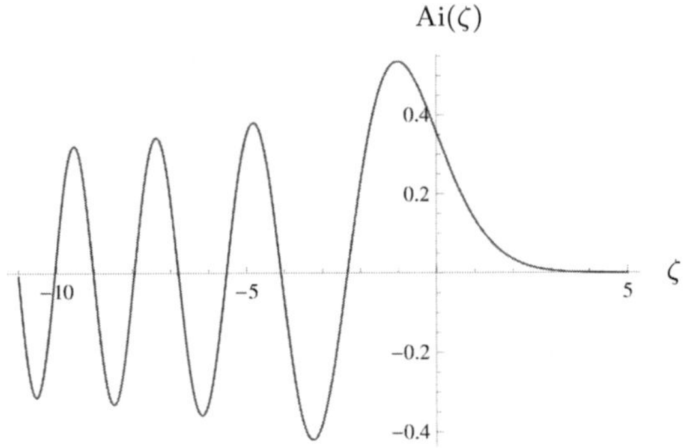

FIGURE 2. Airy function from $\zeta = -10$ to $\zeta = 5$.

A solution to (3.13) that falls off for $\zeta \to \infty$ must be proportional to the Airy function, a plot of which is shown in Figure 2.

As has been recently pointed out in [12], it is remarkable that the penetration depth into the classically forbidden region, which is a simple function of the length κ^{-1}, depends on the *product* of inertial and gravitational mass. Also, suppose we put an infinite potential barrier at $z = 0$. Then the energy eigenstates of a particle in the region $z > 0$ are obtained by the requirement $\psi(0) = 0$, hence $\varepsilon = -z_n$, where $z_n < 0$ is the n-th zero of the Airy function. By (3.15) this gives

$$E_n = \left[\frac{m_g^2 g^2\hbar^2}{2m_i}\right]^{\frac{1}{3}}\cdot(-z_n)\,. \tag{3.16}$$

The energy eigenvalues of this "atom trampoline" [12] depend on the combination m_g^2/m_i. In the gravitational field of the Earth the lowest-lying energies have be realised with ultracold neutrons [17]; these energies are just a few 10^{-12} eV.

In classical physics, the return-time of a body that is projected at level $z = 0$ against the gravitational field $\vec{g} = -g\vec{e}_z$ in positive z-direction, so that it reaches a maximal height of $z = h$, is given by

$$T_{\text{ret}} = 2 \cdot \left[\frac{m_i}{m_g}\right]^{\frac{1}{2}} \cdot \left[\frac{2h}{g}\right]^{\frac{1}{2}} . \tag{3.17}$$

Now, in Quantum Mechanics we may well expect the return time to receive corrections from barrier-penetration effects, which one expects to delay arrival times. Moreover, since the penetration depth is a function of the product rather than the quotient of m_g and m_i, this correction can also be expected to introduce a more complicated dependence of the return time on the two masses. It is therefore somewhat surprising to learn that the classical formula (3.17) can be reproduced as an exact quantum-mechanical result [4]. This is not the case for other shapes of the potential. A simple calculation confirms the intuition just put forward for a step potential, which leads to a positive correction to the classical return which is proportional to the quantum-mechanical penetration depth and also depends on the inertial mass. Similar things can be said of an exponential potential (see [4] for details). Clearly, these results make no more proper physical sense than the notion of *timing* that is employed in these calculations. This is indeed a subtle issue which we will not enter. Suffice it to say that [4] uses the notion of a "Peres clock" [20] which is designed to register and store times of flight without assuming localised particle states. The intuitive reason why barrier penetration does not lead to delays in return time for the linear potential may be read off Figure 2, which clearly shows that the Airy function starts decreasing *before* the classical turning point ($\zeta = 0$) is reached. This has been interpreted as saying that there is also a finite probability that the particle is back-scattered *before* it reaches the classical turning point. Apparently this just cancels the opposite effect from barrier penetration in case of the linear potential, thus giving rise to an unexpectedly close analogue of UFF in Quantum Mechanics.

4. Phase-shift calculation in non-relativistic Quantum Mechanics

We consider the motion of an atom in a static homogeneous gravitational field $\vec{g} = -g\vec{e}_z$. We restrict attention to the motion of the centre of mass along the z-axis, the velocity of which we assume to be so slow that the Newtonian approximation suffices. The centre of mass then obeys a simple Schödinger equation in a potential that depends linearly on the centre-of-mass coordinate. This suggests to obtain (exact) solutions of the time-dependent Schrödinger equation by using the path-integral method; we will largely follow [26].

The time evolution of a Schrödinger wave function in position representation is given by

$$\psi(z_b, t_b) = \int_{\text{space}} dz_a \, K(z_b, t_b\,;\, z_a, t_a)\, \psi(z_a, t_a)\,, \tag{4.1}$$

where

$$K(z_b, t_b\,;\, z_a, t_a) := \langle z_b | \exp\bigl(-iH(t_b - t_a)/\hbar\bigr) | z_a \rangle\,. \tag{4.2}$$

Here z_a and z_b represent the initial and final position in the vertical direction. The path-integral representation of the propagator is

$$K(z_b, t_b\,;\, z_a, t_a) = \int_{\Gamma(a,b)} \mathcal{D}z(t)\, \exp\bigl(iS[z(t)]/\hbar\bigr)\,, \tag{4.3}$$

where

$$\Gamma(a,b) := \bigl\{ z : [t_a, t_b] \to M \mid z(t_{a,b}) = z_{a,b} \bigr\} \tag{4.4}$$

contains all continuous paths. The point is that the path integral's dependence on the initial and final positions z_a and z_b is easy to evaluate whenever the Lagrangian is a potential of at most quadratic order in the positions and their velocities:

$$L(z, \dot z) = a(t)\dot z^2 + b(t)\dot z z + c(t) z^2 + d(t)\dot z + e(t) z + f(t)\,. \tag{4.5}$$

Examples are: 1) The free particle, 2) particle in a homogeneous gravitational field, 3) particle in a rotating frame of reference.

To see why this is true, let $z_* \in \Gamma(a,b)$ denote the solution to the classical equations of motion:

$$\frac{\delta S}{\delta z(t)}\bigg|_{z(t) = z_*(t)} = 0\,. \tag{4.6}$$

We parametrise an arbitrary path $z(t)$ by its difference to the classical solution path; that is, we write

$$z(t) = z_*(t) + \xi(t) \tag{4.7}$$

and regard $\xi(t)$ as path variable:

$$K(z_b, t_b\,;\, z_a, t_a) = \int_{\Gamma(0,0)} \mathcal{D}\xi(t)\, \exp\bigl(iS[z_* + \xi]/\hbar\bigr)\,. \tag{4.8}$$

Taylor expansion around $z_*(t)$ for each value of t, taking into account (4.6), gives

$$\begin{aligned} K(z_b, t_b\,;\, z_a, t_a) =& \exp\left\{ \frac{i}{\hbar} S_*(z_b, t_b\,;\, z_a, t_a) \right\} \\ &\times \int_{\Gamma(0,0)} \mathcal{D}\xi(t) \exp\left\{ \frac{i}{\hbar} \int_{\Gamma(0,0)} dt \left[a(t)\dot\xi^2 + b(t)\dot\xi\xi + c(t)\xi^2 \right] \right\}. \end{aligned} \tag{4.9}$$

Therefore, for polynomial Lagrangians of at most quadratic order, the propagator has the exact representation

$$K(z_b, t_b\,;\, z_a, t_a) = F(t_b, t_a)\, \exp\left\{ \frac{i}{\hbar}\, S_*(z_b, t_b\,;\, z_a, t_a) \right\}, \tag{4.10}$$

where $F(t_b, t_a)$ does not depend on the initial and final position and S_* is the action for the extremising path (classical solution). We stress once more that (4.10) is valid only for Lagrangians of at most quadratic order. Hence we may use it to calculate the exact phase change for the non-relativistic Schrödinger equation in a static and homogeneous gravitational field.

5. Free fall in a static homogeneous gravitational field

We consider an atom in a static and homogeneous gravitational field $\vec{g} = -g\vec{e}_z$. We restrict attention to its centre-of-mass wave function, which we represent as that of a point particle. During the passage from the initial to the final location the atom is capable of assuming different internal states. These changes will be induced by laser interaction and will bring about changes in the inertial and gravitational masses, m_i and m_g. It turns out that for the situation considered here (hyperfine-split ground states of Caesium) these changes will be negligible, as will be shown in footnote 8. Hence we can model the situation by a point particle of fixed inertial and gravitational mass, which we treat as independent parameters throughout. We will nowhere assume $m_i = m_g$.

5.1. Some background from GR

In General Relativity the action for the centre-of-mass motion for the atom (here treated as point particle) is $(-m^2c)$ times the length functional, where m is the mass (here $m_i = m_g = m$):

$$S = -mc \int_{\lambda_1}^{\lambda_2} d\lambda \sqrt{g_{\alpha\beta}(x(\lambda))\dot{x}^\alpha(\lambda)\dot{x}^\beta(\lambda)}\,, \tag{5.1}$$

where, in local coordinates, $x^\alpha(\lambda)$ is the worldline parametrised by λ and $g_{\alpha\beta}$ are the metric components. Specialised to static metrics

$$g = f^2(\vec{x})c^2dt^2 - h_{ab}(\vec{x})\,dx^a dx^b\,, \tag{5.2}$$

we have

$$S = -mc^2 \int_{t_1}^{t_2} dt\, f(\vec{x}(t))\sqrt{1 - \frac{\hat{h}(\vec{v},\vec{v})}{c^2}}\,, \tag{5.3}$$

where $\vec{v} := (v^1, v^2, v^3)$ with $v^a = dx^a/dt$, and where $\hat{h}$ is the "optical metric" of the space sections $t = \text{const}$:

$$\hat{h}_{ab} := \frac{h_{ab}}{f^2}\,. \tag{5.4}$$

This is valid for all static metrics. Next we assume the metric to be spatially conformally flat, i.e., $h_{ab} = h^2\,\delta_{ab}$, or equivalently

$$\hat{h}_{ab} = \hat{h}^2\,\delta_{ab}\,, \tag{5.5}$$

with $\hat{h} := h/f$, so that the integrand (Lagrange function) of (5.3) takes the form

$$L = -mc^2 \, f(\vec{x}(t)) \sqrt{1 - \hat{h}^2(\vec{x}(t)) \frac{\vec{v}^2}{c^2}}\,, \tag{5.6}$$

where $\vec{v}^2 := (v^1)^2 + (v^2)^2 + (v^3)^2$.

We note that spherically symmetric metrics are necessarily spatially conformally flat (in any dimension and regardless of whether Einstein's equations are imposed). In particular, the Schwarzschild solution is of that form, as is manifest if written down in isotropic coordinates:

$$g = \left[\frac{1 - \frac{r_S}{r}}{1 + \frac{r_S}{r}}\right]^2 c^2 dt^2 - \left[1 + \frac{r_S}{r}\right]^4 \left(dx^2 + dy^2 + dz^2\right). \tag{5.7}$$

Here r_S is the Schwarzschild radius:

$$r_s := GM/2c^2\,. \tag{5.8}$$

Hence, in this case,

$$f = \frac{1 - \frac{r_S}{r}}{1 + \frac{r_S}{r}}\,, \quad h = \left[1 + \frac{r_S}{r}\right]^2, \quad \hat{h} = \frac{\left[1 + \frac{r_S}{r}\right]^3}{1 - \frac{r_S}{r}}\,. \tag{5.9}$$

Back to (5.3), we now approximate it to the case of weak gravitational fields and slow particle velocities. For weak fields, Einstein's equations yield to leading order:

$$f = 1 + \frac{\phi}{c^2}\,, \quad h = 1 - \frac{\phi}{c^2}\,, \quad \hat{h} = 1 - \frac{2\phi}{c^2}\,. \tag{5.10}$$

Here ϕ is the Newtonian potential, i.e. satisfies $\Delta\phi = 4\pi G T_{00}/c^2$ where Δ is the Laplacian for the flat spatial metric. Inserting this in (5.6) the Lagrangian takes the leading-order form

$$L = -mc^2 + \tfrac{1}{2}mv^2 - m\phi\,. \tag{5.11}$$

We note that the additional constant m^2c^2 neither influences the evolution of the classical nor of the quantum-mechanical state. Classically this is obvious. Quantum-mechanically this constant is inherited with opposite sign by the Hamiltonian:

$$H = mc^2 + \tfrac{p^2}{2m} + m\phi\,. \tag{5.12}$$

It is immediate that if $\psi(t)$ is a solution to the time-dependent Schrödinger equation for this Hamiltonian, then $\psi(t) := \exp\big(i\alpha(t)\big)\psi'(t)$ with $\alpha(t) = -t(mc^2/\hbar)$, where ψ' solves the time-dependent Schrödinger equation without the term mc^2. But ψ and ψ' denote the *same* time sequence of states (rays). Hence we can just ignore this term.[5]

[5] In [15] this term seems to have been interpreted as if it corresponded to an inner degree of freedom oscillating with Compton frequency, therefore making up a "Compton clock". But as there is no periodic change of state associated to this term, it certainly does not correspond to anything like a clock (whose state changes periodically) in this model.

5.2. Interferometry of freely falling atoms

We now analyse the quantum mechanical coherences of the centre-of-mass motion in a static and homogeneous gravitational field, where we generalise to $m_i \neq m_g$. Hence, instead of (5.11) (without the irrelevant mc^2 term) we take

$$L = \tfrac{1}{2} m_i \dot{z}^2 - m_g g z\,. \tag{5.13}$$

Here we restricted attention to the vertical degree of freedom, parametrised by z, where $\dot{z} := dz/dt$. This is allowed since the equation separates and implies free evolution in the horizontal directions. The crucial difference to (5.11) is that we do not assume that $m_i = m_g$.

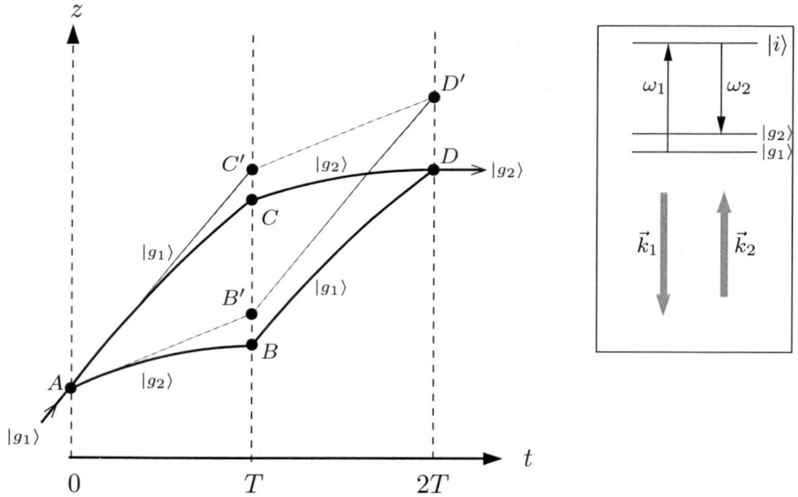

FIGURE 3. Atomic interferometer with beam splitters at A and D and mirrors at B and C, realised by $\pi/2$– and π–pulses of counter-propagating laser beams with $\vec{k}_1 = -k_1\vec{e}_z$, where $k_1 = \omega_1/c$ and $\vec{k}_2 = k_2\vec{e}_z$, where $k_2 = \omega_2/c$. $|g_{1,2}\rangle$ denote the hyperfine doublet of ground states and $|i\rangle$ an intermediate state via which the Raman transitions between the ground states occur. The solid (bent) paths show the classical trajectories in presence of a downward pointing gravitational field $\vec{g} = -g\vec{e}_z$, the dashed (straight) lines represent the classical trajectories for $g = 0$. The figure is an adaptation of Fig. 9 in [26].

The situation described in [15], which is as in [26], is depicted in Fig. 3. A beam of Caesium atoms, initially in the lower state $|g_1\rangle$ of the hyperfine doublet $|g_1\rangle, |g_2\rangle$ of ground states, is coherently split by two consecutive laser pulses which we may take to act simultaneously at time $t = T$: The first downward-pointing pulse with $\vec{k}_1 = -k_1\vec{e}_z$, where $k_1 := \omega_1/c$, elevates the atoms from the ground state $|g_1\rangle$ to an intermediate state $|i\rangle$, thereby transferring momentum $\Delta_{A1}\vec{p} = -\hbar k_1 \vec{e}_z$. The second upward-pointing pulse $\vec{k}_2 = k_2\vec{e}_z$, where $k_2 := \omega_2/c$, induces a transition from $|i\rangle$ to the upper level of the hyperfine-split ground state, $|g_2\rangle$. The emitted photon is pointing upwards so that the atom suffers a recoil by $\Delta_{A2}\vec{p} = -\hbar k_2\vec{e}_z$. The total momentum transfer at A is the sum $\Delta_A\vec{p} := \Delta_{A1}\vec{p} + \Delta_{A2}\vec{p} = -\kappa\vec{e}_z$ with $\kappa = k_1 + k_2 = (\omega_1 + \omega_2)/c$. At A the pulses are so adjusted that each atom has a 50% chance to make this transition $|g_1\rangle \to |i\rangle \to |g_2\rangle$ and proceed on branch AB and a 50% chance to just stay in $|g_1\rangle$ and proceed on branch AC. (So-called $\pi/2$–pulse or atomic beam splitter.) At B and C the pulses are so adjusted that the atoms make transitions $|g_2\rangle \to |i\rangle \to |g_1\rangle$ and $|g_1\rangle \to |i\rangle \to |g_2\rangle$ with momentum changes $\Delta_B\vec{p} = \kappa\vec{e}_z$ and $\Delta_C\vec{p} = -\kappa\vec{e}_z$ respectively. Here the pulses are so adjusted that these transitions occur almost with 100% chance. (So-called π–pulses or atomic mirrors.) Finally, at D, the beam from BD, which is in state $|g_1\rangle$, receives another $\pi/2$-pulse so that 50% of it re-unites coherently with the transmitted 50% of the beam incoming from CD that is not affected by the $\pi/2$–pulse at D. In total, the momentum transfers on the upper and lower paths are, respectively,

$$\Delta_{\text{upper}}\vec{p} = \Delta_C\vec{p} = \underbrace{-\hbar\kappa\vec{e}_z}_{\text{at } C}, \tag{5.14a}$$

$$\Delta_{\text{lower}}\vec{p} = \Delta_A\vec{p} + \Delta_B\vec{p} + \Delta_D\vec{p} = \underbrace{-\hbar\kappa\vec{e}_z}_{\text{at } A}\underbrace{+\hbar\kappa\vec{e}_z}_{\text{at } B}\underbrace{-\hbar\kappa\vec{e}_z}_{\text{at } D}. \tag{5.14b}$$

Following [26] we now wish to show how to calculate the phase difference along the two different paths using (4.10). For this we need to know the classical trajectories.

Remark 5.1. The classical trajectories are parabolic with downward acceleration of modulus $\hat{g}$. If the trajectory is a stationary point of the classical action we would have $\hat{g} = g(m_g/m_i)$. However, the authors of [15] contemplate the possibility that violations of UGR could result in not making this identification.[6] Hence we proceed without specifying $\hat{g}$ until the end, which also has the additional advantage that we can at each stage cleanly distinguish between contributions from the kinetic and contributions from the potential part of the action. But we do keep in mind that (4.10) only represents the right dependence on (z_a, z_b) if $z(t) = z_*(t)$ holds. This will be a crucial point in our criticism to which we return later on.

Now, the unique parabolic orbit with downward acceleration $\hat{g}$ through initial event (t_a, z_a) and final event (t_b, z_b) is

$$z(t) = z_a + v_a(t - t_a) - \tfrac{1}{2}\hat{g}(t - t_a)^2\,, \tag{5.15}$$

where

$$v_a := \frac{z_b - z_a}{t_b - t_a} + \frac{1}{2}\hat{g}(t_b - t_a)\,. \tag{5.16}$$

[6] In [15] the quantity that we call $\hat{g}$ is called g'.

Evaluating the action along this path gives

$$\begin{aligned} S_{\hat{g}}(z_b, t_b; z_a, t_a) = & \frac{m_i}{2}\frac{(z_b - z_a)^2}{t_b - t_a} \\ & - \frac{m_g g}{2}(z_b + z_a)(t_b - t_a) \\ & + \frac{\hat{g}}{24}(t_b - t_a)^3(m_i\hat{g} - 2m_g g)\,. \end{aligned} \tag{5.17}$$

Remark 5.2. Note the different occurrences of g and $\hat{g}$ in this equation. We recall that g is the parameter in the Lagrangian (5.13) that parametrises the gravitational field strength, whereas $\hat{g}$ denotes the modulus of the downward acceleration of the trajectory $z(t)$ along which the atom actually moves, and along which the action is evaluated.

Remark 5.3. In (5.17) it is easy to tell apart those contributions originating from the kinetic term (first term in (5.13)) from those originating from the potential term (second term in (5.13)): The former are proportional to m_i, the latter to m_g.

Remark 5.4. Note also that on the upper path ACD the atom changes the internal state at C and on the lower path ABD it changes the internal state at all three laser-interaction points A, B, and D. According to Special Relativity, the atom also changes its inertial mass at these point by an amount $\Delta E/mc^2$, where $\Delta E \approx 4 \times 10^{-5}\,\text{eV}$, which is a fraction of $3 \cdot 10^{-16}$ of Caesium's rest energy mc^2.[7] Hence this change in inertial mass, as well as a change in gravitational mass of the same order of magnitude, can be safely neglected without making any assumptions concerning the constancy of their quotient m_i/m_g.

Now we come back to the calculation of phase shifts. According to (4.10), we need to calculate the classical actions along the upper and lower paths.

$$\begin{aligned} \Delta S &= S(AC) + S(CD) - \big(S(AB) + S(BD)\big) \\ & \big(S(AC) - S(AB)\big) + \big(S(CD) - S(BD)\big)\,. \end{aligned} \tag{5.18}$$

Remark 5.5. Since the events B and C differ in time from A by the same amount T, and likewise B and C differ from D by the same amount T, we see that the last term on the right-hand side of (5.17) drops out upon taking the differences in (5.18), since it only depends on the time differences but not on the space coordinates. Therefore, the dependence on $\hat{g}$ also drops out of (5.17), which then only depends on g.

[7] Recall that the "second" is defined to be the duration of $\nu = 9,192,631,770$ cycles of the hyperfine structure transition frequency of Caesium-133. Hence, rounding up to three decimal places, the energy of this transition is $\Delta E = h \times \nu = 4.136 \cdot 10^{-15}\,\text{eV} \cdot \text{s} \times 9,193,631,770\,\text{s}^{-1} = 3.802 \cdot 10^{-5}\,\text{eV}$. The mass of Caesium is $m = 132.905\,\text{u}$, where $1\,\text{u} = 931.494 \cdot 10^6\,\text{eV}/\text{c}^2$; hence $mc^2 = 1.238 \cdot 10^{11}\,\text{eV}$ and $\Delta E/mc^2 = 3.071 \cdot 10^{-16}$.

The calculation of (5.18) is now easy. We write the coordinates of the four events A, B, C, D in Fig. 3 as

$$\begin{aligned} A &= (z_A, t_A = 0)\,, & B &= (z_B, t_B = T)\,, \\ C &= (z_C, t_C = T)\,, & D &= (z_D, t_D = 2T) \end{aligned} \tag{5.19}$$

and get

$$\Delta S = \frac{m_i}{T}(z_C - z_B)\big[z_B + z_C - z_A - z_D - g(m_g/m_i)T^2\big]\,. \tag{5.20}$$

Since the curved (thick) lines in Fig. 3 are the paths with downward acceleration of modulus $\hat{g}$, whereas the straight (thin) lines correspond to the paths without a gravitational field, the corresponding coordinates are related as follows[8]:

$$\begin{aligned} A' &= \big(z_{A'} = z_A, \bar{t}_{A'} = 0\big)\,, \\ B' &= \big(z_{B'} = z_B + \tfrac{1}{2}\hat{g}T^2\,,\, t_{B'} = T\big)\,, \\ C' &= \big(z_{C'} = z_C + \tfrac{1}{2}\hat{g}T^2\,,\, t_{C'} = T\big)\,, \\ D' &= \big(z_{D'} = z_D + 2\hat{g}T^2\,,\, t_{D'} = 2T\big)\,. \end{aligned} \tag{5.21}$$

Hence

$$z_B + z_C - z_A - z_D = z_{B'} + z_{C'} - z_{A'} - z_{D'} + \hat{g}T^2 = \hat{g}T^2\,, \tag{5.22}$$

where $z_{B'} + z_{C'} - z_{A'} - z_{D'} = 0$ simply follows from the fact that $A'B'C'D'$ is a parallelogram. Moreover, for the difference $(z_C - z_B)$ we have

$$z_C - z_B = z_{C'} - z_{B'} = \frac{\hbar\kappa}{m_i}T\,, \tag{5.23}$$

[8]If we took into account the fact that the atoms change their inner state and consequently their inertial mass m_i, we should also account for the possibility that the quotient m_i/m_g may change. As a result, the magnitude of the downward acceleration $\hat{g}$ may depend on the inner state. These quantities would then be labelled by indices 1 or 2, according to whether the atom is in state $|g_1\rangle$ or $|g_2\rangle$, respectively. The modulus of the downward acceleration along AC and BD is then $\hat{g}_1$, and $\hat{g}_2$ along AB and CD. Also, the momentum transfers through laser interactions are clearly as before, but if converted into velocity changes by using momentum conservation one has to take into account that the inertial mass changes during the interaction. However, the z-component of the velocity changes at A (for that part of the incoming beam at A that proceeds on AB') and C' will still be equal in magnitude and oppositely directed to that at B', as one can easily convince oneself; its magnitude being $\Delta v = \big(1 - \frac{m_{i1}}{m_{i2}}\big)v_A + \hbar\kappa/m_{i2}$, where v_A is the incoming velocity in A. As a result, it is still true that for laser-induced Raman transitions with momentum transfer $\hbar\kappa$ at $t = 0$ and $t = T$ the two beams from C' and B' meet at time $t = 2T$ at a common point, which is $z_{D'} = z_A + 2v_AT - \frac{\hbar\kappa}{m_{i2}} + \big(\frac{m_{i1}}{m_{i2}} - 1\big)v_AT$. Switching on the gravitational field has the effect that $z_C = z_{C'} - \frac{1}{2}\hat{g}_1T^2$ and $z_B = z_{B'} - \frac{1}{2}\hat{g}_2T^2$, but that the beam from C arrives after time T at $z_D^{(C)} = z_{D'} - \frac{1}{2}(\hat{g}_1 + \hat{g}_2)T^2 - \frac{m_{i1}}{m_{i2}}\hat{g}_1T^2$, whereas the beam from B arrives after time T at $z_D^{(B)} = z_{D'} - \frac{1}{2}(\hat{g}_1 + \hat{g}_2)T^2 - \frac{m_{i2}}{m_{i1}}\hat{g}_2T^2$, which differs from the former by $\Delta z_D := z_D^{(C)} - z_D^{(B)} = T^2\Big(\frac{m_{i2}}{m_{i1}}\hat{g}_2 - \frac{m_{i1}}{m_{i2}}\hat{g}_1\Big)$. Assuming that $\hat{g}_1 = (m_{g1}/m_{i1})g$ and $\hat{g}_2 = (m_{g2}/m_{i2})g$, this is $\Delta z_D = (m_{i2}m_{g2} - m_{i1}m_{g1})gT^2/(m_{i1}m_{i2})$ which, interestingly, vanishes iff the *product* (rather than the quotient) of inertial and gravitational mass stays constant. If the quotient stays approximately constant and so that $\hat{g}_1 = \hat{g}_2 =: \hat{g}$, we write $m_{i2}/m_{i1} = 1 + \varepsilon$, with $\varepsilon = \Delta E/m_{i1}c^2$, and get to first order in ε that $\Delta z_D \approx 2\varepsilon\hat{g}T^2$.

where the last equality follows from the fact that along the path AB' the atoms have an additional momentum of $-\hbar\kappa\vec{e}_z$ as compared to the atoms along the path AC'; compare (5.14b). Using (5.22) and (5.23) in (5.20), we get:

$$\Delta S = \hbar\kappa T^2 \left[\hat{g} - g(m_g/m_i)\right]. \tag{5.24}$$

As advertised in Remarks 5.1, 5.2, and 5.3, we can now state individually the contributions to the phase shifts from the kinetic and the potential parts:

$$(\Delta\phi)_{\text{time}} = +\kappa T^2\, \hat{g}\,, \tag{5.25a}$$

$$(\Delta\phi)_{\text{redshift}} = -\kappa T^2\, g(m_g/m_i)\,. \tag{5.25b}$$

Here "time" and "redshift" remind us that, as explained in Section 5.1, the kinetic and potential energy terms correspond to the leading-order special-relativistic time dilation (Minkowski geometry) and the influence of gravitational fields, respectively.

Finally we calculate the phase shift due to the laser interactions at A, B, C and D. For the centre-of-mass wave function to which we restrict attention here, only the total momentum transfers matter which were already stated in (5.14). Hence we get for the phase accumulated on the upper path ACD minus that on the lower path ABD:

$$\begin{aligned}(\Delta\phi)_{\text{light}} &= \frac{1}{\hbar}\{(\Delta_C\vec{p})\cdot\vec{z}_C - (\Delta_A\vec{p})\cdot\vec{z}_A - (\Delta_B\vec{p})\cdot\vec{z}_B - (\Delta_D\vec{p})\cdot\vec{z}_D\}\\ &= -\kappa(z_B + z_C - z_A - z_D) = -\,\kappa\hat{g}T^2\,, \end{aligned} \tag{5.25c}$$

where we used (5.22) in the last step.

Taking the sum of all three contributions in (5.25) we finally get

$$\Delta\phi = -\frac{m_g}{m_i}\cdot g\cdot\kappa\cdot T^2\,. \tag{5.26}$$

This is fully consistent with the more general formula derived by other methods (no path integrals) in [13], which also takes into account possible inhomogeneities of the gravitational field.

5.3. Atom interferometers testing UFF

Equation (5.26) is the main result of the previous section. It may be used in various ways. For given knowledge of (m_g/m_i) a measurement of $\Delta\phi$ may be taken as a measurement of g. Hence the atom interferometer can be used as a gravimeter. However, in the experiments referred to in [15] there was another macroscopic gravimeter nearby consisting of a freely falling corner-cube retroreflector monitored by a laser interferometer. If M_i and M_g denote the inertial and gravitational mass of the corner cube, its acceleration in the gravitational field will be $\tilde{g} = (M_g/M_i)g$. The corner-cube accelerometer allows to determine this acceleration up to $\Delta\tilde{g}/\tilde{g} < 10^{-9}$ [22]. Hence we can write (5.26) as

$$\Delta\phi = -\frac{m_g}{m_i}\cdot\frac{M_i}{M_g}\cdot\left(\tilde{g}\kappa T^2\right), \tag{5.27}$$

in which the left-hand side and the bracketed terms on the right-hand side are either known or measured. Using the Eötvös ratio (2.1) for the Caesium atom (A) and the reference cube (B)

$$\eta(\text{atom}, \text{cube}) = 2 \cdot \frac{(m_g/m_i) - (M_g/M_i)}{(m_g/m_i) + (M_g/M_i)}, \tag{5.28}$$

we have

$$\frac{m_g M_i}{m_i M_g} = \frac{2+\eta}{2-\eta} = 1 + \eta + \mathcal{O}(\eta^2). \tag{5.29}$$

Hence, to first order in $\eta := \eta(\text{atom}, \text{cube})$, we can rewrite (5.27):

$$\Delta\phi = -(1+\eta) \cdot \left(\tilde{g}\kappa T^2\right). \tag{5.30}$$

This formula clearly shows that measurements of phase shifts can put upper bounds on η and hence on possible violations of UFF.

The experiments [21, 22] reported in [15] led to a measured redshift per unit length (height) which, compared to the predicted values, reads as follows:

$$\zeta_{\text{meas}} := \frac{-\Delta\phi}{\kappa T^2 c^2} = (1.090\,322\,683 \pm 0.000\,000\,003) \times 10^{-16} \cdot \text{m}^{-1}, \tag{5.31a}$$

$$\zeta_{\text{pred}} := \tilde{g}/c^2 = (1.090\,322\,675 \pm 0.000\,000\,006) \times 10^{-16} \cdot \text{m}^{-1}. \tag{5.31b}$$

This implies an upper bound of

$$\eta(\text{atom}, \text{cube}) = \frac{\zeta_{\text{meas}}}{\zeta_{\text{pred}}} - 1 < (7 \pm 7) \times 10^{-9}, \tag{5.32}$$

which is more than four orders of magnitude worse (higher) than the lower bounds obtained by more conventional methods (compare (2.3)). However, it should be stressed that here a comparison is made between a macroscopic body (cube) and a genuine quantum-mechanical system (atom) in a superposition of centre-of-mass eigenstates, whereas other tests of UFF use macroscopic bodies describable by classical (non-quantum) laws.

5.4. Atom interferometers testing URS?

The foregoing interpretation seems straightforward and is presumably uncontroversial; but it is not the one adopted by the authors of [15]. Rather, they claim that a measurement of $\Delta\phi$ can, in fact, be turned into an upper bound on the parameter α_{RS} which, according to them, enters the formula (5.30) for the predicted value of $\Delta\phi$ just in the same fashion as does η. Then, since other experiments constrain η to be much below the 10^{-9} level, the very same reasoning as above now leads to the upper bound (5.32) for α_{RS} rather than η, which now implies a dramatic improvement of the upper bound (2.8) by four orders of magnitude!

However, the reasoning given in [15, 16] for how α_{RS} gets into (5.30) seems theoretically inconsistent. It seems to rest on the observation that (5.25a) cancels with (5.25c) *irrespectively of whether* $\hat{g} = (m_g/m_i)g$ *or not*, so that

$$\Delta\phi = (\Delta\phi)_{\text{redshift}}. \tag{5.33}$$

Then they simply assumed that if violations of UGR existed $(\Delta\phi)_{\text{redshift}}$, and hence $\Delta\phi$, simply had to be multiplied by $(1+\alpha_{\text{RS}})$. (Our α_{RS} is called β in [15].)

Remark 5.6. The cancellation of (5.25a) with (5.25c) for $\hat{g} \neq (m_g/m_i)g$ is formally correct but misleading. The reason is apparent from (4.10): The action has to be evaluated along the solution $z_*(t)$ in order to yield the dynamical phase of the wave function. Evaluating it along any other trajectory will not solve the Schrödinger equation. Therefore, whenever $\hat{g} \neq (m_g/m_i)g$, the formal manipulations performed are physically, at best, undefined. On the other hand, if $\hat{g} = (m_g/m_i)g$, then according to (5.25)

$$(\Delta\phi)_{\text{time}} = -(\Delta\phi)_{\text{redshift}} = -(\Delta\phi)_{\text{light}}\,, \tag{5.34}$$

so that we may just as well say that the total phase is entirely due to the interaction with the laser, i.e. that we have instead of (5.33)

$$\Delta\phi = (\Delta\phi)_{\text{light}} \tag{5.35}$$

and no α_{RS} will enter the formula for the phase.

Remark 5.7. The discussion in Section 2.2 suggests that violations of UGR are quantitatively constrained by violations of UFF if energy conservation holds. If the upper bound for violations of UFF are assumed to be on the 10^{-13} level (compare (2.3)), this means that violations of UGR cannot exceed the 10^{-9} level for magnetic interactions. Since the latter is just the new level allegedly reached by the argument in [15], we must conclude by Nordtvedt's gedanken experiment that the violations of UGR that are effectively excluded by the argument of [15] are those also violating energy conservation.

Remark 5.8. Finally we comment on the point repeatedly stressed in [15] that (4.10) together with the relativistic form of the action (5.1) shows that the phase change due to the free dynamics $(\Delta\phi)_{\text{free}} := \Delta\phi - (\Delta)_{\text{light}}$, which in the leading order approximation is just the dt-integral over (5.11), can be written as the integral of the eigentime times the constant Compton frequency $\omega_C = m_i c^2/\hbar$:

$$(\Delta\phi)_{\text{free}} = \omega_C \int d\tau\,. \tag{5.36}$$

The authors of [15] interpret this as timing the length of a worldline by a "Compton clock" ([15], p. 927). [9] For Caesium atoms this frequency is about $2 \times 10^{26} \cdot \text{s}^{-1}$, an enormous value. However, there is no periodic change of any physical state associated to this frequency, unlike in atomic clocks, where the beat frequency of two stationary states gives the frequency by which the superposition state (ray in Hilbert space) periodically recurs. Moreover, the frequency ω_C apparently plays no rôle in any of the calculations performed in [15], nor is it necessary to express $\Delta\phi$ in terms of known quantities. I conclude that for the present setting this reference to a "Compton clock" is misleading.

---

[9] "The essential realisation of this Letter is that the non-relativistic formalism hides the true quantum oscillation frequency ω_C." ([15], pp. 928 and 930)

6. Conclusion

I conclude from the discussion of the previous section that the arguments presented in [15] are inconclusive and do not provide sufficient reason to claim an improvement on upper bounds on possible violations of UGR—and hence on all of EEP—by four orders of magnitude. This would indeed have been a major achievement, as UCR/UGR is by far the least well-tested part of EEP, which, to stress it once more, is the connector between gravity and geometry. Genuine quantum tests of EEP are most welcome and the experiment described in [15] is certainly a test of UFF, but not of UGR. As a test of UFF it is still more than four orders of magnitude away from the best non-quantum torsion-balance experiments. However, one should stress immediately that it puts bounds on the Eötvös factor relating a classically describable piece of matter to an atom in a superposition of spatially localised states, and as such it remains certainly useful. On the other hand, as indicated in Fig. 3, the atoms are in energy eigenstates $|g_{1,2}\rangle$ between each two interaction points on the upper and on the lower path. Hence we do *not* have a genuine quantum test of UGR where a quantum clock (being in a superposition of energy eigenstates) is coherently moving on two different worldlines. This seems to have been the idea of [15] when calling each massive system a "Compton clock". It remains to be seen whether and how this idea can eventually be realised with real physical quantum clocks, i.e. quantum systems whose state (ray!) is periodically changing in time.[10]

References

[1] Vladimir B. Braginsky and Vladimir I. Panov. The equivalence of inertial and passive gravitational mass. *General Relativity and Gravitation*, 3(4):403–404, 1972.

[2] Luigi Cacciapuoti and Christophe Salomon. Space clocks and fundamental tests: The aces experiment. *The European Physical Journal - Special Topics*, 172(1):57–68, 2009.

[3] Robert Alan Coleman and Herbert Korte. Jet bundles and path structures. *Journal of Mathematical Physics*, 21(6):1340–1351, 1980.

[4] Paul C.W. Davies. Quantum mechanics and the equivalence principle. *Classical and Quantum Gravity*, 21(11):2761–2771, 2004.

[5] Robert H. Dicke. Eötvös experiment and the gravitational red shift. *American Journal of Physics*, 28(4):344–347, 1960.

[6] Jürgen Ehlers and Egon Köhler. Path structures on manifolds. *Journal of Mathematical Physics*, 18(10):2014–2018, 1977.

[7] Anton M. Eisele. On the behaviour of an accelerated clock. *Helvetica Physica Acta*, 60:1024–1037, 1987.

[10] After this contribution had been finished I learned that the third version of [23] makes suggestions in precisely this direction (they call it "clock interferometry").

[8] T.M. Fortier et al. Precision atomic spectroscopy for improved limits on variation of the fine structure constant and local position invariance. *Physical Review Letters*, 98(7):070801 (4 pages), 2007.

[9] Mark P. Haugan. Energy conservation and the principle of equivalence. *Annals of Physics*, 118:156–186, 1979.

[10] Sven Herrmann et al. Test of the isotropy of the speed of light using a continuously rotating optical resonator. *Physical Review Letters*, 95(15):150401 (4 pages), 2005.

[11] Michael A. Hohensee, Brian Estey, Francisco Monsalve, Geena Kim, Pei-Chen Kuan, Shau-Yu Lan, Nan Yu, Achim Peters, Steven Chu, and Holger Müller. Gravitational redshift, equivalence principle, and matter waves. arXiv:1009.2485, 2010.

[12] Endre Kajari et al. Inertial and gravitational mass in quantum mechanics. *Applied Physics B*, 100(1):43–60, 2010.

[13] Claus Lämmerzahl. On the equivalence principle in quantum theory. *General Relativity and Gravitation*, 28(9):1043–1070, 1996.

[14] Alan P. Lightman and David L. Lee. Restricted proof that the weak equivalence principle implies the Einstein equivalence principle. *Physical Review D*, 8(2):364–376, 1973.

[15] Holger Müller, Achim Peters, and Steven Chu. A precision measurement of the gravitational redshift by the interference of matter waves. *Nature*, 463:926–930, 2010.

[16] Holger Müller, Achim Peters, and Steven Chu. Reply to P. Wolf et al. nature doi:10.1038/nature09340 (2010). *Nature*, 467:E2, 2010.

[17] Valery V. Nesvizhevsky et al. Measurement of quantum states of neutrons in the Earth's gravitational field. *Physical Review D*, 67(10):102002 [9 pages], 2003.

[18] Wei-Tou Ni. Equivalence principles and electromagnetism. *Physical Review Letters*, 38(7):301–304, 1977.

[19] Kenneth Nordtvedt. Quantitative relationship between clock gravitational "red-shift" violations and nonuniversality of free-fall rates in nonmetric theories of gravity. *Physical Review D*, 11(2):245–247, 1975.

[20] Asher Peres. Measurement of time by quantum clocks. *American Journal of Physics*, 48(7):552–557, 1980.

[21] Achim Peters, Keng Yeow Chung, and Steven Chu. Measurement of gravitational acceleration by dropping atoms. *Nature*, 400(6747):849–852, 1999.

[22] Achim Peters, Keng Yeow Chung, and Steven Chu. High-precision gravity measurements using atom interferometry. *Metrologia*, 38:25–61, 2001.

[23] Joseph Samuel and Supurna Sinha. Atom interferometers and the gravitational redshift. arXiv:1102.2587.

[24] Leonard I. Schiff. On experimental tests of the general theory of relativity. *American Journal of Physics*, 28(4):340–343, 1960.

[25] S. Schlamminger et al. Test of the equivalence principle using a rotating torsion balance. *Physical Review Letters*, 100(4):041101 (4 pages), 2008.

[26] Pippa Storey and Claude Cohen-Tannoudji. The Feynman path integral approach to atomic interferometry. A tutorial. *Journal de Physique*, 4(11):1999–2027, 1994.

[27] Kip S. Thorne, David L. Lee, and Alan P. Lightman. Foundations for a theory of gravitation theories. *Physical Review D*, 7(12):3563–3578, 1973.

[28] Tim van Zoest. Bose-Einstein condensation in microgravity. *Science*, 328(5985): 1540–1543, 2010.

[29] Robert Vessot et al. Test of relativistic gravitation with a space-borne hydrogen maser. *Physical Review Letters*, 45(26):2081–2084, 1980.

[30] Clifford M. Will. *Theory and Experiment in Gravitational Physics*. Cambridge University Press, Cambridge, 2nd revised edition, 1993. First edition 1981.

[31] Clifford M. Will. The confrontation between general relativity and experiment. *Living Reviews in Relativity*, 9(3), 2006.

[32] Peter Wolf, Luc Blanchet, Serge Bordé, Christian J. Reynaud, Christophe Salomon, and Claude Cohen-Tannoudji. Atom gravimeters and gravitational redshift. *Nature*, 467:E1, 2010.

[33] Peter Wolf, Luc Blanchet, Serge Bordé, Christian J. Reynaud, Christophe Salomon, and Claude Cohen-Tannoudji. Does an atom interferometer test the gravitational redshift at the Compton frequency? arXiv:1012.1194.

Domenico Giulini
Institute for Theoretical Physics
Appelstraße 2
D-30167 Hannover
Germany

Center of Applied Space Technology and Microgravity
Am Fallturm
D-28359 Bremen
Germany
e-mail: `giulini@itp.uni-hannover.de`

Index